Study Guide

JOSEPH P. CHINNICI
VIRGINIA COMMONWEALTH UNIVERSITY

SUSAN M. WADKOWSKI
LAKELAND COMMUNITY COLLEGE

Sixth Edition

Biology
LIFE ON EARTH

Teresa Audesirk

Gerald Audesirk

Bruce E. Byers

Prentice
Hall

Upper Saddle River, NJ 07458

Executive Acquisitions Editor: Teresa Ryu
Project Manager: Travis Moses-Westphal
Executive Managing Editor: Kathleen Schiaparelli
Production Editor: Dinah Thong
Supplement Cover Manager: Paul Gourhan
Supplement Cover Designer: PM Workshop Inc.
Manufacturing Buyer: Lisa McDowell
*Cover Photograph: "Weaver bird building nest,
Maasai Mara national Reserve, Kanya/Manoj Shah"*

© 2002 by Prentice-Hall, Inc.
Upper Saddle River, NJ 07458

Printed in the United States of America

10 9 8 7 6 5 4 3 2

ISBN   0-13-092363-X

Prentice-Hall International (UK) Limited, London
Prentice-Hall of Australia Pty. Limited, Sydney
Prentice-Hall Canada, Inc., Toronto
Prentice-Hall Hispanoamericana, S.A., Mexico City
Prentice-Hall of India Private Limited, New Delhi
Pearson Education Asia Pte. Ltd., Singapore
Prentice-Hall of Japan, Inc., Tokyo
Editora Prentice-Hall do Brazil, Ltda., Rio de Janeiro

CONTENTS

UNIT FIVE: ANIMAL ANATOMY AND PHYSIOLOGY

UNIT SIX: ECOLOGY

A NOTE TO THE STUDENT—PLEASE READ

This study guide is intended be used with the textbook **Biology: Life on Earth**, sixth edition, by Audesirk, Audesirk and Byers. The study guide is set up to help you learn the most from your general biology course (and to earn the highest possible grade). However, working through the study guide is only a portion of the process. You should take advantage of the three interrelated elements of the course: the lectures, the textbook, and the study guide.

1. **Lecture**: Attend all lectures, be attentive in class, and take good and complete notes. Sitting toward the front of the class will help increase your ability to pay attention. You may want to tape the lectures as well. (Always get your instructor's permission to tape lectures.) After each lecture, it's a good idea to recopy your notes into a more readable and organized form, relying on the textbook and study guide for help in understanding confusing and difficult topics and for filling in "gaps" in the notes. (A taped lecture helps here, also.) You'll be pleasantly surprised how much you will learn and remember by simply rewriting and adding to class notes.
2. **Textbook**: Read the textbook chapters before they are covered in lecture. Even if you do not understand what you've read, you have alerted your brain to key words and topics. You will find the lecture more comprehensible since the terminology will be somewhat familiar. If you understand the lecture better, it will be less frustrating and more fun to go to class, and your confidence level about biology will rise. Then re-read the chapters after class.
3. **Study guide**: After reading the textbook and attending lecture, use this study guide to help you prepare for the inevitable exams that go with college life. This study guide has a detailed overview of each chapter in the textbook, along with a set of review exercises with answers.

FORMAT OF THE STUDY GUIDE.

This guide is divided into 41 chapters, corresponding to the organization of the textbook. Each chapter has a similar format, consisting of:

1. A detailed **overview** of the material in the textbook, organized to allow you to gain additional understanding of the chapter and easily find answers to the review questions. All important concepts from the textbook are included in the overview, along with some helpful hints for remembering the material.
2. **Key terms and definitions** for that chapter. It has been suggested that the many new terms encountered in an introductory biology course equals that of a first year course in a foreign language. Thus, the key terms and definitions will prove helpful in understanding the content of each chapter. Refer to them as often as needed.
3. A set of review exercises in a variety of formats, including: **thinking through the concepts**, which may include multiple choice, true false, and fill-in questions, concept maps, tables, or diagrams; and **applying the concepts**, which are essay questions based on ideas covered in the chapter. These questions are designed to increase your ability to think critically about the concepts presented in the chapter. Additionally, questions relating to the Web Investigations in each chapter are included.
4. **Answers to exercises**. Refer to these after answering the review exercises. Looking at the answers before trying to answer them yourself can mislead you into thinking you know the answers when you may not. Use the answers provided in this guide to check your answers for accuracy, not to generate your answers.

HINTS FOR STUDYING.

It is best to first learn the definitions of the key terms (and any others that you are unfamiliar with). Then, you should learn how those terms apply or are related to the topic covered. Try to keep in mind the big picture while you are learning the details of a process. While this can be difficult, it will increase your understanding.

Drawing diagrams of your own to outline a process, will help increase your understanding, especially if you are a more visual learner. Some students use or devise their own mnemonic devices to help them remember terms or concepts in order. (For example: King Philip Came Over For Good Spaghetti is a great way to remember the taxonomic categories Kingdom, Phylum, Class, Order, Family, Genus, Species.) It doesn't matter how ridiculous it sounds, just so it works for you.

HINTS FOR DOING WELL IN CLASS AND ON LECTURE EXAMS.

Whether your college or university has large sections or small classes of general biology, professors may determine grades by giving computer-graded "objective" exams consisting of multiple choice, matching, and true-false questions. If this describes your course (and even if it doesn't), the following hints may help improve your grades, even if you think you don't "test well" on objective exams.

Learn the meaning of biological terms. A listing of key terms with definitions appears in each chapter in the study guide. Learn the meaning of these words. Not only are definitions often asked on objective exams, but it will help in understanding the questions.

Answer questions you know first. If you don't know, or are not sure of, the answer to a question, skip over it and come back to it later. Sometimes you will get a clue to the correct answer from other questions on the exam.

Read all the choices for multiple choice questions. Try to eliminate as many obviously incorrect choices as possible. Choose the choice that appears most correct from those remaining. If your professor says there is no penalty for guessing, answer every question.

If you are sure that two or more choices are correct and another choice is "all of the above," that is usually the correct answer. (But, be careful.)

If part of a statement is false, the entire statement is false.

On **essay exams**, it is always a good idea to write down an outline or a diagram. This will get your thoughts down on the paper and you can reform them into a comprehensive paragraph. Your answer will be more precise and you will be less likely to leave out important points. Begin with a clear, concise introductory statement summarizing your answer. The rest of the answer should be long enough to answer the question fully.

It is always a good idea to look over the entire test before you start; make sure you have all the pages, etc. You might want to start in an area of the test most familliar to you — you don't always have to start at the beginning.

ACKNOWLEDGEMENTS

Joe Chinnici was awarded the 2001 Humanities and Sciences Distinguished Teaching Award at Virginia Commonwealth University. Susan Wadkowski was awarded the 2001 Excellence in Teaching Award at Lakeland Community College. They both thank the many beginning biology students they have taught for providing the inspiration and motivation to develop high quality learning aids such as this study guide.

Joe Chinnici also gives thanks to his wife, Dianna who, in so many ways, helps him persevere not only during the great times but during the other times as well.

Susan Wadkowski also thanks her life partner, Renee Wadkowski, who's patience, dedication, and pre-press proficiency truly makes this a partnership effort.

Chapter 1:   An Introduction To Life On Earth

OVERVIEW

In this chapter, you will learn the characteristics of life and read a brief overview of the vast diversity of living organisms. You will be introduced to basic scientific principles as well as the scientific method. Finally, you will find a brief introduction to the mechanism and evidence for evolution. As you walk across campus, notice the tremendous array of plants and animals. Ever wonder how they all developed and how they interact with each other? What makes living things different from non-living things? This course will help you answer those questions.

1) What Are the Characteristics of Living Things?

The thing we call "life" has **emergent properties**, that is, it has intangible attributes arising from the complex ordered interactions among the individual characteristics of living things. What are some of these emergent properties? Living things display complexity based on highly organized **organic** (carbon-based) **molecules** made up of **elements** of **atoms** with **subatomic particles**. Living things respond to stimuli like light, sound, temperature, hunger, pain, and others in their internal and external environments. Living things display **homeostasis**, which is the active maintenance of an organism's complex structure and internal environment. Living things acquire, convert, and use both materials called **nutrients** and **energy** (the ability to do work) from the environment through **metabolism**, which constitutes the chemical reactions needed to sustain life. Living things grow by converting environmentally-derived materials into specific **molecules** in their bodies. All energy ultimately is trapped by **photosynthesis**, the making of sugar using solar energy. Living things reproduce themselves, using the information in molecules of **DNA** (**deoxyribonucleic acid**) called **genes**. Living things as a whole have the capacity for **evolution** making use of **mutations**, random changes in DNA structure, and usually **natural selection**, the enhanced survival and reproduction of individuals with favorable inherited characteristics.

The hierarchy of life is (from least complex to most complex): subatomic particles, which are grouped into atoms, which are grouped into molecules, which are grouped into **organelles**, which are grouped into **cells** (the smallest unit of life, each with **plasma membrane**, **cytoplasm**, and nucleus), which are grouped into **tissues** like muscles, which are grouped into **organs** like the heart, which are grouped into **organ systems** like the circulatory system, which are grouped into **organisms** like humans. Groups of similar interbreeding organisms constitute a **species**, and members of a species that live in a given area make up a **population**. Populations of several species interact to form a **community**, and a community along with its non-living environment (land, water, and atmosphere) constitutes an **ecosystem**. All the ecosystems on earth make up the **biosphere**.

2) How Do Scientists Categorize the Diversity of Life?

Organisms are classified into three major categories called **domains** based on the traits they exhibit, especially cell type, number of cells in each organism, and the method used to acquire nutrition and energy. These domains are Bacteria, Archaea, and Eukarya. Bacteria and Archaea both have single, simple, smaller (**prokaryotic**) cells lacking nuclei and membrane-bound organelles. The Eukarya organisms each have one or more complex larger (**eukaryotic**) cells with nuclei and organelles, and are divided into four **kingdoms**: Protista, Fungi, Plantae, and Animalia.

Bacteria, Archaea, and members of the kingdom Protista are mostly **unicellular**, while members of the kingdoms Fungi, Plantae, and Animalia are primarily **multicellular**, their lives depending on the intimate cooperation among cells. Members of different kingdoms have different ways of acquiring energy. **Autotrophs** are "self-feeders" and include photosynthetic organisms (plants, some bacteria, and some protists). They capture solar energy and use it to make sugars and fats. Organisms that get energy from the bodies of other organisms are called **heterotrophs** ("other feeders"); these include many archaea, some bacteria, all fungi and animals. Bacteria and fungi absorb predigested food molecules, while animals eat chunks of food and break them down in their digestive tracts, a process called ingestion.

3) What Is the Science of Biology?

As you study biology this semester, you will see that a basic principle of modern biology is that living things obey the same laws of physics and chemistry that govern non-living matter. There are three "scientific principles" (unproven assumptions) essential to biology. The first is **natural causality**, the principle that all earthly events can be traced to preceding natural causes. For instance, an epileptic seizure is caused by misfiring of nerve cells, not the work of the devil. Supernatural intervention has no place in science. The second principle is that natural laws do not change with time or distance. They apply every-where and for all time. For example, scientists assume that gravity always has worked as it does today and works the same way everywhere in the universe. **Creationism** (a supernatural being created each type of organism separately and all at one time) is contrary to both natural causality and uniformity in time. The third principle is common perception, the assumption that all humans individually perceive natural and aesthetic events through their senses in fundamentally the same way. However, our interpretation of such events (like appreciation of rock music or the morality of abortion) may differ.

The **scientific method** is how scientists study the workings of life. It consists of four interrelated operations: observation, hypothesis, experiment, and conclusion. An **observation** is the beginning of a scientific inquiry (for example, maggots appear on fresh meat left uncovered). A **hypothesis** is a tentative testable explanation of an observed event based on an educated guess about its cause (for example, maggots appear on fresh meat left uncovered because flies land on the meat and lay eggs). An **experiment** is a study done under rigidly controlled conditions and based on a prediction stemming from the hypothesis (for example, if the fresh meat is covered with a fine gauze to keep the flies away, no maggots should appear in the meat). Simple experiments test the assertion that a single factor, or **variable**, is the cause of a single observation. Scientists design **controls** into their experiments, in which all variables remain constant. Controls are compared to experiments in which only the variable being tested is changed. A **conclusion** is a judgment about the validity of the hypothesis, based on the results of the experiment (for example, maggots did not appear on the meat covered with gauze but did appear on fresh meat left uncovered in the same place at the same time. Thus, the hypothesis is supported by the results of the experiment).

Science is a human endeavor. While much of human progress can be attributed to the scientific method, real scientific advances often involve accidents and acumen, lucky guesses, controversies among scientists and the unusual insights of brilliant scientists. An example of "real science" was the accidental discovery of penicillin by Alexander Fleming.

When a hypothesis is supported by the results of many different kinds of experiments, scientists are confident enough about its validity to call it a **scientific theory**. Scientific conclusions must always remain tentative and be subject to revision if new observations or experiments demand it. Scientific theories arise through **inductive reasoning**, which is the process of creating a generalization based on many specific observations that support the generalization and the absence of observations that contradict it (for example, the earth exerts a gravitational force on objects near it). A scientific theory can support **deductive reasoning**, which is the process of generating hypotheses about how a specific experiment will turn out based on a well-supported generalization (for example, if a newly discovered organism displays the properties of life, a scientist will deduce that it is made up of cells).

4) Evolution: The Unifying Concept of Biology

Evolution is the theory that present-day organisms descended, with modification, from preexisting life forms; it is the unifying concept in biology. As proposed by the English naturalists Charles Darwin and Alfred Russel Wallace in the mid-1800s, evolution occurs due to three natural processes: (1) genetic variation that exists among members of a population; (2) inheritance of variations from parents to offspring (we now know that inherited variations ultimately arise from gene mutation, changes in DNA structure); and (3) natural selection, the survival and enhanced reproduction of organisms with favorable variations (called **adaptations**) in structure, physiology, or behavior that best meet the challenges of the environment. Ultimately, natural selection has unpredictable results because environments tend to change dramatically (e.g., ice ages). What helps organisms survive today may be a liability tomorrow. For instance, layers of fat and heavy fur allow polar bears to survive in the cold. But if global warming occurs and the climate warms up significantly, fat and fur will no longer be beneficial and might even become a liability to polar bears. All living things within a given geographical area and the interrelationships among them constitute **biodiversity**.

5) How Does Knowledge of Biology Illuminate Everyday Life?

In this text, the authors will be trying to convey to you that an understanding of how nature works will lead to an enhanced appreciation and wonderment about life and how organisms interact with each other and the rest of the world.

KEY TERMS AND CONCEPTS

Fill-In: From the following list of terms, fill in the blanks in the following statements.

autotroph	conclusion	evolution	molecule	organ
atom	DNA	experiment	mutate	plasma membrane
cell	domain	gene	nucleus	theory
community	energy	heterotroph		

1. _____ A relatively stable combination of atoms.

2. _____ The membrane-bound organelle of eukaryotic cells that contains the cell's genetic material.

3. _____ A self-feeder, usually meaning a photosynthetic organism.

4. _____ Any change in the overall genetic composition of a population of organisms from one generation to the next.

5. _____ The smallest unit of life.

6. _____ An organism that eats other organisms.

7. _____ Two or more populations of different species living and interacting in the same area.

8. _____ A structure, usually composed of several tissue types, that acts as a functional unit.

9. _____ A physical unit of inheritance.

10. ___Mutate___ What a gene does when it undergoes a genetic change.

11. _____ A study done under rigidly controlled conditions based on a prediction stemming from the hypothesis.

12. _____ A new major category of organisms, based on comparisons of cellular, molecular and behavioral similarities and differences.

13. _____ A judgment about the validity of the hypothesis, based on the results of the experiment.

14. _____ In science, an explanation for natural events based on a large number of observations.

15. _____ The ability to do work.

16. _____ The outer membrane of a cell, enclosing its contents.

17. _____ The smallest particle of an element that retains all the properties of the element.

18. _____ A type of molecule that encodes the genetic information of all living cells.

Key Terms and Definitions

adaptation: a trait that increases the ability of an individual to survive and reproduce compared to individuals without the trait.

atom: the smallest particle of an element that retains the properties of the element.

autotroph (aw´-tō-trōf): "self-feeder"; normally, a photosynthetic organism; a producer.

biodiversity: the total number of species within an ecosystem and the resulting complexity of interactions among them.

biosphere (bī´-ō-sfēr): that part of Earth inhabited by living organisms; includes both living and nonliving components.

cell: the smallest unit of life, consisting, at a minimum, of an outer membrane that encloses a watery medium containing organic molecules, including genetic material composed of DNA.

community: all the interacting populations within an ecosystem.

conclusion: the final operation in the scientific method; a decision made about the validity of a hypothesis on the basis of experimental evidence.

control: that portion of an experiment in which all possible variables are held constant; in contrast to the "experimental" portion, in which a particular variable is altered.

creationism: the hypothesis that all species on Earth were created in essentially their present form by a supernatural being and that significant modification of those species – specifically, their transformation into new species – cannot occur by natural processes.

cytoplasm (sī´-tō-plaz-um): the material contained within the plasma membrane of a cell, exclusive of the nucleus.

deductive reasoning: the process of generating hypotheses about how a specific experiment or observation will turn out.

deoxyribonucleic acid (dē-ox-ē-rī-bō-noo-klā´-ik; DNA): a molecule composed of deoxyribose nucleotides; contains the genetic information of all living cells.

domain: the broadest category for classifying organisms; organisms are classified into three domains: Bacteria, Archaea, and Eukarya.

ecosystem (ē´kō-sis-tem): all the organisms and their nonliving environment within a defined area.

element: a substance that cannot be broken down, or converted, to a simpler substance by ordinary chemical means.

emergent property: an intangible attribute that arises as the result of complex ordered interactions among individual parts.

energy: the capacity to do work.

eukaryotic (ū-kar-ē-ot´-ik): referring to cells of organisms of the domain Eukarya (kingdoms Protista, Fungi, Plantae, and Animalia). Eukaryotic cells have genetic material enclosed within a membrane-bound nucleus and contain other membrane-bound organelles.

evolution: the descent of modern organisms with modification from preexisting life-forms; strictly speaking, any change in the proportions of different genotypes in a population from one generation to the next.

experiment: the third operation in the scientific method; the testing of a hypothesis by further observations, leading to a conclusion.

heterotroph (het´-er-ō-trōf´): literally, "other-feeder"; an organism that eats other organisms; a consumer.

homeostasis (hōm-ē-ō-stā´sis): the maintenance of a relatively constant environment required for the optimal functioning of cells, maintained by the coordinated activity of numerous regulatory mechanisms, including the respiratory, endocrine, circulatory, and excretory systems.

hypothesis (hī-poth´-eh-sis): the second operation in the scientific method; a supposition based on previous observations that is offered as an explanation for the observed phenomenon and is used as the basis for further observations, or experiments.

inductive reasoning: the process of creating a generalization based on many specific observations that support the generalization, coupled with an absence of observations that contradict it.

kingdom: the second broadest taxonomic category, contained within a domain and consisting of related phyla or divisions. This textbook recognizes four kingdoms within the domain Eukarya: Protista, Fungi, Plantae, and Animalia.

metabolism: the sum of all chemical reactions that occur within a single cell or within all the cells of a multicellular organism.

molecule (mol´-e-kūl): a particle composed of one or more atoms held together by chemical bonds; the smallest particle of a compound that displays all the properties of that compound.

multicellular: many-celled; most members of the kingdoms Fungi, Plantae, and Animalia are multicellular, with intimate cooperation among cells.

mutation: a change in the base sequence of DNA in a gene; normally refers to a genetic change significant enough to alter the appearance or function of the organism.

natural causality: the scientific principle that natural events occur as a result of preceding natural causes.

natural selection: the unequal survival and reproduction of organisms due to environmental forces, resulting in the preservation of favorable adaptations. Usually, natural selection refers specifically to differential survival and reproduction on the basis of genetic differences among individuals.

nucleus (atomic): the central region of an atom, consisting of protons and neutrons.

nutrient: a substance acquired from the environment and needed for the survival, growth, and development of an organism.

observation: the first operation in the scientific method; the noting of a specific phenomenon, leading to the formulation of a hypothesis.

organ: a structure (such as the liver, kidney, or skin) composed of two or more distinct tissue types that function together.

organelle (or-guh-nel´): a structure, found in the cytoplasm of eukaryotic cells, that performs a specific function; sometimes refers specifically to membrane-bound structures, such as the nucleus or endoplasmic reticulum.

organic/organic molecule: describing a molecule that contains both carbon and hydrogen.

organism (or´-guh-niz-um): an individual living thing.

organ system: two or more organs that work together to perform a specific function; for example, the digestive system.

photosynthesis: the complete series of chemical reactions in which the energy of light is used to synthesize high-energy organic molecules, normally carbohydrates, from low-energy inorganic molecules, normally carbon dioxide and water.

plasma membrane: the outer membrane of a cell, composed of a bilayer of phospholipids in which proteins are embedded.

population: all the members of a particular species within an ecosystem, found in the same time and place and actually or potentially interbreeding.

prokaryotic (prō-kar-ē-ot´-ik): referring to cells of the domains Bacteria or Archaea. Prokaryotic cells have genetic material that is not enclosed in a membrane-bound nucleus; they lack other membrane-bound organelles.

scientific method: a rigorous procedure for making observations of specific phenomena and searching for the order underlying those phenomena; consists of four operations: observation, hypothesis, experiment, and conclusion.

scientific theory: a general explanation of natural phenomena developed through extensive and reproducible observations; more general and reliable than a hypothesis.

species (spē´-sēs): the basic unit of taxonomic classification, consisting of a population or series of populations of closely related and similar organisms. In sexually reproducing organisms, a species can be defined as a population or series of populations of organisms that interbreed freely with one another under natural conditions but that do not interbreed with members of other species.

spontaneous generation: the proposal that living organisms can arise from nonliving matter.

subatomic particle: the particles of which atoms are made: electrons, protons, and neutrons.

tissue: a group of (normally similar) cells that together carry out a specific function; for example, muscle; may include extracellular material produced by its cells.

unicellular: single-celled; most members of the domains Bacteria and Archaea and the kingdom Protista are unicellular.

variable: a condition, particularly in a scientific experiment, that is subject to change.

THINKING THROUGH THE CONCEPTS

True or False: Determine if the statement given is true or false. If it is false, change the underlined word(s) so that the statement reads true.

19. _____   Biology is basically different from other sciences.

20. _____   The basic assumptions of science can be proven.

21. _____   The conclusions of science are permanent.

22. _____   Science accepts only natural explanations for natural processes.

23. _____   Creationism is not a science.

24. _____   Organisms that produce their own food are heterotrophic.

25. _____   Bacteria are eukaryotic organisms.

26. _____   Prokaryotic forms do not possess distinct nuclei.

27. _____   Fungi are autotrophic organisms.

28. _____   Redi's experiments supported the theory that life can occur by spontaneous generation.

Matching: The kingdoms.

29. _____   multicellular and autotrophic

30. _____   multicellular, heterotrophic, ingestive

31. _____   unicellular and eukaryotic

32. _____   multicellular, heterotrophic, absorptive

Choices:

   a. Protista

   b. Fungi

   c. Plantae

   d. Animalia

Matching: Which scientific principles do each of the following violate?

33. _____   in biblical times, humans lived to be 900 years old

34. _____   God created all life on earth in six days

35. _____   a six-foot man may look ten feet tall to you

36. _____   gravity did not affect the dinosaurs as much as it does us

37. _____   miracles

38. _____   an anorexic woman sees herself as fat

Choices:

a. natural causality

b. uniformity in time and space

c. common perception

Multiple Choice: Pick the most correct choice for each question.

39. Which of the following are not characteristics used to categorize organisms into kingdoms?
    a. types of cells present
    b. numbers of cells present
    c. presence or absence of cell walls
    d. how the organisms acquire energy
    e. how the organisms move
    f. choices a and b
    g. choices c and e
    h. none of the above

40. What is the ultimate source of genetic variation?
    a. mutations in DNA
    b. adaptations to a changing environment
    c. natural selection
    d. spontaneous generation
    e. homeostasis

41. The basic difference between a prokaryotic cell and a eukaryotic cell is that the prokaryotic cell
    a. possesses membrane-bound organelles
    b. lacks DNA
    c. lacks a nuclear membrane
    d. is considerably larger
    e. is in multicellular organisms

42. Why can scientists NOT accept creationism as fact?
    a. The U.S. government requires a separation of church and state matters.
    b. The tenets of creationism cannot be explained by natural laws or tested scientifically.
    c. They would not have any research to do if they accept the fact that religion has already answered questions about the origin of life.
    d. Scientists come from so many different religious backgrounds that they could never agree on one creation story.

43. Which of the following is TRUE?
    a. Science and religion discover knowledge in fundamentally the same way.
    b. Both science and religion need proof in order to accept something as valid.
    c. Science discovers knowledge by observation and experimentation, while religion relies on revelation and faith.
    d. Creationism and evolution are equally valid from a scientific point of view.

44. The scientific principle of "common perception" does NOT mean that
    a. all people hear music in a common manner with their ears
    b. all people see a work of art in a common manner with their eyes
    c. all people appreciate opera in a common manner with their minds
    d. all people feel heat in a common manner with their nervous systems

45. A hypothesis is simply
    a. a test devised to gain information
    b. an initial observation of some event
    c. a possible explanation for something observed
    d. drawing a conclusion
    e. designing an experiment

APPLYING THE CONCEPTS

These practice questions are intended to sharpen your ability to apply critical thinking and analysis to biological concepts covered in this chapter.

46. Last summer, Zachary grew pepper plants in his garden and decided to use a new fertilizer called UltraGrow. He claims that his plants produced a larger quantity of peppers that were larger in size than those he harvested the year before, and he credited the use of UltraGrow for the improvement. Scientifically speaking, has he proven the effectiveness of the fertilizer?

_____

_____

_____

_____

_____

_____

47. How could Zachary more validly test his hypothesis that UltraGrow works to produce larger plants that yield more fruit?

_____

_____

_____

_____

_____

48. Suppose while walking across campus, you spotted a robin's nest with several baby robins being fed worms by their mother. Describe some elements of an ecosystem containing the things you observed.

_____

_____

_____

_____

_____

49. Describe several ways that you may have displayed, within the past few months, some of the properties characteristic of a living organism.

_____

_____

_____

_____

_____

50. Which characteristics do you possess that would allow you to determine the domain and kingdom to which you belong?

_____

_____

_____

_____

51. Many department stores sell objects called "banana holders." These are stands with a hook at the top where bunches of bananas are supposed to be hung up so that they are not in contact with the kitchen counter top where they might become bruised. Those who make banana holders claim that their use will allow bananas to stay fresh longer, and retard the development of dark spots on their surface. Design an experiment to determine whether these claims are scientifically valid.

_____

_____

_____

_____

_____

52. Some small plants, called "stoneplants," look like pebbles. Discuss several ways that these plants differ from the pebbles that they mimic.

_____

_____

_____

_____

_____

53. Discuss several ways that a fungus, a plant, and an animal differ from each other.

_____

_____

_____

_____

_____

54. Some people like classical music while others don't. Rap music appeals to some students but not to others. Modern art is adored by many but there are some who don't appreciate it as art. Aren't these differences among people an example of why the scientific notions of "common perception" is not a valid basic principle of science? Please explain your answer.

_____

_____

_____

_____

_____   _____

Use the following information to answer questions 55–63.

A study has been done to determine whether the toxic material known as "DDE" is affecting the reproductive tracts of alligators living in contaminated lakes. It is known that incubation of alligator eggs at higher temperatures results in normal male sexual development, whereas incubation of eggs at lower temperatures produces normal females.

Identify: Determine which part of the scientific method each of the following statements about the study represents: **observation**, **hypothesis**, **experimentation**, **results**, **interpretation of results**, **repeatability of results**, **conclusion**, or **new hypothesis**.

55. Investigators collected alligator eggs from a clean lake and incubated them at high temperatures to ensure that the hatchlings would be male. Just before incubation began, the researchers painted some of the eggs with estrogen, some with DDE, and some with an inert substance.

_____

56. Alligators living in a lake containing DDE, which is a breakdown product of the pesticide DDT, have a high rate of reproductive tract problems.

_____

57. The egg-painting experiments also were done with eggs from two other lakes, and the results were similar.

_____

58. If exposure to large amounts of estrogen causes reproductive problems in animals, then animal eggs intentionally exposed to estrogen or similar substances should not hatch, or they will hatch to yield abnormal offspring.

_____

59. Among the estrogen-treated eggs, only 20% produced male hatchlings (dramatically lower than the expected 100%). The other estrogen-treated eggs either died or yielded abnormal alligators. Only 40% of the eggs painted with DDE hatched normal males. Another 40% of the DDE-treated eggs hatched intersexes, which had both male and female reproductive structures (intersexes are not found among alligators in clean lakes). The remaining 20% of the DDE-treated eggs produced female hatchlings. All of the eggs painted with the inert substance hatched normal male alligators.

_____

60. Both estrogen and DDE appear to alter the reproductive tract of developing male alligators in similar ways.

_____

61. Since the effects of exposure to estrogen and exposure to DDE are similar in altering the reproductive tracts of developing alligators (that should develop as males), DDE acts as an "estrogen mimic."

_____

62 If "estrogen mimics" like DDE produce the same or similar effects as estrogen, then the mimics should bind to cellular estrogen receptor molecules.

_____

63. Using the description in question 55, what did the researchers use as their control group in their experiments?

_____

Use the Case Study and the Web sites for this chapter to answer the following question.

64. The Tree of Life Project Hominidae has constructed a phylogenetic tree for humans and their closest living and extinct relatives. Is the orangutan, the chimpanzee and bonobo (pygmy chimpanzee), or the gorilla our closest living relative? How reliable do you think this phylogenetic tree is?

_____

_____

_____

_____

_____

_____

_____

_____

_____

_____

_____

_____

_____

_____

_____

_____

_____

_____

ANSWERS TO EXERCISES

1. molecule	13. conclusion	24. false, autotrophic	35. c
2. nucleus	14. theory	25. false, prokaryotic	36. b
3. autotroph	15. energy	26. true	37. a
4. evolution	16. plasma membrane	27. false, heterotrophic	38. c
5. cell	17. atom	28. false, disproved	39. g
6. heterotroph	18. DNA	29. c	40. a
7. community	19. false, essentially	30. d	41. c
8. organ	similar to	31. a	42. b
9. gene	20. false, cannot	32. b	43. c
10. mutate	21. false, tentative	33. b	44. c
11. experiment	22. true	34. a	45. c
12. domain	23. true		

46. Scientifically speaking, Zachary did not prove his hypothesis that the fertilizer works. He did not perform a controlled experiment, meaning that there could be a number of reasons why he got more and larger peppers last year than the year before: differences in the amounts of sunlight, rainfall, richness of the soil, the types of seeds used, differences in plant pest activity each year, and perhaps others.

47. What Zachary must do is plant two groups of pepper plants side by side at the same time. To test the effect of UltraGrow, only one factor (the amount of UltraGrow) must differ between the experimental (fed UltraGrow) and the control (not fed UltraGrow) plants. If there is a difference between the fruits produced from the experimental and the control plants, then Zachary may validly conclude that the fertilizer was responsible for the difference.

48. An ecosystem consists of a community of living organisms along with their non-living environment. In this case, the robin's nest is probably built of twigs and grass and located in a tree that provides some safety from predators, as well as shade. The robin is feeding her young with worms that live in the soil and feed there on microorganisms and moisture, and grass roots.

49. You may have responded to an overly warm environment by sweating to cool off, or you may have responded to an overly cool environment by shivering to keep warm. These are examples of your body responding to stimuli and displaying homeostasis. You have also acquired and used materials and energy. Whenever you eat, you are ingesting food molecules that contain chemical energy. Your digestive system digests the food into smaller molecules, some of which are used to maintain body parts. The digestive process also releases chemical energy from the food, and some of that energy is stored in your cells so that it is available when you need energy to keep your body alive (run, breathe, sleep, etc.).

50. Since your body is composed of cells that contain nuclei (eukaryotic cells), you must belong in the domain Eukarya. Because your body consists of many cells and you acquire energy by ingesting food produced by other organisms, you belong in the kingdom Animalia.

51. To perform a valid experiment testing the claims of the banana holder proponents, you need to test their product under normal conditions while at the same time comparing the changes that occur in bananas hung on a banana holder (the experimental group) with very similar bananas not in contact with the banana holder (the control group). Both groups of bananas, however, should be exposed to an extremely similar environment, with similar sunlight, temperature, humidity, hours of daylight and darkness, etc. (you need to eliminate extraneous variables). So, you should buy a large bunch of fresh bananas, divide the bunch equally, count the number of spots and record the size of the spots on each at the beginning on the experiment. Then hang one bunch on the banana holder and place the other bunch right beside the banana holder on the counter top. Check each bunch each morning and evening

for a week and keep track of the numbers and sizes of spots on the bananas in each bunch. If the banana holder works as claimed, the numbers and sizes of surface spots should increase in the control group of bananas faster than in the experimental group. However, if the spots show no difference in the two groups, the claims made by the banana holder manufacturers would not be supported.

52. The plants will respond to stimuli in their eternal and internal environments, such as direction of sunlight and availability of water; pebbles do not. The plants are cellular in nature and the pebbles are not. The plants acquire energy by photosynthesis, making sugar molecules for food through metabolism; pebbles do not acquire energy (except passively by growing warm in the sun) or make food. The plants grow larger by dividing their cells internally in an orderly predictable fashion, while pebbles might increase in size simply by adding more material to their surface in a random process. The plants reproduce themselves by passing on cells that contain the genetic material that controls the process; pebbles do not reproduce themselves, although they may be split haphazardly by environmental events. The cells of plants have DNA as the genetic material and this DNA has the ability to control all cellular processes, replicate itself precisely, and mutate occasionally to produce variation necessary for evolution; pebbles do not possess cells or DNA and cannot mutate or evolve.

53. Fungi, plants, and animals are similar in that they all possess eukaryotic cells, and tend to be multi-cellular. They differ from each other in the ways they acquire food. Plants generally make their own food through the processes of photosynthesis, trapping solar energy within sugar molecules. Thus, plants are autotrophic. Fungi lack chlorophyll and obtain food by absorbing food that has already been digested into their bodies. Animals tend to ingest undigested food into their bodies and digest the food internally. Thus, fungi and animals both are heterotrophic, with fungi having absorptive digestion and animals having ingestive digestion. Animals have active means of moving their bodies from place to place, while plants and fungi do not.

54. These differences among people in their appreciation of music or art do not violate the scientific principle of common perception. Common perception refers to the manner that the sounds of music or the images of art are taken into the body by our senses of hearing and sight. Normally, the ears and eyes of all people respond to stimulation in the same manner, sending signals to the brain to be processed. The personal differences among people who prefer one type of music over another or one expression of art over alternatives are beyond the scope of the science of biology.

55. experimentation
56. observation
57. repeatability
58. hypothesis
59. results
60. interpretation of results
61. conclusion
62. new hypothesis

63. Treatment with the inert substance.

64. From these data, it appears that the chimpanzees and bonobos are the closest relatives of humans. This phylogenetic tree appears to be quite reliable because the information for constructing it was obtained by highly trained scientists whose experimental work was done recently and published in peer-reviewed journals, and is based on a comprehensive review of the literature. Also, the phylogenetic tree appears on the official web pages of a respected university (University of Arizona). So, the data used to make this tree is recent, objective, peer-reviewed, university-sanctioned, and based on a comprehensive review of the literature.

Chapter 2: Atoms, Molecules, and Life

OVERVIEW

In this chapter, you will learn about matter and energy, as well as the structure of atoms and molecules. The three major types of chemical bonding are also discussed here, along with important inorganic molecules like water. As new information is discovered by scientists, old ideas about the healthfulness of foods changes. For instance, the latest research indicates that chocolate may protect cells from harmful "free radicals" containing a destructive type of oxygen. Free radicals have been implicated in aging, cancer, heart disease, and nervous system disorders.

1) What Are Atoms?

Atoms are the fundamental structural units of matter. An element is a substance containing the same kind of atoms and cannot be broken down or converted into another substance under ordinary conditions. Each atom has a central dense nucleus (the **atomic nucleus**) which contains positively charged **protons** and uncharged **neutrons**. Atoms also have negatively charged **electrons** that spin about the nucleus in paths called energy levels or shells. Protons and neutrons are roughly the same weight, but electrons are much lighter. There are 92 naturally occurring types of atoms, and each type of atom forms the structural unit of a different **element**. An atom's **atomic number** is the number of protons present in its nucleus and this is a constant feature of all atoms of a particular type. For instance, every hydrogen atom has one proton. Similar atoms that differ in the number of neutrons they possess are called **isotopes**. Some, but not all, isotopes are **radioactive**, which means they break apart spontaneously, forming different types of atoms and giving off energy. Radioactive isotopes are used to "label" biological materials used in certain research studies.

Electrons orbit the nucleus at fixed distances, forming **electron shells** that correspond to different energy levels. Up to two electrons are found in the electron shell nearest the nucleus, while the next shell may contain up to eight electrons. Normally, electrons fill the shell closest to the nucleus, then begin filling the next shell. For instance, a carbon has six electrons, two of which are found in the innermost shell and four are in the second shell. Electrically neutral, atoms have equal numbers of protons and electrons. For instance, a carbon atom has six protons and six electrons. Whereas the nuclei of atoms are stable and resistant to change, the electron shells are dynamic, and atoms interact with each other by gaining, losing, or sharing electrons, thus forming chemical bonds.

2) How Do Atoms Interact to Form Molecules?

A **molecule** consists of two or more atoms of either the same or of different elements. A substance whose molecules are formed of different types of atoms is called a **compound**. Atoms react with other atoms when there are vacancies in their outermost electron shells. Inert or stable atoms have completely filled or empty outer electron shells (like helium) while reactive atoms have partially filled outer shells (like hydrogen). An atom with a partially full outer electron shell is reactive and can become more stable through filling the outer shell by losing, gaining, or sharing electrons with other atoms. These electron interactions with other atoms create attractive forces called **chemical bonds**. **Chemical reactions**, the making and breaking of chemical bonds to form new substances, are essential for the maintenance of life. Free radicals are atoms or molecules which lack one or more electrons in their outer shells, making them highly reactive.

Charged atoms, called **ions**, interact to form **ionic bonds**. Atoms with almost full or almost empty outer electron shells will interact by gaining or by losing electrons, respectively, forming charged ions that will attract each other (e.g., sodium [Na^+] and chloride [Cl^-] ions), forming molecules by the formation of ionic bonds, like NaCl (table salt). Ionic molecules tend to form crystals, although ionic bonds are relatively weak and easily broken.

An atom with a partially full outermost electron shell can also become stable by sharing electrons with another atom, forming a **covalent bond**. An atom of hydrogen has one electron in a shell that can hold two. Two hydrogen atoms each share an electron with the other, forming a molecule of hydrogen gas (H_2) held together by one covalent bond. Oxygen atoms need two electrons to fill their outer shells; thus, two oxygen atoms form a molecule of O_2 gas by forming two covalent bonds. All covalent bonds are strong compared with ionic bonds, but some covalent bonds are stronger than others depending on the types of atoms involved. Water and carbon dioxide have strong covalent bonds (it takes a lot of energy to break them apart) while oxygen gas and gasoline have weaker covalent bonds. Most biological molecules are held together by covalent bonding. The atoms C,H,N,O,P,S (pronounced "chanops") are often found in cellular molecules. Hydrogen forms one covalent bond with another atom, oxygen and sulfur form two, nitrogen forms three, and carbon forms four bonds. Phosphorus is unusual in that it can form up to five covalent bonds with up to four other atoms.

Polar covalent bonds form when atoms share electrons unequally. A molecule having only one type of atom, like H_2 gas, is held together by a **nonpolar covalent bond** since each hydrogen nucleus exerts equal attraction on the shared electrons. In molecules made of different atoms, like water (H_2O), polar covalent bonds may form since the oxygen nucleus has more protons than the hydrogen nuclei and exerts a greater attraction on the electrons, which spend more time orbiting the stronger nucleus. Thus, water is electrically neutral overall, but the oxygen end is more negatively charged than the hydrogen end, making water a polar molecule. Consequently, the negative end of one water attracts the positive end of another, forming an electrical attraction called a **hydrogen bond**. Although individual hydrogen bonds are quite weak, many of them working together are quite strong. Many other molecules in cells form hydrogen bonds, including proteins and DNA.

3) Why Is Water So Important to Life?

Living organisms contain about 60% to 90% water. Water interacts with many other molecules. Since water is a polar molecule, it can dissolve many other substances (it is a good **solvent**). Water containing dissolved substances forms a solution. Water will surround positive and negative ions, dissolving crystals of polar molecules. Water is attracted to and dissolves molecules containing polar covalent bonds, called **hydrophilic** ("water-loving") molecules, such as sugars and amino acids. Uncharged and nonpolar molecules, like fats and oils, are **hydrophobic** ("water-fearing") and do not dissolve in water. Oil molecules dropped into water will nestle close together, forming **hydrophobic interactions**, and will be surrounded by water molecules attracted to each other through hydrogen bonding. The membranes of living cells owe much of their structure to hydrophobic interactions of the lipids they contain.

Due to hydrogen bonding, water molecules stick together (**cohesion**), producing the **surface tension** at the surface of lakes and pools that allows light insects to walk on water. The cohesion of water allows trees to pull water from the roots up into the leaves.

About two out of a billion water molecules at any given time become ionized, that is, they break apart into H^+ (hydrogen ions) and OH^- (hydroxide ions). A volume of pure water contains equal amounts of these ions and is said to have a value of 7 on the **pH scale** (a measure of the degree of acidity). Solutions with $H^+ > OH^-$ is **acidic** (the pH value is less than 7) and a solution with $OH^- > H^+$ is **basic** (pH greater than 7). An **acid** is a substance that releases hydrogen ions when it is dissolved in water. Hydrochloric acid (HCl) is an example of an acid. A **base** is a substance that combines with hydrogen ions, decreasing their number in solution; sodium hydroxide (NaOH) is an example of a base. Each pH scale unit represents a 10-fold increase or decrease in the concentration of H^+ ions. A cola drink has a pH of 3; thus it has 10 x 10 x 10 x 10 = 10,000 times more hydrogen ions than does pure water with a pH of 7. A **buffer** compound, like bicarbonate, helps organisms maintain a constant pH in their cells and bloodstream by accepting or releasing H^+ ions in response to small changes in pH.

Water moderates the effects of temperature changes due to three properties. Water has high specific heat (it takes a lot of energy to raise the temperature of water) due to the presence of hydrogen bonds between water molecules. The amount of energy needed to heat 1 gram of water $1°$ C is measured as a **calorie**. Water has high heat of vaporization (it takes a lot of heat to evaporate a molecule of water, leaving the remaining water cooler). Water also has a high heat of fusion (much energy is removed from water as it forms ice, thus heating up its surroundings). Since ice floats on cold liquid water, lakes remain liquid on the bottom in winter, allowing aquatic life to continue.

Health Foods revisited. Normal cellular processes produce molecules containing oxygen atoms with unfilled outer electron shells (called free radicals). Since these oxygen atoms are very reactive, they cause damage (called "oxidative stress") to other cellular molecules and structures (akin in concept to the damage oxygen in the atmosphere does when it causes rusting). Substances that react with oxygen-containing free radicals, rendering them harmless, are called antioxidants. Many foods we eat contain antioxidants (orange, yellow, and red fruits and vegetables), along with vitamins C and E. Surprisingly, even chocolate contains antioxidants called flavenoids.

KEY TERMS AND CONCEPTS

Fill-In: From the following list of terms, fill in the blanks in the following statements.

acidic	chemical bonds	hydrogen	ionic	pH	single covalent
atomic number	covalent	hydrophilic	ions	polar	specific heat
atoms	electrons	hydrophobic	neutrons	protons	triple covalent
basic	electron shells	ion	nonpolar	released	vaporization
buffers	fusion				

1. _____ contain a central dense nucleus in which are found positively charged _____ and electrically neutral _____.

2. Atoms also have negatively charged _____, which spin about the nucleus in paths called _____.

3. The _____ of an atom is the number of protons present in its nucleus and this is constant for all atoms of a particular type.

4. Atoms enter into chemical reactions when there are vacancies in their outermost _____.

5. _____ are attractive forces between atoms due to interactions of their _____.

6. _____ result when atoms lose or gain electrons. The attraction between a Na^+ _____ and a Cl^- _____ is called _____ bonding.

7. Atoms that interact by sharing electrons form molecules by _____ bonding. If two atoms share one electron each, they form a _____ bond; if two atoms share three electrons each, they form a _____ bond.

8. If two similar atoms share electrons, they form a _____ covalent bond, but if dissimilar atoms share electrons, as in water, they may form _____ covalent bonds.

9. When two water molecules electrically attract each other, the attraction is called a _____ bond. Charged molecules that attract water are called _____ molecules, while uncharged molecules that do not attract water are called _____ molecules.

10. A solution with equal amounts of H^+ and OH^- _____ has a _____ of 7.

11. _____ solutions have a pH that is less than 7, while _____ solutions have a pH greater than 7. _____ are important because they help maintain the pH of a cell at approximately 7.

12. Water moderates temperature changes because it has a high _____ (a lot of energy is needed to increase the temperature of water), it has a high heat of _____ (a lot of energy is needed to evaporate water), and it has high heat of _____ (a lot of energy is _____ when water becomes ice).

Key Terms and Definitions

acid: a substance that releases hydrogen ions (H^+) into solution; a solution with a pH of less than 7.

acidic: with an H^+ concentration exceeding that of OH^-; releasing H^+.

atom: the smallest particle of an element that retains the properties of the element.

atomic nucleus: the central dense area of an atom which contains positively charged protons and uncharged neutrons.

atomic number: the number of protons in the nuclei of all atoms of a particular element.

base: a substance capable of combining with and neutralizing H^+ ions in a solution; a solution with a pH of more than 7.

basic: with an H^+ concentration less than that of OH^-; combining with H^+.

buffer: a compound that minimizes changes in pH by reversibly taking up or releasing H^+ ions.

calorie (kal´-ō-rē): the amount of energy required to raise the temperature of 1 gram of water by 1 degree Celsius.

chemical bond: the force of attraction between neighboring atoms that holds them together in a molecule.

chemical reaction: the process that forms and breaks chemical bonds that hold atoms together.

cohesion: the tendency of the molecules of a substance to stick together.

compound: a substance whose molecules are formed by different types of atoms; can be broken into its constituent elements by chemical means.

covalent bond (kō-vā´-lent): a chemical bond between atoms in which electrons are shared.

electron: a subatomic particle, found in an electron shell outside the nucleus of an atom, that bears a unit of negative charge and very little mass.

electron shell: a region within which electrons orbit that corresponds to a fixed energy level at a given distance from the atomic nucleus of an atom.

element: a substance that cannot be broken down, or converted, to a simpler substance by ordinary chemical means.

energy level: the specific amount of energy characteristic of a given electron shell in an atom.

hydrogen bond: the weak attraction between a hydrogen atom that bears a partial positive charge (due to polar covalent bonding with another atom) and another atom, normally oxygen or nitrogen, that bears a partial negative charge; hydrogen bonds may form between atoms of a single molecule or of different molecules.

hydrophilic (hī-drō-fil´-ik): pertaining to a substance that dissolves readily in water or to parts of a large molecule that form hydrogen bonds with water.

hydrophobic (hī-drō-fō´-bik): pertaining to a substance that does not dissolve in water.

hydrophobic interaction: the tendency for hydrophobic molecules to cluster together when immersed in water.

ion (ī´-on): a charged atom or molecule; an atom or molecule that has either an excess of electrons (and hence is negatively charged) or has lost electrons (and is positively charged).

ionic bond: a chemical bond formed by the electrical attraction between positively and negatively charged ions.

isotope: one of several forms of a single element, the nuclei of which contain the same number of protons but different numbers of neutrons.

molecule (mol´-e-kūl): a particle composed of one or more atoms held together by chemical bonds; the smallest particle of a compound that displays all the properties of that compound.

neutron: a subatomic particle that is found in the nuclei of atoms, bears no charge, and has a mass approximately equal to that of a proton.

nonpolar covalent bond: a covalent bond with equal sharing of electrons.

pH scale: a scale, with values from 0 to 14, used for measuring the relative acidity of a solution; at pH 7 a solution is neutral, pH 0 to 7 is acidic, and pH 7 to 14 is basic; each unit on the scale represents a tenfold change in H^+ concentration.

polar covalent bond: a covalent bond with unequal sharing of electrons, such that one atom is relatively negative and the other is relatively positive.

proton: a subatomic particle that is found in the nuclei of atoms, bears a unit of positive charge, and has a relatively large mass, roughly equal to the mass of the neutron.

radioactive: pertaining to an atom with an unstable nucleus that spontaneously disintegrates, with the emission of radiation.

solvent: a liquid capable of dissolving (uniformly dispersing) other substances in itself.

surface tension: the property of a liquid to resist penetration by objects at its interface with the air, due to cohesion between molecules of the liquid.

THINKING THROUGH THE CONCEPTS

Identify the types of chemical bonding in the following diagram of water molecules:

13. _____

14. _____

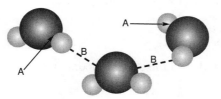

Refer to the figure to the right when answering the following questions.

15. Which is more chemically stable, a sodium atom or a sodium ion?

_____

16. Which has the higher atomic number, a chlorine atom or a chloride ion?

_____

17. The type of chemical bond depicted in part (b) is called a(n)

_____ bond.

18. The structure shown in part (c) is called a

_____.

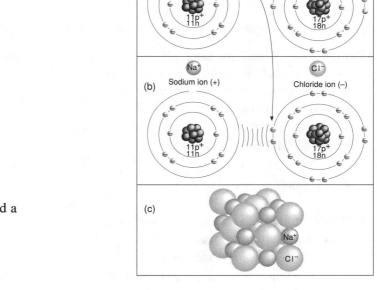

Matching: Chemical bonding.

19. _____ Two atoms share a pair of electrons.

20. _____ One atom donates an electron and another atom accepts it.

21. _____ It holds a crystal of salt together.

22. _____ It is the attraction between the polar regions of molecules.

23. _____ It results in atoms with unbalanced electrical charges.

24. _____ It holds a molecule of water together.

25. _____ It is the attraction between two water molecules.

26. _____ It is the electrical attraction between oppositely charged ions.

27. _____ It holds most biological molecules together.

28. _____ It can result in polar molecules.

Choices:

a. ionic bonding

b. covalent bonding

c. hydrogen bonding

Matching: Properties of water.

29. _____ why leaves "float" on water

30. _____ why a lot of energy is needed to convert liquid water to water vapor

31. _____ why water is such a good solvent

32. _____ why a lot of energy is needed to raise the temperature of water

33. _____ why water can break down salt crystals

34. _____ why fish can live in lakes in the wintertime

35. _____ why a lot of heat must be lost before water turns to ice

36. _____ why evaporation cools us in the summertime

37. _____ why belly-flopping dives hurt

Choices:

a. a polar molecule

b. high specific heat

c. high heat of vaporization

d. high heat of fusion

e. solid less dense than liquid

f. cohesion

Multiple Choice: Pick the most correct choice for each question.

38. A glass of lemon juice has a pH of 2 while a glass of grapefruit juice has a pH of 3. How much higher is the H^+ ion concentration in lemon juice than in grapefruit juice?

a. ten times as much

b. two-thirds as much (2/3)

c. one and a half times as much (3/2)

d. one-tenth as much

39. Which of the following is an example of hydrogen bonding? The bond between

a. O and H in a single molecule

b. O of one water molecule and the H of a second water molecule

c. O of one water molecule and the O of a second water molecule

d. H of one water molecule and the H of a second water molecule

40. What determines the atomic number of an atom?

a. the number of electrons in its outermost energy level

b. the total number of energy levels of electrons

c. the arrangement of neutrons in the atomic nucleus

d. the number of protons in the atomic nucleus

41. The nucleus of an atom never contains

a. protons

b. neutrons

c. electrons

42. For an atom to achieve maximum stability and become chemically unreactive, what must occur?

a. Its outermost energy level must be filled with electrons.

b. The number of electrons must equal the number of protons.

c. Sharing of electrons between atoms must occur.

d. Ionization of atoms is required.

e. Hydrogen bonds must form.

43. How is the formation of ions explained?

a. Different atoms share electrons.

b. Different atoms gain and lose electrons.

c. Different atoms gain and lose protons.

d. Different atoms share protons.

e. Different atoms share neutrons.

44. If a substance measures 7.0 on the pH scale, that substance
    a. has equal concentrations of H⁺ and OH⁻ ions
    b. may be very acidic
    c. has greater concentration of H⁺ and less of OH⁻ ions
    d. probably lacks OH⁻ ions
    e. may be very basic

45. As ice melts, it
    a. releases heat into its surroundings
    b. absorbs heat from its surroundings
    c. increases its property of cohesion
    d. increases its heat of vaporization
    e. immediately vaporizes

46. Why are "free radical" molecules harmful to cells?
    a. they contain poisons that kill cells
    b. they contain hydrogen atoms that attract water
    c. they contain oxygen atoms in a form that reacts with other molecules in cells
    d. they destroy the water molecules in a cell
    e. they attack food molecules, rendering them useless

47. Name some vitamins that act as antioxidants in cells.
    a. vitamins A and B
    b. vitamins B and C
    c. vitamins B and E
    d. vitamins C and E
    e. vitamins K and C

48. What is a beneficial function of antioxidants in cells?
    a. they react to form free radicals in cells
    b. they react with free radicals to render them less active
    c. they activate free radical molecules
    d. they allow reactive oxygen to increase in activity
    e. they cause free radical molecules to leave the cells

49. What does chocolate contain that may be beneficial to cells?
    a. sugar
    b. flavenoids
    c. vitamin C
    d. free radicals
    e. water

True or False: Determine if the statement given is true or false. If it is false, change the underlined word(s) so that the statement reads true.

50. _____ The smallest unit of matter in an element is the molecule.

51. _____ A positive unit in an atom is the proton.

52. _____ An electron is heavier than a proton.

53. _____ The number of protons in an atom determines its atomic weight.

54. _____ Helium is more likely to explode than hydrogen.

55. _____ In salt, sodium and chlorine atoms attract each other by forming covalent bonds.

56. _____ The sodium atoms in salt tend to take on electrons.

57. _____ In a polar water molecule, the hydrogen region has a positive electrical charge.

58. _____ When atoms share electrons, they form ionic bonds.

59. _____ Atomic reactivity depends on the number of electrons in the innermost electron shell.

60. Fill in the blanks in the following table.

Atom/Ion	Atomic number	Number of protons	Number of electrons	Number of electrons in the outermost energy level
hydrogen (H)	1	1	1	1
oxygen (O)	8		8	
carbon (C)		6		
chloride ion (Cl⁻)	17		18	

APPLYING THE CONCEPTS

These practice questions are intended to sharpen your ability to apply critical thinking and analysis to biological concepts covered in this chapter.

61. Why are winter temperatures near a large lake often a few degrees warmer than in a land-locked city in the same state, while during the summer the reverse is true?

_____

_____

_____

_____

_____

_____

62. Why does water bead up when it is spilled onto the surface of a newly waxed automobile or counter top, whereas alcohol spreads out without beading up?

_____

_____

_____

_____

_____

_____

63. Why does it hurt when you belly-flop into a swimming pool but it doesn't hurt when you make a clean dive into the pool hands first?

_____

_____

_____

_____

_____

_____

_____

64. Explain why a molecule of oxygen gas contains two covalent bonds between the pair of oxygen atoms in its structure. Remember that an oxygen atom contains 6 protons and 6 electrons.

_____

_____

_____

_____

65. Explain what a buffer is and how a buffer like bicarbonate works.

_____

_____

_____

_____

66. Mr. Smith sells his house to Mr. Jones. Mr. Brown and Mr. Green jointly buy a condominium under a time-sharing agreement in which Mr. Brown lives in the condo eight months of the year and Mr. Green lives there four months a year. Explain how these transactions are analogous to the covalent bonding of some atoms and ionic bonding of other types of atoms.

_____

_____

_____

_____

_____

67. When scuba divers swim underwater, they release into the water gas bubbles that rise to the surface. Why do you think that the bubbles formed are always spherical and never shaped like cubes or pyramids? Hint: the answer has to do with the water and not the gas.

_____

_____

_____

_____

68. A carbon atom has 6 protons and 6 electrons. How many covalent bonds do you think a carbon atom is most likely to make in order to become more stable? What would be the chemical formula of methane gas, a molecule containing one carbon along with a number of hydrogen atoms.

_____

_____

_____

_____

69. One way of expressing what happens during photosynthesis in plants is the following: carbon dioxide (CO_2) and water (H_2O) produce sugars and oxygen gas (O_2). How could a scientist use radioactive isotopes of oxygen to determine the source of the oxygen gas produced by photosynthesis?

_____

_____

_____

_____

_____

70. Explain the connection between the rusting of iron nails left outside all summer and the aging that occurs in animals like cats, dogs, and people.

_____

_____

_____

_____

_____

_____

Use the Case Study and the Web sites for this chapter to answer the following question.

71. Chocolate is not just a single molecular entity. Chocolate contains many compounds, including phenylethylamine, theobromine, anandamine, caffeine, and flavenoids. Determine the effects of these molecules on the human body.

_____

_____

_____

_____

_____

_____

ANSWERS TO EXERCISES

1. Atoms	6. Ions	9. hydrogen	vaporization
protons	ion	hydrophilic	fusion
neutrons	ion	hydrophobic	released
2. electrons	ionic	10. ions	13. polar covalent
electron shells	7. covalent	pH	14. hydrogen
3. atomic number	single covalent	11. Acidic	15. sodium ion
4. energy levels	triple covalent	basic	16. neither
5. chemical bonds	8. non-polar	Buffers	17. ionic
electrons	polar	12. specific heat	18. crystal

19. b	30. c	41. c	51. true
20. a	31. a	42. a	52. false, lighter
21. a	32. b	43. b	53. false, number
22. c	33. a	44. a	54. false,
23. a	34. e	45. b	hydrogen, helium
24. b	35. d	46. c	55. false, ionic
25. c	36. c	47. d	56. false, give up
26. a	37. f	48. b	57. true
27. b	38. a	49. b	58. false, covalent
28. b	39. b	50. false, atom	59. false, outermost
29. f	40. d		

60.

Atom/Ion	Atomic number	Number of protons	Number of electrons	Number of electrons in the outermost energy level
hydrogen (H)	1	1	1	1
oxygen (O)	8	8	8	6
carbon (C)	6	6	6	4
chloride ion (Cl⁻)	17	17	18	8

61. During the winter, the water on the surface of the lake freezes into ice, giving off heat into the surrounding atmosphere (heat of fusion). This raises the temperature around the lake in winter. During the summer, the water on the surface of the lake evaporates into the atmosphere, absorbing heat from the surrounding atmosphere (heat of vaporization). This lowers the temperature around the lake in summer.

62. Water molecules are polar and alcohol molecules are nonpolar. A waxed surface also is nonpolar. Thus, water molecules are attracted to each other by cohesion. The positive H ends of some water molecules are attracted to negative O ends of other water molecules, forming spheres of water with high surface tension (the water molecules on the surface of the sphere are not attracted to the waxy surface, only to other water molecules). Alcohol molecules spread out on the waxy surface because they are not attracted to each other, so that gravity forces them to spread out on the surface.

63. When you belly-flop into a pool, you come into contact with many millions of water molecules at the surface of the pool all at once. These water molecules at the surface have formed many hydrogen bonds with each other, resulting in a collectively strong surface tension which holds you back for a split second, causing noise and pain. However, when you do a clean dive, your hands encounter only a relatively few water molecules at the surface and it is easier to break the few hydrogen bonds attracting these few water molecules to each other.

64. Each oxygen atom has 2 electrons in the first electron shell and 6 electrons in the second shell. In order to become more stable, each oxygen seeks 2 additional electrons in the second shell. So, in order for both oxygen atoms to become more stable, they share two electrons each with each other, so that each has a filled second electron shell half of the time. Since two pairs of electrons are shared between the two oxygen atoms, a molecule of oxygen gas has two covalent bonds.

65. A buffer is a compound that minimizes changes in pH by reversibly taking up or releasing H^+ ions. If the H^+ ion concentration in a solution rises, buffers combine with them, and if the H^+ ion concentration falls, buffers release H^+ ions. Bicarbonate (HCO_3^-) is an example of a buffer. If the blood becomes too acidic, bicarbonate combines with H^+ ions to form carbonic acid [$HCO_3^- + H^+ \rightarrow H_2CO_3$]. If the blood becomes too basic, carbonic acid releases hydrogen ions, which combine with the excess hydroxide ions to form water [$H_2CO_3 + OH^- \rightarrow HCO_3^- + H_2O$].

66. Selling a house is analogous to two atoms interacting to become oppositely charged ions. When sodium (Na) and chlorine (Cl) interact, for instance, Na gives one electron permanently to Cl, so that

Na^+ and Cl^- ions result. The Na^+ no longer "owns" the electron and the Cl^- "owns" the electron completely. Time sharing a house is like atoms forming covalent bonds between them. In a water molecule, for instance, the oxygen atom and the two hydrogen atoms "time share" electrons with each other. The two pairs of shared electrons spend most of the time circling the oxygen nucleus and less time circling the hydrogen nuclei because the oxygen nucleus has more protons and, thus, exerts a stronger "pull" on the shared electrons.

67. Since water is a polar covalent molecule, the positive and negative regions of nearby water molecules will attract each other, forming hydrogen bonds. This attraction among water molecules will "push away" other molecules that are nonpolar (lack electrically charged regions). So, when scuba divers release gas into the water, the water molecules attempt to "compress" the gas into the smallest possible area. Since a sphere has less surface area per unit of internal volume than do cubes or pyramids, gas bubbles assume spherical shapes as the water molecules that surround them exert pressure.

68. Since carbon has 6 electrons, they are distributed so that 2 are in the first electron shell and 4 are in the second shell. Therefore, in order for a carbon atom to have 8 electrons in the second electron shell and become more stable, it must share its four outer electrons with four electrons from other atoms. So, each carbon tends to form 4 covalent bonds within molecules. In the case of methane gas, the carbon would form one covalent bond with each of 4 hydrogen atoms, forming a CH_4 molecule.

69. A radioactive isotope is an atom with an unstable nucleus that spontaneously disintegrates with the emission of energy. Radioactive isotopes of oxygen could be used to determine the source of the oxygen gas produced during photosynthesis. Theoretically, the O_2 could come from the CO_2 or from the H_2O molecules that react together to form the sugar and O_2 during photosynthesis. In one experiment, the H_2O could be normal and the CO_2 could contain radioactive oxygen, while in another separate experiment the CO_2 could be normal while the H_2O could contain radioactive oxygen. In each case, one could then check the O_2 produced to see in which case the gas is radioactive. When this experiment was actually performed, it was determined that the O_2 is radioactive only when the H_2O contained radioactive oxygen, thus providing evidence that the oxygen gas from photosynthesis comes from the breakdown of water molecules used during the process.

70. Iron nails left exposed to water and oxygen gas undergo "oxidation" (reacting with oxygen) to produce rust (iron oxide), destroying the structural integrity of the nail. Free radicals containing reactive oxygen cause "biological rusting" in cells, since the oxygen reacts with cellular components, oxidizing them and causing a degradation in their structures. Many scientists think the process of aging seen in cells and organisms is at least in part caused by structural damage due to oxidation by free radical molecules. For this reason, many health food products containing "antioxidant" compounds are touted as having anti-aging properties.

71. According to the authors of the "Composition of Chocolate" web site, phenylethylamine is a natural chemical in our body. It is similar to an amphetamine. Scientists say that our body releases phenylethylamine when we are in love. It has been rumored that chocolate produces the same feelings because of the phenylethylamine it contains, but it isn't true. Phenylethylamine in chocolate is broken down before it can affect the brain. A one-ounce portion of chocolate contains 20 mg of caffeine. A cup of coffee contains 135 mg. Researchers say that caffeine can cause anxiety and sleep problems. The effects of theobromine are similar to those of caffeine. Cocoa contains about seven times more theobromine than it does caffeine. Theobromine can increase the pulse rate and can produce migraines. It's important to know that theobromine is toxic to dogs. An old study found that anandamine, contained in chocolate, could act in the brain in a manner similar to tetrahydrocannabinol (THC), found in marihuana. Tyramine, an amino acid found in chocolate, is known to produce headaches. According to L. Calderon at Cal State, Los Angeles, flavenoids are phytochemicals which are non-nutritive substances in plants that possess health protective effects. Flavenoids have anti-inflammatory effects, inhibit the formation of blood clots, and have strong antioxidant activity.

Chapter 3:   Biological Molecules

OVERVIEW

In this chapter, you will learn about the basic molecules that make up living things, especially the structure, synthesis, and function of proteins, carbohydrates, fats, and sugars. These molecules are the major building blocks of our bodies. In affluent countries, obesity is a serious health problem due to overeating. Scientists are attempting to modify biological food molecules to make them less nutritious or even non-nutritious, as a means of combating obesity. For instance, sugar substitutes (like aspartame and sucralose) provide a sweet taste with few or no calories. The U.S. Food and Drug Administration (FDA) must test and approve all non-natural food molecules for safety. Olestra, for instance, mimics the culinary properties of natural oils but is completely indigestible. Though approved by the FDA, some people's digestive systems have difficulty coping with olestra, making this product controversial.

1) Why Is Carbon So Important in Biological Molecules?

Organic molecules contain carbon skeletons along with some hydrogen, while **inorganic** molecules include carbon dioxide and all molecules lacking carbon. Since each carbon can form four covalent bonds, organic molecules with many carbons can form complex shapes including chains, rings, and branches. Organic molecules also contain **functional groups** of atoms including hydroxyl (–OH), carboxyl (–COOH), amino (–NH$_2$), phosphate (–H$_2$PO$_4$), and methyl (–CH$_3$) groups. These groups help determine the chemical characteristics and chemical reactivity of organic molecules. Organic molecules are similar because they have similar functional groups, and cells use a "modular approach" to make large organic molecules.

2) How Are Organic Molecules Synthesized?

Small organic molecules (sugars, for instance) are used as **subunits** to make larger ones (like starch), by cells hooking them together like cars in a train. Many **monomers** ("one part") are joined to make a **polymer** ("many parts"). Biological molecules are built up or broken down by adding or removing water molecules. Subunits are linked together to form large molecules by a chemical reaction called **dehydration synthesis** ("to form by removing water"). One subunit loses a hydrogen (–H) and another loses a hydroxyl (–OH) group; the two subunits form a covalent bond linking them together, and the (–H) and (–OH) join to form water (H$_2$O). The reverse reaction is called **hydrolysis** ("to break apart with water"). Water splits into (–H) and (–OH) and each covalently bonds to one or another subunit of a polymer, resulting in the breakdown of the polymer by removal of individual monomers.

3) What Are Carbohydrates?

Carbohydrates have carbon, hydrogen, and oxygen in an approximate 1:2:1 ratio. Carbohydrates are single **sugars** (like glucose and fructose) called **monosaccharides**, double sugars called **disaccharides**, or longer chains of sugars (like starch and cellulose) called **polysaccharides**. Carbohydrates (sugars and starches) are used to provide energy to cells or to provide structural support. Most monosaccharides have a backbone of between 3 and 7 carbons, and have a formula of (CH$_2$O)$_n$, where n = the number of carbons in the backbone. Different monosaccharides have slightly different structures even though they contain the same types of atoms. For instance, glucose (the most common monosaccharide in living organisms), fructose (corn sugar), and galactose (part of lactose, or milk sugar) each are C$_6$H$_{12}$O$_6$. Monosaccharides may exist in either a linear or a circular (ring) form. It is in the ring form that sugars link together to make disaccharides and polysaccharides. Most small carbohydrates are soluble in water.

Disaccharides consist of two single sugars linked by dehydration synthesis. Disaccharides such as **sucrose** (table sugar made up of glucose + fructose), **lactose** (milk sugar: glucose + galactose), and **maltose** (glucose + glucose; made from starch as it breaks down in the digestive tract) often are used for short-term energy storage in plants. When energy is required, the disaccharides are broken apart into their monosaccharides by hydrolysis.

Polysaccharides are chains of single sugars, mostly glucose. Some polysaccharides are used for long-term energy storage in plants (**starch**) and animals (**glycogen**), while others provide structural support for plants (**cellulose**) and certain insects and fungi (**chitin**).

dehydration synthesis

4) What Are Lipids?

Lipid molecules are insoluble in water and contain mainly carbons and hydrogens. Some lipids (**fats** and **oils**) store energy, some (**waxes**) form waterproof coatings on plants and animals, some (phospholipids) are found in cell membranes, and some (steroids) act as hormones (made in one part of an organism and used in another). There are three major groups of lipids: oils, fats, and waxes which are simple in structure and contain only C, H, and O; phospholipids, structurally similar to oils but also containing P and N; and steroids made up of fused rings of atoms.

Oils, fats, and waxes contain only carbon, hydrogen, and oxygen. They contain **fatty acid** subunits (long chains of carbons with a –COOH (carboxyl group) at one end, and do not form ringed structures. Fats and oils form from **glycerol** and three fatty acid molecules through dehydration synthesis and are called **triglycerides**. Fats and oils store a much higher concentration of chemical energy (9.3 Calories per gram) than do sugars (4.1 Calories per gram) and proteins. One Calorie equals 1000 calories. Bears build up fat during the summer to store energy for use during the winter months while they hibernate. Fats are solid and oils are liquid at room temperature, because fats contain fatty acids **saturated** with H (making them more compact) and oils contain fatty acids **unsaturated** with less H (making them more kinky due to the presence of double covalent bonds between carbons, and thus less compact). Waxes are chemically similar to saturated fats.

Phospholipids have water-soluble "heads" and water insoluble "tails" and are found in high concentration in cell membranes. Phospholipids are similar to oils but with one fatty acid "tail" replaced by a phosphate "head" group attached to a polar charged functional group containing nitrogen. Thus, phospholipids have dissimilar ends: the "tail" is nonpolar (not soluble in water) while the "head" is polar (soluble in water).

Steroids consist of four rings of carbons fused together, with various functional groups attached. One type of steroid is cholesterol, which is found in cell membranes and is used to make male and female sex hormones and bile that assists in fat digestion.

5) What Are Proteins?

Proteins are composed of one or more chains of **amino acids**. Proteins perform many functions. Depending on the sequences of amino acids in proteins, they may function as **enzymes** (guide all chemical reaction within cells) or as structural components (elastin in skin, or keratin in hair, horns and claws), energy storage (albumin in egg white), transport (hemoglobin to carry oxygen in the blood), cell movement (muscle proteins), hormones (insulin and growth hormone), antibodies to fight infection, and poisons (snake venom).

Proteins are polymers of amino acids. All amino acids have a similar structure: a central carbon bonded to an amino group ($-NH_2$), a carboxyl group ($-COOH$), a hydrogen, and a variable (or R) group which differs among the 20 types of amino acids and gives each its distinctive properties. Some amino acids are hydrophilic because their R groups are polar and soluble in water. Other amino acids are hydrophobic, with nonpolar R groups that are insoluble in water. The amino acid "cysteine" has sulfur in its R group and can form bonds with other cysteines, linking protein chains together through the **disulfide bridges** their R groups form. Amino acids differ in their chemical and physical properties because of their different R groups. Thus, the exact sequence of amino acids dictates the function of each protein; one misplaced amino acid can cause a protein to function incorrectly.

Amino acids are joined to form protein chains by dehydration synthesis. The $-NH_2$ of one amino acid is joined to the $-COOH$ of the next by a covalent bond called a **peptide bond**, resulting in a **peptide** molecule with two amino acids. As more amino acids are added one by one, a polypeptide develops until the protein is complete.

A protein may have up to four levels of three-dimensional structure since they are highly organized molecules. The **primary structure** is the sequence of amino acids in a protein and is coded by the genes. The **secondary structure** is a coiled or **helix** structure caused by hydrogen bonding between the C=O and N-H regions of different amino acids in the sequence. Some proteins like silk consist of many protein chains lying side-by-side, with hydrogen bonds holding adjacent chains together in a **pleated sheet** arrangement (another type of secondary structure). The helical coil or pleated sheet is distorted into the **tertiary structure** when the R groups of different amino acids interact, particularly when covalent bonds (called disulfide bridges) form between the R groups of different cysteine amino acids in a protein. Interactions between the R groups of different polypeptides can form huge proteins (like hemoglobin) with two or more polypeptide subunits and such complex proteins have **quaternary structure**. Within a protein, the exact type, position, and number of amino acids bearing specific R groups determines the three-dimensional structure of the protein which in turn determines its biological function. The function of proteins is linked to its three-dimensional structure. A protein may lose its three dimensional shape, and its biological function, if it becomes **denatured**. A protein may become denatured due to exposure to high heat or an acid solution.

6) What Are Nucleic Acids?

The genetic material is composed of **nucleic acids** which are long chains of subunits called **nucleotides**. Each nucleotide has a five-carbon sugar (ribose in RNA or deoxyribose in DNA), a phosphate group, and a variable nitrogen-containing base. The four types of **ribonucleic acid** (**RNA**) nucleotides contain ribose and either adenine (A), cytosine (C), guanine (G), or uracil (U) bases, whereas the four types of **deoxyribonucleic acid** (**DNA**) nucleotides contain deoxyribose and either A, C, G, or thymine (T) bases. Nucleotides are covalently bonded together into long chains to form DNA and RNA molecules. DNA is found in the chromosomes of all living things, with the sequence of its bases providing the genetic information needed for cells to make specific proteins. RNA molecules are copied from DNA in the nucleus, move into the cytoplasm (carrying the DNA's genetic code sequence), and direct the construction of proteins there.

Other nucleotides exist singly in the cell or occur as parts of other molecules. **Cyclic nucleotides**, such as cyclic adenosine monophosphate (cyclic AMP), act as intracellular messengers that carry information from the plasma membrane to other molecules in the cell. Cyclic AMP is made when certain hormones contact the plasma membrane. The cyclic AMP then stimulates reactions in the cell. Some nucleotides (like **adenosine triphosphate** or ATP) have extra phosphate groups and carry energy from one place to another within cells. Some nucleotides (called **coenzymes**) usually combine with vitamins to assist enzymes in their functions.

KEY TERMS AND CONCEPTS

Fill-In: From the following list of terms, determine the correct answer to each of the following statements

amino acid	glucose	lactose	peptide
ATP	enzyme	lipid	protein
chitin	fat	nucleic acid	sugar
DNA	helix	nucleotides	wax

1. A _____ is a simple carbohydrate, such as a monosaccharide. _____ is the most common monosaccharide, with the formula $C_6H_{12}O_6$. _____ is a disaccharide found in mammalian milk and _____ is a polysaccharide used by fungi and some animals for structural support.

2. A _____ is a water-insoluble organic molecule such as a wax. Specifically, a _____ is a triglyceride that is solid at room temperature

3. An _____ is a molecule with a central carbon joined to –H, –NH_2, –COOH, and a variable R group. A chain made of several of these molecules joined covalently forms a _____.

4. The abbreviation for the genetic material is _____; it is made up of many _____ each containing a phosphate group, a five-carbon sugar, and a nitrogen-containing base.

5. A _____ molecule may have primary, secondary, and tertiary structure. A specific shape of the secondary structure would be a _____.

6. A _____ is a specific lipid coating that plants use to repel water.

7. A protein that guides chemical reactions in a cell is called an _____.

8. An organic molecule composed of many nucleotides is a _____; a special nucleotide with three phosphate groups for energy transfer is _____.

Key Terms and Definitions

adenosine triphosphate (a-den´-ō-sēn trī-fos´-fāt; ATP): a molecule composed of the sugar ribose, the base adenine, and three phosphate groups; the major energy carrier in cells. The last two phosphate groups are attached by "high-energy" bonds.

amino acid: the individual subunit of which proteins are made, composed of a central carbon atom bonded to an amino group (–NH_2), a carboxyl group (–COOH), a hydrogen atom, and a variable group of atoms denoted by the letter *R*.

carbohydrate: a compound composed of carbon, hydrogen, and oxygen, with the approximate chemical formula $(CH_2O)_n$; includes sugars and starches.

cellulose: an insoluble carbohydrate composed of glucose subunits; forms the cell wall of plants.

chitin (kī´-tin): a compound found in the cell walls of fungi and the exoskeletons of insects and some other arthropods; composed of chains of nitrogen-containing, modified glucose molecules.

coenzyme: an organic molecule that is bound to certain enzymes and is required for the enzymes' proper functioning; typically, a nucleotide bound to a water-soluble vitamin.

cyclic nucleotide (sik´-lik noo´-klē-ō-tī d): a nucleotide in which the phosphate group is bonded to the sugar at two points, forming a ring; serves as an intracellular messenger.

dehydration synthesis: a chemical reaction in which two molecules are joined by a covalent bond with the simultaneous removal of a hydrogen from one molecule and a hydroxyl group from the other, forming water; the reverse of hydrolysis.

denature: to disrupt the secondary and/or tertiary structure of a protein while leaving its amino acid sequence intact. Denatured proteins can no longer perform their biological functions.

deoxyribonucleic acid (dē-ox-ē-rī-bō-noo-klā´-ik; DNA): a molecule composed of deoxyribose nucleotides; contains the genetic information of all living cells.

disaccharide (dī-sak´-uh-rīd): a carbohydrate formed by the covalent bonding of two monosaccharides.

disulfide bridge: the covalent bond formed between the sulfur atoms of two cysteines in a protein; typically causes the protein to fold by bringing otherwise distant parts of the protein close together.

enzyme (en´zīm): a protein catalyst that speeds up the rate of specific biological reactions.

fat (molecular): a lipid composed of three saturated fatty acids covalently bonded to glycerol; solid at room temperature.

fatty acid: an organic molecule composed of a long chain of carbon atoms, with a carboxylic acid (COOH) group at one end; may be saturated (all single bonds between the carbon atoms) or unsaturated (one or more double bonds between the carbon atoms).

functional group: one of several groups of atoms commonly found in an organic molecule, including hydrogen, hydroxyl, amino, carboxyl, and phosphate groups, that determine the characteristics and chemical reactivity of the molecule.

glucose: the most common monosaccharide, with the molecular formula $C_6H_{12}O_6$; most polysaccharides, including cellulose, starch, and glycogen, are made of glucose subunits covalently bonded together.

glycerol (glis´-er-ol): a three carbon alcohol to which fatty acids are covalently bonded to make fats and oils.

glycogen (glī´-kō-jen): a long, branched polymer of glucose that is stored by animals in the muscles and liver and metabolized as a source of energy.

helix (hē´-liks): a coiled, springlike secondary structure of a protein.

hydrolysis (hī-drol´-i-sis): the chemical reaction that breaks a covalent bond by means of the addition of hydrogen to the atom on one side of the original bond and a hydroxyl group to the atom on the other side; the reverse of dehydration synthesis.

inorganic: describing any molecule that does not contain both carbon and hydrogen.

lactose (lak´-tōs): a disaccharide composed of glucose and galactose; found in mammalian milk.

lipid (li´-pid): one of a number of organic molecules containing large nonpolar regions composed solely of carbon and hydrogen, which make lipids hydrophobic and insoluble in water; includes oils, fats, waxes, phospholipids, and steroids.

maltose (mal´-tōs): a disaccharide composed of two glucose molecules.

monomer (mo´-nō-mer): a small organic molecule, several of which may be bonded together to form a chain called a *polymer*.

monosaccharide (mo-nō-sak´-uh-rīd): the basic molecular unit of all carbohydrates, normally composed of a chain of carbon atoms bonded to hydrogen and hydroxyl groups.

nucleic acid (noo-klā´-ik): an organic molecule composed of nucleotide subunits; the two common types of nucleic acids are ribonucleic acid (RNA) and deoxyribonucleic acid (DNA).

nucleotide: a subunit of which nucleic acids are composed; a phosphate group bonded to a sugar (deoxyribose in DNA), which is in turn bonded to a nitrogen-containing base (adenine, guanine, cytosine, or thymine in DNA). Nucleotides are linked together, forming a strand of nucleic acid, as follows: Bonds between the phosphate of one nucleotide link to the sugar of the next nucleotide.

oil: a lipid composed of three fatty acids, some of which are unsaturated, covalently bonded to a molecule of glycerol; liquid at room temperature.

organic/organic molecule: describing a molecule that contains both carbon and hydrogen.

peptide (pep´-tīd): a chain composed of two or more amino acids linked together by peptide bonds.

peptide bond: the covalent bond between the amino group's nitrogen of one amino acid and the carboxyl group's carbon of a second amino acid, joining the two amino acids together in a peptide or protein.

phospholipid (fos-fō-li´-pid): a lipid consisting of glycerol bonded to two fatty acids and one phosphate group, which bears another group of atoms, typically charged and containing nitrogen. A double layer of phospholipids is a component of all cellular membranes.

pleated sheet: a form of secondary structure exhibited by certain proteins, such as silk, in which many protein chains lie side-by-side, with hydrogen bonds holding adjacent chains together.

polymer (pah´-li-mer): a molecule composed of three or more (perhaps thousands) smaller subunits called *monomers*, which may be identical (for example, the glucose monomers of starch) or different (for example, the amino acids of a protein).

polysaccharide (pahl-ē-sak´-uh-rīd): a large carbohydrate molecule composed of branched or unbranched chains of repeating mono saccharide subunits, normally glucose or modified glucose molecules; includes starches, cellulose, and glycogen.

primary structure: the amino acid sequence of a protein.

protein: polymer of amino acids joined by peptide bonds.

quaternary structure (kwat´-er-nuh-rē): the complex three-dimensional structure of a protein composed of more than one peptide chain.

ribonucleic acid (rī-bō-noo-klā´-ik; RNA): a molecule composed of ribose nucleotides, each of which consists of a phosphate group, the sugar ribose, and one of the bases adenine, cytosine, guanine, or uracil; transfers hereditary instructions from the nucleus to the cytoplasm; also the genetic material of some viruses.

saturated: referring to a fatty acid with as many hydrogen atoms as possible bonded to the carbon backbone; a fatty acid with no double bonds in its carbon backbone.

secondary structure: a repeated, regular structure assumed by protein chains held together by hydrogen bonds; for example, a helix.

starch: a polysaccharide that is composed of branched or unbranched chains or glucose molecules; used by plants as a carbohydrate-storage molecule.

steroid: see *steroid hormone*.

steroid hormone: a class of hormone whose chemical structure (four fused carbon rings with various functional groups) resembles cholesterol; steroids, which are lipids, are secreted by the ovaries and placenta, the testes, and the adrenal cortex.

subunit: a small organic molecule, several of which may be bonded together to form a larger molecule. See also *monomer*.

sucrose: a disaccharide composed of glucose and fructose.

sugar: a simple carbohydrate molecule, either a monosaccharide or a disaccharide.

tertiary structure (ter´-shē-er-ē): the complex three-dimensional structure of a single peptide chain; held in place by disulfide bonds between cysteines.

triglyceride (trī-glis´-er-īd): a lipid composed of three fatty acid molecules bonded to a single glycerol molecule.

unsaturated: referring to a fatty acid with fewer than the maximum number of hydrogen atoms bonded to its carbon backbone; a fatty acid with one or more double bonds in its carbon backbone.

wax: a lipid composed of fatty acids covalently bonded to long-chain alcohols.

THINKING THROUGH THE CONCEPTS

True or False: Determine if the statement given is true or false. If it is false, change the <u>underlined</u> word(s) so that the statement reads true

9. _____   Sucrose is a <u>monosaccharide</u>.

10. _____   In carbohydrates, the amount of C equals the amount of <u>H</u>.

11. _____   When two monosaccharides become a disaccharide, a water molecule is <u>added</u>.

12. _____   Two glucose molecules may join together to form a <u>polysaccharide</u>.

13. _____   Animals store their food in the form of <u>glycogen</u>.

14. _____   Carbohydrates have <u>more</u> energy per gram than fats.

15. _____   Fats are made of fatty acids and <u>cholesterol</u>.

16. _____   The water-attracting portion of a phospholipid is located <u>in the middle</u> of the cell membrane.

17. _____   The acid portion of an amino acid is the <u>NH_2</u> group.

18. _____   The sequence of amino acids is the <u>secondary</u> structure of a protein.

Matching: Subunits making up large organic molecules

19. _____   nitrogen bases

20. _____   amino acids

21. _____   glycerol

22. _____   phosphate group

23. _____   five-carbon sugars

24. _____   monosaccharides

25. _____   fatty acids

Choices:

   a. carbohydrates

   b. nucleic acids

   c. proteins

   d. fats

Matching: Carbohydrates

26. _____   one sugar molecule per carbohydrate

27. _____   two sugar molecules per carbohydrate

28. _____   many sugar molecules per carbohydrate

29. _____   glucose, fructose, and galactose

30. _____   starch, glycogen, and cellulose

31. _____   sucrose, lactose, and maltose

32. _____   ribose and deoxyribose

Choices:

   a. disaccharide

   b. monosaccharide

   c. polysaccharide

Matching: Specific carbohydrates

33. _____ energy storage polysaccharide in plants

34. _____ plant cell wall component

35. _____ milk sugar

36. _____ found in insect skeletons

37. _____ most common sugar

38. _____ table sugar

39. _____ found in some nucleic acids

40. _____ energy storage polysaccharide in animals

Choices:

a. glucose

b. sucrose

c. lactose

d. deoxyribose

e. starch

f. glycogen

g. cellulose

h. chitin

Matching: Lipids

41. _____ chemically similar to fats, but is not a fat

42. _____ cholesterol

43. _____ liquid at room temperature

44. _____ contains all saturated fatty acids

45. _____ contains unsaturated fatty acids

46. _____ energy storage for plant embryos

47. _____ have hydrophilic and hydrophobic ends

48. _____ have four fused rings of carbon atoms

49. _____ a major component of cell membranes

Choices:

a. fats

b. oils

c. waxes

d. phospholipids

e. steroids

Matching: Protein structure

50. _____ interaction of R groups of different amino acids, causing contortions in shape

51. _____ sequence of amino acids in a polypeptide chain

52. _____ joining of several polypeptide chains

53. _____ helical shape of a peptide chain due to hydrogen bonding

Choices:

a. primary structure

b. secondary structure

c. tertiary structure

d. quaternary structure

Multiple Choice: Pick the most correct choice for each question

54. What type of chemical reaction results in the breakdown of organic polymers into their respective subunits?
    a. dehydration synthesis
    b. oxidation
    c. hydrolysis

55. Which of the following reactions requires the removal of water to form a covalent bond?
    a. glycogen → glucose subunits
    b. dipeptide → two amino acids
    c. cellulose → glucose
    d. glucose + galactose → lactose
    e. triglyceride → 3 fatty acids + glycerol

56. What maintains the secondary structure of a protein?
    a. peptide bonds
    b. disulfide bonds
    c. hydrogen bonds
    d. ionic bonds
    e. covalent bonds

57. What determines the specific function of a protein?
    a. the exact sequence of its amino acids
    b. the number of disulfide bonds
    c. having a hydrophilic head and a hydrophobic tail region
    d. having fatty acids as monomers
    e. the length of the molecule

58. Complex 3-dimensional tertiary structures of proteins are characterized by:
    a. an absence of hydrophilic amino acids
    b. a helical shape
    c. a lack of cysteines in the amino acid sequence
    d. the presence of disulfide bridges
    e. a pleated sheet shape

59. Which of the following is a sugar substitute?
    a. glucose
    b. aspartame
    c. fructose
    d. olestra
    e. glycerol

60. Which of the following mimics the culinary properties of natural oils?
    a. glucose
    b. aspartame
    c. fructose
    d. olestra
    e. glycerol

61. Which of the following is NOT a way that scientists recommend to combat obesity?
    a. trying a variety of "fad" diets
    b. eating less
    c. using sugar substitutes
    d. using indigestible mimics of natural oils
    e. using modified foods that are less nutritious in the diet

APPLYING THE CONCEPTS

These practice questions are intended to sharpen your ability to apply critical thinking and analysis to biological concepts covered in this chapter.

62. A bear wandered into town and died, apparently of natural causes, and the forensics lab wants to determine the types of food the bear ate just before it died. When the contents of its stomach are analyzed, the following types of molecules are found: glucose, adenine, galactose, long chains of carbon with –COOH at one end, ribose, fragments with both phosphate and nitrogen components, deoxyribose, cytosine, and chitin. Determine which of the following types of large molecules the bear consumed: lipids, proteins, carbohydrates, nucleic acids. Also, indicate whether specific dietary products such as beef or milk were present, and what types of plants or animals might have contributed to the stomach contents. Explain your choices.

_____

_____

_____

_____

_____

_____

_____

_____

63. There are literally millions of different kinds of proteins possible in living cells. However, there are very few potentially different types of starch molecules found in cells. Why do proteins and starches differ in this regard?

_____

_____

_____

_____

_____

_____

64. A trinucleotide is a small nucleic acid consisting of three nucleotide subunits. Knowing that any nucleotide can contain either an adenine, a guanine, a cytosine, or a thymine base, explain how you could determine how many different trinucleotides there could be. What is the total number of trinucleotides possible?

_____

_____

_____

_____

65. People who have difficulty digesting the lactose sugar in whole milk suffer from a condition called "lactose intolerance" often manifested as gastrointestinal distress. Scientists have discovered a substance, marketed under the name "lactaid," which can be added to a glass of milk to cause the lactose sugar to be broken down into two smaller sugars called glucose and galactose. This treated milk does not cause discomfort in lactose intolerant people. Most likely, "lactaid" is what type of molecule?

_____

_____

_____

66. Plants store most of their energy in carbohydrates, especially starch. Animals store some energy in glycogen (animal starch), but store most of their energy in lipids, especially fats. Why is using fat as a storage molecule important in animals but not in plants?

_____

_____

_____

_____

67. One technique for studying mixtures of proteins is "gel electrophoresis." A sample of various proteins is placed on an appropriate carrier (usually a rectangular slab of starch gel) and subjected to an electrical field. Some of the proteins will move towards the positively charged electrode at different speeds, others will move towards the negative electrode at different speeds, and still others will not move at all. What is it about the structure of proteins that allows gel electrophoresis to separate different types of proteins?

_____

_____

_____

_____

_____

_____

_____

_____

68. Why is obesity considered a problem in American society, and what can be done about it?

_____

_____

_____

_____

_____

_____

_____

Use the Case Study and the Web sites for this chapter to answer the following question.

69. Let's take a closer look at olestra. While most artificial sweeteners, food colors, and fat replacers, can be used for any food product, olestra is currently approved by U.S. Food and Drug Administration (FDA) for use in "savory snacks," e.g., potato chips only. Some experts say that it is safe. Others think the side effects outweigh the benefits. The Center for Science in the Public Interest (CSPI) opposes the use of olestra. What are their reasons? What are the proported benefits of olestra?

_____

_____

_____

_____

_____

_____

_____

_____

_____

_____

ANSWERS TO EXERCISES

1. sugar	16. false, on the edges	39. d
Glucose	17. false, COOH	40. f
Lactose	18. false, primary	41. c
chitin	19. b	42. e
2. lipid	20. c	43. b
fat	21. d	44. a
3. amino acid	22. b	45. b
peptide	23. b	46. b
4. DNA	24. a	47. d
nucleotides	25. d	48. e
5. protein	26. b	49. d
helix	27. a	50. c
6. wax	28. c	51. a
7. enzyme	29. b	52. d
8. nucleic acid	30. c	53. b
ATP	31. a	54. c
9. false, disaccharide	32. b	55. d
10. false, oxygen	33. e	56. c
11. false, taken away	34. g	57. a
12. false, disaccharide	35. c	58. d
13. true	36. h	59. b
14. false, less	37. a	60. d
15. false, glycerol	38. b	61. a

62. The bear had eaten lipids (long chains of carbon with –COOH at one end are fatty acids), carbohydrates (glucose and galactose are monosaccharides; chitin is a polysaccharide), and nucleic acids (ribose and deoxyribose are sugars in nucleic acids, adenine and cytosine are nitrogen bases in nucleic acids, and nucleic acid fragments have both phosphate and nitrogen components). No proteins were consumed since no amino acids are present. Milk might have been consumed since galactose sugar is present, and some insects (indicated by the presence of chitin) were eaten.

63. Starch contains only one type of subunit, the small sugar called glucose. Therefore, all starch molecules are essentially very similar, being polymers of glucose. Protein molecules, however, contain many amino acids, of which there are 20 kinds. Therefore, even small proteins can show remarkable variation in the sequence of their amino acids. For instance, if we consider proteins containing only four amino acids, there could be $(20)^4 = 20 \times 20 \times 20 \times 20 = 160,000$ varieties, since any one of 20 kinds of amino acids could be in position 1, position 2, position 3, and position 4.

64. For a trinucleotide, any of the four possible bases could be position 1, position 2, and position 3. Therefore, the number of possible trinucleotides is $(4)^3 = 4 \times 4 \times 4 = 64$.

65. "Lactaid" is a protein molecule that acts as an enzyme, breaking down the lactose disaccharide into the glucose and galactose monosaccharides.

66. Lipids are capable of storing more energy per gram of weight than are starches, and are, therefore, more efficient for organisms that must move around, like animals. Since land plants do not move around, the excess weight encountered by using starch as an energy storage molecule is of little consequence.

67. Proteins are linear polymers of 20 different kinds of amino acids. Because each amino acid has a variable side group, different individual amino acids may have either a positive charge, a negative charge, or no charge, depending on the structure of the side group. The sum of the charges on the amino acids is the "net charge" on the protein molecule. Two proteins with sufficiently different primary structures will carry different net charges, and will be attracted to one or the other electrode to a greater or lesser extent and, therefore, may be separated by gel electrophoresis.

68. Obesity is considered a problem because of the many health-related issues arising from obesity, leading to poor health and shortened life spans, and the accompanying expenses, both business related and personal, associated with those health problems. Good nutrition habits, moderate eating patterns, eating foods lower in fat content and calories (including sugar substitutes and less digestible lipids), more exercise, and generally more active lifestyles can prevent obesity in the vast majority of people.

69. The Center for Science in the Public Interest (CSPI) opposes the use of olestra for the following reasons (see the web-site for more details). (1) Olestra rapidly depletes blood levels of many valuable fat-soluble substances, including carotenoids. (2) Supplementing olestra with selected vitamins will not solve all of olestra's nutrient-depletion problems. (3) Olestra causes gastrointestinal disturbances, which are sometimes severe, including diarrhea, fecal urgency, and more frequent and looser bowel movements. (4) Olestra sometimes causes underwear staining associated with "anal leakage." (5) Data are lacking on the health effects of olestra on potentially vulnerable segments of the population. Key tests were unacceptably brief. (6) Olestra's possible carcinogenicity needs to be better resolved. (7) Procter & Gamble's claim that olestra's gastrointestinal effects are similar to those caused by high-fiber diets is not true. (8) It is not possible to set an Acceptable Daily Intake (ADI) for olestra use in snack foods. (9) Any benefits of olestra do not outweigh the risks. Some promoted benefits of olestra, according to Proctor & Gamble, include olestra has great taste and satisfaction, low fat foods form the foundation for a healthy diet, and calorie reduction is healthy (olestra is a smart tool in the fight against fat).

Chapter 4:   Cell Membrane Structure and Function

OVERVIEW

This chapter introduces you to the characteristics and functions of cell walls and cell membranes. You will learn about the transport of molecules across cell membranes, especially the processes of diffusion and osmosis. You also will learn how cells are connected and how cells communicate with each other. The chapter begins with an account of two young hikers, one of which is bitten by a Western Diamondback rattlesnake, with initial symptoms of rapidly spreading bruising in the bitten area, dropping blood pressure, shortness of breath, dizziness and nausea.

1) How Is the Structure of a Membrane Related to Its Function?

The **plasma membrane** acts as a cellular gatekeeper, surrounding, protecting and isolating the cell while allowing it extensive communication with its surroundings. The three basic functions of a plasma membrane are to selectively isolate the cell's contents from the external environment, to regulate the exchange of essential substances between the cell's contents and the surrounding environment, and to communicate with other cells. The basic structure of membranes is a collection of proteins moving within a double layer of phospholipids (and, in animal cells, cholesterol). This is called the **fluid mosaic model** of cellular membrane structure; the proteins are like "tiles" moving within a double layer of phospholipids which acts as a viscous, fluid "grout." The overall distribution of proteins and various types of phospholipids can change over time within a plasma membrane.

Within membranes, phospholipids arrange themselves into a double layer called the **phospholipid bilayer** with the hydrophilic head forming the outer borders facing the watery extracellular fluid or the internal watery **cytoplasm**, and the hydrophobic tails facing each other inside the layers. Polar water-soluble molecules (salts, amino acids, sugars) cannot pass easily through the phospholipid bilayer. A variety of proteins are embedded within or attached to the phospholipid bilayer to regulate molecular movement through the membrane and to communicate with the environment. Some have carbohydrates attached, forming **glycoproteins**, which aid in cell communication. In most animal cells, the phospholipid bilayer also contains cholesterol, which makes the bilayer stronger, more flexible but less fluid, and less permeable to water-soluble substances such as ions and monosaccharides.

The three major categories of membrane proteins are transport proteins, receptor proteins, and recognition proteins. **Transport proteins** regulate movement of water-soluble molecules through the plasma membrane. Specifically, **channel proteins** form pores to allow small molecules and ions (Ca^{++}, K^+, Na^+) to pass through and **carrier proteins** bind molecules and, by changing shape, pass them across the membrane. **Receptor proteins** trigger cell responses and/or communication between cells when certain molecules (hormones or nutrients) bind to them. **Recognition proteins** often are glycoproteins on the outer membrane surface of certain cells (immune system cells, for instance) and serve as identification tags and attachment sites for other cells and molecules.

2) How Do Substances Move Across Membranes?

A **fluid** is any substance that can move or change shape in response to external forces without breaking apart. The **concentration** of molecules in a fluid is the number of molecules in a given amount of volume. A **gradient** is a physical difference between two regions of space that causes molecules to move from one region to the other. Cells frequently generate or encounter gradients of concentration, pressure, and/or electrical charge. Molecules in fluids move randomly in response to **concentration gradients**, from regions of greater concentration (numbers of molecules in a given volume) to regions of lower concentration. Over time, random movement will produce a net movement of molecules from regions of high concentration to regions of low concentration, a process called **diffusion**. Eventually, diffusion will result in molecules becoming evenly dispersed throughout a fluid or the air, creating a dynamic equilibrium, since the molecules continue to move randomly. You can watch diffusion in action by pacing a drop of food coloring in a glass of water and looking at it every five minutes.

Movement across membranes occurs by both passive and active transport. During **passive transport**, substances move into and out of cells down concentration gradients, from regions of high concentration to regions of low concentration by diffusion and require no cell energy. The plasma membrane's lipid and protein pores regulate which molecules can cross, but they do not influence the direction of flow. During **active transport**, cells use energy to move substances against the concentration gradient, using membrane transport proteins to control the direction of flow. For example, when you ride a bike, if you don't pedal, you can only go downhill (like passive transport), but if you use energy to pedal, you can go uphill (like active transport). The greater the concentration gradient, the faster diffusion occurs, but diffusion cannot move molecules rapidly over long distances.

Passive transport includes simple diffusion, facilitated diffusion, and osmosis. Plasma membranes are **differentially permeable** to diffusion of molecules since they allow some molecules to pass across but prevent other molecules from passing across. Different molecules cross the plasma membrane at different locations and at different rates. In **simple diffusion**, water, dissolved gases, or lipid-soluble molecules pass freely through the phospholipid bilayer. In **facilitated diffusion**, most water-soluble molecules such as ions, amino acids, and small sugars, cross the membrane in a way that doesn't use energy; they are assisted by channel proteins (form pores or channels) and carrier proteins (that bind specific molecules, then change shape to allow the molecules to pass through). Facilitated transport is a slower process than simple diffusion.

Diffusion of water across differentially permeable membranes is called **osmosis**. Extracellular fluids in animals usually are equal in water concentration (**isotonic** or "having the same strength") to cellular fluids, so water diffuses equally into and out of cells. Pure water has the highest water concentration (100%). If a cell is placed in a concentrated salt solution, the salt solution has a lower water/higher dissolved salt concentration (is **hypertonic** or "having greater strength of dissolved molecules") than the cell (higher water/lower dissolved molecules concentration or **hypotonic**), water will diffuse down its concentration gradient and leave the cell faster than it enters and the cytoplasm will shrink. If, however, a solution is hypotonic to cells that are hypertonic, water will enter the cells faster than it leaves and the cytoplasm will expand. In protists that live in fresh water, cells have contractile vacuoles that pump the excess water out, while plant cells have central vacuoles which expand and allow those cells to become rigid. Examples of osmosis include water uptake by plant roots, absorption of dietary water in animal intestines, and reabsorption of water and minerals in kidneys.

Active transport uses energy to move substances against their concentration gradients into or out of cells. Digestive cells concentrate nutrients and brain cells get rid of excess ions by active transport. Active transport proteins, often called "pumps," span plasma membranes and use energy (usually from breaking down ATP molecules) to transport molecules across the membrane against the concentration gradient.

Many cells acquire particles too large to pass through membranes by **endocytosis** (using energy to surround the substance with plasma membrane and pinching it off internally to form a **vesicle**). In **pinocytosis**, a small area of membrane pinches inward to surround extracellular fluid and buds off into the cytoplasm to form a tiny vacuole. In **receptor-mediated endocytosis**, depressed areas of membrane called coated pits contain many copies of a receptor protein. These attach to specific extracellular molecules and the coated pit deepens into a U-shaped area that pinches off into the cytoplasm forming a coated vesicle. In **phagocytosis**, cells (such as *Amoeba* and white blood cells) can ingest entire microorganisms or large molecules by extending sections of plasma membrane to form **pseudopods** that surround the object and enclose it within a food vacuole in the cytoplasm for digestion. Through **exocytosis**, cells eliminate unwanted materials (like digestive waste) or secrete molecules (like hormones) into the extracellular fluid. A membrane-bound vesicle moves within the cytoplasm to the cell surface, where its membrane fuses with the plasma membrane, excreting the contents.

3) How Are Cells Specialized?

A variety of junctions allow cells to connect and communicate. **Desmosomes** attach cells together, particularly within animal tissues, by gluing together the plasma membranes of adjacent cells with proteins or carbohydrates. Protein filaments run from the desmosomes into the interiors of each cell, adding additional strength to the attachment. **Tight junctions** waterproof the membranes of cells forming tubes or sacs that must remain watertight (like the urinary bladder). The membranes of adjacent cells nearly fuse along a series of ridges, forming leakproof gaskets. **Gap junctions** are clusters of protein channels directly connecting the cytoplasm of adjacent cells to help communication (via the flow of hormones, nutrients, ions, or electrical signals) among heart muscle cells, gland cells, or brain cells, for instance. **Plasmodesmata** are cytoplasmic strands, surrounded by plasma membrane, that pass through openings in the walls of adjacent plant cells, allowing water, nutrients, and hormones to pass freely from one cell to another.

Some cells are supported by **cell walls**. Cell walls cover the outer surfaces of many cells. In protists, cell walls are made of cellulose, protein, or glassy silica. In plants, they are made of cellulose and other polysaccharides, in fungi they are made of chitin, and in bacteria they are made of a chitin-like material. In plants, cells secrete a **primary cell wall** of cellulose through their plasma membranes, and later secrete a thicker **secondary cell wall** of cellulose and other polysaccharides beneath the primary wall. The primary cell walls of adjacent cells are joined by a **middle lamella** layer made of a polysaccharide called pectin. Cells walls are strong yet porous.

Case study revisited. The reason that Western Diamondback rattlesnake venom is dangerous, even deadly, is that it contains enzymes called phospholipases, which break down the phospholipids of the plasma membranes of cells, causing the cells to rupture and die. All of Karl's symptoms were the result of the action of phospholipases on his skin, blood cells and blood vessels, muscle, and blood clotting. Antivenom contains proteins that bind to and neutralize the various toxins in the snake venom.

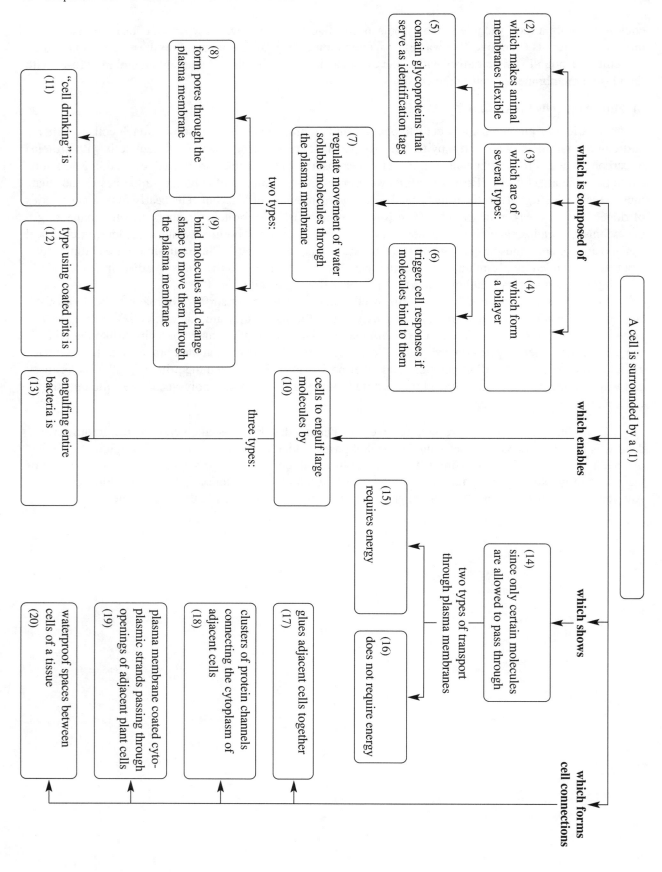

KEY TERMS AND CONCEPTS

Fill-In: From the following list of key terms, fill in all the boxes in the preceding concept map.

active transport
carrier proteins
channel proteins
cholesterol
desmosomes
differential permeability
endocytosis

gap junctions
passive transport
phagocytosis
phospholipid
pinocytosis
plasma membrane
plasmodesmata

proteins
receptor proteins
receptor-mediated endocytosis
recognition proteins
tight junctions
transport proteins

Key Terms and Definitions

active transport: the movement of materials across a membrane through the use of cellular energy, normally against a concentration gradient.

carrier protein: a membrane protein that facilitates the diffusion of specific substances across the membrane. The molecule to be transported binds to the outer surface of the carrier protein; the protein then changes shape, allowing the molecule to move across the membrane through the protein.

cell wall: a layer of material, normally made up of cellulose or cellulose-like materials, that is outside the plasma membrane of plants, fungi, bacteria, and some protists.

channel protein: a membrane protein that forms a channel or pore completely through the membrane and that is usually permeable to one or to a few water-soluble molecules, especially ions.

concentration: the number of particles of a dissolved substance in a given unit of volume.

concentration gradient: the difference in concentration of a substance between two parts of a fluid or across a barrier such as a membrane.

cytoplasm (sī´-tō-plaz-um): the material contained within the plasma membrane of a cell, exclusive of the nucleus.

desmosome (dez´-mō-sōm): a strong cell-to-cell junction that attaches adjacent cells to one another.

differentially permeable: referring to the ability of some substances to pass through a membrane more readily than can other substances.

diffusion: the net movement of particles from a region of high concentration of that particle to a region of low concentration, driven by the concentration gradient; may occur entirely within a fluid or across a barrier such as a membrane.

endocytosis (en-dō-sī-tō´-sis): the process in which the plasma membrane engulfs extracellular material, forming membrane-bound sacs that enter the cytoplasm and thereby move material into the cell.

exocytosis (ex-ō-sī-tō´-sis): the process in which intracellular material is enclosed within a membrane-bound sac that moves to the plasma membrane and fuses with it, releasing the material outside the cell.

facilitated diffusion: the diffusion of molecules across a membrane, assisted by protein pores or carriers embedded in the membrane.

fluid: a liquid or gas.

fluid mosaic model: a model of membrane structure; according to this model, membranes are composed of a double layer of phospholipids in which various proteins are embedded. The phospholipid bilayer is a somewhat fluid matrix that allows the movement of proteins within it.

gap junction: a type of cell-to-cell junction in animals in which channels connect the cytoplasm of adjacent cells.

glycoprotein: a protein to which a carbohydrate is attached.

gradient: a difference in concentration, pressure, or electrical charge between two regions.

hypertonic (hī-per-ton´-ik): referring to a solution that has a higher concentration of dissolved particles (and therefore a lower concentration of free water) than has the cytoplasm of a cell.

hypotonic (hī-pō-ton´-ik): referring to a solution that has a lower concentration of dissolved particles (and therefore a higher concentration of free water) than has the cytoplasm of a cell.

isotonic (ī-sō-ton´-ik): referring to a solution that has the same concentration of dissolved particles (and therefore the same concentration of free water) as has the cytoplasm of a cell.

middle lamella: a thin layer of sticky polysaccharides, such as pectin, and other carbohydrates that separates and holds together the primary cell walls of adjacent plant cells.

osmosis (oz-mō´-sis): the diffusion of water across a differentially permeable membrane, normally down a concentration gradient of free water molecules. Water moves into the solution that has a lower concentration of free water from a solution with the higher concentration of free water.

passive transport: the movement of materials across a membrane down a gradient of concentration, pressure, or electrical charge without using cellular energy.

phagocytosis (fa-gō-sī-tō´-sis): a type of endocytosis in which extensions of a plasma membrane engulf extracellular particles and transport them into the interior of the cell.

phospholipid bilayer: a double layer of phospholipids that forms the basis of all cellular membranes. The phospholipid heads, which are hydrophilic, face the water of extracellular fluid or the cytoplasm; the tails, which are hydrophobic, are buried in the middle of the bilayer.

pinocytosis (pi-nō-sī-tō´-sis): the nonselective movement of extracellular fluid, enclosed within a vesicle formed from the plasma membrane, into a cell.

plasma membrane: the outer membrane of a cell, composed of a bilayer of phospholipids in which proteins are embedded.

plasmodesma (plaz-mō-dez´-muh; pl., plasmodesmata): a cell-to-cell junction in plants that connects the cytoplasm of adjacent cells.

primary cell wall: cellulose and other carbohydrates secreted by a young plant cell between the middle lamella and the plasma membrane.

pseudopod (sood´-ō-pod): an extension of the plasma membrane by which certain cells, such as amoebae, locomote and engulf prey.

receptor-mediated endocytosis: the selective uptake of molecules from the extracellular fluid by binding to a receptor located at a coated pit on the plasma membrane and pinching off the coated pit into a vesicle that moves into the cytoplasm.

receptor protein: a protein, located on a membrane (or in the cytoplasm), that recognizes and binds to specific molecules. Binding by receptor proteins typically triggers a response by a cell, such as endocytosis, increased metabolic rate, or cell division.

recognition protein: a protein or glycoprotein protruding from the outside surface of a plasma membrane that identifies a cell as belonging to a particular species, to a specific individual of that species, and in many cases to one specific organ within the individual.

secondary cell wall: a thick layer of cellulose and other polysaccharides secreted by certain plant cells between the primary cell wall and the plasma membrane.

simple diffusion: the diffusion of water, dissolved gases, or lipid-soluble molecules through the phospholipid bilayer of a cellular membrane.

tight junction: a type of cell-to-cell junction in animals that prevents the movement of materials through the spaces between cells.

transport protein: a protein that regulates the movement of water-soluble molecules through the plasma membrane.

vesicle (ves´-i-kul): a small, membrane-bound sac within the cytoplasm.

THINKING THROUGH THE CONCEPTS

True or False: Determine if the statement given is true or false. If it is false, change the <u>underlined</u> word(s) so that the statement reads true.

21. _____   As a cell increases in size, its surface area increases <u>more rapidly</u> than its internal volume.

22. _____   Red blood cells will <u>burst</u> when placed in fresh water.

23. _____   The water-loving portion of a compound is <u>hydrophobic</u>.

24. _____   The rate of diffusion is increased by <u>decreasing</u> the temperature.

25. _____   In diffusion, molecules move toward regions of <u>higher</u> concentration.

26. _____   More water will enter a cell if it is placed in a <u>hypotonic</u> solution.

27. _____   Solutions with higher salt concentrations than a cell are <u>hypertonic</u> when compared to the cell.

28. _____   Freshwater organisms deal with the tendency of their cells to <u>gain</u> water.

29. _____   Endocytosis is the movement of substances <u>into</u> cells.

30. _____   The movement of a solid substance into a cell is <u>pinocytosis</u>.

Identify: Determine whether the following statements refer to **cell walls** or **plasma membranes**.

31. _____   contains cellulose in plants

32. _____   isolates the cytoplasm from the external environment

33. _____   regulates flow of materials into and out of cells

34. _____   contains chitin in fungi

35. _____   communicates with other cells

36. _____   stiff, porous, and non-living

37. _____   "fluid-mosaic model"

38. _____   has a lipid bilayer

Identify: Determine whether the following statements refer to **diffusion** or specifically to **osmosis**.

39. _____   effect of movement of all molecules down the concentration gradient

40. _____   effect of water moving down its concentration gradient across a differentially permeable membrane

41. _____   movement of O_2 into a cell and CO_2 out of a cell

42. _____   a cell expands when placed in pure water

43. _____   can cause cells to shrink

Identify: Determine whether the following statements refer to effect of osmosis on cells placed in a **hypertonic**, **hypotonic**, or **isotonic** solution.

44. _____   Animal cells will expand.

45. _____   Animal cells will shrivel up.

46. _____   Red blood cells will burst.

47. _____   Celery will wilt.

48. _____   Lettuce leaves will become turgid (rigid, crisp) in fluid.

49. _____   Red blood cells will neither shrivel up nor swell up.

Matching: Cell connections and communications

50. _____   clusters of protein channels for communication between cells

51. _____   waterproof gaskets between cells

52. _____   attachments between cells that are stretched, compressed, or bent as organisms move

53. _____   large, membrane-bound tubes for water passage between cells

Choices:

   a. tight junctions

   b. plasmodesmata

   c. gap junctions

   d. desmosomes

Multiple Choice: Pick the most correct choice for each question.

54. The hydrophobic tails of a phospholipid bilayer are oriented toward the
    a. interior of the plasma membrane
    b. extracellular fluid surrounding the cell
    c. cytoplasm of the cell
    d. nucleus of the cell

55. Molecules that permeate a plasma membrane by facilitated diffusion
    a. require the use of energy
    b. require the aid of transport proteins
    c. move from areas of low concentration to areas of high concentration
    d. do so much more quickly than those crossing by simple diffusion
    e. are water molecules

56. A molecule that can diffuse freely through a phospholipid bilayer is probably
    a. water-soluble
    b. positively charged
    c. nonpolar
    d. negatively charged
    e. a membrane-spanning protein

57. The preferential movement of water molecules across a differentially permeable membrane is termed
    a. facilitated diffusion
    b. osmosis
    c. active transport
    d. exocytosis
    e. a concentration gradient

58. If red blood cells are placed in a hypotonic solution, what happens?
    a. The cells swell and burst.
    b. The cells shrivel up and shrink.
    c. The cells remain unchanged in volume.
    d. The cells take up salt molecules from the hypotonic solution.
    e. The cells release salt molecules into the hypotonic solution.

59. Solutions that cause water to preferentially enter cells by osmosis are called
    a. hypertonic
    b. isotonic
    c. hypotonic
    d. endosmotic
    e. exosmotic

60. What does snake antivenom do to alleviate the effects of a snakebite?
    a. it kills the snake
    b. it cuts off circulation to the bitten body part
    c. it neutralizes the various toxins in the snake venom
    d. it contains new cells to take the place of the affected ones
    e. it provides vitamins to strengthen the victim

61. Most snakes are aggressive and seek out humans to attack.
    a. true
    b. false

62. Which of the following is NOT an initial symptom of being bitten by a Western Diamondback rattlesnake?
    a. a sudden burst of energy
    b. rapidly spreading bruising in the bitten area
    c. dropping blood pressure
    d. shortness of breath
    e. dizziness and nausea

63. Which enzymes contained in rattlesnake venom are especially dangerous?
    a. polymerases
    b. endonucleases
    c. lipases
    d. phospholipases
    e. lactases

APPLYING THE CONCEPTS

These practice questions are intended to sharpen your ability to apply critical thinking and analysis to biological concepts covered in this chapter.

64. Suppose you are taking a cruise from San Francisco to Hawaii. About halfway there, the ship begins to sink and all passengers and crew board lifeboats and are floating around in the ocean waiting to be rescued. After several days, you are so thirsty that you bend over the side of your life boat and drink some of the seawater. Did you do a wise thing? Explain what you think will happen to your body within a few hours of drinking the ocean water, and explain the biological basis for your reactions.

_____

_____

_____

_____

_____

_____

_____

65. Usually, lipids and water do not mix. However, soap allows us to remove greasy dirt from our hands by washing it down the drain with water. What molecules in membranes allow lipids and water to "mix" and what does that suggest to us about the chemical composition of soap?

_____

_____

_____

_____

_____

_____

_____

66. An organism that lives in fresh water is almost always hypertonic to its watery environment. Why is this a serious problem for such organisms? How do these organisms cope with the problem?

_____

_____

_____

_____

_____

_____

67. An organism that lives in the ocean is almost always hypotonic to its salt water environment. Why is this a serious problem for such organisms? How do these organisms cope with the problem?

_____

_____

_____

_____

_____

_____

68. Membranes have phospholipid, protein, and carbohydrate components. Discuss how their chemical properties are related to their functions within cell membranes.

_____

_____

_____

_____

_____

_____

_____

_____

69. Compare simple diffusion, facilitated diffusion, and active transport, including the source of energy that drives each process and use the terms "concentration gradient" and "carrier proteins" in your response.

_____

_____

_____

_____

_____

_____

_____

70. In the human body, the ways cells are attached to each other usually indicates something about the function of the cells involved. Relate the three ways that cells are attached to each other with a specific function in each case.

_____

_____

_____

_____

_____

71. Describe the differences in membrane structure in the upper and lower parts of the legs of arctic caribou that allow their lower legs to reach freezing temperatures in the winter without loss of membrane function.

_____

_____

_____

_____

_____

_____

72. Both humans and snakes have phospholipase enzymes. Explain their functions in these organisms.

_____

_____

_____

_____

_____

_____

73. Why is it important for cells to be capable of active transport?

_____

_____

_____

_____

_____

Use the Case Study and Web sites for this chapter to answer the following question.

74. Many people are afraid of snakes. A person bitten by a poisonous snake might die without anti-venom treatment, yet venom components can also be life-saving drugs. This exercise takes a closer look at snake bites and venom. What are the three major functions of snake venom? What are the major components of venom? Why does it takes several weeks to produce snake anti-venom, used to treat snake bites?

_____

_____

_____

_____

_____

_____

_____

ANSWERS TO EXERCISES

1. plasma membrane
2. cholesterol
3. proteins
4. phospholipid
5. recognition proteins
6. receptor proteins
7. transport proteins
8. channel proteins
9. carrier proteins
10. endocytosis
11. pinocytosis
12. receptor-mediated endocytosis
13. phagocytosis
14. differential permeability
15. active transport
16. passive transport
17. desmosomes
18. gap junctions
19. plasmodesmata
20. tight junctions
21. false, less rapidly

22. true
23. false, hydrophilic
24. false, increasing
25. false, lower
26. true
27. true
28. true
29. true
30. false, phagocytosis
31. cell wall
32. plasma membrane
33. plasma membrane
34. cell wall
35. plasma membrane
36. cell wall
37. plasma membrane
38. plasma membrane
39. diffusion
40. osmosis
41. diffusion
42. osmosis

43. osmosis
44. hypotonic
45. hypertonic
46. hypotonic
47. hypertonic
48. hypotonic
49. isotonic
50. c
51. a
52. d
53. b
54. a
55. b
56. c
57. b
58. a
59. c
60. c
61. b
62. a
63. d

64. Although you were thirsty and your cells craved water, drinking the salty seawater was unwise because seawater is hypertonic (has a higher concentration of salts) to hypotonic cellular cytoplasm. So, in your stomach, cells will begin to lose water due to osmosis, since water flows through a selectively permeable membrane from regions of greater water concentration (the hypotonic cytoplasm) toward regions of lesser water concentration (the hypertonic seawater). Soon, your stomach cells will have lost so much water that they will begin to die, causing you to go into convulsions and, perhaps, die as well.

65. Phospholipid molecules in cell membranes interact with both nonpolar lipids (at their fatty acid ends) and polar water molecules (at their charged ends). Soaps work in a similar fashion. One way to make soap is to boil animal fat with sodium hydroxide, creating molecules of soap with fatty acid ends (nonpolar) and ends containing sodium ions (charged polar ends). When we apply soap to our hands, the nonpolar ends attract the greasy dirt. Then, when we rinse our hands, the soap molecules are attracted to the running water, pulling the greasy dirt off our hands as the soap is washed down the drain with the water.

66. An organism living in fresh water is hypertonic to its environment and constantly will be taking in excess water due to osmosis. This could lead to the cells in the organism bursting due to the increased amount of water. Aquatic plants counteract this by having rigid cell walls around their cells that prevent the cells from expanding to the point that they burst. Aquatic animals prevent bursting by constantly pumping the excess water out through contractile vacuoles in their cells and specialized organs in their bodies.

67. An organism living in salty water is hypotonic to its environment and constantly will be losing water due to osmosis. This could lead to the cells in the organism shriveling up due to the loss of water. Marine plants prevent water loss by storing water in their central vacuoles and adding salts to the water to slow or stop the effect of osmosis. Marine animals must constantly drink water to replace that which is lost. Such animals often have specialized organs, called "salt glands" to remove salt from the ingested water and pump it out of the organism.

68. Phospholipids have both nonpolar hydrophobic regions and polar hydrophilic regions. This allows them to associate spontaneously to form a bilayer pattern in membranes. The phospholipid bilayer not only determines the permeability properties of the membrane but also determines the nature of the proteins that can be associated with the membrane. The primary structure of the proteins determines their tertiary structures, which determines where in the membrane they will reside and what their functions will be. The ability of carbohydrates to form branched patterns confers on them the specificity needed to function as molecules involved in cellular recognition reactions.

69. In simple diffusion, all substances move in the direction of the concentration gradient, and no outside energy source is needed since the process is driven by the energy of molecular movement. In facilitated transport, specific types of molecules are transported by carrier proteins in the plasma membrane. Energy for facilitated transport comes from molecular movement of molecules moving in the direction of the concentration gradient. Facilitated transport can be in either direction across the membrane, depending on the concentration gradient. Active transport involves carrier proteins and is specific in the types of molecules affected. The major difference between active transport and the other two is the use of cellular energy to drive transport against a concentration gradient. Active transport is in one direction only.

70. Desmosome connections produce very tight adhesion between cells. Desmosomes are found attaching skin cells, where protection from physical stress is required to keep the skin intact. Tight junctions prevent movement of liquids between cells. Tight junctions are found between cells lining the urinary bladder to prevent leakage of urine into adjacent tissues. Gap junctions allow for communication between the cells involved. Gap junctions are found in cardiac muscle cells needed to contract in synchrony, since gap junctions allow for rapid transfer of electrical and chemical signals in the heart.

71. During the long arctic winters, temperatures fall far below freezing. For caribou to keep their legs and feet really warm would waste energy. Specialized arrangements of arteries and veins in caribou legs allow the temperature of the lower legs to drop almost to freezing, thus conserving body heat while the upper legs and trunk remain at body temperature. To remain fluid at these different temperatures requires the phospholipids in the membranes of cells in the upper and lower legs to be very different. The fluidity of a membrane is a function of the fatty acid tails of its phospholipids. Unsaturated fatty acids remain more fluid at lower temperatures than do saturated fatty acids. In caribou, the membranes of cells near the chilly hoof have lots of unsaturated fatty acids, whereas the membranes of cells near the warmer trunk have more saturated ones.

72. Poisonous snake venom contains phospholipase enzymes that attack the cell membranes of their prey, breaking down the phospholipids of the plasma membranes of cells, causing the cells to rupture and die. Thus, snake bites cause death of tissues around the bite, rupture of red blood cells in the bloodstream leading to oxygen loss, and rupture of the blood vessels as well as muscles. All animals, including snakes and humans, have phospholipase enzymes in their digestive tracts, where they aid in the breakdown of ingested food particles. The lining of the digestive tract protects the cells of the digestive organs from being attacked by the phospholipase enzymes used to digest food.

73. Due to active transport, cells use energy to move substances into or out of cells against a concentration gradient. For instance, this allows the kidney cells to reabsorb water and send it back into the blood stream. This also allows the root cells of plants to take up minerals from the soil and make them more concentrated in the cells. Without active transport, cells could never have concentrations of molecules different from their surroundings, making life impossible.

74. According to the authors of TheSnake.org web site, snake venom is used to immobilize prey, to begin the digestion of the prey, and as a defense against predators. Venoms are highly toxic secretions produced in special oral glands. Because these oral glands are related to the salivary glands of other vertebrates, venom can be considered a modified saliva. Venoms are at least 90% protein (by dry weight), and most of the proteins in venoms are enzymes. About 25 different enzymes have been isolated from snake venom, ten of which occur in the venoms of most snakes. Proteolytic enzymes, phospholipases, and hyaluronidases are the most common types. Proteolytic enzymes catalyze the breakdown of tissue proteins. Phospholipases, which occur in almost all snakes, vary from mildly toxic to highly destructive of musculature and nerves. The hyaluronidases dissolve intercellular materials and speed the spread of venom through the prey's tissue. Other enzymes include collagenases, which occur in the venom of vipers and pitvipers and promote the breakdown of a key structural component of connective tissues (the protein collagen). Ribonucleases, deoxyribonucleases, nucleotidases, amino acid oxidases, lactate dehydrogenases, and acidic and basic phosphatases all disrupt normal cellular function, causing the collapse of cell metabolism, shock, and death. Not all toxic chemical compounds in snake venoms are enzymes. Polypeptide toxins, glycoproteins, and low-molecular-weight compounds are also present in some snakes. The roles of the other components of venom are largely unknown. Every snake's venom contains more than one toxin, and in combination the toxins have a more potent effect than the sum of their individual effects. In general, venoms are described as either neurotoxic (affecting the nervous system) or hemotoxic (affecting the circulatory system), although the venoms of many snakes contain both neurotoxic and hemotoxic components.

According to researchers at the National Aquarium in Baltimore, anti-venom is made in a complex and time-consuming process, with the following steps. (1) Venom is extracted with gentle pressure on the venom glands, or mild electric shock to constrict muscles around the gland. (2) Venom is purified and freeze-dried. Dried venom can be stored for years in a cool place. (3) A weak venom solution is injected into a horse or other animal. The concentration is increased as the animal builds immunity. The host animal makes blood rich in antibodies. (4) Blood from the animal donor is separated into dark red cells and clear serum. The serum has antibodies against venom. (5) Clear serum is further purified, stabilized for storage and standardized for effectiveness.

Chapter 5:   Cell Structure and Function

OVERVIEW

In this chapter, you will learn about the structure and function of the more primitive prokaryotic cells and the more evolutionarily advanced eukaryotic cells. In particular, the authors cover the organization of eukaryotic cells and emphasize the various organelles found in these cells. Although structurally simple in comparison to eukaryotic cells, bacterial cells have been found to withstand nearly three years exposure to the harsh environment on the surface of the moon.

1) What Are the Basic Features of Cells?

All living things are composed of one or more cells. In the 1850s, Rudolf Virchow proclaimed "All cells come from cells." Modern cell theory principles are: (1) every organism is made of at least one cell; (2) cells are the functional units of life; and (3) all cells arise from preexisting cells. All cells obtain energy and nutrients from their environment, make molecules necessary for growth and repair, get rid of wastes, interact with other cells, and reproduce. All cells share certain features.

The **plasma membrane** (phospholipid bilayer with embedded proteins) encloses the cell and mediates interactions between the cell and its environment. It (1) isolates cytoplasm from external environment, (2) regulates flow of materials between cytoplasm and its environment, and (3) allows interactions with other cells.

Cells use **DNA** as a hereditary blueprint to determine cell structure and function and allow cells to reproduce. In eukaryotic cells (plants, animals, fungi, and protists), DNA is found within the membrane-bound **nucleus**. In prokaryotic cells (bacteria and archaeans), DNA is in a non-membrane enclosed region of the cell.

All cells contain **cytoplasm** (all material inside the plasma membrane and outside the DNA-containing region). Cytoplasm includes water, salts, and organic molecules. Most of the cell's metabolic activities (biochemical reactions that underlie life) occur in the cell's cytoplasm. For instance, protein synthesis takes place on special cytoplasmic structures called **ribosomes**.

All cells obtain energy and nutrients from their environment, in order to maintain their incredible complexity. Cells that harness solar energy directly and incorporate it into high-energy molecules provide the source of energy for nearly all other life forms. The building blocks of biological molecules ultimately come from the environment.

Cell function limits cell size. Most cells are small (1 to 100 micrometers in diameter) because they need to exchange nutrients and wastes through their plasma membranes mainly by diffusion, a slow process especially over long distances. For instance, in a cell 8.5 inches in diameter, it would take over 200 days for oxygen molecules to diffuse to the middle of the cell. Larger cells have greater needs for exchange of molecules with the environment, but they have smaller surface area/internal volume ratio than do smaller cells. A cell that doubles in size becomes eight times greater in volume but only four times greater in surface area. Thus, cells tend to remain small.

2) What Are the Features of Prokaryotic ("before the nucleus") Cells?

Prokaryotic cells are small (less than 5 micrometers long) with simple internal features (no nucleus or membrane-bound organelles). Most are surrounded by a stiff cell wall that confers shape and protection. Some move using simple flagella. Surface protein projections, called **pili** (singular is **pilus**) are used to attach some bacteria surfaces or to exchange genetic material. **Capsules** or **slime layers** are polysaccharide or protein coatings that some disease-causing bacteria secrete outside their cell wall to attach to their hosts and perhaps evade attack by immune cells. Prokaryotic cells have a single, circular strand of DNA in a cytoplasmic region called the **nucleoid**. Prokaryotic cells lack nuclei as well as other membrane-enclosed organelles, but some photosynthetic bacteria have cytoplasmic membranes which contain the light-capturing proteins and enzymes that catalyze the making of high-energy molecules. Bacterial cytoplasm contains ribosomes, made of **ribonucleic acid** (**RNA**) and proteins, on which proteins are made. The cytoplasm also may contain food granules.

3) What Are the Features of Eukaryotic Cells?

Eukaryotic cells are larger (greater than 10 micrometers in diameter) and contain **organelles** that perform specific functions within cells. They also have a **cytoskeleton** that provides a network of protein fibers for cellular shape and organization.

The nucleus is usually the largest organelle in the cell, and functions as the cellular control center, containing genetic material (DNA). The DNA is used selectively by eukaryotic cells, depending on their stage of development and environmental conditions. Nuclear components are: (1) the **nuclear envelope** (two membranes perforated with pores to control flow of informational molecules), which separates nuclear material from the cytoplasm. Ribosomes line the outer nuclear membrane, which is continuous with membranes of the rough endoplasmic reticulum; (2) **chromatin** (DNA and associated proteins organized into **chromosomes**). DNA stays in the nucleus, but makes information molecules of RNA that are sent into the cytoplasm to direct the synthesis of cellular proteins; and (3) one or more **nucleoli** ("little nuclei"), containing ribosomal RNA, ribosomes in various stages of assembly, and DNA genes for making ribosomal RNA. Nucleoli are the sites of ribosome synthesis.

Eukaryotic cells contain a complex system of membranes, including the plasma membrane and several organelles including the nuclear envelope. The plasma membrane both isolates a cell and allows selective interactions between a cell and its environment. The **endoplasmic reticulum** (**ER**) forms interconnected membrane-bound tubes and channels within the cytoplasm and is continuous with the nuclear membrane. Numerous ribosomes stud the outside of the **rough ER** (site of protein synthesis), while **smooth ER** (major site of lipid synthesis) lacks ribosomes. Proteins made by ribosomes in rough ER move through ER channels and accumulate in regions that bud off to form vesicles (membrane-bound cytoplasmic sacs) that carry their protein cargo to the Golgi complex.

The **Golgi complex** (membranous sacs derived from smooth ER) has three functions: (1) it separates out lipids and proteins obtained from ER according to their destinations; (2) it chemically alters some molecules, for instance adding sugars to proteins to make glycoproteins; and (3) it packages molecules into vesicles for transport. **Lysosomes** are cellular digestive centers containing digestive enzymes to break down proteins, fats, and carbohydrates taken into cells as food. Many cells eat by phagocytosis, engulfing extracellular particles using extensions of the plasma membrane which form **food vacuoles**. Lysosomes fuse with food vacuoles and lysosomal enzymes then digest the food into small molecules. Lysosomes also digest defective organelles. Membranes flow through a cell in an orderly way, for example from ER to the Golgi complex to a vesicle which fuses with plasma membrane.

Vacuoles, which are fluid-filled membrane-bound sacs, serve many functions, including water regulation, support, and storage. **Contractile vacuoles**, found in freshwater microorganisms, use energy to pump out water constantly entering due to osmosis. **Central vacuoles**, found in plant cells, may: (1) collect cellular wastes; (2) store poisons that deter feeding animals; (3) store sugars and amino acids for cellular use; and (4) collect pigments that give flowers their colors. Central vacuole contents become hypertonic to cytoplasm and take in water through osmosis. The pressure of the expanding central vacuole, called **turgor pressure**, stiffens the cell, providing support for nonwoody plant parts. Houseplants stiffen when watered and wilt when sufficient water is lacking.

Mitochondria extract energy from food molecules and **chloroplasts** capture solar energy. Biologists believe that both mitochondria and chloroplasts evolved from prokaryotic ancestors that took up residence within the cytoplasm of ancestral eukaryotic cells. This is explained in the **endosymbiotic theory**. Both chloroplasts and mitochondria (1) are oblong and about 1–5 micrometers in diameter; (2) are surrounded by a double membrane; (3) have DNA; and (4) make **ATP**.

Mitochondria, the "powerhouses of the cell," are found in all eukaryotic cells and make ATP using energy stored in food molecules. **Anaerobic** (without oxygen) metabolism of sugar in the cytoplasm produces little ATP energy. Mitochondria use **aerobic** (with oxygen) metabolism to generate about 18 or 19 times as much ATP. The inner mitochondrial membrane loops back and forth to form deep folds (**cristae**) so that there are two regions: the **intermembrane compartment** between outer and inner membrane and the inner **matrix** region.

Chloroplasts (specialized **plastids**) are the sites of photosynthesis. Their inner membranes enclose semifluid **stroma**. The stroma contains **thylakoids** (interconnected stacks of hollow membranous discs) containing **chlorophyll** and other pigments; a stack of thylakoids is a **granum**. Chlorophyll captures sunlight energy and transfers it to other molecules that make ATP and other energy-carrier molecules. In the stroma, these energy molecules are used to combine carbon dioxide and water into sugars. Plastids store various types of molecules, including pigments in fruits and starch in potatoes.

The cytoskeleton provides shape, support, and movement. It includes several types of protein fibers: thin **microfilaments**, medium-sized **intermediate filaments**, and thick **microtubules**. Cytoskeleton functions include: (1) cell shape (especially in animal cells); (2) cell movement (through assembly, disassembly, and sliding of microfilaments and microtubules); (3) organelle movement (especially vesicles, by microfilaments and microtubules); and (4) cell division (microtubules move chromosomes into daughter nuclei, and division of the cytoplasm in animal cells results from contraction of a ring of microfilaments).

Cilia (short, with many per cell) and **flagella** (long, with few per cell) move cells or move fluid past cells. These are slender extensions of plasma membrane containing a ring of nine fused pairs of microtubules with an unfused pair in the center of the ring (a "9+2" arrangement). They are powered by ATP made by many mitochondria at their bases. Some prokaryotic cells have "flagella" but these do not contain microtubules. A **centriole** is a short, barrel-shaped ring consisting of nine microtubule triplets, but with no microtubules in the center (a "9+0" arrangement). Centrioles provide a basis for the formation of cilia or flagella. Once it begins forming the cilium or flagellum, the centriole is referred to as a **basal body**, since it is located at their base, anchoring them to the plasma membrane.

Case study revisited. Bacteria from humans have survived on the moon surface for several years in a vacuum, at temperatures near absolute zero (-273 degrees, Celsius), without nutrients, energy, or water. Bacteria form sturdy resting structures, called endospores, and may survive in that state for millions of years. Perhaps they are hardy because their simple structure is similar to forms that first colonized Earth about 3.5 billion years ago under extremely harsh conditions.

KEY TERMS AND CONCEPTS

Fill-In: Fill in the names of the structures indicated in this diagram of an animal cell, identifying the following:

cytoplasm
chromatin
Golgi complex
lysosome

mitochondrion
nucleolus
nucleus

plasma membrane
rough endoplasmic reticulum
smooth endoplasmic reticulum

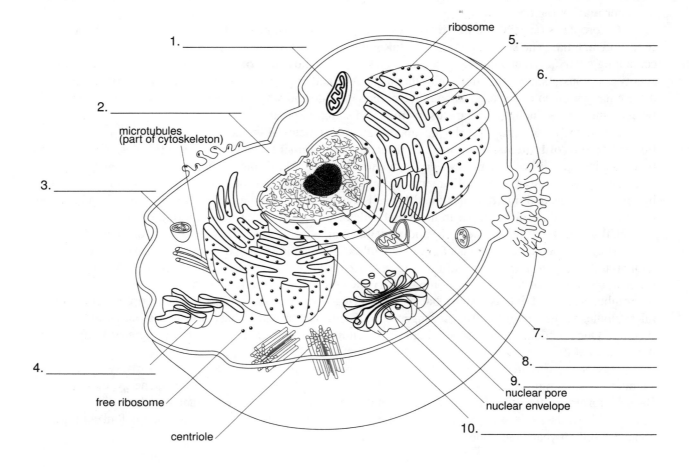

1. _____

2. _____

microtubules
(part of cytoskeleton)

3. _____

4. _____

free ribosome

centriole

ribosome

5. _____

6. _____

7. _____

8. _____

9. _____
nuclear pore
nuclear envelope

10. _____

Key Terms and Definitions

aerobic: using oxygen.

anaerobic: not using oxygen.

basal body: a structure resembling a centriole that produces a cilium or flagellum and anchors this structure within the plasma membrane.

capsule: a polysaccharide or protein coating that some disease-causing bacteria secrete outside their cell wall.

central vacuole: a large, fluid-filled vacuole occupying most of the volume of many plant cells; performs several functions, including maintaining turgor pressure.

centriole (sen´-trē-ōl): in animal cells, a short, barrel-shaped ring consisting of nine microtubule triplets; a microtubule-containing structure at the base of each cilium and flagellum; gives rise to the microtubules of cilia and flagella and is involved in spindle formation during cell division.

chlorophyll (klor´-ō-fil): a pigment found in chloroplasts that captures light energy during photosynthesis; absorbs violet, blue, and red light but reflects green light.

chloroplast (klor´-ō-plast): the organelle in plants and plantlike protists that is the site of photosynthesis; surrounded by a double membrane and containing an extensive internal membrane system that bears chlorophyll.

chromatin (krō´-ma-tin): the complex of DNA and proteins that makes up eukaryotic chromosomes.

chromosome (krō´-mō-sōm): a single DNA double helix together with proteins that help to organize the DNA.

cilium (sil´-ē-um; pl., cilia): a short, hairlike projection from the surface of certain eukaryotic cells that contains microtubules in a 9 + 2 arrangement. The movement of cilia may propel cells through a fluid medium or move fluids over a stationary surface layer of cells.

contractile vacuole: a fluid-filled vacuole in certain protists that takes up water from the cytoplasm, contracts, and expels the water outside the cell through a pore in the plasma membrane.

crista (kris´-tuh; pl., cristae): a fold in the inner membrane of a mitochondrion.

cytoplasm (sī´-tō-plaz-um): the material contained within the plasma membrane of a cell, exclusive of the nucleus.

cytoskeleton: a network of protein fibers in the cytoplasm that gives shape to a cell, holds and moves organelles, and is typically involved in cell movement.

deoxyribonucleic acid (dē-ox-ē-rī-bō-noo-klā´-ik; DNA): a molecule composed of deoxyribose nucleotides; contains the genetic information of all living cells.

endoplasmic reticulum (ER) (en-dō-plaz´-mik re-tik´-ū-lum): a system of membranous tubes and channels within eukaryotic cells; the site of most protein and lipid syntheses.

endosymbiont hypothesis: the hypothesis that certain organelles, especially chloroplasts and mitochondria, arose as mutually beneficial associations between the ancestors of eukaryotic cells and captured bacteria that lived within the cytoplasm of the pre-eukaryotic cell.

eukaryotic (ū-kar-ē-ot´-ik): referring to cells of organisms of the domain Eukarya (kingdoms Protista, Fungi, Plantae, and Animalia). Eukaryotic cells have genetic material enclosed within a membrane-bound nucleus and contain other membrane-bound organelles.

flagellum (fla-jel´-um; pl., flagella): a long, hairlike extension of the plasma membrane; in eukaryotic cells, it contains microtubules arranged in a 9 + 2 pattern. The movement of flagella propel some cells through fluids.

food vacuole: a membranous sac, within a single cell, in which food is enclosed. Digestive enzymes are released into the vacuole, where intracellular digestion occurs.

Golgi complex (gōl´-jē): a stack of membranous sacs, found in most eukaryotic cells, that is the site of processing and separation of membrane components and secretory materials.

granum (gra´-num; pl., grana): a stack of thylakoids in chloroplasts.

intermediate filament: part of the cytoskeleton of eukaryotic cells that probably functions mainly for support and is composed of several types of proteins.

intermembrane compartment: the fluid-filled space between the inner and outer membranes of a mitochondrion.

lysosome (lī´-sō-sōm): a membrane-bound organelle containing intracellular digestive enzymes.

matrix: the fluid contained within the inner membrane of a mitochondrion.

microfilament: part of the cytoskeleton of eukaryotic cells that is composed of the proteins actin and (in some cases) myosin; functions in the movement of cell organelles and in locomotion by extension of the plasma membrane.

microtubule: a hollow, cylindrical strand, found in eukaryotic cells, that is composed of the protein tubulin; part of the cytoskeleton used in the movement of organelles, cell growth, and the construction of cilia and flagella.

mitochondrion (mī-tō-kon´-drē-un): an organelle, bounded by two membranes, that is the site of the reactions of aerobic metabolism.

nuclear envelope: the double-membrane system surrounding the nucleus of eukaryotic cells; the outer membrane is typically continuous with the endoplasmic reticulum.

nucleoid (noo-klē-oid): the location of the genetic material in prokaryotic cells; not membrane-enclosed.

nucleolus (noo-klē´-ō-lus): the region of the eukaryotic nucleus that is engaged in ribosome synthesis; consists of the genes encoding ribosomal RNA, newly synthesized ribosomal RNA, and ribosomal proteins.

nucleus (cellular): the membrane-bound organelle of eukaryotic cells that contains the cell's genetic material.

organelle (or-guh-nel´): a structure, found in the cytoplasm of eukaryotic cells, that performs a specific function; sometimes refers specifically to membrane-bound structures, such as the nucleus or endoplasmic reticulum.

pilus (pil´-us; pl., pili): a hairlike projection that is made of protein, located on the surface of certain bacteria, and is typically used to attach a bacterium to another cell.

plasma membrane: the outer membrane of a cell, composed of a bilayer of phospholipids in which proteins are embedded.

plastid (plas´-tid): in plant cells, an organelle bounded by two membranes that may be involved in photosynthesis (chloroplasts), pigment storage, or food storage.

prokaryotic (prō-kar-ē-ot´-ik): referring to cells of the domains Bacteria or Archaea. Prokaryotic cells have genetic material that is not enclosed in a membrane-bound nucleus; they lack other membrane-bound organelles.

ribonucleic acid (rī¯-bo¯-noo-kla¯´-ik; RNA): a molecule composed of ribose nucleotides, each of which consists of a phosphate group, the sugar ribose, and one of the bases adenine, cytosine, guanine, or uracil; transfers hereditary instructions from the nucleus to the cytoplasm; also the genetic material of some viruses.

ribosome: an organelle consisting of two subunits, each composed of ribosomal RNA and protein; the site of protein synthesis, during which the sequence of bases of messenger RNA is translated into the sequence of amino acids in a protein.

rough endoplasmic reticulum: endoplasmic reticulum lined on the outside with ribosomes.

slime layer: a sticky polysaccharide or protein coating that some disease-causing bacteria secrete outside their cell wall; helps the cells aggregate and stick to smooth surfaces.

smooth endoplasmic reticulum: endoplasmic reticulum without ribosomes.

stroma (strō´-muh): the semi-fluid material inside chloroplasts in which the grana are embedded.

thylakoid (thī´-luh-koid): a disk-shaped, membranous sac found in chloroplasts, the membranes of which contain the photosystems and ATP-synthesizing enzymes used in the light-dependent reactions of photosynthesis.

turgor pressure: pressure developed within a cell (especially the central vacuole of plant cells) as a result of osmotic water entry.

vacuole (vak´-ū-ōl): a vesicle that is typically large and consists of a single membrane enclosing a fluid-filled space.

THINKING THROUGH THE CONCEPTS

True or False: Determine if the statement given is true or false. If it is false, change the underlined word(s) so that the statement reads true.

11. _____   More primitive types of cells are called <u>eukaryotic</u> cells.

12. _____   The presence of large numbers of ribosomes is characteristic of <u>rough</u> endoplasmic reticulum.

13. _____   In eukaryotic cells, the majority of the hereditary material is found in the <u>cytoplasm</u>.

14. _____   <u>Mitochondria</u> are associated with release of energy from sugar.

15. _____   <u>Mitochondria</u> are associated with the storage of energy in sugar.

16. _____   Lipid-producing enzymes are more common in <u>rough</u> endoplasmic reticulum.

17. _____   <u>Animals</u> store food in the form of starch.

18. _____   <u>Animal</u> cells are more likely to have vacuoles.

19. _____   Higher <u>plants</u> lack ciliated or flagellated cells.

20. _____   Cilia are <u>longer</u> than flagella.

Identify: Determine whether the following statements refer to **eukaryotic** or **prokaryotic** cells or **both**.

21. _____   lack a membrane-bound nucleus

22. _____   have many chromosomes with DNA and protein

23. _____   lack most cytoplasmic organelles

24. _____   larger and more complex

25. _____   have nucleoid regions in the cytoplasm

26. _____   have DNA

27. _____   have flagella with 9+2 structure

Identify: Determine whether the following statements refer to **mitochondria**, **chloroplasts**, or **both**.

28. _____   make ATP using solar energy

29. _____   capture sunlight energy to make sugar

30. _____   convert food energy into ATP energy

31. _____   have DNA

32. _____   have thylakoid membranes and semi-fluid stroma

33. _____   extract energy from food molecules

34. _____   found in plants

35. _____   have cristae membranes and semi-fluid matrix

36. _____   have chlorophyll

37. _____   function in photosynthesis

38. _____   use oxygen

Identify: Determine whether the following statements refer to **cilia**, **flagella**, or **both**.

39. _____ microtubular extensions through the cell membrane

40. _____ shorter and more numerous per cell

41. _____ have a "9 + 2" arrangement of microtubular pairs

42. _____ move in a wavelike motion

43. _____ move substances across a surface

44. _____ used for food gathering and for movement

Fill-In: Fill in the names of the structures indicated in the following figure.

cristae   intermembrane compartment   outer membrane
inner membrane   matrix

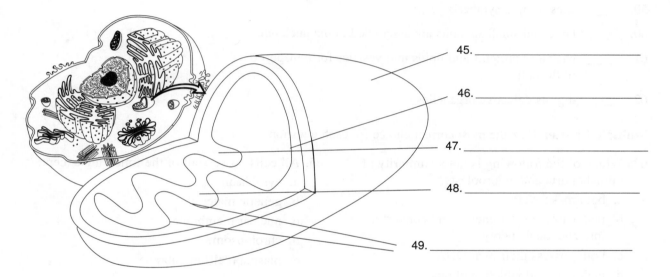

45. _____

46. _____

47. _____

48. _____

49. _____

Matching: Organelles that manufacture or digest proteins and lipids.

50. _____ digest food particles

51. _____ interconnected membrane tubes and channels
in the cytoplasm

52. _____ stacks of membranes in the cytoplasm

53. _____ made of RNA and proteins

54. _____ function in the nucleus

55. _____ membrane-bound vesicles

56. _____ "workbenches" for protein synthesis

57. _____ sort out various lipids and proteins

58. _____ may be rough or smooth in appearance

59. _____ sites of lipid synthesis

60. _____ large and small subunits are assembled in the nucleolus

61. _____ packages proteins and lipids into vesicles for transport out
of the cell

62. _____ digests defective organelles

Choices:

a. ribosomes

b. endoplasmic reticulum

c. Golgi complexes

d. lysosomes

e. none of these

Multiple Choice: Pick the most correct choice for each question.

63. Which of the following is not a similarity of
mitochondria and chloroplasts?
a. both make ATP
b. both capture solar energy and convert it
into chemical energy
c. both possess their own DNA
d. both have a double membrane
e. both probably evolved from bacteria long
ago

64. Which type of organism is most likely to
survive on the lunar surface for years at a
time?
a. plants
b. bacteria
c. fungi
d. animals
e. algae

65. All cells possess all of the following except:
a. cytoplasm
b. genetic material
c. nuclear membrane
d. chromosome
e. plasma (cell) membrane

66. Both prokaryotic and eukaryotic cells possess:
a. mitochondria
b. chloroplasts
c. cytoskeleton
d. ribosomes
e. lysosomes

67. Which organelle extracts energy from food
molecules and uses it to make ATP?
a. mitochondrion
b. chloroplast
c. ribosome
d. centriole
e. nucleus

APPLYING THE CONCEPTS

These practice questions are intended to sharpen your ability to apply critical thinking and analysis to biological concepts covered in this chapter.

68. Why do cells seldom grow large enough to be seen without the aid of a microscope?

_____

_____

_____

_____

_____

69. Suppose you discovered a chemical that had all of the following effects on a cell: cell growth slowed down, the movement of cilia slowed down, cell divisions occurred less frequently, proteins were made less often, and endocytosis occurred less often. Which of these organelles do you think the chemical affected most severely: lysosomes, Golgi complex, or mitochondria? Briefly explain your answer.

_____

_____

_____

70. Why is it not surprising that simple bacteria have been shown to be able to survive on the lunar surface for long periods of time?

_____

_____

_____

71. Although it appears that plants and animals are very different types of organisms, their cells have much more in common than you would guess. List the organelles common in plants and animals, and the organelles unique to each.

_____

_____

_____

_____

72. Why do scientists think that chloroplasts and mitochondria arose from prokaryotic cells?

_____

_____

_____

_____

73. Vigorous exercise, repeated regularly, may result in larger and firmer muscles. What organelles within the muscle cells might also increase (in number) as a result of the exercise regimen?

_____

_____

_____

_____

74. Suppose a cell is shaped like a cube, 4 micrometers long on each side. Calculate the surface area (length x width x 6 sides) and the volume (length x width x height). Also, compare the surface area to the volume of the cell (divide the surface area by the volume). Now, compare those calculations to similar ones for a cube-shaped cell 8 micrometers long on each side. What does this comparison indicate about the functioning of smaller and larger cells? What conclusion can you draw about why cells seldom grow very large?

_____

_____

_____

_____

_____

_____

_____

75. Estimate the number of cells in your little finger. Assume your finger is a cylinder, for which the volume is 3 x [(radius)^2] x height; assume that average size of a body cell is 1000 μm^3 (cubic micrometers).

_____

_____

_____

_____

_____

76. Why are scientists so interested in studying lysosomes?

_____

_____

_____

_____

_____

Use the Case Study and the Web sites for this chapter to answer the following question.

77. Bacteria that survive on the moon may get all the headlines, but conditions can be harsh on the Earth too. "Extremophiles" are organisms that not only survive but thrive under some unusual conditions on Earth. Most extremophiles are Archaea, single celled organisms that do not have nuclei but share other features with eukaryotes. There are 5 major groups of extremophiles. What are they?

_____

_____

_____

_____

_____

_____

_____

ANSWERS TO EXERCISES

1. mitochondrion	22. eukaryotic	45. outer membrane
2. cytoplasm	23. prokaryotic	46. inner membrane
3. lysosome	24. eukaryotic	47. intermembrane compartment
4. smooth endoplasmic reticulum	25. prokaryotic	48. matrix
5. rough endoplasmic reticulum	26. both	49. cristae
6. plasma membrane	27. eukaryotic	50. d
7. nucleus	28. chloroplasts	51. b
8. nucleolus	29. chloroplasts	52. c
9. chromatin	30. mitochondria	53. a
10. Golgi complex	31. both	54. e
11. false, prokaryotic	32. chloroplasts	55. d
12. true	33. mitochondria	56. a
13. false, nucleus	34. both	57. c
14. true	35. mitochondria	58. b
15. false, chloroplasts	36. chloroplasts	59. b
16. false, smooth	37. chloroplasts	60. a
17. false, plants	38. mitochondria	61. c
18. false, plant	39. both	62. d
19. true	40. cilia	63. b
20. false, shorter	41. both	64. b
21. prokaryotic	42. flagella	65. c
	43. cilia	66. d
	44. both	67. a

68. Cells seldom grow this large because of two factors. First, a cell this size would have too much volume for the amount of surface area present, so that not enough plasma membrane area would be present for the movement of molecules into and out of such a giant cell. Second, the time it would take for molecules deep inside such a cell to move to the surface or vice versa would be extremely long because diffusion over relatively long distances is too slow to support life processes.

69. Mitochondria, because all the processes affected require energy and mitochondria are the chief organelles where the energy in sugar is converted into cellular energy in the form of ATP molecules.

70. Since bacteria can produce resting cells called endospores with thick coatings they have the capacity to withstand extremely harsh conditions, it is not very surprising that they could survive on the lunar surface.

71. Plant and animal cells share the following organelles: nucleus with chromosomes and nucleolus, plasma membranes, rough and smooth endoplasmic reticulum, ribosomes, vesicles, Golgi complexes, lysosomes, food vacuoles, mitochondria, and the cytoskeleton. Plants alone have plastids (including chloroplasts), cell walls, and central vacuoles; animal cells alone have cilia and flagella, contractile vacuoles, and centrioles.

72. Chloroplasts and mitochondria are about the size and shape of large bacteria. Each has circular chromosomes made up of DNA only, like bacteria. Each has simple ribosomes resembling those of bacteria.

73. The growth of muscle cells caused by repeated, vigorous exercise would be accompanied by increased amounts of proteins and cell membranes, produced by more ribosomes and Golgi complexes. Microfilaments, made of actin protein, would increase as well. The number of mitochondria also should increase to provide more energy for muscle contraction and other cellular activities.

74. For the smaller cell, the surface area is 144 square micrometers, the volume is 64 cubic micrometers, and the ratio of surface area to volume is 2.25. For the larger cell, the surface area is 384 square micrometers, the volume is 512 cubic micrometers, and the ratio of surface area to volume is 0.75. The smaller cell has the greater surface-to-volume ratio, which would allow it to exchange materials more readily with its surroundings. As cells grow larger, the volume grows more quickly than the surface area, putting larger cells at a severe disadvantage.

75. The number of cells in a little finger is surprisingly high. Let's say a little finger is 55 millimeters tall and 10 millimeters in radius (half of its diameter). Multiply these numbers by 1000 to convert them into micrometers. So, the volume of this little finger is $3 \times [(10,000)^2] \times 55,000 = 3 \times 100,000,000 \times 55,000 = 16,500,000,000,000$, or 16.5 trillion cubic micrometers. If each cell is 1000 cubic micrometers, there are 16,500,000,000 or 16.5 billion cells in a little finger. Imagine how many cells there are in an adult human body!

76. Scientists think that lysosomes may be important in understanding cellular (and organismal) deterioration as aging occurs. As cells age, they may by subject to degeneration due to the breakdown of the lysosome membrane. If the lysosome membrane begin to leak, digestive enzymes may be released into the cytoplasm, slowly killing the cell by digesting its organelles. Being able to control the stability of the lysosome membrane may be the key to slowing the aging process.

77. According to the European Network Project in Biotechnologie, environments that are considered by humans to be extreme, are colonized by special microorganisms which are adapted to these ecological niches. These organisms are called extremophiles and may be divided into five categories: thermophiles, acidophiles, alkaliphiles, halophiles and psychrophiles, clearly indicating the nature of habitats used by these microorganisms. These habitats include hot springs, shallow submarine hydrothermal systems or abyssal hot-vent systems where microorganisms can be found at temperatures above 100° C. Extremophiles are also found in highly saline lakes, or salterns, sometimes at salt conditions near that of saturation, and in environments with extreme pH values, either acidic (acidic solfatara fields and acidic sulfur pyrite areas), or alkaline (freshwater, alkaline hot springs, carbonate springs, alkaline soils and soda lakes). The habitats of psychrophilic organisms include the cold polar seas and soils, and Alpine glaciers, as well as deep-sea sediments which are not only permanently cold but at high pressure.

Chapter 6:   Energy Flow in the Life of a Cell

OVERVIEW

In this chapter, you will learn about the flow of energy through the universe and particularly through living cells and organisms. You will read about the basic laws of thermodynamics and the basic types of chemical reactions that occur in cells and organisms. The authors also explain how cells use enzymes to control chemical reactions. The runners in the New York Marathon expend lots of energy running the 26-mile course, and more energy traveling between their homes and the race venue. What is energy? Do machines and humans follow the same principle regarding energy? What causes us to heat up when exercising? How do we store energy and use it when we need to? How do our cells "burn" sugar to release its energy?

1) What Is Energy?

Energy is the capacity to do work like making molecules, moving them around, and generating light and heat. **Kinetic energy** is the energy of movement including light, heat, and electricity. **Potential energy** is stored energy, including chemical energy stored in the bonds of molecules. Potential energy can be converted to kinetic energy and vice versa. To understand energy flow, we need to know how much energy is available, and how useful the energy is.

The **laws of thermodynamics** define the basic properties and behavior of energy. The **first law of thermodynamics** (called the law of conservation of energy) is that energy cannot be created or destroyed, although it can be changed from one form to another (chemical energy in gasoline can become the heat and movement of cars, for instance). The **second law of thermodynamics** states that when energy is converted from one form to another, the amount of useful energy decreases, since less useful heat usually is given off. Energy is automatically converted from more useful into less useful forms. Spontaneous energy conversions in nature produce an increase in randomness and disorder, called **entropy**. No process is 100% efficient. Disorder spreads through the universe, and life alone battles against it by using energy from the sun to maintain orderliness within cells.

2) How Does Energy Flow in Chemical Reactions?

A **chemical reaction** converts **reactant** substances into **products**. If the reactant energy is greater than the product energy, the reaction is **exergonic** ("energy out") because energy is released during the chemical reaction. For example, when sugar is heated (**activation energy** is added) with oxygen until it burns, chemical energy within the sugar molecules is then released as heat and light (fire), and the molecules produced (carbon dioxide and water) have less energy. Exergonic reactions release energy.

If the product energy is greater than the reactant energy, the reaction is **endergonic** ("energy in") because energy is added to the reaction as it occurs. For example, green plants use solar energy to make high-energy sugar and oxygen from low-energy water and carbon dioxide. Cells make complex biological molecules like proteins using endergonic reactions.

Coupled reactions link exergonic and endergonic reactions. In a coupled reaction, an exergonic reaction provides the energy for an endergonic reaction. The exergonic reaction of burning gasoline provides the energy for the endergonic reaction of starting a stationary car into motion and keeping it moving, even though much heat energy is lost. In photosynthesis, the exergonic reaction occurs in the sun and the endergonic reaction occurs in the plant. Most of the solar energy is lost as heat, so the second law still applies since useable energy decreases. If the coupled reactions occur in different places within cells, the energy usually is transferred from place to place by energy-carrier molecules like ATP.

63

3) How Is Cellular Energy Carried Between Coupled Reactions?

The energy from glucose (exergonic reactions) is transferred to reusable **energy-carrier molecules** for transfer to the muscle protein that uses energy to contract (endergonic reactions). Since energy-carrying molecules are somewhat unstable, they are not used for long-term energy storage. They are used only to carry energy from place to place within a cell, not between cells. **Adenosine triphosphate (ATP)** is the most common energy-carrier molecule in cells. Energy from exergonic reactions is used to make ATP from **adenosine diphosphate (ADP)** and phosphate (P). ATP carries the energy to various cellular sites where energy-requiring reactions occur. The ATP then is broken down into ADP and P, releasing energy to drive the endergonic reactions. Heat is released during these energy transfers, resulting in a loss of usable energy (an increase in entropy). Energy may be transported within a cell by other carrier molecules as well as by ATP. Some energy may be captured by electrons, which are carried by molecules called electron carriers to other parts of the cell to be released to drive endergonic reactions. Common electron carriers include nicotinamide adenine dinucleotide (NAD^+) and flavin adenine dinucleotide (FAD).

4) How Do Cells Control Their Metabolic Reactions?

Cell **metabolism** refers to the sum of all chemical reactions within cells; these often occur in sequences called **metabolic pathways**. The biochemistry of cells is controlled in three ways: (1) Cells regulate chemical reactions by using proteins called enzymes, which act as catalysts; (2) Cells couple endergonic and exergonic reactions together; and (3) Cells use energy-carrier molecules to transfer energy from exergonic reactions to endergonic ones

At body temperature, spontaneous reactions proceed too slowly to sustain life. Most reactions can be accelerated by raising the temperature, thus supplying more activation energy, but high temperatures would kill cells. Molecules called **catalysts** reduce activation energy, allowing spontaneous reactions to occur at normal body temperature at rates needed for life. All catalysts speed up spontaneous reactions, but are not permanently changed in the reactions they promote. **Enzymes** are biological catalysts made by living organisms. Enzymes lower the activation energy needed to begin exergonic chemical reactions. Most enzymes are proteins, but some ribosomal RNA molecules also act as biological catalysts. Protein enzymes are quite specific in the type of reactions they catalyze, and their activity is regulated (enhanced or suppressed) by other molecules in cells.

Due to its three-dimensional shape, a particular protein enzyme is very specific, catalyzing at most only a few types of reactions. The **active site** region of an enzyme has a distinctive shape and distribution of electrical charges that is complementary to those of its reactants (called **substrates**). Several steps occur when an enzyme catalyzes a reaction. In step 1, substrates enter active sites in specific orientations. In step 2, both substrate and active site change shape, promoting the specific chemical reaction catalyzed by the particular enzyme. In step 3, after the final reaction between the substrates is finished, the product(s) no longer fit properly into the active site and are expelled. The enzyme resumes its original configuration and, thus, is not permanently changed by the reaction it catalyzes.

Cells regulate the amount and the activity of their enzymes in several ways. Cells regulate the synthesis of enzymes to meet their changing needs. Cells make some enzymes in inactive form and activate them only when needed (some digestive enzymes are active only in the acidic environment of the stomach). Cells inhibit enzymes when adequate amounts of the enzyme's product are available (in **feedback inhibition**, an enzyme's activity is inhibited by its own product or by a subsequent product of the metabolic pathway). Certain enzymes are subject to **allosteric regulation**, when the enzyme's action is enhanced or inhibited by small organic molecules (not substrates or products) that act as regulators, binding to special allosteric regulatory sites and causing a change in the structure of the active site and the enzyme becomes either more or less able to bind its substrate. Finally, **competitive inhibition** may occur when two or more molecules somewhat similar in structure compete for the active site of an enzyme. Some poisons are competitive inhibitors that keep an enzyme from breaking down its normal substrate.

The activity of enzymes is influenced by their environment, such as pH, temperature, salt concentration, or the availability of **coenzyme** molecules (often derived from water-soluble vitamins) necessary to aid enzymes in interacting with substrates.

KEY TERMS AND CONCEPTS

Fill-In: From the following list of terms, fill in the blanks below.

active site	entropy	metabolic pathway
adenosine triphosphate (ATP)	enzyme	metabolism
catalyst	exergonic	potential energy
coenzyme	feedback inhibition	product
coupled reaction	1st law of thermodynamics	reactant
endergonic	kinetic energy	2nd law of thermodynamics
energy		

1. A substance that speeds up a chemical reaction without itself being permanently changed in the process is a _____. Specifically, a protein that speeds up the rate of a particular biological reaction is an _____. The region of this protein that binds to substrates is called the _____.

2. The _____ states that within any isolated system, energy can be neither created nor destroyed, however, energy can be converted from one form to another. The _____ states that any change in an isolated system causes the amount of useful energy to decrease, and the amount of randomness and disorder to increase. The measure of the amount of randomness and disorder in a system is _____.

3. _____ is the capacity to do work. When it involves movement, it is called _____ such as light, heat, and mechanical movement. When it is stored energy, such as in the bonds of molecules, it is called _____.

4. A molecule composed of ribose sugar, adenine, and three phosphate groups is an _____ molecule. It is the major energy carrier in cells.

5. A molecule bound to an enzyme that is required for its proper functioning is called a _____. These molecules are made from water-soluble vitamins.

6. A _____ is a molecule used up in a chemical reaction to form the _____, the molecule that results from a chemical reaction.

7. A _____ is a sequence of chemical reactions within a cell. However, _____ refers to the total of all chemical reactions occurring within a cell or organism.

8. When the product of a reaction inhibits an enzyme involved in making the product, _____ is occurring.

9. When a pair of reactions, one exergonic and one endergonic, are linked together so that the energy produced by one provides the energy needed for the other, this pair is called a _____.

10. An _____ reaction is a chemical reaction requiring an input of energy to proceed. An _____ reaction is a chemical reaction that liberates energy and increases entropy.

Key Terms and Definitions

activation energy: in a chemical reaction, the energy needed to force the electron shells of reactants together, prior to the formation of products.

active site: the region of an enzyme molecule that binds substrates and performs the catalytic function of the enzyme.

adenosine diphosphate (a-den´-ō-sēn dī-fos´-fāt; ADP): a molecule composed of the sugar ribose, the base adenine, and two phosphate groups; a component of ATP.

adenosine triphosphate (a-den´-ō-sēn trī-fos´-fāt; ATP): a molecule composed of the sugar ribose, the base adenine, and three phosphate groups; the major energy carrier in cells. The last two phosphate groups are attached by "high-energy" bonds.

allosteric regulation: the process by which enzyme action is enhanced or inhibited by small organic molecules that act as regulators by binding to the enzyme and altering its active site.

catalyst (kat´-uh-list): a substance that speeds up a chemical reaction without itself being permanently changed in the process; lowers the activation energy of a reaction.

chemical reaction: the process that forms and breaks chemical bonds that hold atoms together.

coenzyme: an organic molecule that is bound to certain enzymes and is required for the enzymes' proper functioning; typically, a nucleotide bound to a water-soluble vitamin.

competitive inhibition: the process by which two or more molecules that are somewhat similar in structure compete for the active site of an enzyme.

coupled reaction: a pair of reactions, one exergonic and one endergonic, that are linked together such that the energy produced by the exergonic reaction provides the energy needed to drive the endergonic reaction.

electron carrier: a molecule that can reversibly gain or lose electrons. Electron carriers generally accept high-energy electrons produced during an exergonic reaction and donate the electrons to acceptor molecules that use the energy to drive endergonic reactions.

endergonic (en-der-gon´-ik): pertaining to a chemical reaction that requires an input of energy to proceed; an "uphill" reaction.

energy: the capacity to do work.

energy-carrier molecule: a molecule that stores energy in "high-energy" chemical bonds and releases the energy to drive coupled endothermic reactions. In cells, ATP is the most common energy-carrier molecule.

entropy (en´-trō-pē): a measure of the amount of randomness and disorder in a system.

enzyme (en´zīm): a protein catalyst that speeds up the rate of specific biological reactions.

exergonic (ex-er-gon´-ik): pertaining to a chemical reaction that liberates energy (either as heat or in the form of increased entropy); a "downhill" reaction.

feedback inhibition: in enzyme-mediated chemical reactions, the condition in which the product of a reaction inhibits one or more of the enzymes involved in synthesizing the product.

first law of thermodynamics: the principle of physics that states that within any isolated system, energy can be neither created nor destroyed but can be converted from one form to another.

kinetic energy: the energy of movement; includes light, heat, mechanical movement, and electricity.

laws of thermodynamics: the physical laws that define the basic properties and behavior of energy.

metabolic pathway: a sequence of chemical reactions within a cell, in which the products of one reaction are the reactants for the next reaction.

metabolism: the sum of all chemical reactions that occur within a single cell or within all the cells of a multicellular organism.

potential energy: "stored" energy, normally chemical energy or energy of position within a gravitational field.

product: an atom or molecule that is formed from reactants in a chemical reaction.

reactant: an atom or molecule that is used up in a chemical reaction to form a product.

second law of thermodynamics: the principle of physics that states that any change in an isolated system causes the quantity of concentrated, useful energy to decrease and the amount of randomness and disorder (entropy) to increase.

substrate: the atoms or molecules that are the reactants for an enzyme-catalyzed chemical reaction.

THINKING THROUGH THE CONCEPTS

True or False: Determine if the statement given is true or false. If it is false, change the underlined word(s) so that the statement reads true.

11. _____   Within a closed system, the amount of energy is <u>variable</u> over time.

12. _____   A fire <u>creates</u> energy.

13. _____   The <u>second</u> law of thermodynamics is concerned with entropy.

14. _____   Photosynthesis and similar reactions <u>decrease</u> entropy.

15. _____   Eventually, all molecules in the universe will become <u>randomly dispersed</u>.

16. _____   Reactions that release energy are <u>endergonic</u>.

17. _____   Enzymes <u>increase</u> the activation energy needed for chemical reactions to occur.

18. _____   ATP contains a <u>six-carbon</u> sugar.

19. _____   Conversion of ATP to ADP <u>releases</u> energy.

20. _____   In coupled reactions, the "downhill" reaction liberates <u>less</u> energy than the "uphill" reaction.

Matching: Chemical reactions.

21. _____   Once started, these will continue by themselves.   Choices:

22. _____   These need "activation energy" to get started.   a. exergonic reactions

23. _____   Reactants have more energy than products.   b. endergonic reactions

24. _____   Photosynthesis is classified as this.   c. both of these

25. _____   Products have more energy than reactants.   d. coupled reactions

26. _____   Energy is released from the reaction.

27. _____   This usually involves energy-carrier molecules.

28. _____   Burning wood in a fireplace is classified as this.

Identify: Determine whether the following statements refer to **catalysts** in general, specifically to **enzymes**, or to **both**.

29. _____   Are protein molecules.

30. _____   Can speed up chemical reactions.

31. _____   Are not changed in the reactions they affect.

32. _____   Generally, these are very specific as to the reaction they affect.

33. _____   Cannot cause energetically unfavorable reactions to occur.

34. _____   Their activity can be regulated.

35. _____   Does not have to be a protein molecule.

Multiple Choice: Pick the most correct choice for each question.

36. In exergonic chemical reactions
    a. reactants have more energy than products
    b. reactants have less energy than products
    c. reactants and products have equal amounts of energy
    d. energy is stored in the reactions
    e. enzymes are not necessary

37. Which of the following statements about catalysts is false?
    a. Biological catalysts usually are enzymes.
    b. Catalysts increase energy of activation requirements.
    c. Catalysts often increase the rate of reaction.
    d. Catalysts are not permanently altered during the reaction.
    e. Enzymes affect the amount of activation energy required in reactions.

38. Cells regulate enzyme activity in all the following ways except
    a. the amount of enzyme manufactured may be regulated
    b. enzymes may be synthesized in an inactive form
    c. feedback inhibition may occur
    d. energy carrier molecules may be used to regulate enzyme activity

39. The second law of thermodynamics states that
    a. light can be converted into heat
    b. within an isolated system, the total amount of energy remains constant
    c. energy always flows from an area of higher concentration to an area of lower concentration
    d. useful energy increases within an isolated system
    e. useful energy decreases within an isolated system

40. When a muscle cell requires energy for contraction, what happens to ATP?
    a. ATP makes more ATP.
    b. ATP enters a metabolic pathway.
    c. ATP is broken down.
    d. ATP is phosphorylated.
    e. ATP is synthesized.

41. Which is the most common short-term energy-storage molecule?
    a. glycogen
    b. fat
    c. sucrose
    d. adenosine triphosphate

42. The statement that "energy is neither created nor destroyed" is part of
    a. entropy
    b. first law of thermodynamics
    c. second law of thermodynamics
    d. allosteric inhibition

43. What causes us to heat up and sweat when exercising?
    a. The clothes we wear make us feel hot.
    b. The room heats up and makes us feel hot.
    c. As we break down sugar, half the energy released is heat.
    d. The people exercising near us give off heat and make us feel hot.
    e. The sugar we break down turns to warm water that is given off as sweat.

44. Do the same laws of thermodynamics hold true for humans and for machines?
    a. Both laws do.
    b. The first law does, but the second law does not.
    c. The first law does not, but the second law does.
    d. Neither law does.
    e. Humans can circumvent the physical laws of nature.

Refer to the figure below to answer the following questions.

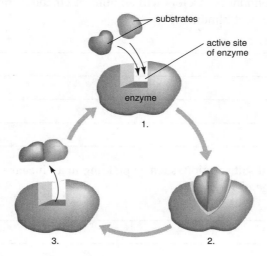

45. Which of the following statements are true?
    a. Substrates enter an enzyme's active site in a random orientation.
    b. While in the active site, substrates change their shape.
    c. The active site changes its shape once the substrates are present.
    d. Products fit the active site just as well as substrates do.
    e. Enzymes are permanently changed during chemical reactions.

46. If the figure depicts an endergonic reaction, which of the following statements are true?
    a. The substrates have more energy than the products.
    b. The products have more energy than the substrates.
    c. ATP could be produced from energy released by this reaction
    d. ATP could be used up to provide energy for this reaction
    e. The enzyme will be destroyed by the reaction.

APPLYING THE CONCEPTS

These practice questions are intended to sharpen your ability to apply critical thinking and analysis to biological concepts covered in this chapter.

47. When paper burns, it gives off both heat and light. Thus, the reaction is exergonic. Why, then, doesn't the paper this book contains spontaneously burst into flames? If you touched the book with a burning match, you could set it on fire. What role does the energy supplied by the match play in this process?

_____

_____

_____

_____

_____

_____

48. When a person drinks methanol (wood alcohol), it is broken down into formaldehyde, which causes blindness. To prevent the blindness, doctors will administer ethanol (grain alcohol) to the patient. How does this treatment prevent blindness?

_____

_____

_____

_____

_____

49. Why is using a concentrated salt solution, such as pickling in a vinegar-salt solution, effective in preventing bacterial spoilage of foods?

_____

_____

_____

_____

50. Marathon runners often will take a cup of water and pour it over their heads during the race to "cool off" as they continue to run. Why are they "heating up" during the race, and why does the water soaking help?

_____

_____

_____

_____

_____

51. Because of the phenomenon associated with the second law of thermodynamics, most automobiles need radiators and engine fans. Explain.

_____

_____

_____

_____

52. Why would you quickly die if not for the action of enzymes in your cells?

_____

_____

_____

_____

_____

Use the Case Study and the Web sites for this chapter to answer the following question.

53. Dieters focus on two simple rules. Take in too many calories, and you gain weight. Take in too few calories and weight loss, disability, and eventually death will result. But how much is too much? Use online calorie calculators to explore how different factors affect energy needs. [Remember this is only an exercise. The data obtained should not be used as the basis for any dietary or exercise program.] Let's say you are a 25-year-old male, are 6 feet tall, and weigh 250 pounds. First, enter these data into the "Basal Metabolism Calculator" to calculate your basic (non-exercising) energy needs. Now, let's say you need to start an exercise program. Enter the same numbers into the "Energy Calculator." Be patient, this calculator loads slowly. Choose "running, cross country, 20 minutes" as your activity. Calculate! How many calories did you burn while exercising? Now, return to the "Energy Calculator" and change the weight value to 200 lbs. Calculate. Did you use more or fewer calories than before?

_____

_____

_____

_____

_____

_____

_____

_____

_____

ANSWERS TO EXERCISES

1. catalyst
   enzyme
   active site
2. 1st law of thermodynamics
   2nd law of thermodynamics
   entropy
3. Energy
   kinetic energy
   potential energy
4. adenosine triphosphate (ATP)
5. coenzyme
6. reactant
   product
7. metabolic pathway
   metabolism
8. feedback inhibition
9. coupled reaction
10. endergonic
    exergonic

11. false, constant
12. false, changes
13. true
14. true
15. true
16. false, exergonic
17. false, decrease
18. false, five-carbon
19. true
20. false, more
21. a
22. c
23. a
24. d
25. b
26. a
27. d
28. a
29. enzymes

30. both
31. both
32. enzymes
33. both
34. enzymes
35. catalyst
36. a
37. b
38. d
39. e
40. c
41. d
42. b
43. c
44. a
45. b and c
46. b and d

47. The paper in this book doesn't spontaneously burst into flames because a sufficient amount of activation energy is not present to begin the intense exergonic reaction called burning. Holding a lit match to the paper would supply sufficient activation energy to begin the burning process. Often, as paper ages, it turns brown and becomes brittle. This occurs because the paper is reacting with oxygen in much the same way as in burning, except that the process is very slow due to insufficient activation energy.

48. Methanol is broken down in the body through the action of the enzyme alcohol dehydrogenase. To prevent the methanol from being broken down, administer a large quantity of ethanol, which is the normal form of alcohol broken down by alcohol dehydrogenase. The ethanol acts as a competitive inhibitor of methanol, competing for the enzyme's active site and preventing much methanol from being broken down into the harmful formaldehyde.

49. Before refrigerators were invented, meats were preserved mainly by impregnating the meats in concentrated salt solutions. These solutions kill bacteria, particularly by interfering with the functioning of their enzymes. Molecules in salts are in the form of electrically charged ions which bond with enzymes and change their three dimensional shapes, thus destroying enzyme activity.

50. As the runner's muscles contract, lots of ATP energy is used up. As each ATP is broken down to power muscle contraction, some of the energy is converted to kinetic energy as movement, and some is lost as heat. It is the accumulated heat released by the breakdown of many ATP molecules that causes the runner to sweat and feel hot. By pouring water over their heads and skin, the water absorbs a lot of heat as it warms up and evaporates (water has a high heat of evaporation). This action of water draws off heat from the runner's body and, thus, allows the runner to cool down during the race.

51. During the conversion of the chemical energy of gasoline into the kinetic energy used to move the engine parts and move the automobile, the reactions are not completely efficient, and a considerable amount of energy is released as heat, which quickly heats up the engine parts. This heat energy is dissipated by the radiator and the engine fan to prevent damage to the engine if the engine overheats.

52. The chemical reactions that need to occur in your cells to keep you alive normally happen at rates far too slow to be effective in our bodies under normal conditions. One way to speed up these reactions is to raise the body's temperature so that the molecules will move faster, hit each other at a faster rate, and speed up the reactions. But the higher temperatures needed are incompatible with life. Enzymes allow reactions to occur faster at normal body temperature. Because of the tertiary structure of the active site, the enzyme brings the reacting molecules together in a favorable orientation for the reaction to occur, thus lowering the needed activation energy. Often the interaction between enzyme and substrate results in the stretching of bonds in the substrate, allowing it to break apart without the addition of heat energy.

53. Using the "Basal Metabolism Calculator" from Room 42 Software for a 25-year-old male who is 6 feet tall and weighs 250 pounds, the metabolism results are: 2166.8 calories per day is his Basal Caloric Rate. This is: no more then 72.229 grams of fat (30%) for his Basal Caloric Rate, 81.258 grams of protein (15%) for his Basal Caloric Rate, and 297.94 grams of carbohydrate (55%) for his Basal Caloric Rate. Running cross-country for 20 minutes would use up 180 kilocalories of energy for a 250 pound man, and also 180 kcal for a 200 (or even a 300) pound man.

Chapter 7: Capturing Solar Energy: Photosynthesis

OVERVIEW

You will learn about photosynthesis in this chapter. The authors outline the light-dependent and light-independent steps of photosynthesis, and present the two major ways that plants trap carbon dioxide. About 65 million years ago, a 6-mile in diameter meteorite crashed into the Earth near the Yucatan Peninsula, creating a hole one mile deep and 120 miles across. Trillions of tons of debris were thrown up into the stratosphere, and heat from the impact caused massive fires. Ashes, smoke and dust blocked sunlight from reaching Earth for months, causing the demise of the dinosaurs.

1) What Is Photosynthesis?

Around 2 billion years ago, mutations caused some cells to gain the ability to combine simple molecules into glucose by harvesting solar energy, a process called **photosynthesis**. Other mutations caused cells to use oxygen to break down glucose more efficiently, a process called cellular respiration.

Starting with carbon dioxide and water, photosynthesis converts sunlight energy into chemical energy stored in the bonds of glucose and oxygen ($6 CO_2 + 6 H_2O + \text{light energy} \rightarrow C_6H_{12}O_6 + 6 O_2$). It occurs in plants, algae, and some bacteria. Photosynthetic organisms are called autotrophs ("self- feeders"). In plants, leaves and chloroplasts are adaptations for photosynthesis. Leaves obtain CO_2 from the air through adjustable pores (called **stomata**; singular is **stoma**) in the outer transparent cells of the epidermis. **Mesophyll** ("middle of the leaf") cells within leaves contain most of the chloroplasts. Vascular bundles (veins) carry water and minerals to the mesophyll cells and carry sugars to other plant parts.

Chloroplast organelles consist of a double outer membrane enclosing a semifluid medium, the **stroma**, which contains disk-shaped membranous sacs called **thylakoids**. Thylakoids tend to be stacked atop each other in stacks called **grana** (singular is **granum**).

The dozens of reactions in photosynthesis can be grouped into two series of reactions. In the **light-dependent** reactions, chlorophyll and other molecules in the thylakoids capture sunlight energy and convert it into chemical energy stored in ATP and NADPH (nicotinamide adenine dinucleotide phosphate) energy-carriers. Oxygen gas is released. In the **light-independent** reactions, enzymes in the stroma use the chemical energy from ATP and NADPH to make glucose or other organic molecules.

2) Light-Dependent Reactions: How Is Light Energy Converted to Chemical Energy?

Light is first captured by pigments in chloroplasts. Wavelengths of light are composed of **photons** (individual packets of energy). Short wavelength photons are very energetic, whereas long wavelength photons have lower energies. Visible light is energetic enough to alter the shape of certain pigment molecules but not strong enough to cause mutations in DNA. When light strikes a leaf, it is either absorbed, reflected (bounced back again), or transmitted (passed through). Reflected and transmitted light reaches our eyes, and gives an object its color.

Chloroplasts contain several pigments that absorb different wavelengths of light: **chlorophyll** absorbs violet, blue, and red light and reflects green (this is why leaves are green) and **accessory pigments** (**carotenoids** absorb blue and green and reflect yellow, orange, and red, and **phycocyanins** absorb green and yellow and reflect blue and purple) absorb light energy and transfer it to chlorophyll.

Light-dependent reactions occur in clusters of molecules called **photosystems** (proteins including chlorophyll, accessory pigments, and electron-carrying molecules) within the thylakoid membranes. Each thylakoid contains thousands of copies of two types of photosystems, named photosystem I and photosystem II. Each photosystem has two major parts: (1) a **light-harvesting complex** (300 pigment molecules that absorb light and pass the energy to a specific chlorophyll called the **reaction center**); and (2) an **electron transport system** (ETS). The ETS is a series of electron carrier molecules embedded in the thylakoid membranes. Photosystem II generates ATP, and photosystem I generates NADPH. When the reaction center chlorophyll receives energy, one of its electrons enters the ETS and moves from one carrier to the next, releasing energy that allows ADP to form ATP (through a hydrogen ion gradient called **chemiosmosis** in photosystem II) and $NADP^+$ to form NADPH (in photosystem I). Electrons from the ETS of photosystem II replenish those lost by the reaction center chlorophyll of photosystem I.

Splitting water maintains the flow of electrons through the photosystems. Electrons flow one way: from splitting water through reaction center of system II through ETS of system II to reaction center of system I through ETS of system I to NADPH.

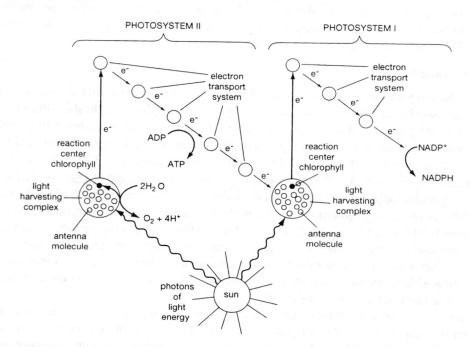

3) Light Independent Reactions: How Is Chemical Energy Stored in Glucose Molecules?

As long as sufficient ATP and NADPH are available, the light-independent reactions do not require sunlight. The **C_3 cycle** (three-carbon or **Calvin-Benson cycle**) captures carbon dioxide. It requires (1) CO_2 from the air; (2) CO_2-capturing sugar, ribulose bisphosphate (RuBP); (3) stroma enzymes; and (4) energy from ATP and NADPH (from light-dependent reactions).

The three steps of the C_3 cycle are **carbon fixation** (CO_2 attaches to RuBP), using PGA to make G3P intermediate molecules using energy from ATP and NADPH, and regeneration of RuBP. Carbon fixed during the C_3 cycle is used to make glucose.

4) How Are the Light-Dependent and Light-Independent Reactions Related?

The "photo" part of photosynthesis is the capture of light energy by the light-dependent reactions, using chlorophyll in the thylakoids to "charge up" ADP and $NADP^+$ to form ATP and NADPH. The "synthesis" part refers to the making of glucose that occurs in the light-independent reactions using enzymes in the stroma and energy from ATP and NADPH captured by the light-dependent reactions.

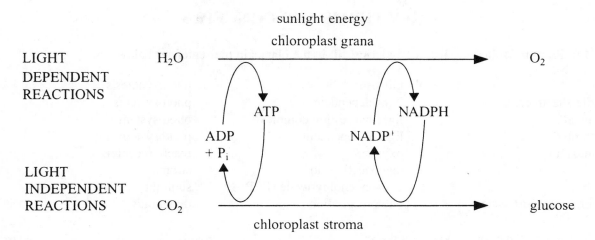

sunlight energy

chloroplast grana

LIGHT
DEPENDENT
REACTIONS

H_2O

ATP

NADPH

O_2

ADP
$+ P_i$

$NADP^+$

LIGHT
INDEPENDENT
REACTIONS

CO_2

glucose

chloroplast stroma

5) Water, CO_2, and the C_4 Pathway.

For land plants, having leaves porous to CO_2 also allows water to evaporate. Large waterproof leaves with adjustable pores (stomata) help compensate for this. When stomata close to reduce water loss, however, less CO_2 enters and less O_2 leaves.

When stomata are closed to conserve water, wasteful **photorespiration** occurs. During photorespiration, oxygen combines with RuBP and no useful cellular energy results, preventing the C_3 pathway from making glucose. During hot dry weather, plants may die due to lack of sufficient glucose.

C_4 plants (plants that have chloroplasts in mesophyll cells and in **bundle-sheath cells** around their leaf veins) reduce photorespiration using a two-stage carbon fixation process called the **C_4 pathway**. In C_4 plants, CO_2 reacts with PEP (phosphoenolpyruvate) instead of RuBP in the mesophyll, making oxaloacetate (a 4-carbon molecule). This reaction is highly specific for CO_2 and is not hindered by high O_2 concentrations. Oxaloacetate travels to bundle sheath cells where it breaks down to release CO_2 there, where the regular C_3 cycle occurs. The remnant molecule returns to the mesophyll where ATP energy is used to regenerate PEP.

C_3 and C_4 plants are each adapted to different environmental conditions. C_4 plants use more energy to make glucose than do C_3 plants. Thus, C_4 plants (crabgrass, for instance) thrive in deserts and in midsummer (much light energy, scarce water) but C_3 plants (Kentucky bluegrass, for instance) thrive in cool, wet, cloudy climates.

Case study revisited. Did dinosaurs die from lack of sunlight? About 70% of all species became extinct around 65 million years ago (the end of the Cretaceous period). Clay deposited at that time contains soot, and has 30 times the normal level of a rare element called iridium, often found in some meteorites. The Yucatan meteorite could have led to the extinction of the dinosaurs, but increased volcanic activity, spewing iridium from deep within the Earth, also could be to blame. Either a meteorite impact or increased volcanic activity could have significantly reduced the amount of sunlight and impacted the rate of photosynthesis. Less vegetation for large herbivorous dinosaurs could have reduced their numbers and had an impact on the predatory dinosaurs as well, leading to the extinction of both.

KEY TERMS AND CONCEPTS

Fill-In: From the following list of key terms, fill in the blanks in the sentences below.

ATP	glucose	photosynthesis
bundle-sheath cells	light-dependent	photosystems
chemical	light-harvesting complex	photosystem I
chlorophyll	light-independent	photosystem II
chemiosmosis	oxygen	reaction center
C_3	oxaloacetic acid	stroma
C_4	phosphoenolpyruvate (PEP)	sunlight
electron transport system	photorespiration	thylakoids

1. Starting with carbon dioxide and water, _photosynthesis_ converts _sunlight_ energy into _chemical_ energy stored in the bonds of glucose and oxygen.

2. During the _light dependent_ reactions of photosynthesis, _chlorophyll_ in the thylakoid membranes captures sunlight energy to split water and make _oxygen_ gas and some _ATP_ and NADPH energy-carriers.

3. During the _light dependent_ reactions, _stroma_ enzymes use chemical energy in ATP and NADPH to make _glucose_ from CO_2.

4. Light-dependent reactions occur in clusters of molecules called _photosystems_ consisting of proteins including chlorophyll, accessory pigments, and electron-carrying molecules in the _thylakoids_.

5. Each photosystem has two major parts. The _light harvesting complex_ has 300 pigment molecules that absorb light and pass the energy to a specific chlorophyll called the _reaction center_. The _electron transport system_ is a series of electron carrier molecules embedded in the thylakoid membrane.

6. In _photosystem II_, ADP becomes ATP through a hydrogen ion gradient called _chemiosmosis_. In _photosystem I_, NADP$^+$ becomes NADPH. Electrons from the electron transport system leave _photosystem 2_ and replenish those lost by the reaction center chlorophyll of _photo 1_.

7. Plants with chloroplasts in both mesophyll and _bundle sheath_ around the leaf veins reduce _photorespiration_ using a two-stage carbon fixation process called the _C4_ pathway.

8. In these plants, CO_2 reacts with _PEP_ instead of RuBP in the mesophyll, making _oxaloacetic acid_ (4-carbon molecule) which travels to the _bundle sheath cell_ cells where it breaks down to release CO_2.

9. In the bundle sheath cells, the regular _C_3_ cycle occurs. The remnant molecule returns to the mesophyll where _ATP_ energy is used to regenerate _PEP_.

Key Terms and Definitions

accessory pigments: colored molecules other than chlorophyll that absorb light energy and pass it to chlorophyll.

bundle-sheath cell: one of a group of cells that surround the veins of plants; in C_4 (but not in C_3) plants, bundle-sheath cells contain chloroplasts.

C_3 cycle: the cyclic series of reactions whereby carbon dioxide is fixed into carbohydrates during the light-independent reactions of photosynthesis; also called *Calvin-Benson cycle.*

C_4 pathway: the series of reactions in certain plants that fixes carbon dioxide into oxaloacetic acid, which is later broken down for use in the C_3 cycle of photosynthesis.

Calvin-Benson cycle: see *C_3 cycle.*

carbon fixation: the initial steps in the C_3 cycle, in which carbon dioxide reacts with ribulose bisphosphate to form a stable organic molecule.

carotenoid (ka-rot´-en-oid): a red, orange, or yellow pigment, found in chloroplasts, that serves as an accessory light-gathering molecule in thylakoid photosystems.

chemiosmosis (ke-mē-oz-mō´-sis): a process of ATP generation in chloroplasts and mitochondria. The movement of electrons down an electron transport system is used to pump hydrogen ions across a membrane, thereby building up a concentration gradient of hydrogen ions across the membrane; the hydrogen ions diffuse back across the membrane through the pores of ATP-synthesizing enzymes; the energy of their movement down their concentration gradient drives ATP synthesis.

chlorophyll (klor´-ō-fil): a pigment found in chloroplasts that captures light energy during photosynthesis; absorbs violet, blue, and red light but reflects green light.

electron transport system: a series of electron carrier molecules, found in the thylakoid membranes of chloroplasts and the inner membrane of mitochondria, that extract energy from electrons and generate ATP or other energetic molecules.

granum (gra´-num; pl., grana): a stack of thylakoids in chloroplasts.

light-dependent reactions: the first stage of photosynthesis, in which the energy of light is captured as ATP and NADPH; occurs in thylakoids of chloroplasts.

light-harvesting complex: in photosystems, the assembly of pigment molecules (chlorophyll and accessory pigments) that absorb light energy and transfer that energy to electrons.

light-independent reactions: the second stage of photosynthesis, in which the energy obtained by the light-dependent reactions is used to fix carbon dioxide into carbohydrates; occurs in the stroma of chloroplasts.

mesophyll (mez´-ō-fil): loosely packed parenchyma cells beneath the epidermis of a leaf.

photon (fō´-ton): the smallest unit of light energy.

photorespiration: a series of reactions in plants in which O_2 replaces CO_2 during the C_3 cycle, preventing carbon fixation; this wasteful process dominates when C_3 plants are forced to close their stomata to prevent water loss.

photosynthesis: the complete series of chemical reactions in which the energy of light is used to synthesize high-energy organic molecules, normally carbohydrates, from low-energy inorganic molecules, normally carbon dioxide and water.

photosystem: in thylakoid membranes, a light-harvesting complex and its associated electron transport system.

phycocyanin (fī-kō-sī´-uh-nin): a blue or purple pigment that is located in the membranes of chloroplasts and is used as an accessory light-gathering molecule in thylakoid photosystems.

reaction center: in the light-harvesting complex of a photosystem, the chlorophyll molecule to which light energy is transferred by the antenna molecules (light-absorbing pigments); the captured energy ejects an electron from the reaction center chlorophyll, and the electron is transferred to the electron transport system.

stoma (stō´-muh; pl., stomata): an adjustable opening in the epidermis of a leaf, surrounded by a pair of guard cells, that regulates the diffusion of carbon dioxide and water into and out of the leaf.

stroma (strō´-muh): the semi-fluid material inside chloroplasts in which the grana are embedded.

thylakoid (thī´-luh-koid): a disk-shaped, membranous sac found in chloroplasts, the membranes of which contain the photosystems and ATP-synthesizing enzymes used in the light-dependent reactions of photosynthesis.

THINKING THROUGH THE CONCEPTS

True or False: Determine if the statement given is true or false. If it is false, change the <u>underlined</u> word(s) so that the statement reads true.

10. __F__   Cellular respiration probably evolved <u>before</u> photosynthesis.

11. __F__   The light-dependent reactions occur in the <u>stroma</u>. *GRANA*

12. __F__   Glucose is synthesized in the <u>grana</u>. *STROMA*

13. __T__   Blue is a <u>more</u> energetic wavelength of light than red.

14. __F__   Light energy is first captured in <u>photosystem I</u>. *PHOTO II*

15. __F__   As electrons are transferred from one carrier to another, the electrons <u>gain</u> energy. *lose*

16. __F__   Photosynthesis <u>uses</u> O_2 and <u>produces</u> CO_2. *PRODUCES* *lose*

17. __F__   In the light dependent reactions, chlorophyll captures sunlight energy and uses it to make <u>glucose</u>. *ATP & NADPH*

18. __T__   The electron transport system is part of the light-<u>dependent</u> process of photosynthesis.

Identify: Determine whether the following statements refer to the light **dependent** reactions or the light **independent** reactions of photosynthesis.

19. _____ CO_2 is captured and converted into sugars.

20. _____   Light energy is converted into chemical energy of ATP and NADPH.

21. _____   Occurs in chloroplast grana.

22. _____   Uses chemical energy to make glucose.

23. _____   Uses chlorophyll, carotenoids, and phycocyanins to trap light energy.

24. _____   Calvin-Benson, or C_3 cycle.

25. _____   Energy obtained from NADPH and ATP.

26. _____   Produces oxygen gas.

27. _____   Thylakoid membranes.

28. _____   Photosystems I and II.

29. _____   Carbon fixation occurs.

30. _____   Involves electron transport.

31. _____   Occurs in chloroplast stroma.

32. _____   Water is split into oxygen and hydrogen.

Multiple Choice: Pick the most correct choice for each question.

33. Molecules of chlorophyll are located in the membranes of sacs called
    a. cristae
    b. thylakoids
    c. stroma
    d. grana
    e. chloroplasts

34. A pigment that absorbs red and blue light and reflects green light is
    a. phycocyanin
    b. carotenoid
    c. chlorophyll
    d. melanin
    e. colored orange

35. Light dependent photosynthetic reactions produce
    a. ATP, NADPH, oxygen gas
    b. ATP, NADPH, carbon dioxide gas
    c. glucose, ATP, oxygen gas
    d. glucose, ATP, carbon dioxide gas
    e. ADP, NADP, glucose

36. Where does the oxygen gas produced during photosynthesis come from?
    a. carbon dioxide
    b. water
    c. ATP
    d. glucose
    e. the atmosphere

37. Carbon fixation requires which of the following?
    a. sunlight
    b. products of energy-capturing reactions
    c. high levels of oxygen gas and low levels of carbon dioxide gas
    d. water, ADP, and NADP

38. The immediate source of hydrogen atoms for the production of sugar during photosynthesis comes from
    a. ATP
    b. water
    c. NADPH
    d. glucose
    e. chlorophyll

APPLYING THE CONCEPTS

These practice questions are intended to sharpen your ability to apply critical thinking and analysis to biological concepts covered in this chapter.

39. Suppose you wanted to devise an experiment to determine whether the oxygen gas generated by photosynthesis came from oxygen molecules released from water or from carbon dioxide, each of which are broken down during photosynthesis. Briefly describe an idea you might have for such a study. Hint: perhaps you might want to use a radioactive isotope of oxygen in your study.

_____

_____

_____

_____

_____

40. Suppose you wanted to determine which color of light had the greatest effect on the amount of photosynthesis occurring in plants. How would you set up such an experiment? What variables would you use and what aspects would serve as controls?

_____

_____

_____

_____

_____

41. The leaves of plants contain many types of pigments. One group of pigments present in leaves is the carotenoids, which are yellow or orange in color. Carotenoids function during photosynthesis by absorbing certain wavelengths of light energy. What colors of light do carotenoids absorb and reflect, and why is that useful to the plant?

_____

_____

_____

_____

_____

42. What are the three physical things that could happen to a photon of light when it interacts with a pigment granule, and which of the three is most important in terms of photosynthesis?

_____

_____

_____

_____

_____

_____

43. Crabgrass, a C_4 plant, grows well in the summer but not in the spring. The reverse is true for Kentucky bluegrass, a C_3 plant. Describe the process used by crabgrass to trap carbon dioxide and explain why this process is more advantageous in hot, dry weather.

_____

_____

_____

_____

_____

_____

_____

44. How would humans be different if we had all the enzymes necessary to carry out photosynthesis?

_____

_____

_____

_____

_____

_____

_____

_____

_____

45. Suppose more plants were C_4 rather than C_3. What would be the economic impact of this change?

_____

_____

_____

_____

46. Explain why the meat-eating dinosaurs would suffer as a consequence of the impact of a huge meteorite like the one that hit the Earth 65 million years ago.

_____

_____

_____

_____

Use the Case Study and the Web sites for this chapter to answer the following questions.

47. What killed the dinosaurs? There are two current theories. What are they?

_____

_____

_____

_____

_____

_____

_____

_____

_____

_____

_____

_____

_____

_____

_____

48. The Dodo bird was hunted to extinction by men, pigs, and rats less than 100 years after it was first discovered but it has taken over 300 years for a secondary biological repercussion to become apparent. What is it?

_____

_____

_____

_____

_____

_____

ANSWERS TO EXERCISES

1. photosynthesis
   sunlight
   chemical
2. light-dependent
   chlorophyll
   oxygen
   ATP
3. light-independent
   stroma
   glucose
4. photosystems
   thylakoids
5. light-harvesting complex
   reaction center
   electron transport system
6. photosystem II
   chemiosmosis
   photosystem I
   photosystem II
   photosystem I

7. bundle-sheath cells
   photorespiration
   C_4
8. phenylenolpyruvate (PEP)
   oxaloacetic acid
   bundle-sheath cells
9. C_3
   ATP
   phenylenolpyruvate (PEP)
10. false, after
11. false, grana
12. false, stroma
13. true
14. false, photosystem II
15. false, lose
16. false, produces, uses
17. false, NADPH & ATP
18. true
19. independent

20. dependent
21. dependent
22. independent
23. dependent
24. independent
25. independent
26. dependent
27. dependent
28. dependent
29. independent
30. dependent
31. independent
32. dependent
33. b
34. c
35. a
36. b
37. b
38. c

39. You could set up one study where the plants got normal CO_2 and water containing radioactive oxygen (H_2O^*), and another study where the plants got normal (H_2O) and carbon dioxide containing radioactive oxygen (CO_2^*). In each study, you would collect the oxygen gas and the sugars produced by photosynthesis and determine where the radioactive oxygen atoms show up.

40. After keeping a group of plants in the dark for several days, expose them, in separate experiments, to light from bulbs of different colors but equal intensities for equal amounts of time, keeping the plants equally watered and at constant temperatures, and compare the amounts of oxygen gas each plant produces.

41. Since carotenoids are yellow and orange, those wavelengths are reflected by the pigments, which absorb all the other wavelengths including green. Since chlorophyll reflects green light, the carotenoids are able to capture light energy not absorbed by chlorophyll and channel this energy into the system used for photosynthesis.

42. A photon of light may be reflected off the pigment granule, giving the molecule a corresponding color when that photon hits our eyes. Or, a photon could be transmitted through the pigment molecule, passing through it virtually unchanged as though the pigment were transparent. Finally, a photon and its energy could be absorbed by a pigment molecule. If the photon energy is absorbed, the amount of energy in the pigment is increased from a lower to a higher state. In the example of chlorophyll, such "excitation" of the pigment makes it capable of initiating the chain of events in photosynthesis leading to the production of ATP and the splitting of water.

43. Crabgrass, a C_4 plant, has an enzyme, PEP carboxylase, which catalyzes the formation of oxaloacetate from carbon dioxide and phosphoenolpyruvate (PEP). Because PEP carboxylase has a higher affinity for CO_2 than does RuBP carboxylase, such plants can trap CO_2 more efficiently than C_3 plants. Since C_4 plants are found in dry environments, they keep their stomata closed much of the time to conserve water, and this results in a depletion of CO_2 in the leaves. However, since the C_4 plants trap CO_2 more efficiently, they continue to carry out the light independent reactions of photosynthesis at rates surpassing that of C_3 plants in dry, hot environments.

44. Photosynthetic humans would be different in anatomical, physiological, and behavioral ways. Since we would need to absorb light energy through our skin, we would be thinner and flatter, and probably have more than four appendages, so that our surface area would be greater relative to our volume. Our skin color would be transparent, so that more sunlight energy could pass through to the pigment granules beneath our skin. Since we would make our own food by photosynthesis, our digestive system would become much simpler. We wouldn't need a mouth or teeth, or stomachs filled with acid and specialized enzymes to digest complex animal bodies. Our excretory system would become simpler, since we wouldn't have to excrete fecal matter or urine. Our respiratory system would be reversed, since we would have to inhale carbon dioxide and exhale oxygen to keep photosynthesis going. Our behavior would change, since instead of avoiding the sun, we would seek it out as the source of energy needed to make our photosynthetic food. Can you think of other changes?

45. The C_4 plants are more competitive than the C_3 plants in adverse growing conditions such as low light intensities, very hot climates, and dry conditions. This means that humans could generate a greater yield of food production in the less than optimal farming areas of the world. This could, realistically, help to provide more food for a rapidly growing world population.

46. If a large meteorite struck the Earth and the resulting dust clouds lasted for months, the amount of sunlight reaching the surface would dwindle significantly, causing most vegetation to die off. This in turn would cause starvation and death of all the large herbivore dinosaurs. Without the large herbivores to feed upon, the large carnivorous dinosaurs also would find little food and would die off.

47. According to J.R. Hutchinson at Univ. of CA, Berkeley, the current theories about what killed the dinosaurs are called the "intrinsic gradualist" and "extrinsic catastrophist" theories. The intrinsic gradualists believe that the ultimate cause of the extinction was intrinsic (meaning of an Earthly nature) and gradual, taking some time to occur (several million years). The two main hypotheses resulting in intrinsic gradualism are increased volcanic activity (over a period of several million years, increased volcanism could have created enough dust and soot to block out sunlight, producing the climatic change) and plate tectonics (major changes in the organization of the continental plates were occurring at the time). The extrinsic catastrophists believe that the ultimate cause of the extinction was extrinsic (of an extraterrestrial nature) and catastrophic (fairly sudden and punctuated). The main hypothesis was proposed in 1980 by (among others) Luis and Walter Alvarez of the University of California at Berkeley. A large extraterrestrial object collided with the Earth, its impact throwing up enough dust to cause the climatic change. The iridium layer is what prompted the Alvarez team to blame an asteroid impact for the extinction — asteroids and similar extraterrestrial bodies are higher in iridium content than the Earth's crust, so they figured that the iridium layer must be composed of the dust from the vaporized meteor. No crater was found, but it was assumed that one existed that was about 65 million years old and 100 kilometers (about 65 miles) in diameter. Later research found a likely candidate for the crater at Chicxulub, on the Yucatan Peninsula of Mexico.

48. According to the author's of tombtown.com, the Dodo bird of Mauritius became extinct in 1681. Recently, a scientist noticed that a certain species of tree was becoming quite rare on Mauritius. In fact, all 13 of the remaining trees of this species are about 300 years old. No new trees had germinated since the late 1600s. Since the average life span of this tree was about 300 years, the last members of the species were extremely old. They would soon die, and the species would be extinct. It is not just a coincidence that the tree had stopped reproducing 300 years ago and that the Dodo had become extinct 300 years ago. The Dodo ate the fruit of this tree, and it was only by passing through the Dodo's digestive system that the seeds became active and could grow. Now, more than 300 years after one species became extinct, another was to follow.

Chapter 8: Harvesting Energy: Glycolysis and Cellular Respiration

OVERVIEW

You will learn about the processes of glycolysis and cellular respiration in this chapter. The authors explain glycolysis and fermentation and discuss the role of mitochondria in converting the chemical energy of organic molecules, especially glucose, into the usable energy of ATP during aerobic respiration. Hummingbirds have extraordinary energy demands, since their wings beat 60 times a second. They burn calories 50 times faster than do humans. Consequently, hummingbirds must eat frequently, gaining energy from sugar-rich nectar from flowers and protein from eating insects. How do hummingbirds extract energy from sugar and then store the energy in their cells?

1) How Is Glucose Metabolized?

Virtually all cells break down glucose for energy. The metabolism of glucose is relatively simple compared to other molecules. When cells use other molecules for energy, they usually convert those molecules to glucose or glucose-like molecules.

The chemical equations for glucose formation by photosynthesis and the complete metabolism of glucose are nearly symmetrical:

Photosynthesis:
6 CO_2 + 6 H_2O + solar energy → $C_6H_{12}O_6$ + 6 O_2

Complete glucose metabolism:
$C_6H_{12}O_6$ + 6 O_2 → 6 CO_2 + 6 H_2O + chemical energy and heat energy

The conversion of energy into different forms always results in the decrease of the amount of concentrated, useful energy, as explained by the second law of thermodynamics. So, lots of heat energy is produced when glucose is metabolized. The major steps in glucose metabolism in eukaryotic cells are: (1) glycolysis, which does not require oxygen and splits apart a glucose molecule into two three-carbon molecules of **pyruvate**, releasing a small amount of energy used to make 2 molecules of ATP. If no oxygen is present, the pyruvate is converted by fermentation into lactate or ethanol. (2) If oxygen is present, cellular respiration occurs in the mitochondria, breaking down the pyruvate into carbon dioxide and water, generating an additional 34–36 molecules of ATP.

2) How Is the Energy of Glucose Harvested During Glycolysis?

In all living cells, the first step of glucose metabolism is called **glycolysis** ("to break apart a sweet thing"), and glycolysis proceeds the same either in the presence (aerobic) or absence (anaerobic) of oxygen. Glycolysis consists of two major steps: glucose activation and energy harvest.

Glycolysis breaks down glucose to pyruvate, releasing chemical energy. In glucose activation, energy from two ATPs is used to convert stable glucose into unstable fructose bisphosphate. During energy harvest, fructose bisphosphate is split apart into two three-carbon molecules of glyceraldehyde 3-phosphate (G3P). The G3P molecules are then converted into pyruvate, during which two ATP are generated for each G3P (4 ATP molecules made in all), so that there is a net gain of two ATPs per glucose molecule. In addition, two NAD^+ molecules are converted into NADH electron carriers using energy released when fructose bisphosphate is converted into pyruvate. So, each molecule of glucose is broken down into two molecules of pyruvate and two ATP molecules, and two NADH electron carriers are formed.

Some cells ferment pyruvate to form lactate. Under anaerobic conditions in animal muscle, **fermentation** of pyruvate to produce lactate occurs. The lactate gains electrons and hydrogen ions when NADH is converted into NAD^+. Lactate is toxic when concentrated, causing discomfort and fatigue. When oxygen is present, lactate is converted back into pyruvate which enters cellular respiration.

Other cells ferment pyruvate to alcohol. Anaerobic conditions in many microorganisms produce alcoholic fermentation: pyruvate is converted into ethanol and CO_2, gaining electrons and hydrogen ions when NADH is converted into NAD^+. Alcoholic fermentation in yeast is useful in the brewing (ethanol) and baking (CO_2 makes bread rise) industries.

3) How Does Cellular Respiration Generate Still More Energy from Glucose?

Cellular respiration is a series of reactions, occurring under aerobic conditions, during which large amounts of ATP are produced. During aerobic cellular respiration in the mitochondria of eukaryotic cells, pyruvate is converted into $CO_2 + H_2O$, plus many ATP molecules. The final reaction requires oxygen because oxygen is the final acceptor of electrons.

Mitochondria have two membranes that produce two compartments: an inner compartment enclosed by the inner membrane and that contains the fluid **matrix**, and an **intermembrane compartment** between the two membranes. Most ATP made during cellular respiration is generated by reactions catalyzed by enzymes in the mitochondrial matrix, electron transfer proteins in the inner membrane, and movement of hydrogen ions through ATP-synthesizing proteins in the inner membrane. The steps involved are as follows:

(1) The two molecules of pyruvate are transported across both membranes into the mitochondrial matrix;
(2) Each pyruvate is split into CO_2 and a two-carbon acetyl group which enters the Kreb's cycle, during which the remaining carbons become CO_2, one ATP is produced, and energetic electrons are donated to several electron-carrying molecules;
(3) The electron carriers donate their energetic electrons to the electron transport system (ETS) of the inner membrane where the energy is used to transport hydrogen ions from the matrix to the intermembrane compartment where electrons combine with hydrogen ions and oxygen to make H_2O;
(4) In chemiosmosis, the hydrogen ion gradient (high in the intermembrane compartment, low in the matrix) created by the ETS discharges through pores in the ATP-synthesizing enzymes located in the inner membrane, and the energy is used to make ATP;
(5) The ATP is transported out of the mitochondria into the cytoplasm.

Pyruvate diffuses down its concentration gradient into the mitochondrial matrix through pores in the membranes. Then pyruvate reacts with coenzyme A molecules, and each pyruvate is split into CO_2 and a two-carbon acetyl group, which attaches to coenzyme A (CoA), forming an acetyl-coemzyme A complex (*acetyl CoA*). During this reaction, two energetic electrons and a hydrogen ion are transferred to NAD^+, forming NADH. Then a cyclic pathway called the **Krebs Cycle** (or *citric acid cycle*) occurs. Each acetyl CoA combines with *oxaloacetate*. The two-carbon acetyl group bonds to the four-carbon oxaloacetate to form the six-carbon *citrate*. CoA is released to be reused. The citrate then is rearranged to regenerate oxaloacetate, give off two CO_2 molecules, release energy to make one ATP, and four electron carriers (one $FADH_2$, flavin adenine dinucleotide, and three NADH).

Up to this point, from the energy of each glucose broken down, the cell has made 4 ATP, 10 NADH, and 2 $FADH_2$ molecules. The carriers deposit their electrons in **electron transport systems** (ETSs) located in the inner mitochondrial membrane. The energetic electrons move from molecule to molecule along the ETSs, releasing energy used to pump in hydrogen ions from the matrix across the inner membrane during **chemiosmosis**. At the end of the ETSs, oxygen and hydrogen ions accept the low energy electrons to form water. Without oxygen, the electrons would "pile up" in the ETS, stopping the reactions.

Chemiosmosis captures energy stored in a hydrogen ion gradient and produces ATP. Hydrogen ion pumping produces a large concentration gradient (high in the intermembrane compartment and low in the matrix). Energy is released when the ions move down their concentration gradient and the energy is captured to make 32 to 34 ATP molecules. The ATP diffuses out of the mitochondria into the cytoplasm.

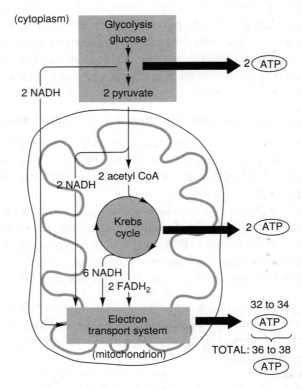

Glycolysis and cellular respiration influence the way entire organisms function. Cyanide kills quickly because it reacts with one of the proteins of the ETS, blocking the movement of electrons and halting cellular respiration. Now, consider Olympic runners. During the 100 meter dash, runners' leg muscles use far more ATP than cellular respiration can supply because their bodies cannot supply oxygen to the muscles fast enough. Glycolysis and lactate fermentation can supply a little more ATP, but lactate buildup causes fatigue and cramps. Runners in the 100 meter dash can rely on anaerobic respiration for that short time, but long distance runners must pace themselves so that cellular respiration can occur, saving the anaerobic sprint for the finish. Marathon runners practice running 50–100 miles a week to build up their respiratory and circulatory systems to deliver enough oxygen to their muscles. Sustaining life depends on efficiently obtaining, storing, and using energy.

Case study revisited. To fly over 600 miles of open water, the ruby throated hummingbird (weighing as much as a penny) must store a great deal of energy as fat (another penny's worth by weight). The cells of a hummer's wings are packed with mitochondria to produce the maximum amount of ATP, and its respiratory system is designed to extract oxygen from the air even while exhaling, to supply the wing cells with oxygen for cellular respiration

KEY TERMS AND CONCEPTS

Fill-In: Write the answers in the numbered blanks in the table below.

1.

Name of metabolic process:	Is oxygen necessary?	Part of a cell where it occurs:	Number of ATP molecules produced:	Types of molecules produced:
glycolysis				
alcoholic fermentation				
lactate fermentation				
Krebs cycle				
electron transport system and chemiosmosis				

Key Terms and Definitions

cellular respiration: the oxygen requiring reactions, occurring in mitochondria, that break down the end products of glycolysis into carbon dioxide and water while capturing large amounts of energy as ATP.

chemiosmosis (ke-mē-oz-mō´-sis): a process of ATP generation in chloroplasts and mitochondria. The movement of electrons down an electron transport system is used to pump hydrogen ions across a membrane, thereby building up a concentration gradient of hydrogen ions across the membrane; the hydrogen ions diffuse back across the membrane through the pores of ATP-synthesizing enzymes; the energy of their movement down their concentration gradient drives ATP synthesis.

citric acid cycle: see *Krebs cycle*.

electron transport system: a series of electron carrier molecules, found in the thylakoid membranes of chloroplasts and the inner membrane of mitochondria, that extract energy from electrons and generate ATP or other energetic molecules.

fermentation: anaerobic reactions that convert the pyruvic acid produced by glycolysis into lactic acid or alcohol and CO_2.

glycolysis (glī-kol´-i-sis): reactions, carried out in the cytoplasm, that break down glucose into two molecules of pyruvic acid, producing two ATP molecules; does not require oxygen but can proceed when oxygen is present.

intermembrane compartment: the fluid-filled space between the inner and outer membranes of a mitochondrion.

Krebs cycle: a cyclic series of reactions, occurring in the matrix of mitochondria, in which the acetyl groups from the pyruvic acids produced by glycolysis are broken down to CO_2, accompanied by the formation of ATP and electron carriers; also called *citric acid cycle*.

matrix: the fluid contained within the inner membrane of a mitochondrion.

pyruvate: a three-carbon molecule that is formed by glycolysis and then used in fermentation or cellular respiration.

THINKING THROUGH THE CONCEPTS

True or False: Determine if the statement given is true or false. If it is false, change the underlined word(s) so that the statement reads true.

2. _____   Aerobic forms of life evolved before anaerobic forms.

3. _____   Aerobic (cellular) respiration uses O_2 and produces CO_2.

4. _____   Glycolysis requires oxygen in order to function.

5. _____   Glycolysis occurs in the mitochondria of a cell.

6. _____   Pyruvic acid (pyruvate) is produced by glycolysis.

7. _____   The chemical energy in sugar is used to make O_2.

8. _____   When NADH becomes NAD^+, the hydrogens are used to make sugar.

9. _____   Lactic acid (lactate) fermentation occurs when oxygen is abundant in muscle cells.

10. _____   When each pyruvic acid is completely broken down, six CO_2 molecules are released.

11. _____   Each glucose molecule releases enough energy to make 100 molecules of ATP.

Matching: Glucose metabolism.

12. _____   most of the ATP is made

13. _____   occurs only under anaerobic conditions

14. _____   occurs only under aerobic conditions

15. _____   occurs under either anaerobic or aerobic conditions

16. _____   glucose is split into 2 pyruvate molecules

17. _____   occurs in mitochondria

18. _____   lactate is formed

19. _____   occurs in the cytoplasm

20. _____   produces CO_2 and ATP

21. _____   requires some ATP energy to get started

22. _____   produces ethanol

23. _____   acetyl-CoA is used

24. _____   Krebs cycle occurs

25. _____   fructose diphosphate is produced

Choices:

a. glycolysis

b. fermentation

c. both a. and b.

d. cellular respiration

Short answer.

26. Explain, using chemical equations, how photosynthesis and aerobic cellular respiration are "complementary" processes.

_____

_____

_____

_____

_____

_____

_____

_____

Multiple choice: Pick the most correct choice for each question:

27. During glycolysis, what provides the initial energy to break down glucose?
    a. ATP
    b. pyruvate
    c. NADH
    d. cytoplasmic enzymes
    e. mitochondria

28. At the end of glycolysis, where are the original carbons of the glucose molecule located?
    a. in six molecules of carbon dioxide
    b. in two molecules of NADH
    c. in two molecules of pyruvate
    d. in two molecules of citric acid

29. When oxygen is present
    a. most eukaryotic cells utilize aerobic cellular respiration
    b. most animal cells carry out lactate fermentation
    c. most bacteria and yeast carry out alcoholic fermentation
    d. glucose is broken down to produce 2 ATP molecules
    e. mitochondria are less likely to function normally

30. The anaerobic breakdown of glucose is called
    a. artificial respiration
    b. glycolysis
    c. photosynthesis
    d. fermentation
    e. b and d

31. What happens when pyruvate is converted into lactate?
    a. the lactate enters the Krebs cycle
    b. the mitochondria are activated
    c. NAD^+ is regenerated for use in glycolysis
    d. oxidation of pyruvate occurs
    e. oxygen gas is liberated

32. Oxygen is necessary for cellular respiration because oxygen
    a. combines with electrons and hydrogen ions to form water
    b. combines with carbon to form carbon dioxide
    c. combines with carbon dioxide and water to form glucose
    d. breaks down glucose into carbon dioxide and water
    e. allows glucose to be converted into pyruvic acid

APPLYING THE CONCEPTS

These practice questions are intended to sharpen your ability to apply critical thinking and analysis to biological concepts covered in this chapter.

33. Why can drowning, suffocation, or carbon monoxide poisoning lead to death? The obvious initial response is that they prevent oxygen from reaching our cells, but go beyond that to explain why this can cause death.

_____

_____

_____

_____

_____

_____

34. Some animals that live in deserts survive without actually drinking water. They do eat food containing a little water, but most of the water they need is made within their cells and called "metabolic water." From what you have read in this chapter, explain one way by which metabolic water is produced.

_____

_____

_____

35. Why are runners in the 100 meter dash and marathon runners forced to employ different running strategies in order to win their respective races?

_____

_____

_____

_____

_____

_____

36. "Have you thanked a green plant today?" is a common bumper sticker. Why are plants necessary in order for animals to survive?

_____

_____

_____

_____

_____

_____

37. In the wine-making process, why do yeast consume glucose quicker in the absence of oxygen, and why do yeast produce alcohol only in the absence of oxygen?

_____

_____

_____

_____

38. What would a scientist look for in various types of cells that might accurately predict the rates of cellular respiration?

_____

_____

_____

_____

39. Why couldn't the ruby throated hummingbird fly 600 miles to Guatemala if it stored energy as glycogen instead of as fat?

_____

_____

_____

_____

_____

40. The electron carriers used in cellular respiration, FAD and NAD^+, are derived from the B vitamins riboflavin and niacin, respectively. Why is it that a person needs to consume lots of food each day, but only tiny amounts of these B vitamins?

_____

_____

_____

_____

41. Explain how the mitochondria of plants and animals and the chloroplasts of plants "cooperate" to create cycles of carbon dioxide, oxygen, and energy throughout the biosphere.

_____

_____

_____

_____

_____

Use the Case Study and the Web sites for this chapter to answer the following question.

42. Two myths about hummingbirds are: Hummingbirds live on a strict diet of sugar water (nectar), and they obtain this nectar by sucking it out of flowers using their beaks as straws. The fact is that the feeding behaviors and diet of hummingbirds are perfectly adapted to their energy needs. What and how do hummingbirds eat?

_____

_____

_____

_____

ANSWERS TO EXERCISES

1.

Name of metabolic process:	Is oxygen necessary?	Part of a cell where it occurs:	Number of ATP molecules produced:	Types of molecules produced:
glycolysis	no	cytoplasm	2	ATP, NADH, pyruvate
alcoholic fermentation	no	cytoplasm	0	CO_2, NAD^+, ethanol
lactate fermentation	no	cytoplasm	0	NAD^+, lactate
Krebs cycle	yes	mitochondrial matrix	1 ATP per pyruvate	CO_2, ATP, NADH, $FADH_2$
electron transport system and chemiosmosis	yes	inner mitochondrial membrane and intermembrane compartment	32 to 34	H_2O, ATP, FAD, NAD^+

2. false, after	8. false, water	14. d	20. d
3. true	9. false, absent	15. a	21. a
4. false,	10. false, three	16. a	22. b
does not require	11. false, 36	17. d	23. d
5. false, cytoplasm	12. d	18. b	24. d
6. true	13. b	19. c	25. a
7. false, ATP			

26. Photosynthesis

$$6\ CO_2 + 6\ H_2O + \text{solar energy} \rightarrow C_6H_{12}O_6 + 6\ O_2$$

used in photosynthesis   made in photosynthesis
made in cell respiration   used in cell respiration

Aerobic Cell Respiration

$$C_6H_{12}O_6 + 6\ O_2 \rightarrow 6\ CO_2 + 6\ H_2O + \text{energy}$$

made in photosynthesis   used in photosynthesis
used in cell respiration   made in cell respiration

27. a	29. a	31. c
28. c	30. e	32. a

33. Oxygen is the final electron acceptor in aerobic cellular respiration, receiving electrons and hydrogen ions from the electron transport chain and forming water. Without sufficient oxygen, the electrons and hydrogen ions clog up the electron transport chain, not allowing NADH and $FADH_2$ molecules to give off their electrons and hydrogen ions, and as cells run out of these molecules, mitochondria shut down. This forces cells to rely on glycolysis and fermentation to produce very little ATP from glucose metabolism, and without sufficient energy from ATP, cells cannot continue functioning and die.

34. When oxygen combines with electrons and hydrogen ions at the end of the electron transport chain, "metabolic water" is produced. Some animals can use this cell-made water to help meet the water demands of their cells, and thus survive.

35. Humans can run at full speed for short distances (100 meters, for example) because they have stored up enough ATP molecules for relatively short intense bursts of energy release without running out of ATP. For marathon runners, however, stored ATP would run out long before the 26 miles were run if they ran at full speed, and muscle mitochondria would not be able to make more ATP fast enough to have a constant supply since the respiratory system cannot supply oxygen to muscle cells fast enough. What would happen is the mitochondria would shut down due to inadequate oxygen availability, and muscles would produce limited ATP by lactic acid fermentation. The buildup of lactic acid would cause fatigue and muscle cramping, forcing the runner to stop. Consequently, a marathon runner must run at a slower pace to allow the rate of oxygen supply to more nearly keep up with ATP utilization during the race.

36. Animals need oxygen in order for cellular respiration to occur in mitochondria. The principle source of oxygen on Earth is supplied as a by-product of photosynthesis in plants and algae. Without this source of atmospheric oxygen, animals would die very quickly. Also, plants and algae are the ultimate sources of all the food on Earth, produced as a product of photosynthesis, which traps solar energy in sugar molecules as chemical energy. Animals (and plants) convert the chemical energy in food into the chemical energy of ATP.

37. In the absence of oxygen, yeast cannot perform cellular respiration using their mitochondria. Instead they use alcohol fermentation to produce a smaller amount of ATP from partial sugar metabolism, yielding carbon dioxide and ethanol as by-products.

38. A scientist might look for the numbers of mitochondria in various types of cells as an indicator of the relative rates of cellular respiration in those cell types. Also the scientist could measure the amounts of oxygen consumed and the rates of carbon dioxide produced as other indicators of the relative rates of cellular respiration.

39. Before a ruby-throated hummingbird begins its journey, it doubles its body weight (2 grams) by storing up about 2 grams of fat. With that additional weight, it barely can take off, and it barely reaches its destination as it totally depletes all its fat supply. Since fat contains twice as much energy per unit weight as do proteins or carbohydrates, the hummer would be too heavy to take off if it stored enough glycogen for the journey.

40. The energy derived from food is converted into ATP which is constantly being utilized to do cellular work and ultimately converted into less useful forms such as heat. Thus, energy from food must be constantly supplied to cells. However, because FAD and NAD^+ are constantly being recycled, cells require only a small supply of them at any particular time.

41. Chloroplasts utilize the energy in sunlight to convert inorganic carbon dioxide gas into sugar, releasing oxygen gas. Mitochondria break down the sugar using oxygen as the ultimate electron acceptor, releasing carbon dioxide and energy. The actions of chloroplasts and mitochondria jointly produce a continuous cycle of use and production of carbon dioxide and oxygen gas, driven by energy from the sun.

42. According to "The Birds of North America," hummingbird diets include nectar from various flower species, various insects (mosquitoes, spiders, gnats, fruit flies, and small bees), and butterfly larvae, aphids, and insect eggs from leaves and bark of various tree species. According to the Richmond Audubon Society, a hummingbird's tongue is long and tubelike and darts deep into flowers for nectar, taking up liquid by capillary action. The tongue also has a brushy tip that traps insects.

Chapter 9:   DNA: The Molecule of Heredity

OVERVIEW

In this chapter, you will learn about the molecular basis of inheritance, namely DNA. The authors describe the structure of DNA and explain how DNA makes accurate copies of itself. Melanoma is a relatively common skin cancer that begins in pigmented cells of the lower parts of the skin. Melanoma can spread to other body organs, frequently leading to death. The frequency of melanoma is increasing, and often is caused by increased exposure to sunlight, which can cause changes in the genetic material.

1) How Did Scientists Discover that Genes Are Made of DNA?

Fifty years ago, no one knew that deoxyribonucleic acid, or **DNA**, is the molecule that carries the blueprints for all forms of life on Earth. DNA enables organisms and their cells to transmit information accurately from one generation to the next. Learning DNA's structure and how it worked required the incremental advances of dozens of scientists working for decades. Starting in the late 1800s, scientists learned that heritable information exists in discrete units called **genes**. Studies of dividing cells provided strong evidence that genes are located in **chromosomes**, which are made of both DNA and **protein**, indicating that genes are made of either DNA or protein. Experiments using bacteria eventually provided clear proof that genes are made of DNA.

Transformed bacteria revealed the link between genes and DNA. In the 1920s Griffith was trying to make a vaccine to prevent pneumonia infections, using two strains of *Streptococcus pneumonia* bacteria. The R-strain did not cause pneumonia when injected into mice, and the S-strain caused pneumonia. Heat killed S-strain bacteria did not cause disease. He mixed living R-strain with heat-killed S-strain bacteria, and injected them into mice, which sickened and died, yielding living S-strain bacteria from their organs. Some substance in the heat-killed S-strain transformed the living, harmless R-strain into a deadly S-strain. Later, Avery, MacLeod, and McCarty purified the molecules from the S-strain bacteria that could transform the R-strain into a deadly S-strain. The transforming molecules were DNA.

2) What Is the Structure of DNA?

DNA is made of four types of small subunits called **nucleotides**, each with a sugar (S) called deoxyribose, a phosphate group (P), and one of four possible nitrogen-containing **bases** (B): **adenine** (A), **thymine** (T), **guanine** (G), or **cytosine** (C).

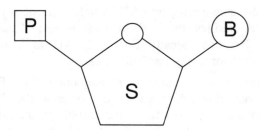

In the 1940s, Chargaff discovered that the DNA of any species contains equal amounts of adenine and thymine, as well as equal amounts of cytosine and guanine. Wilkins and Franklin used X-ray diffraction to study DNA structure, finding that the molecule is helical (twisted like a corkscrew), has a uniform diameter (2 nanometers), and consists of repeating subunits.

Watson and Crick suggested that a DNA molecule consists of two DNA **strands** of linked nucleotides. Within each DNA strand, the phosphate group of one nucleotide bonds to the sugar of the next, producing a "backbone" of alternating, covalently bonded sugars and phosphates. The bases protrude from the **sugar-phosphate backbone**.

$$-P\text{-}S\ -P\text{-}S\ -P\text{-}S\ -P\text{-}S\ -P\text{-}S\ -P\text{-}S\ -$$
$$\ \ \ \ B\ \ \ \ \ \ B\ \ \ \ \ \ B\ \ \ \ \ \ B\ \ \ \ \ \ B\ \ \ \ \ \ B$$

Watson and Crick said that hydrogen bonds between complementary bases hold the DNA strands together. Hydrogen bonds form between the protruding bases of the two separate DNA **strands**, giving DNA a ladder-like structure with the sugar-phosphate backbones on the outside and the nucleotide bases on the inside. The DNA strands are twisted about each other to form a **double helix**, like a ladder twisted lengthwise into a circular staircase. In the double helix, the two DNA strands are oriented in opposite directions. At each end of the double helix, one DNA strand ends with a free phosphate and the other ends with a free sugar. Within DNA, adenine forms hydrogen bonds only with thymine, and guanine forms bonds only with cytosine. These A-T and G-C linkages are called **complementary base pairs**, and explain Chargaff's results.

It is not the number of different subunits, but their order that is important. The order of nucleotides in DNA can encode vast amounts of information. Within a DNA strand, the four types of bases can be arranged in any linear order, with each sequence representing a unique set of genetic instructions, like a biological Morse code. A stretch of DNA just 10 nucleotides long can have more than a million possible sequences of the four bases, and a typical plant or animal chromosome has billions of nucleotides.

3)   How Does DNA Replication Ensure Genetic Constancy?

DNA replication produces two DNA double helices, each with one old strand and one new strand. Each chromosome has a single long DNA double helix together with proteins that help organize and fold up the DNA. DNA duplication (replication) produces two identical double helices of DNA, each of which will be passed, within its chromosome, to one of the new daughter cells. DNA replication begins when enzymes pull the parental DNA double helix apart, so that the parental strands no longer form base pairs with each other. Other enzymes move along each separated DNA strand, selecting free nucleotides with complementary bases to the parental strand. The enzymes join these **free nucleotides** together to form two new DNA strands, each complementary to one of the original parental strands. When replication is complete, one original strand and one new complementary strand wind together into one double helix. The other original strand and new complementary strand do likewise. This process is called **semiconservative replication**. The new double helices are held together while the cell prepares for division. The chromosome at this stage is called a duplicated chromosome with two "sister" chromatids.

More details of the process follow. **DNA helicase** (an enzyme that breaks apart the helix) separates the parental DNA strands, using ATP energy to break the hydrogen bonds between complimentary base pairs. This activity separates and unwinds the parental DNA double helix, forming a **replication "bubble"** containing two replication "forks" where the two parental DNA strands have not yet unwound. In eukaryotic cells, many replication bubbles occur simultaneously on each chromosome to speed up the process. The bubbles grow and meet up.

DNA polymerase (an enzyme that makes a DNA polymer) makes new DNA strands. At each replication fork, DNA polymerase makes two new DNA strands that are complementary to the two parental strands by matching up unpaired parental bases with free nucleotides with the correct complementary bases. Then DNA polymerase catalyzes formation of new covalent bonds that link the phosphate of the incoming free nucleotide to the sugar of the previously added one, making the sugar-phosphate backbone of the daughter strand.

DNA polymerase can only add new nucleotides onto the free sugar end of the new DNA strand that it is making. But the two strands of the parental DNA molecule are oriented in opposite directions. So, as

DNA helicase separates parental strands, one DNA polymerase moves in the same direction as the helicase, adding nucleotides to form a long continuous daughter DNA strand. The second DNA polymerase must move in the opposite direction, making a daughter strand in small discontinuous segments. These segments are subsequently connected by **DNA ligase**. DNA ligase also plays a role in repairing DNA damaged by sunlight.

Proofreading produces almost error-free replication of DNA. Nearly 700 nucleotides per second are replicated by DNA polymerase, leading to an error about once in every 10,000 base pairs. But most of these errors are repaired by a variety of DNA repair enzymes that "proofread" each daughter strand during and after it is made and make any necessary repairs. Ultimately, the completed DNA strands contain about one mistake in every billion base pairs. But these mistakes do happen. Also, the DNA in each human cell loses about 10,000 base pairs daily by spontaneous chemical breakdown due to our body temperature of 98.6° F. Environmental agents, like ultraviolet radiation, also can damage DNA. Deterioration in the accuracy of DNA replication as people get older may contribute to the aging process.

Case study revisited. Thymine and cytosine are particularly prone to UV damage. Two adjacent thymines in a DNA strand can be improperly linked together by UV radiation. Unless the damage is repaired, DNA polymerase may be unable to properly replicate this region of DNA and may insert incorrect nucleotides into the daughter strand. If the involved gene controls cell division, it may lead to skin cancer. When you get a sunburn, many skin cells sustain more damage than can be repaired, and the cells die and peel off. Less damaged cells are repaired, but people who lack functional repair enzymes needed to repair UV-damage can have a variety of disorders, including xeroderma pigmentosum, being very sensitive to sunlight and prone to cancer. Getting sunburned as a child raises the risk of developing melanoma about three-fold. Recognizing a melanoma is as easy as ABCD: look for moles with Asymmetry, irregular Border, irregular Color, or a Diameter larger than the eraser at the end of a pencil.

KEY TERMS AND CONCEPTS

Fill-In: From the following list of terms, fill in the blanks below.

AATC	cytosine	GC and AT	helicase	nucleotides
adenine	DNA	GTTAC	helix	polymerase
chromosomes	four	guanine	ligase	semiconservative

1. Genes are made of molecules of _____ which is made of _____ that have a phosphate group, a sugar, and a nitrogenous base. The genetic molecule has a double-_____ 3-dimensional structure.

2. There are _____ different bases that are used to build a DNA molecule.

3. _____ is the base that pairs with thymine, _____ is the base that pairs with guanine, and _____ is the base that pairs with cytosine in DNA. Therefore, _____ are complementary base pairs.

4.   DNA replication is _____.

5. _____ is the enzyme that unwinds DNA during replication. New strands of DNA are made using the enzyme _____, and _____ is the enzyme that connects the short segments of nucleotides in a newly made DNA strand.

6.   The complimentary base sequence for TTAG is _____ and the complementary base sequence for CAATG is _____

7.   _____ are the cell structures in the nucleus that contain the genetic material.

Key Terms and Definitions

adenine: a nitrogenous base found in both DNA and RNA; abbreviated as *A*.

base: in molecular genetics, one of the nitrogen-containing, single- or double-ringed structures that distinguish one nucleotide from another. In DNA, the bases are adenine, guanine, cytosine, and thymine.

chromosome (krō´-mō-sōm): a single DNA double helix together with proteins that help to organize the DNA.

complementary base pair: in nucleic acids, bases that pair by hydrogen bonding. In DNA, adenine is complementary to thymine and guanine is complementary to cytosine; in RNA, adenine is complementary to uracil, and guanine to cytosine.

cytosine: a nitrogenous base found in both DNA and RNA; abbreviated as C.

deoxyribonucleic acid (dē-ox-ē-rī-bō-noo-klā´-ik; **DNA**): a molecule composed of deoxyribose nucleotides; contains the genetic information of all living cells.

DNA helicase: an enzyme that helps unwind the DNA double helix during DNA replication.

DNA ligase: an enzyme that joins the sugars and phosphates in a DNA strand to create a continuous sugar-phosphate backbone.

DNA polymerase: an enzyme that bonds DNA nucleotides together into a continuous strand, using a preexisting DNA strand as a template.

DNA replication: the copying of the double-stranded DNA molecule, producing two identical DNA double helices.

double helix (hē´-liks): the shape of the two-stranded DNA molecule; like a ladder twisted lengthwise into a corkscrew shape.

free nucleotides: nucleotides that have not been joined together to form a DNA or RNA strand.

gene: a unit of heredity that encodes the information needed to specify the amino acid sequence of proteins and hence particular traits; a functional segment of DNA located at a particular place on a chromosome.

guanine: a nitrogenous base found in both DNA and RNA; abbreviated as *G*.

nucleotide: a subunit of which nucleic acids are composed; a phosphate group bonded to a sugar (deoxyribose in DNA), which is in turn bonded to a nitrogen-containing base (adenine, guanine, cytosine, or thymine in DNA). Nucleotides are linked together, forming a strand of nucleic acid, as follows: Bonds between the phosphate of one nucleotide link to the sugar of the next nucleotide.

protein: polymer of amino acids joined by peptide bonds.

replication bubble: the unwound portion of the two parental DNA strands, separated by DNA helicase, in DNA replication.

semiconservative replication: the process of replication of the DNA double helix; the two DNA strands separate, and each is used as a template for the synthesis of a complementary DNA strand. Consequently, each daughter double helix consists of one parental strand and one new strand.

strand: a single polymer of nucleotides; DNA is composed of two strands.

sugar-phosphate backbone: a major feature of DNA structure, formed by attaching the sugar of one nucleotide to the phosphate from the adjacent nucleotide in a DNA strand.

thymine: a nitrogenous base found only in DNA; abbreviated as *T*.

THINKING THROUGH THE CONCEPTS

True or False: Determine if the statement given is true or false. If it is false, change the underlined word(s) so that the statement reads true.

8.   _____   A molecule of DNA is <u>single</u> stranded.

9.   _____   DNA contains four types of <u>sugars</u>.

10.   _____   Sugars found in DNA have <u>five</u> carbons each.

11.   _____   DNA is found in cellular <u>chromosomes</u>.

12.   _____   DNA contains sugars, bases, and <u>sulfur</u> groups.

13.   _____   The concentration of DNA is <u>constant</u> for different body cells of the same species.

14.   _____   Adenine pairs with <u>guanine</u> in DNA.

15.   _____   The duplication of DNA is called <u>fully</u> conservative replication.

16.   _____   The building blocks of nucleic acids are <u>amino acids</u>.

Multiple Choice: Pick the most correct choice for each question.

17. If amounts of bases in a DNA molecule are measured, we find
    a. A = C and G = T
    b. A = G and C = T
    c. T = A and C = G
    d. that no two bases would be equal in amount
    e. that all bases are equal in amount

18. The DNA of a certain organism has guanine as 30% of its bases. What percentage of its bases would be adenine?
    a. 0
    b. 10%
    c. 20%
    d. 30%
    e. 40%

19. The correct structure of a nucleotide is
    a. phosphate–ribose–adenine
    b. phospholipid–sugar–base
    c. phosphate–sugar–phosphate–sugar
    d. adenine–thymine and guanine–cytosine
    e. phosphate–sugar–base

20. The two polynucleotide chains in a DNA molecule are attracted to each other by
    a. covalent bonds between carbon atoms
    b. hydrogen bonds between bases
    c. peptide bonds between amino acids
    d. ionic bonds between "R" groups in amino acids
    e. covalent bonds between phosphates and sugars

21. Using an analogy of DNA as a twisted ladder, the rungs (steps) of the ladder are
    a. phosphate groups
    b. sugar groups
    c. paired nitrogenous bases
    d. oxygen-carbon double bond

22. All the cells of a specific organism contain equal amounts of
    a. adenine and guanine
    b. guanine and cytosine
    c. adenine and cytosine
    d. thymine and cytosine

23. The sequence of subunits in the DNA backbone is
    a. –base–phosphate–sugar–base– phosphate–sugar–
    b. –base–phosphate–base–phosphate– base–phosphate–
    c. –phosphate–sugar–phosphate–sugar– phosphate–sugar–
    d. –sugar–base–sugar–base–sugar–base– sugar–base–
    e. –base–sugar–phosphate–base– sugar–phosphate–

24. Figuratively speaking, a double helix is comparable to
    a. coiled rope
    b. stacked up plates
    c. braided hair
    d. twisted ladder
    e. tangled threads

25. Melanoma is a cancer caused by exposure to
    a. X rays
    b. harmful chemicals
    c. sea water
    d. ultraviolet radiation
    e. skin cream

26. Excessive amounts of sunlight will cause which adjacent DNA bases to form improper bonding?
    a. adjacent thymine bases
    b. adjacent adenine bases
    c. adjacent cytosine bases
    d. adjacent guanine bases
    e. any pair of adjacent bases

27. The genetic condition "xeroderma pigmentosum" occurs due to the lack of which functional enzymes?

    a. DNA polymerase

    b. DNA repair enzymes

    c. DNA ligase

    d. DNA helicase

    e. DNA replicase

28. Getting severely sunburned as a child raises the risk of developing melanoma later in life about three-fold.

    a. true

    b. false

APPLYING THE CONCEPTS

These practice questions are intended to sharpen your ability to apply critical thinking and analysis to biological concepts covered in this chapter.

29. Instead of DNA being a double-stranded molecule, suppose DNA was a single-stranded molecule, so that each complete molecule of DNA consisted of one chain of nucleotides. If DNA were single-stranded, describe how DNA replication could take place, using the same DNA polymerase enzyme that creates complimentary base pairing, so that exact copies of genes could be made. Would this be more efficient or less efficient than replication involving double-stranded DNA? Briefly explain.

_____

_____

_____

_____

_____

_____

_____

_____

_____

_____

_____

_____

30. Look at figure 9.3 in your textbook. Can you briefly explain why cytosine can pair only with guanine, and not with adenine or thymine?

_____

_____

_____

_____

_____

_____

_____

31. Analysis of the relative amounts of the four types of bases in DNA from tuberculosis bacterium shows that it contains roughly 15% adenine, 15% thymine, 35% cytosine, and 35% guanine. A similar analysis of human DNA shows more adenine (30%) and thymine (30%), and less cytosine (20%) and guanine (20%). However, both human DNA and tuberculosis bacterial DNA contain 50% of the one-ring bases thymine and cytosine, and 50% of the two-ring bases adenine and guanine. Explain the differences and similarities of the DNA from the two sources.

    _____

    _____

    _____

    _____

    _____

    _____

32. Suppose a genetic transformation experiment is performed between two types of bacteria, where one strain (strain A) can make its own vitamin B_6 due to the presence of an active gene, and the other strain (strain B) cannot make its own vitamin B_6 due to the presence of a defective gene. Describe an experiment that might determine whether genes are made of DNA or of proteins.

    _____

    _____

    _____

    _____

    _____

    _____

    _____

33. In DNA molecules, when base pairing occurs between adenine and thymine, two hydrogen bonds form between them. However, when base pairing occurs between cytosine and guanine, three hydrogen bonds form between them. Let's say that we compared DNA from two different species. Species A's DNA has 40% A-T base pairs and 60% G-C base pairs, while Species B's DNA has 70% A-T base pairs and 30% G-C base pairs. Which type of DNA would have to be heated to a higher temperature in order to get the double stranded DNA to dissociate into single strands?

    _____

    _____

    _____

    _____

    _____

    _____

34. Suppose someone presented data from an analysis of the DNA of a newly discovered species showing that the relative amounts of the DNA bases were 30% for adenine, 30% for guanine, 20% for thymine, and 20% for cytosine. In what ways would these data not make sense in relation to what we know about the structure of DNA?

_____

_____

_____

_____

_____

_____

_____

_____

_____

35. Predict the complementary base sequence for each of the following short DNA molecules. Also predict which molecule would be more stable at a temperature 15 degrees above normal body temperature.

Molecule One: 5'-CGTAGTAGTAGAATATTGCTGCACC-3'

Molecule Two: 5'-AGTGCGAAGGCTCCTTGGGAACGTG-3'

_____

_____

_____

_____

_____

_____

36. Suppose one strand of a DNA molecule contains the following proportions of bases: 10% T, 20% A, 30% C, and 40% G. What proportions of the four types of bases would you expect in the double-stranded form of this DNA molecule?

_____

_____

_____

_____

_____

37. What are the things to look for when examining your skin for the presence of melanoma cancer?

_____

_____

Use the Case Study and the Web sites for this chapter to answer the following questions.

38. There are three kinds of skin cancer: basal cell carcinoma, squamous cell carcinoma and melanoma. What is the difference between a carcinoma and a melanoma? Which is the most invasive?

_____

_____

_____

39. The number of skin cancer cases increases every year. What is the best way to avoid skin cancer?

_____

_____

_____

_____

ANSWERS TO EXERCISES

1. DNA	5. Helicase	11. true	20. b
nucleotides	polymerase	12. false, phosphorus	21. c
helix	ligase	13. true	22. b
2. four	6. AATC	14. false, thymine	23. c
3. Adenine	GTTAC	15. false, semi	24. d
cytosine	7. Chromosomes	16. false, nucleotides	25. d
guanine	8. false, double	17. c	26. a
GC and AT	9. false, bases	18. c	27. b
4. semiconservative	10. true	19. e	28. a

29. Suppose a hypothetical single-stranded DNA molecule has the base sequence AAAAAAAAAA. During replication, the DNA polymerase would make a complimentary DNA molecule with the base sequence TTTTTTTTTT. Then, the DNA polymerase would have to make a complementary copy of the TTTTTTTTTT DNA molecule, which would be AAAAAAAAAA, the same as the original gene. Then, the TTTTTTTTTT molecule would have to be broken down, since it isn't a normal DNA molecule for that organism. This scheme, where two replications yield one new DNA molecule, is much less efficient than replicating a double-stranded DNA molecule, where one round of semiconservative replication yields two DNA molecules.

30. Hydrogen bonding between specific base pairs in the center of the helix holds the two strands together. Three hydrogen bonds hold guanine to cytosine; two hydrogen bonds hold adenine to thymine. Cytosine, theoretically, could pair with either guanine or adenine. Of the two possibilities, cytosine always pairs with guanine because they have complementary chemical side groups positioned so that three hydrogen bonds link them.

31. Within a double helix of DNA, whether from bacterial or human sources, adenine base pairs with thymine, and cytosine base pairs with guanine (A-T and C-G base pairs). Thus, within a DNA molecule, the amounts of A and T must be equal, as well as the amounts of C and G. However, since human DNA has more A-T pairs and fewer C-G base pairs than tuberculosis bacterial DNA, this accounts for the differing percentages when the DNAs are compared. Also, since each base pair contains a base with two rings of atoms (A and G) and a base with one ring of atoms (C and T), there has to be a total of 50% one-ring bases and 50% two-ring bases within any DNA molecule.

32. You might extract DNA from heat-killed strain A bacteria and mix it with living type B bacteria. If type B bacteria can absorb DNA fragments and use them as genes, the type B cells that absorb the normal vitamin B_6 gene could then begin to make the vitamin and behave like living type A cells. This would demonstrate that DNA is the genetic material. In a second experiment, you might extract protein from the heat-killed strain A cells and mix it with living type B bacteria. If type B bacteria can absorb the protein fragments but not use them as genes, the type B cells still could not make the vitamin. This would demonstrate that protein is not the genetic material.

33. The greater the number of hydrogen bonds attracting different molecules to each other, the greater will be the amount of energy needed to break all the hydrogen bonds and allow the molecules to separate from each other. Due to this, it takes more heat energy to separate G from C (3 hydrogen bonds involved) than to separate A from T (2 bonds). Thus, the higher the percentage of G-C pairings within a DNA molecule, the higher the DNA must be heated before the double stranded molecule separates into a pair of single stranded entities. It will take more heat to separate double stranded DNA from Species A (with 60% G-C base-pairs) than to separate double stranded DNA from Species B (with 30% G-C base pairs).

34. What the scientist's data would suggest is that in this peculiar type of DNA, A pairs with G and T pairs with C. There would be two problems with these data. The chemical structures of A and G make them incapable of forming hydrogen bonds with each other; likewise, the chemical structures of T and C make them incapable of forming hydrogen bonds with each other. In addition, even if A-G and C-T pairings were possible, this would mean that some pairings would involve a two-ring base pairing with a two-ring base (A-G), and some pairings would involve a one-ring base pairing with a one-ring base (C-T). That would mean the diameters of the G-C and A-T paired bases would be of different widths, neither of which would fit properly in the space available in a standard double stranded DNA molecule.

35. Molecule One: 5'-CGTAGTAGTAGAATATTGCTGCACC-3'
   3'-GCATCATCATCTTATAACGACGTGG-5'
   Molecule Two: 5'-AGTGCGAAGGCTCCTTGGGAACGTG-3'
   3'-TCACGCTTCCGAGGAACCCTTGCAC-5'
   The numbers of GC base pairs in molecule one is 11, while the number of AT base pairs in molecule one is 14. The numbers of GC base pairs in molecule two is 15, while the number of AT base pairs in molecule one is 10. Therefore, molecule two has more hydrogen bonds holding the two strands together than does molecule one, and molecule two is more stable at a higher temperature (see Question 33).

36. The double stranded DNA molecule would have the following proportions of bases:
   one strand: 10% T, 20% A, 30% C, and 40% G
   other strand: 10% A, 20% T, 30% G, and 40% C
   Collectively: 60 ÷ 200 = 30% A-T and 140 ÷ 200 = 70% G-C, which means that A = 15%, T = 15%, G = 35%, and C = 35%

37. To recognize a possible melanoma in your skin, remember to check moles for "ABCD": Asymmetrical shape, irregular Border, irregular Color, or a Diameter bigger than the eraser on the end of a pencil.

38. According to the National Cancer Institute, a carcinoma is cancer that begins in the cells that cover or line an organ. Another type of cancer that occurs in the skin is melanoma, which begins in the melanocytes. Melanomas are much more dangerous due to their tendency to spread to other parts of the body.

39. Protecting a baby's skin will prevent sunburn now and also may guard against a more serious problem in the future. "The damaging effects of the sun's rays are cumulative," says Paul J. Honig, M.D., chief of Dermatology, The Children's Hospital of Philadelphia. "Although the incidence of skin cancer is low in young people, severe sunburn or continued exposure to the sun early on in life may be correlated with skin cancer later in life."

Chapter 10: Gene Expression and Regulation

OVERVIEW

In this chapter you will learn how genes are expressed and regulated. The authors introduce the "one-gene, one-protein" hypothesis. They discuss the processes of transcription, during which information in DNA molecules makes RNA, and translation, during which the information in RNA molecules makes proteins. You also will learn about the three types of RNA and their functions, as well as transcriptional regulation of genes in eukaryotic cells and the effects of mutation.

Boys and girls have many physical differences that are biologically determined. However, the genes of men and women do not differ so dramatically. In fact, boys have all the genes needed to make female genitalia, and girls have all the genes needed to make male genitalia. In boys, the action of a single gene activates the male developmental pathway and deactivates the female pathway. How does this gene work?

1) How Are Genes and Proteins Related?

The information in DNA must be linked to proteins, since proteins are responsible for building cell components and carrying out biochemical reactions. Within a biochemical pathway, the product of one enzyme becomes the substrate of the next enzyme in the pathway. Using mold (Neurospora) with single doses of chromosomes and **genes** in its cells, Beadle and Tatum used X-rays to cause mutations (changes in the base sequence of DNA). Each mutation caused the loss of the ability of Neurospora to make one enzyme. Thus, a mutation in a single gene affected only a single enzyme within a single biochemical pathway. Data supported the hypothesis that each gene encodes information (as sequence of bases) needed for making one specific protein (an amino acid sequence): the "one-gene, one-enzyme" relationship. Most genes contain the information for the synthesis of a single protein. Some genes code for structural proteins or for types of **RNA (ribonucleic acid)**, but most code for enzymes. Some functional proteins have more than one subunit, each subunit made by a different gene, so the "one-gene, one-enzyme" relationship has been clarified to become the "one-gene, one-**polypeptide**" relationship.

DNA in the nucleus provides instructions for protein synthesis in the cytoplasm via RNA inter-mediaries. RNA differs from DNA in three ways: RNA is single stranded, RNA has the sugar ribose instead of deoxyribose in its backbone, and RNA has the base uracil instead of thymine. DNA codes for three kinds of RNA: **messenger RNA (mRNA)**, **ribosomal RNA (rRNA)**, and **transfer RNA (tRNA)**.

Overview: Genetic information is transcribed into RNA and translated into protein. During **transcription**, the information in DNA is copied into either mRNA, rRNA, or tRNA. Thus, a gene is a segment of DNA that can be transcribed into RNA. **Translation** is catalyzed by RNA polymerase and occurs in the nucleus. During protein synthesis (translation), tRNA and rRNA, together with proteins, use the nucleotide sequence in mRNA to make a specific amino acid sequence within a protein. The rRNA assembles together with dozens of proteins to form **ribosomes**. Translation is catalyzed by ribosomes and occurs in the cytoplasm.

In cells, the language of nucleotide sequences in DNA and mRNA must be translated into the language of amino acid sequences in proteins. The dictionary for translation is the **genetic code**. In the genetic code, a sequence of three nucleotides specifies an amino acid or a "stop." Four different nucleotide bases must code for 20 different amino acids in protein. A three-base sequence gives 64 possible combinations, more than enough to code for 20 different amino acids. "Words" in the genetic code are three bases long and most signify an amino acid. An RNA with only U bases (UUUUUUUUU…) makes a protein with only phenylalanine (phe-phe-phe…). So, the triplet UUU codes for phenylalanine. The 64 mRNA triplets are called **codons**. All amino acids begin with the same amino acid methionine, specified by the codon AUG. Three codons, UAG, UAA, and UGA, are **stop codons**, which signal the completion of protein synthesis. Each codon specifies one, and only one, amino acid (except the stop codons).

2) How Is Information in a Gene Transcribed into RNA?

Transcription consists of initiation, elongation, and termination. Initiation of transcription begins when **RNA polymerase** binds to the promoter of a gene. A different version of the enzyme RNA polymerase makes each type of RNA (m, r, and tRNA). RNA polymerase must select the appropriate genes to transcribe in each cell type and at each stage of a cell's life. The enzyme uses a gene's **promoter** region, a non-transcribed sequence of DNA that marks the beginning of the gene. When RNA polymerase binds to the promoter, transcription of that gene begins.

When RNA polymerase binds to the promoter region, it begins to unwind the gene and then travels in one direction along one of the DNA strands, making an RNA molecul that is complementary to that strand of DNA (the template strand). After about 10 nucleotides have been added to the growing RNA chain, the first nucleotides of the RNA molecule separate from the DNA **template strand**, and that region of the gene rewinds. When the RNA polymerase reaches the termination signal, it releases the completed RNA molecule and detaches from the DNA. Transcription is selective, being restricted to only those genes needed by that particular cell type at that particular time. And, transcription copies only the template strand of most genes.

3) How Is the Sequence of a Messenger RNA Molecule Translated into Protein?

Messenger RNA carries the code for the amino acid sequence of a protein from the nucleus into the cytoplasm, while the gene remains safely in the nucleus. Ribosomal RNA forms an important part of the protein-synthesizing machinery of a ribosome. Each ribosome has a large and a small subunit, which remain separate unless the ribosome is making proteins. When active, the small and large ribosomal subunits come together with an mRNA molecule in between them. The large subunit has binding sites for tRNA and a catalytic site for joining the amino acids brought in by the tRNAs.

Transfer RNA molecules decode the sequence of bases in mRNA into the amino acid sequence of a protein. Cells make 61 types of tRNA, one type for each different amino-acid encoding mRNA codon. Each type of tRNA is capable of picking up only one type of amino acid. Each tRNA has three exposed bases, called the **anticodon**, that form base pairs with the mRNA codon. For example, the mRNA codon AUG will attract the tRNA with anticodon UAC carrying the amino acid methionine.

During translation, mRNA, tRNA, and ribosomes cooperate to make proteins. Translation has three steps: initiation (protein synthesis begins when tRNA and mRNA bind to a ribosome), elongation, and termination. In initiation, the first AUG codon in a eukaryotic mRNA sequence specifies where translation will begin. An "initiation complex" (the small ribosomal subunit and a methionine-tRNA) binds to the end of an mRNA and moves along until it encounters the first AUG codon, which base pairs with the UAC anticodon of the methionine tRNA. The large ribosomal subunit then attaches, sandwiching the mRNA between the small and large subunits.

During elongation and termination, protein synthesis proceeds one amino acid at a time until a stop codon is reached. The assembled ribosome holds two mRNA codons in alignment with the two tRNA binding sites in the large subunit. With one tRNA in place, a second tRNA with the complementary anticodon and amino acid moves into the second tRNA binding site on the large subunit, which then breaks the bond holding the first amino acid to its tRNA and forms a peptide bond between the two amino acids. The first tRNA is empty and leaves the ribosome, and the second tRNA has two amino acids bound. The ribosome shifts to the next mRNA codon, moving the second tRNA into the first binding site and allowing the third tRNA to move into place. The third amino acid forms a peptide bond with the second one, which detaches from its tRNA. The empty second tRNA leaves, the ribosome shifts to read the fourth codon, shifting the third tRNA with attached amino acids into the first binding site, and the process continues.

When the ribosome reaches a stop codon, special proteins bind to the ribosome, forcing it to release the finished protein chain and the mRNA, and the ribosome disassembles. A protein 100 amino acids long can be assembled by translation in about 6 seconds.

4) How Do Mutations in DNA Affect the Function of Genes?

Mutations are changes in the sequence of bases in DNA, often through a mistake in base pairing during DNA replication, or triggered by an environmental chemical or radiation. In a **nucleotide substitution** (or **point mutation**), a pair of bases becomes incorrectly matched and the cell replaces the correct base with another incorrect one. An **insertion mutation** occurs when one or more new nucleotide pairs are added into a gene. A **deletion mutation** occurs when one or more nucleotide pairs are removed from a gene. Deletions and insertions can have quite harmful effects on a gene because all the codons that follow the deletion or insertion will be misread.

Mutations have different effects on protein structure and function. Four types of effects may result from mutations: (1) The protein is unchanged. If the gene's base sequence is changed from GAG to GAA (a point mutation), the same amino acid will be incorporated into the protein this gene codes for. (2) The new protein is equivalent to the original one if the active site is unchanged and the rest of the molecule is changed in an insignificant way. This is a **neutral mutation**. For instance, a change from GAG to CAG replaces one hydrophilic amino acid with another. (3) Protein function is changed by an altered amino acid sequence. If GAG changes to GTG, a hydrophilic amino acid is replaced by a hydrophobic one. (4) Protein function is destroyed by a misplaced stop codon, resulting in a much shortened protein chain. If AAG is replaced by TAG, this creates a new stop codon, halting translation prematurely.

Mutations provide the raw material for evolution. Mutation rates vary from 1 in 100,000 to 1 in 1,000,000 gametes. So, a human male releases about 600 mutant sperm during each ejaculation of 300–400 million sperm. If mutations in gametes are not lethal, they may be passed on to future generations. Mutation is the source for genetic variation and thus is essential for evolution.

5) How Are Genes Regulated?

How do cells control the presence, concentration, and activity of the proteins encoded by genes? Proper regulation of gene expression is critical for an organism's development and health. Most cells in the body have identical DNA (30,000 to 50,000 genes) but they don't use all the DNA all the time. Gene expression changes over time, and an organism's environment can determine which genes are translated. Four of the steps at which the rate of gene activity may be regulated are: (1) transcription of individual genes; (2) translation of various mRNAs; (3) modification of inactive proteins into their active forms; and (4) regulation of a protein's lifespan.

Eukaryotic cells may regulate the transcription of (1) individual genes, through the action of regulatory proteins such as steroid hormones; (2) regions of chromosomes with several genes, by condensing those regions into compact DNA that is inaccessible to RNA polymerase; or (3) entire chromosomes with thousands of genes, such as one of the X chromosomes in mammalian females to form a **Barr Body**. Which X chromosome is inactivated in any cell is random, but all its daughter cells will have the same condensed chromosome. For example, separate patches of orange and black fur in female calico cats are due to fur-color genes on the X chromosome.

Case study revisited: boy or girl? The Y chromosome of mammals contains the SRY (sex-determining region on the Y) gene. An XX mouse embryo given the SRY gene develops into a sterile male, indicating that an XX female has all the genes to be a male. Likewise, an XY mouse embryo lacking the SRY gene develops into a female, indicating that an XY male has all the genes to be a female. The SRY gene normally is transcribed for only a short time in embryonic development and only in the cells that will become the testes. The SRY protein initiates the expression of many other genes, the proteins from which are needed for testes development. The testes secrete testosterone, which activates other genes leading to development of the penis and scrotum. So, the SRY gene is the initial genetic switch to activate male development.

KEY TERMS AND CONCEPTS

Fill-In: From the following list of key terms, fill in the blanks in the following statements.

anticodon	messenger RNA	RNA polymerase
codon	mutation	stop codons
deletion mutation	one-gene, one-protein hypothesis	transcription
genetic code	point mutation	transfer RNA
insertion mutations	ribosome	translation

1. The _____ is the amino acid translation of all the codons, each of which directs the incorporation of an amino acid during protein synthesis.

2. The enzyme that catalyzes the covalent bonding of free RNA nucleotides into a continuous strand, using the sequence of bases in DNA as a template, is called _____.

3. The hypothesis that each gene encodes information (as a sequence of bases) needed for making one specific protein (amino acid sequence) is the _____.

4. An _____ is a sequence of three nucleotides in transfer RNA that is complementary to the three nucleotides in messenger RNA.

5. The process in which a sequence of nucleotide bases in mRNA is converted into a sequence of amino acids in a protein is called _____.

6. The mRNA "start codon" is AUG, coding for the first amino acid in a protein. Three codons (UAG, UAA, UGA) are mRNA _____, signaling that the protein's amino acid sequence is completed.

7. A molecule of _____ is a strand of nucleotides, complementary to the DNA of a gene, that conveys genetic information to ribosomes to be used to sequence amino acids during protein synthesis.

8. _____ is the synthesis of an RNA molecule from a DNA template.

9. A change in the base sequence of DNA is called a _____.

10. A molecule that binds to a specific amino acid and has a set of three nucleotides complementary to the codon for that amino acid is known as _____.

11. In a _____, a pair of bases becomes incorrectly matched; _____ occur when one or more new nucleotide pairs are added into a gene; and a _____ occurs when one or more nucleotide pairs are removed from a gene.

12. A _____ is a sequence of three nucleotides of mRNA that specifies a particular amino acid to be incorporated into a protein.

13. An organelle with two subunits, each composed of RNA and protein, that serves as the site of protein synthesis is a _____.

Key Terms and Definitions

androgen insensitivity: a rare condition in which an individual with XY chromosomes is female in appearance because the body's cells don't respond to the male hormones that are present.

anticodon: a sequence of three bases in transfer RNA that is complementary to the three bases of a codon of messenger RNA.

Barr body: an inactivated X chromosome in cells of female mammals, which have two X chromosomes; normally appears as a dark spot in the nucleus.

codon: a sequence of three bases of messenger RNA that specifies a particular amino acid to be incorporated into a protein; certain codons also signal the beginning or end of protein synthesis.

deletion mutation: a mutation in which one or more pairs of nucleotides are removed from a gene.

exon: a segment of DNA in a eukaryotic gene that codes for amino acids in a protein (see also *intron*).

gene: a unit of heredity that encodes the information needed to specify the amino acid sequence of proteins and hence particular traits; a functional segment of DNA located at a particular place on a chromosome.

genetic code: the collection of codons of mRNA, each of which directs the incorporation of a particular amino acid into a protein during protein synthesis.

insertion mutation: a mutation in which one or more pairs of nucleotides are inserted into a gene.

intron: a segment of DNA in a eukaryotic gene that does not code for amino acids in a protein.

messenger RNA (mRNA): a strand of RNA, complementary to the DNA of a gene, that conveys the genetic information in DNA to the ribosomes to be used during protein synthesis; sequences of three bases (codons) in mRNA specify particular amino acids to be incorporated into a protein.

mutation: a change in the base sequence of DNA in a gene; normally refers to a genetic change significant enough to alter the appearance or function of the organism.

neutral mutation: a mutation that has little or no effect on the function of the encoded protein.

nucleotide substitution: a mutation that replaces one nucleotide in a DNA molecule with another; for example, a change from an adenine to a guanine.

point mutation: a mutation in which a single base pair in DNA has been changed.

polypeptide: a short polymer of amino acids; often used as a synonym for protein.

promoter: a specific sequence of DNA to which RNA polymerase binds, initiating gene transcription.

ribonucleic acid (rī -bō-noo-klā´-ik; RNA): a molecule composed of ribose nucleotides, each of which consists of a phosphate group, the sugar ribose, and one of the bases adenine, cytosine, guanine, or uracil; transfers hereditary instructions from the nucleus to the cytoplasm; also the genetic material of some viruses.

ribosomal RNA (rRNA): a type of RNA that combines with proteins to form ribosomes.

ribosome: an organelle consisting of two subunits, each composed of ribosomal RNA and protein; the site of protein synthesis, during which the sequence of bases of messenger RNA is translated into the sequence of amino acids in a protein.

RNA polymerase: in RNA synthesis, an enzyme that catalyzes the bonding of free RNA nucleotides into a continuous strand, using RNA nucleotides that are complementary to those of a strand of DNA.

stop codon: a codon in messenger RNA that stops protein synthesis and causes the completed protein chain to be released from the ribosome.

template strand: the strand of the DNA double helix from which RNA is transcribed.

transcription: the synthesis of an RNA molecule from a DNA template.

transfer RNA (tRNA): a type of RNA that binds to a specific amino acid by means of a set of three bases (the anticodon) on the tRNA that are complementary to the mRNA codon for that amino acid; carries its amino acid to a ribosome during protein synthesis, recognizes a codon of mRNA, and positions its amino acid for incorporation into the growing protein chain.

translation: the process whereby the sequence of bases of messenger RNA is converted into the sequence of amino acids of a protein.

uracil: a nitrogenous base found in RNA; abbreviated as *U*.

Werner syndrome: a rare condition in which a defective gene causes premature aging; caused by a mutation in the gene that codes for DNA replication/repair enzymes.

THINKING THROUGH THE CONCEPTS

True or False: Determine if the statement given is true or false. If it is false, change the underlined word(s) so that the statement reads true.

14. _____ Genes are made of <u>RNA</u> in human cells.

15. _____ <u>Transfer RNA</u> carries amino acids to the ribosomes.

16. _____ Protein synthesis occurs in the <u>ribosome</u>.

17. _____ Messenger RNA is <u>double</u> stranded.

18. _____ Messenger RNA is manufactured in the <u>cytoplasm</u>.

19. _____ The triplets of bases in messenger RNA are called <u>anticodons</u>.

20. _____ Proteins are made during <u>transcription</u>.

21. _____ Proteins contain many <u>nucleotide</u> subunits.

22. _____ Barr bodies are found in normal mammalian <u>females</u>.

23. _____ Barr bodies are <u>active</u> X chromosomes found in mammals.

Identify: Determine whether the following statements refer to **mRNA**, **tRNA**, or **rRNA**.

24. _____ has anticodons

25. _____ deciphers the genetic code

26. _____ carries the genetic code to make proteins

27. _____ picks up and transports amino acids

28. _____ part of ribosomes

29. _____ has codons

30. _____ twisted into a cloverleaf shape

31. _____ fits into binding sites in ribosomes

Identify: Determine whether the following statements refer to **transcription** or **translation**.

32. _____ information from DNA makes RNA

33. _____ information from RNA makes protein

34. _____ occurs in the nucleus of eukaryotic cells

35. _____ occurs in the cytoplasm of eukaryotic cells

36. _____ involves RNA polymerase

37. _____ involves amino acids

38. _____ involves ribosomes

39. _____ involves codon-anticodon interactions

40. _____ involves copying the genetic code

41. _____ involves deciphering the genetic code

Multiple Choice: Pick the most correct choice for each question.

42. Inherited disorders induced by X-rays in red bread mold by Beadle and Tatum
    a. are caused by errors in mitosis
    b. are related to enzyme deficiencies
    c. can always be cured by dietary restrictions
    d. are environmental and not genetic in origin
    e. can never be cured by supplying the missing end product

43. If a bacterial protein has 30 amino acids, how many nucleotides are needed to code for it?
    a. 30
    b. 60
    c. 90
    d. 120
    e. 600

44. Which of these choices is coded for by the shortest piece of DNA?
    a. a tRNA having 75 nucleotides
    b. an mRNA having 50 codons
    c. a protein having 40 amino acids
    d. a protein with 2 polypeptides, each having 35 amino acids
    e. an mRNA having 100 bases

45. Blood cells and muscle cells make different enzymes because
    a. blood cells contain only genes for blood cell proteins and muscle cells contain only muscle protein genes
    b. all cells of an organism have all genes
    c. not every gene acts in every type of cell
    d. blood cells have hemoglobin while muscle cells have microtubules
    e. adult red blood cells lack nuclei in mammals

46. Because of random X chromosome inactivation, one of the X chromosomes of a mammalian female
    a. is functionally inactive
    b. is present in each cell in three doses
    c. does not divide during meiosis
    d. disappears from each cell early during development
    e. is genetically identical to the other X chromosome

Fill-In: Based on the figure to the right, answer the following questions.

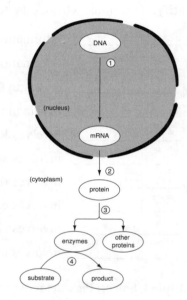

47. The step indicated by (1) in the figure is commonly known as

    _____.

48. The step indicated by (2) in the figure is commonly known as

    _____.

49. The step indicated by (3) in the figure is commonly known as

    _____.

50. The step indicated by (4) in the figure is commonly known as

    _____.

Fill-In: Fill in the blanks in the following table.

51.

Type of molecule	Sequences of bases or amino acids See text Table 10-3 for genetic code.					
DNA template strand	_____	TAG	_____	AGC	_____	TCA
DNA non-template strand	GAA	_____	TTA	_____	CCG	_____
messenger RNA codons	_____	_____	_____	_____	_____	_____
transfer RNA anticodons	_____	_____	_____	_____	_____	_____
protein amino acid sequence	_____	_____	_____	_____	_____	_____

APPLYING THE CONCEPTS

These practice questions are intended to sharpen your ability to apply critical thinking and analysis to biological concepts covered in this chapter.

Short Answer.

Suppose a section of DNA from a normal gene has the following sequences of bases in one of its polynucleotide strands:

    normal base sequence: TACTTTACGTCGTGAAAACGGTAT

    If this strand is used to make an mRNA molecule:

52. the base sequence in the mRNA is _____

53. the normal amino acid sequence in the polypeptide is (use text Table 10-3)

    _____

Now, suppose a point mutation occurs and a single base (C*) is added (an insertion mutation) to the gene sequence, causing a new altered sequence to occur:

    mutant base sequence: TACC*TTTACGTCGTGAAAACGGTAT

    If this mutant strand is used to make an mRNA molecule:

54. the base sequence in the mRNA is _____

55. the abnormal amino acid sequence in the polypeptide is (use text Table 10-3)

    _____

56. Briefly explain why, in questions 52–55, the addition of a single base in the DNA caused so many amino acids to change in the polypeptide.

    _____

    _____

    _____

57. What sort of evidence exists to back up the claim that mammalian females have all the genes to develop into a male?

    _____

    _____

    _____

    _____

    _____

58. What sort of evidence exists to back up the claim that mammalian males have all the genes to develop into a female?

    _____

    _____

    _____

    _____

59. Are male calico cats expected in a normal cat population? How could they occur?

_____

_____

_____

_____

_____

_____

_____

_____

Use the Case Study and the Web sites for this chapter to answer the following questions.

60. The SRY region of the Y chromosome has been shown to code for a factor that is "necessary and sufficient" for male physiological development. What is the experimental evidence for the essential role of SRY?

_____

_____

_____

_____

_____

_____

_____

_____

61. Are individuals with XXY (Klinefelter syndrome) male or female? How about individuals with Turner syndrome (XO), XYY individuals, or XXX individuals? Are they capable of reproducing?

_____

_____

_____

62. What biological errors of metabolism result in XY females? Consider gene rearrangements and the SRY gene. How could an XX individual be physiologically male?

_____

_____

_____

_____

_____

ANSWERS TO EXERCISES

1. genetic code
2. RNA polymerase
3. "one-gene, one-protein" hypothesis
4. anticodon
5. translation
6. stop codons
7. messenger RNA (mRNA)
8. Transcription
9. mutation
10. transfer RNA (tRNA)
11. point mutation
    insertion mutations
    deletion mutation
12. codon
13. ribosome
14. false, DNA
15. true

16. true
17. false, single
18. false, nucleus
19. false, codons
20. false, translation
21. false, amino acid
22. true
23. false, inactive
24. tRNA
25. tRNA
26. mRNA
27. tRNA
28. rRNA
29. mRNA
30. tRNA
31. tRNA
32. transcription
33. translation

34. transcription
35. translation
36. transcription
37. translation
38. translation
39. translation
40. transcription
41. translation
42. b
43. c
44. a
45. c
46. a
47. transcription
48. translation
49. modification
50. catalysis

51.

Type of molecule	Sequences of bases or amino acids See text Table 10-3 for genetic code.					
DNA template strand	CTT	TAG	AAT	AGC	GGC	TCA
DNA non-template strand	GAA	ATC	TTA	TCG	CCG	AGT
messenger RNA codons	GAA	AUC	UUA	UCG	CCG	AGU
transfer RNA anticodons	CUU	UAG	AAU	AGC	GGC	UCA
protein amino acid sequence	glutamic acid	isoleucine	leucine	serine	proline	serine

52. AUG-AAA-UGC-AGC-ACU-UUU- GCC-AUA

53. Methionine-lysine-cysteine-serine-threonine-phenylalanine-alanine-isoleucine

54. AUG-GAA-AUG-CAG-CAC-UUU-UGC-CAU-A

55. Methionine-glutamic acid-methionine-glutamine-histidine-phenylalanine-cysteine-histidine-

56. The reason so many amino acids are changed is that the addition of one base causes all the codons to become different because they are read by ribosomes as three consecutive mRNA bases and the addition of one base results in a shift in the "reading frame" the ribosome uses to determine the codons.

57. If a mouse embryo with two X chromosomes is given a copy of the SRY gene normally found only in males, the embryo develops completely male characteristics, including a penis and testes, and the juvenile and adult mouse behaves as a male. However, it is sterile because other genes on the Y chromosome apparently are needed for production of functional sperm. Thus, an XX mouse has all the genes necessary to develop into a male if the SRY gene is present.

58. A mouse embryo, either XX or XY in sex chromosome content, will develop into a female if the SRY gene is lacking or not functional. Such XY females are sterile, however, since two X chromosomes are needed for completely normal ovary development. Thus, an XY mouse has all the genes necessary to develop into a female but usually doesn't because the SRY gene is present.

59. Normally, all calico cats are females, because the genes for black and orange fur color are on the X chromosome in cats. In mammalian females, only one X is active in body cells, determined randomly early in development. So, in the hair cell population of a female cat who has an orange color gene in one X and a black color gene in the other X, half the hair cells would make black hair and the other half would make orange hair. Male cats, on the other hand, with only one X chromosome per cell, would have either all black or all orange hair. However, once in a while, a calico male cat is born. This type of cat has an XXY sex chromosome constitution, with the orange color gene on one X and the black color gene on the other X. Thus, the X chromosomes behave as they do in females, while the cat is a male because of the action of the Y chromosome.

60. According to McKusick et. al. at the National Center for Biotechnology Information, three lines of evidence indicate that the SRY gene is necessary and sufficient for male sex determination and is the testis-determining factor. First, the SRY gene maps to the smallest region of the Y chromosome known to be male determining in humans and in the mouse; namely, the 35-kb interval just proximal to the pseudoautosomal boundary of the Y chromosome. Second, XY females with gonadal dysgenesis and with the Y chromosome intact have mutations in the SRY gene. Third, a 14-kb genomic fragment carrying the mouse SRY gene and no other coding sequence was sufficient to induce testis differentiation and subsequent male development when introduced into chromosomally female mouse embryos.

61. According to R. Bock at National Institute of Health, XXY individuals are male and sterile. According to the Turner's Syndrome Society, XO individuals are female and sterile. And According to J. Nielsen at the Turner Center, XYY individuals are male and fertile, and XXX individuals are female and fertile. Any human with a Y chromosome containing a functional SRY gene is male, and all other humans are female.

62. According to researchers at the John Hopkins University School of Medicine, Division of Pediatric Endocrinology, an XY human could be physiologically female if the SRY gene on the Y chromosome is abnormal and non-functional due to mutation. Another possibility is that during meiosis in the father of the XY "female," a rare crossover occurred between the X and the Y, moving the SRY gene from the Y chromosome over to the X chromosome. If that occurs, the father's Y chromosome without the SRY could produce an XY "female" offspring, and the father's X chromosome containing the SRY gene could produce an XX "male" offspring.

Chapter 11:   The Continuity of Life: Cellular Reproduction

OVERVIEW

In this chapter, you will learn about mitosis and meiosis. The authors begin with a brief discussion of the cell cycle in prokaryotes, then discuss mitosis, a basic type of eukaryotic cell division. They describe the structure of eukaryotic chromosomes and review the typical chromosome numbers in body cells and sex cells. You will read about the eukaryotic cell cycle as well as interphase, mitosis, and cytokinesis. Then, the authors contrast asexual and sexual reproduction, and focus on meiosis. They describe the stages of meiosis I and meiosis II as they relate to the production of sex cells. They compare meiosis I and II, as well as mitosis and meiosis. This chapter concludes with descriptions of the three basic types of life cycles among eukaryotic organisms. Female rattlesnakes can sometimes give birth to male offspring without having mated with a male rattlesnake at all. How can this happen?

1) What Are the Functions of Cellular Reproduction?

Cellular reproduction enables a parent cell to accurately distribute both genes and cell components to its daughter cells in a process called **cell division**: binary fission in prokaryotic cells, and mitosis and meiosis in eukaryotic cells. **Binary fission** ("splitting in two") is the cell division process in prokaryotes, producing two cells which are genetically identical. First, the cell replicates its DNA double helix, a closed circular structure with about 4000 genes. The two identical double helices attach to the plasma membrane at nearby, separate points. The plasma membrane expands, pushing the double helices apart. The plasma membrane around the middle of the cell grows inward between the two DNA attachment sites until the cell is split into two daughter cells. After binary fission, each daughter cell contains one DNA double helix and about half of the original cell's cytoplasm. Under ideal conditions, the common intestinal bacterium *Escherichia coli* (*E. coli*) can complete binary fission in about 20 minutes.

Mitotic cell division allows development, growth, maintenance, and repair of body tissues in eukaryotes. Mitotic cell division involves a process of nuclear divisions called mitosis followed by a single cell division, producing two genetically identical daughter cells. In conjunction with the differential expression of genes in different cells, mitotic cell division allows a fertilized egg to produce cells needed in a new-born organism, and allows the organism to grow into an adult. It also allows an organism to maintain its tissues, many of which may require frequent replacement of cells. It allows for repair or even regeneration of cells following injury or surgery. For instance, your liver has the ability to regenerate.

Mitotic cell division forms the basis of **asexual reproduction** in eukaryotes, in which offspring are formed from a single parent without the uniting of male and female gametes. This is routine for many unicellular organisms and some multicellular organisms as well. For instance, in *Hydra*, a small replica of the parent grows by mitotic cell division, and this bud separates from its parent to live independently. The bud is genetically identical to the parent; they are called **clones**. Asexual reproduction is not responsible for the baby rattlesnake described above. Many plants and fungi reproduce both asexually and sexually. Some organisms produce sex cells by mitotic cell division. Mitosis also gave rise to the nucleus used to produce Dolly, the cloned sheep.

Sexual reproduction in eukaryotes produces offspring with a mixture of genetic material from two parents. This is possible by a process called **meiotic cell division**. In mammals, it occurs only in the ovaries and testes. Meiotic cell division involves a nuclear division called meiosis, and cellular divisions to produce daughter cells that can become specialized **gametes** (egg and sperm) that carry half of the genetic material of the parent. The cells produced by meiosis are not identical to each other or to the original cell.

2) How Is DNA in Eukaryotic Cells Organized into Chromosomes?

Cell division enables accurate passage of chromosomes from one generation to the next. Eukaryotic cellular DNA is packaged in **chromosomes** (darkly staining nuclear rods of condensed protein and DNA). Every species has a particular number of chromosomes, and consequently, a specific number of DNA double helices. Human body cells have 46 chromosomes, the largest of which (chromosome 1) contains about 3,000 genes and one of the smallest of which (chromosome 22) contains about 600 genes.

The eukaryotic chromosome consists of a linear DNA double helix bound to proteins, including histones that organize and compact the DNA so that it can fit into the nucleus. The total length of DNA in a single cell is about 6 feet long (about 2 meters), while the nucleus is a million times smaller. The degree of DNA compaction, or **condensation**, varies with the stage of the cell cycle. During cell growth, the DNA is maximally dispersed and active, and individual chromosomes are too thin to be visible in light microscopes. During cell division, chromosomes are condensed and shortened for easier transport, and proteins fold up the DNA into compact structures that can be seen under the microscope. At the time it condenses, the chromosomal DNA has already replicated, forming two double helices attached to each other at the **centromere** ("middle body") region. Each double helical chromosome of the attached pair is called a sister **chromatid**. Thus, the DNA replication has produced a **duplicated chromosome** with two identical sister chromatids. During mitotic cell division, the two sister chromatids separate and each chromatid becomes an independent chromosome.

Eukaryotic chromosomes usually occur in homologous pairs with similar genetic information. An entire set of stained chromosomes from a single nonreproductive cell (the **karyotype**) shows pairs of chromosomes. Typically, the members of each pair are of the same length and have the same staining pattern, and carry the same genes arranged in the same order. Chromosomes containing the same genes are called homologous chromosomes or **homologues** ("to say the same thing"). Cells with pairs of homologues are called **diploid** ("times two"). A human skin cell, for instance, has 46 chromosomes, organized into 23 pairs of homologues. One pair is the **sex chromosomes**, two X chromosomes (XX) in females and an X and Y chromosome (XY) in males. Most body cells are diploid. During sexual reproduction, cells in the ovaries and testes undergo meiotic cell division to produce gametes (eggs or sperm) with only 23 chromosomes, one of each type. Such cells are called **haploid**. Fusion of two haploid cells produces a diploid cell. The number of different types of chromosomes in a species is the haploid number and is designated n. Diploid cells are $2n$. In humans, $n=23$ (in gametes) and $2n=46$ (in body cells). In male mammals, one pair of chromosomes, the X and Y, have very different sizes and staining patterns, but behave like homologues during cell division.

3) What Are the Events of the Eukaryotic Cell Cycle?

The eukaryotic **cell cycle** has two phases: interphase and mitotic cell division. Acquiring nutrients, growth, replication of DNA, and most cell functions occur during **interphase**, the relatively long time period between cell divisions. Interphase contains three subphases. During the G_1 phase (gap or growth phase 1), the cell acquires nutrients, performs specialized functions, and grows. The cell is sensitive to internal and external signals that help the cell decide whether to divide. Active cells no longer dividing (heart, brain, and eye cells, for instance) enter the G_0 phase. Alternately, cells enter the S phase (DNA synthesis phase), during which DNA and chromosome replication occurs. Next, cells enter the G_2 phase (gap or growth phase 2), during which cells make molecules required for cell division. Progression through the cell cycle is carefully regulated. At key times called "checkpoints," molecular signals ensure that the cell has completed all necessary processes before entering the next step. One such checkpoint occurs at the end of G_1.

Mitotic cell division consists of nuclear division (mitosis) and cytoplasmic division (cytokinesis).

Mitosis produces two nuclei, each containing a copy of every chromosome present in the original nucleus. In most cells, the cytoplasm divides following mitosis to form two daughter cells each with a nucleus and about half the cytoplasm, a process called **cytokinesis**. The two cells produced by mitotic cell division are essentially identical to each other cytoplasmically, and genetically identical to each other and to the parent cell. Some cells undergo mitosis without cytokinesis, producing single cells with many nuclei. All your different types of body cells (brain and liver, for example) are genetically identical but undergo **differentiation**, the process whereby cells assume specialized functions because they *use* different genes as they develop.

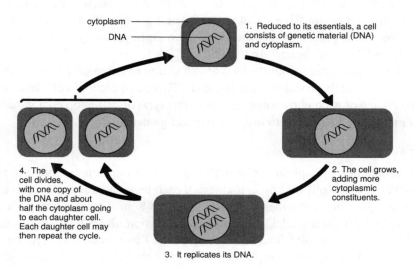

cytoplasm

DNA

1. Reduced to its essentials, a cell consists of genetic material (DNA) and cytoplasm.

2. The cell grows, adding more cytoplasmic constituents.

3. It replicates its DNA.

4. The cell divides, with one copy of the DNA and about half the cytoplasm going to each daughter cell. Each daughter cell may then repeat the cycle.

4) What Are the Phases of Mitosis?

Mitosis has four phases, based on the appearance and behavior of the chromosomes, within a continuous process: prophase, metaphase, anaphase, and telophase.

During **prophase** ("the stage before"), the duplicated chromosomes condense, **spindle microtubules** form, and the chromosomes are captured by the spindle. Also, the nucleolus disappears and the nuclear membrane disintegrates. The spindle microtubules assemble, originating in animal cells from a region containing a pair of **centrioles**. The centrioles replicate and each pair migrates to opposite sides of the nucleus forming the "poles." From the poles, spindle microtubules radiate towards the plasma membrane and towards the nucleus, forming a basket around the nucleus. Some of the spindle microtubules attach to the chromatids at protein-containing structures called **kinetochores**. One kinetochore of each duplicated chromosome is tethered to one spindle pole and the other kinetochore is tethered to the other spindle pole.

During **metaphase** ("the middle stage"), chromosomes become aligned along the equator of the cell through the interactions of the spindle fibers and the kinetochores. During **anaphase** ("the moving away stage"), sister chromatids separate from each other and the protein "motors" in the kinetochores pull the chromosomes to opposite poles along the spindle microtubules. The unattached spindle microtubules lengthen to push the poles of the cell apart. Because the sister chromatids are identical copies of the original chromosomes, both clusters of chromosomes moving to opposite ends of the cell contain one copy of every chromosome in the original cell. During **telophase** (the "end stage"), nuclear envelopes form around each group of chromosomes, the chromosomes revert to their extended form, and nucleoli reappear.

5) What Are the Events of Cytokinesis?

In most cells, cytokinesis occurs during telophase, enclosing each daughter nucleus into a separate cell. In animal cells, microfilaments attached to the plasma membrane around the equator contract and constrict the equator region, pinching the cell in two. In plant cells, the Golgi complex buds off carbohydrate-filled vesicles along the equator. The vesicles fuse, forming the **cell plate** that expands to fuse with the plasma membrane, eventually forming a new cell wall. Cytokinesis is followed by G_1 of interphase.

6) What Are Some Advantages of Sexual Reproduction?

Mitosis can only produce clones, genetically identical offspring. The reshuffling of genes among individuals to create genetically unique offspring results from **sexual reproduction**. DNA mutations are the ultimate source of genetic variability and the raw material for evolution. Mutations form **alleles**, alternate forms of a gene that confer variability on individuals (black, brown or blonde hair, for instance). Reshuffling genes may combine different alleles in beneficial ways through sexual reproduction. For instance, if a parent with good camouflage ("Coloration" gene) mated with a parent with motionless behavior ("Freeze" gene) when predators are nearby, some offspring with good behavior and camouflage (both the "Coloration" and "Freeze" genes) may result.

Meiotic cell division produces haploid cells that can merge to combine genetic material from two parents. The first eukaryotic cells probably were haploid. Then, two haploid cells fused, resulting in a diploid cell with two copies of each chromosome. Then, this type of cell evolved a variation of mitotic cell division, called meiotic cell division, to produce haploid gametes.

7) What Are the Events of Meiosis?

Meiosis is the production of haploid nuclei with unpaired chromosomes from diploid parent nuclei. In meiotic cell division (meiosis followed by cytokinesis), each new daughter cell receives one member from each pair of homologous chromosomes. Meiosis separates homologous chromosomes in a diploid nucleus, producing haploid daughter nuclei. In meiotic cell division, the cell undergoes one round of DNA replication followed by two nuclear divisions known as meiosis I and meiosis II. During meiosis, the following events occur: (1) Chromosomes replicate before meiosis begins; (2) During meiosis I, duplicated homologous chromosomes separate, one duplicated homologue moving into each of the two daughter cells (a *2n* cell produces a pair of *n* cells); (3) In meiosis II, sister chromatids in each daughter cell separate and cytokinesis may occur to produce four haploid cells; each with one set of unduplicated chromosomes (each *n* cell with duplicated chromosomes becomes a pair of *n* cells with unduplicated chromosomes).

Meiosis I separates homologous chromosomes into two daughter nuclei. During *prophase* I, homologous chromosomes pair up and exchange DNA, a process called **crossing over**, involving the formation of **chiasmata** (singular is **chiasma**, regions of exchange), and resulting in **genetic recombination** (formation of new combinations of different alleles on a chromosome). During *metaphase* I, paired homologous chromosomes move to the equator of the spindle. Different pairs of chromosomes align themselves randomly at the equator, allowing independent assortment to occur. During *anaphase* I, homologous chromosomes separate, with one duplicated chromosome of each pair moving to opposite poles. During *telophase* I, two haploid clusters of duplicated chromosomes are formed.

Meiosis II is very similar to mitosis; it separates sister chromatids into four daughter nuclei. You should note that there is *no* duplication of the chromosomes between meiosis I and meiosis II. During *prophase* II, the spindle microtubules re-form and the duplicated chromosomes attach to spindle microtubules as they did in mitosis. During *metaphase* II, the duplicated chromosomes line up at cell's equator. During *anaphase* II, the centromeres holding the sister chromatids together split, and spindle microtubules pull each chromatid (unduplicated daughter chromosome) to opposite poles. *Telophase* II and cytokinesis complete meiosis II; the nuclear envelopes re-form, the chromosomes unwind, and the cytoplasm divides.

The life cycles of almost all eukaryotic organisms on Earth have a common overall pattern, usually including meiosis and mitosis. First, two haploid cells fuse, endowing the resulting diploid cell with new gene combinations. Second, at some point in the life cycle, meiosis occurs, producing haploid cells. Third, at some point, mitosis of either haploid or diploid cells, or both, results in the growth of multicellular bodies and/or asexual reproduction.

8) How Do Meiosis and Sexual Reproduction Produce Genetic Variability?

Shuffling of homologues creates new combinations of chromosomes. During metaphase of the first meiotic division, random alignment of homologous chromosomes at the equator creates new combinations of chromosomes. Crossing over creates chromosomes with novel combinations of genes. Fusion of gametes adds further genetic variability to the offspring. In humans there are 2^{23} or about 8 million different types of gametes based on random alignment of homologous chromosomes at metaphase I. Fusion of gametes from two people makes 8 million x 8 million (64 trillion) possible genetically different children. Crossing over increases this number substantially.

Case study revisited. Females in many species of reptiles and birds can produce offspring without mating with males. When meiosis II is not properly completed in females, the result is two cells that have identical copies of each type of chromosome. The resulting egg can sometimes develop into a viable, diploid embryo in a process called parthenogenesis ("virgin birth"). One breed of turkey produces some 40% of its offspring by parthenogenesis.

KEY TERMS AND CONCEPTS

Fill-In: From the following list of key terms, fill in the blanks below.

asexual	crossing over	genetic recombination	meiosis
binary fission	cytokinesis	homologous	mitosis
chiasmata	gametes	interphase	sexual
chromosomes			

1. Eukaryotic cellular DNA is packaged in _____ (darkly staining nuclear rods of condensed protein and DNA).

2. Cell division in bacteria is called _____, and can occur in 30 minutes or less.

3. Eukaryotic chromosomes usually occur in _____ pairs with similar sizes, staining patterns, and genetic information in nonreproductive diploid cells (*2n*).

4. At some point in the life cycle of sexually reproducing organisms, _____ produces haploid cells (*n*) called _____ (sperm and egg), with one copy of each type of chromosome.

5. Growth, replication of chromosomes, and most cell functions occur during _____ , the period between cell divisions.

6. Mitotic cell division consists of nuclear division or _____ and cytoplasmic division called _____.

7. _____ reproduction (formation of genetically identical offspring without fusion of eggs and sperm) occurs in simple organisms and plants. However, the reshuffling of genes among individuals to create genetically unique offspring results from _____ reproduction.

8. _____ separates homologous chromosomes in a diploid nucleus, producing haploid daughter nuclei.

9. Meiosis I separates homologous chromosomes into two daughter nuclei. During prophase I, homologous chromosomes pair up and exchange DNA through _____ and the formation of _____ (regions of exchange), resulting in _____ (formation of new combinations of different alleles on a chromosome).

Key Terms and Definitions

allele (al-ēl´): one of several alternative forms of a particular gene.

anaphase (an´-a-fāz): in mitosis, the stage in which the sister chromatids of each chromosome separate from one another and are moved to opposite poles of the cell; in meiosis I, the stage in which homologous chromosomes, consisting of two sister chromatids, are separated; in meiosis II, the stage in which the sister chromatids of each chromosome separate from one another and are moved to opposite poles of the cell.

asexual reproduction: reproduction that does not involve the fusion of haploid sex cells. The parent body may divide and new parts regenerate, or a new, smaller individual may form as an attachment to the parent, to drop off when complete.

binary fission: the process by which a single bacterium divides in half, producing two identical offspring.

cell cycle: the sequence of events in the life of a cell, from one division to the next.

cell division: splitting of one cell into two; the process of cellular reproduction.

cell plate: in plant cell division, a series of vesicles that fuse to form the new plasma membranes and cell wall separating the daughter cells.

centriole (sen´-trē-ōl): in animal cells, a short, barrel-shaped ring consisting of nine microtubule triplets; a microtubule-containing structure at the base of each cilium and flagellum; gives rise to the microtubules of cilia and flagella and is involved in spindle formation during cell division.

centromere (sen´-trō-mēr): the region of a replicated chromosome at which the sister chromatids are held together until they separate during cell division.

chiasma (kī-as´-muh; pl., chiasmata): a point at which a chromatid of one chromosome crosses with a chromatid of the homologous chromosome during prophase I of meiosis; the site of exchange of chromosomal material between chromosomes.

chromatid (krō´-ma-tid): one of the two identical strands of DNA and protein that forms a replicated chromosome. The two sister chromatids are joined at the centromere.

chromosome (krō´-mō-sōm): a single DNA double helix together with proteins that help to organize the DNA.

clone: offspring that are produced by mitosis and are therefore genetically identical to each other.

cloning: the process of producing many identical copies of a gene; also the production of many genetically identical copies of an organism.

condensation: compaction of eukaryotic chromosomes into discrete units in preparation for mitosis or meiosis.

crossing over: the exchange of corresponding segments of the chromatids of two homologous chromosomes during meiosis.

cytokinesis (sī-tō-ki-nē´-sis): the division of the cytoplasm and organelles into two daughter cells during cell division; normally occurs during telophase of mitosis.

differentiation: the process whereby relatively unspecialized cells, especially of embryos, become specialized into particular tissue types.

diploid (dip´-loid): referring to a cell with pairs of homologous chromosomes.

duplicated chromosome: a eukaryotic chromosome following DNA replication; consists of two sister chromatids joined at the centromeres.

gamete (gam´-ēt): a haploid sex cell formed in sexually reproducing organisms.

haploid (hap´-loid): referring to a cell that has only one member of each pair of homologous chromosomes.

homologue (hō-´mō-log): a chromosome that is similar in appearance and genetic information to another chromosome with which it pairs during meiosis; also called *homologous chromosome.*

interphase: the stage of the cell cycle between cell divisions; the stage in which chromosomes are replicated and other cell functions occur, such as growth, movement, and acquisition of nutrients.

karyotype: a preparation showing the number, sizes, and shapes of all chromosomes within a cell and, therefore, within the individual or species from which the cell was obtained.

kinetochore (ki-net´-ō-kor): a protein structure that forms at the centromere regions of chromosomes; attaches the chromosomes to the spindle.

meiosis (mī-ō´-sis): a type of cell division, used by eukaryotic organisms, in which a diploid cell divides twice to produce four haploid cells.

meiotic cell division: meiosis followed by cytokinesis.

metaphase (met´-a-fāz): the stage of mitosis in which the chromosomes, attached to spindle fibers at kinetochores, are lined up along the equator of the cell.

mitosis (mī-tō´-sis): a type of nuclear division, used by eukaryotic cells, in which one copy of each chromosome (already duplicated during interphase before mitosis) moves into each of two daughter nuclei; the daughter nuclei are therefore genetically identical to each other.

mitotic cell division: mitosis followed by cytokinesis.

prophase (pro´-fāz): the first stage of mitosis, in which the chromosomes first become visible in the light microscope as thickened, condensed threads and the spindle begins to form; as the spindle is completed, the nuclear envelope breaks apart, and the spindle fibers invade the nuclear region and attach to the kinetochores of the chromosomes. Also, the first stage of meiosis: In meiosis I, the homologous chromosomes pair up and exchange parts at chiasmata; in meiosis II, the spindle reforms and chromosomes attach to the microtubules.

recombination: the formation of new combinations of the different alleles of each gene on a chromosome; the result of crossing over.

sex chromosomes: the pair of chromosomes that usually determines the sex of an organism; for example, the X and Y chromosomes in mammals.

sexual reproduction: a form of reproduction in which genetic material from two parent organisms is combined in the offspring; normally, two haploid gametes fuse to form a diploid zygote.

spindle microtubules: microtubules organized in a spindle shape that separate chromosomes during mitosis or meiosis.

telophase (tēl´-ō-fāz): in mitosis, the final stage, in which a nuclear envelope reforms around each new daughter nucleus, the spindle fibers disappear, and the chromosomes relax from their condensed form; in meiosis I, the stage during which the spindle fibers disappear and the chromosomes normally relax from their condensed form; in meiosis II, the stage during which chromosomes relax into their extended state, nuclear envelopes reform, and cytokinesis occurs.

THINKING THROUGH THE CONCEPTS

True or False: Determine if the statement given is true or false. If it is false, change the underlined word(s) so that the statement reads true.

10. _____   Asexually produced organisms are similar but not identical to their parents.

11. _____   Prokaryotes have many chromosomes per cell.

12. _____   The body cells of a human are haploid.

13. _____   DNA replication occurs during the G_2 portion of interphase.

14. _____   Cell plates form during metaphase of the cell cycle.

15. _____   Diploid cells produce haploid cells by the process of meiosis.

16. _____   Crossing over occurs during meiosis II.

17. _____   Sister chromatids become daughter chromosomes during anaphase I of meiosis.

18. _____   Meiosis II resembles mitosis.

19. _____   DNA replicates between meiosis I and meiosis II.

20. _____   Reduction of chromosome number occurs during meiosis I.

Identify: Determine whether the following statements refer to **mitosis, meiosis, both,** or **neither.**

21. _____   production of haploid cells from haploid cells

22. _____   the mechanism by which eukaryotic unicellular organisms reproduce

23. _____   produces genetically variable cells

24. _____   produces sperm and egg cells in animals

25. _____   allows multicellular organisms to grow

26. _____   cells divide twice

27. _____   ensures that each body cell gets a complete set of genes

28. _____   can produce haploid cells

29. _____   chromosomes replicate once

30. _____   produces genetically identical cells

31. _____   production of haploid cells from diploid ones

32. _____   occurs in the human body

33. _____   maintains the same number of chromosomes

34. _____   doubles the number of chromosomes

35. _____   reduces the number of chromosomes by half

Matching: Interphase

36. _____   phase in cells that will no longer divide   Choices:

37. _____   period after DNA synthesis occurs   a. S phase

38. _____   period before DNA synthesis occurs   b. G_0 phase

39. _____   period of DNA replication or synthesis   c. G_1 phase

40. _____   period of most cell growth and metabolic activity   d. G_2 phase

Identify: Determine whether the following statements refer to **prophase**, **metaphase**, **anaphase**, or **telophase** of mitosis.

41. _____   Independent chromosomes have reached the opposite spindle poles.

42. _____   Duplicated chromosomes migrate to the cell's equator.

43. _____   Replicated chromosomes coil up and condense.

44. _____   The centromeres divide.

45. _____   Independent chromosomes move to the poles.

46. _____   The spindle breaks down.

47. _____   The nuclear envelope disintegrates.

48. _____   Sister chromatids become independent chromosomes.

49. _____   The spindle forms.

50. _____   Cytokinesis begins.

Matching: Meiosis.

51. _____   Individual chromosomes migrate to the equator.   Choices:

52. _____   Chiasmata form.   a. prophase I

53. _____   Homologous chromosomes move towards opposite poles.   b. metaphase I

54. _____   Centromeres divide.   c. anaphase I

55. _____   Homologous chromosomes pair up.   d. telophase I

56. _____   Independent chromosomes migrate towards opposite poles.   e. prophase II

57. _____   Homologous pairs of chromosomes are together at the   f. metaphase II
             equator.
                                                                                 g. anaphase II
58. _____   Crossing over occurs.
                                                                                 h. telophase II

Multiple Choice: Pick the most correct choice for each question.

59. The genetic material in bacteria consists of
    a. several circular DNA molecules
    b. one circular RNA molecule
    c. many rod-like DNA molecules with protein
    d. one circular DNA molecule
    e. DNA in mitochondria

60. The daughter cells of binary fission are
    a. structurally identical
    b. chromosomally different
    c. genetically identical
    d. structurally similar and genetically identical
    e. not genetically the same as the parent cell

61. Cell reproduction in prokaryotic cells differs from eukaryotic cells in that
    a. prokaryotic cells reproduce asexually but eukaryotic cells do not.
    b. each prokaryotic cell has a circular chromosome but the chromosomes of eukaryotic cells are linear.
    c. prokaryotic cells lack nuclei and do not replicate their DNA before dividing but eukaryotic cells have nuclei and replicate their DNA before dividing.
    d. prokaryotic chromosomes have DNA and protein but eukaryotic chromosomes are made of only DNA.
    e. they do not differ significantly.

62. A diploid cell contains six chromosomes. After meiosis I, each of the cells contains
    a. three maternal and three paternal chromosomes each time
    b. mixture of maternal and paternal chromosomes totaling three
    c. six maternal or six paternal chromosomes each time
    d. a mixture of maternal and paternal chromosomes totaling six
    e. three pairs of chromosomes

63. A region of attachment for two sister chromatids is the
    a. centriole
    b. centromere
    c. equator
    d. microtubule
    e. spindle fiber

64. Which of the following does not occur during prophase?
    a. the nuclear membrane disintegrates
    b. nucleoli break up
    c. the spindle apparatus forms
    d. the chromosomes condense
    e. DNA replicates

65. In sexually reproducing organisms, the source of chromosomes in the offspring is
    a. almost all from one parent, usually the father
    b. almost all from one parent, usually the mother
    c. half from the father and half from the mother
    d. a random mixing of chromosomes from both parents

66. Meiosis results in the production of
    a. diploid cells with unpaired chromosomes
    b. diploid cells with paired chromosomes
    c. haploid cells with unpaired chromosomes
    d. haploid cells with paired chromosomes
    e. none of the above choices is correct

67. Which occurs in meiosis I but not in meiosis II?
    a. diploid daughter cells are produced
    b. chromosomes without chromatids line up at the equator
    c. centromeres divide
    d. pairing of homologous chromosomes occurs
    e. the spindle apparatus forms

68. The process of females giving birth to offspring without the genetic contribution of males is called
    a. sexual reproduction
    b. alternation of generations
    c. typical asexual reproduction
    d. mitosis
    e. parthenogenesis

69. When snakes give birth to offspring without a genetic contribution from a male, the offspring are usually
    a. females
    b. males
    c. either sex
    d. intersex
    e. stillborn

APPLYING THE CONCEPTS

These practice questions are intended to sharpen your ability to apply critical thinking and analysis to biological concepts covered in this chapter.

70. Suppose you are working in a lab that grows various types of human body cells in petri dishes, a technique known as "cell culturing." While you are looking at one particular cell culture under the microscope, you decide to count up the numbers of cells you see in the various phases of the mitotic cell cycle. You count 45 cells in anaphase, 34 cells in prophase, 23 cells in telophase, 11 cells in metaphase, and 1008 cells in interphase. Could you use these numbers to determine the relative amounts of time this cell type spends in each phase of the mitotic cell cycle? Explain.

_____

_____

_____

_____

_____

_____

_____

_____

_____

_____

_____

_____

71. Suppose that the cloning process used to make the ewe named Dolly becomes widely used to clone human beings. What do you suppose the effect of widespread cloning might be on human populations?

_____

_____

_____

_____

_____

72. Why do you think that many treatments for cancer cause the loss of hair and severe nausea brought on by loss of the gastrointestinal lining?

_____

_____

_____

_____

_____

_____

_____

_____

73. There are three elements involved in sexual reproduction that lead to the production of genetically diverse offspring. What are those three elements?

_____

_____

_____

_____

_____

_____

_____

_____

_____

74. Why is there so much scientific interest in trying to understand the mechanisms controlling the process of mitosis and cell differentiation?

_____

_____

_____

_____

_____

75. What are some characteristics shared by chromosomes which are considered to be homologous, and what makes a cell diploid or haploid?

_____

_____

_____

_____

76. If a diploid cell has 100 units of DNA at the G_1 stage of interphase before meiosis begins, how much DNA would a cell have during the following stages of meiosis: prophase I, prophase II, and sperm cell?

_____

_____

_____

_____

_____

_____

_____

_____

Use the Case Study and the Web sites for this chapter to answer the following questions.

77. Give the sex chromosomes for males and females in each of the following organisms: humans, fruit flies, honey bees, and birds.

humans _____   fruit flies _____

birds_____   honey bees_____

78. There are no male *C. uniparens* (whiptail lizards). How does this species reproduce? What are the hypothesized origins of the species?

_____

_____

_____

_____

_____

_____

_____

_____

_____

79. How is the automictic parthenogenesis observed in turkeys and the occasional rattlesnake different from that of the parthenogenic whiptails? (Hint: check the chromosome numbers.) Why are all the offspring male?

_____

_____

_____

ANSWERS TO EXERCISES

1. chromosomes	14. false, cytokinesis	32. both	51. f
2. binary fission	15. true	33. mitosis	52. a
3. homologous	16. false, meiosis I	34. neither	53. c
4. meiosis	17. false, anaphase II	35. meiosis	54. g
gametes	18. true	36. b	55. a
5. interphase	19. false, does not	37. d	56. g
6. mitosis	replicate	38. c	57. b
cytokinesis	20. true	39. a	58. a
7. Asexual	21. mitosis	40. c	59. d
sexual	22. mitosis	41. telophase	60. d
8. Meiosis	23. meiosis	42. metaphase	61. b
9. crossing over	24. meiosis	43. prophase	62. b
chiasmata	25. mitosis	44. anaphase	63. b
genetic	26. meiosis	45. anaphase	64. e
recombination	27. mitosis	46. telophase	65. d
10. false, identical	28. both	47. prophase	66. c
11. false, one	29. both	48. anaphase	67. d
12. false, diploid	30. mitosis	49. prophase	68. e
13. false, S	31. meiosis	50. telophase	69. b

70. If the cells are dividing randomly in the culture dish, you can relate the percentage of time the cells spend in each phase of the cell cycle with the percentage of cells observed at each stage of the cell cycle. For instance, if cells spend 50% of the time in interphase, you would expect to find about half of the actively dividing cells in interphase at any particular time. So, if you saw 1121 cells and 34 (34 ÷ 1121 = 3.0%) were in prophase, 11 (1.0%) were in metaphase, 45 (4.0%) were in anaphase, 23 (2.1%) were in telophase, and 1008 (89.9%) were in interphase, the percentages would indicate the relative amounts of time the cells spend in each phase of the cell cycle. If the average time for a complete mitotic division is known for the cells growing in the culture, these percentages could be used to determine the amounts of time cells spend in each phase. For instance, if it takes 24 hours (1140 minutes) for a complete cell cycle, these cells spend about 34 minutes (3% of 1140) in prophase, 11 minutes in metaphase, 46 minutes in anaphase, 24 minutes in telophase, and 1025 minutes in interphase.

71. One effect might be a reduction in genetic variation in human populations as fewer types of people produce more and more copies of themselves. There is a danger of eugenic manipulation of human traits, since those with "superior" traits might be cloned and those with "inferior" traits might be prohibited from being cloned (or even from reproducing at all). Another effect might be a skewing of sex ratios as more and more males are cloned and fewer and fewer females, especially in countries like China where males are considered more "valuable" to society than females. A much higher male sex ratio could, in turn, lead to other societal changes such as legalized prostitution.

72. Since cancer cells characteristically reproduce rapidly, chemicals that differentially attack actively dividing cells kill cancer cells. Also, chromosomes in their condensed condition during cell division are more vulnerable to damage from radiation treatment, often leading to cell death, so that radiation is quite effective against cancer cells. Unfortunately, these cancer treatments will also target rapidly and constantly dividing normal body cells as well. So, during cancer treatment, hair cells and cells lining the gastrointestinal tract also die off, since they are normally rapidly dividing.

73. The three elements leading to genetically diverse offspring are: (1) crossing over occurs early in meiosis, allowing homologous chromosomes to exchange genes and produce new combinations of genes; (2) the independent assortment of paired homologous maternally-derived and paternally-derived chromosomes, each with many different versions of the genes they carry, produce many different combinations of chromosomes in the sex cells produced by meiosis; and (3) since fertilizations are random combinations from among the variety of eggs and sperm available, this further increases the genetic variation among offspring.

74. Since cancer occurs as the result of uncontrolled cell division, gaining knowledge of the mechanisms controlling cell division might provide a way to control the abnormal growth. In addition, the aging process is also affected by cell division. As organisms age, the rate of replacement of worn out cells decreases. Being able to control cell division and differentiation so that needed cell types could be produced might be a way to counteract the aging process.

75. Homologous chromosomes share several properties, including similar lengths, similar positions of the centromere, similar types and locations of genes, and the ability to pair with each other during meiosis. A cell is diploid when both members of each homologous pair of chromosomes are present. Haploid cells contain one chromosome of each homologous pair of chromosomes.

76. If we think of the amount of DNA in the chromosomes, at G_1 of interphase there are 100 units of DNA in the unreplicated chromosomes. By the end of interphase, the chromosomes have replicated, so that during prophase I, there would be 200 units of DNA in the pairs of replicated chromosomes. As a result of the first meiotic division, the pairs of replicated chromosomes move away from each other and into separate cells, so that during prophase II, there would be 100 units of DNA in the single replicated chromosomes. During the second meiotic division, the centromeres holding the sister chromatids of the replicated chromosomes together divide, so that the sister chromatids become unreplicated chromosomes as they move towards opposite poles, so that the sperm cells produced would each have 50 units of DNA in their single unreplicated chromosomes.

77. According to W. Richardson and L. Dale, human males are XY and females are XX, and fruit fly males are XY and females are XX. According to J. Bell of Cal State Univ., Chico, honey bee males are haploid with one X and females are diploid with XX (in bees, if a queen lays a fertilized egg, it develops into a diploid XX female, but if the queen lays an unfertilized egg, it develops into a haploid X male). And, according to S. Carr, bird sex determination is the opposite of humans; XY is female (sometimes, the sex chromosomes are called WZ in birds), and XX (ZZ) is male.

78. According to reports from Discovery.com, the New Mexico Whiptail is an all-female species that is actually a mixture of two other species, the Western Whiptail, which lives in the desert, and the Little Striped Whiptail, a denizen of grasslands. Most products of crossbreeding, such as the mule, are sterile. But the New Mexico Whiptail, as well as several other all-female species of whiptail lizard, does reproduce, and all of its offspring are female. Moreover, it reproduces by parthenogenesis — its haploid eggs require no fertilization, and its offspring are exact and complete genetic duplicates of the mother. Scientists understand only partially how this reproductive mode developed, and it raises many questions. One of the most intriguing is how this cloning affects the lizard's ability to adapt to environmental changes. Since there is no genetic variation except that which occurs through mutation, the New Mexico Whiptail cannot evolve as other species do.

79. According to G. Schuett et. al. and D. Blanchard, automictic parthenogenesis occurs when an egg with the X (also called the Z) chromosome fuses with a polar body also with an X (Z) chromosome to form a male offspring (XX = ZZ) without any sperm involved. These parthenogenetic rattlesnake and turkey offspring are diploid, whereas parthenogenetic whiptails are haploid.

Chapter 12: Patterns of Inheritance

OVERVIEW

In this chapter, you will learn about Gregor Mendel's concepts of inheritance, namely the segregation and independent assortment of genes, the dominance and recessiveness of different alleles, and the randomness of fertilization. You will also read about the relationship of genes to chromosomes, especially sex-linkage, sex determination, linkage, and crossing over. This chapter also presents variations on the Mendelian theme, including mutation, incomplete dominance, multiple alleles, codominance, polygenic inheritance, gene interactions, and environmental influences. In addition, the authors discuss chromosomal nondisjunction as a cause of humans with abnormal numbers of chromosomes, as well as the more common results of nondisjunction: XO, XXX, XXY, XYY, and Down syndrome. Genes control many traits in humans, including narcolepsy, a strange condition of the central nervous system causing uncontrolled sleepiness even when well-rested, and periods of paralysis called cataplexy brought on by being startled. Several colonies of dogs at Stanford also have narcolepsy, and recent studies have identified a gene associated with this strange malady. How did the Stanford scientists find those genes?

1) How Did Gregor Mendel Lay the Foundations for Modern Genetics?

A gene's specific location on a chromosome is its **locus**. Homologous chromosomes carry similar **genes** at similar loci. Slightly different DNA nucleotide sequences at the same gene locus on two homologous chromosomes produce alternate forms of the gene, called **alleles**. If both homologous chromosomes in an organism have identical alleles, the organism is **homozygous** ("same pair"); if the alleles are similar but not identical, the organism is **heterozygous** ("different pair"), and is called a **hybrid**. All gametes produced by an individual homozygous at a particular locus will contain the same allele. Gametes produced by heterozygous individuals at a locus are of two kinds: half of the gametes contain one allele, and half contain the other allele.

Mendel was successful because: (1) he chose the right organism to work with, garden peas which reproduce by **self-fertilization** (the same flower produced the egg and sperm cells) and are thus **true-breeding** or homozygous, and not by **cross-fertilization** (gametes come from different parents); (2) he designed and performed the experiment carefully, choosing easy to distinguish traits and counting up all the offspring from each mating; and (3) he analyzed the data properly, using simple statistics such as determining the ratios of offspring with differing traits.

2) How Are Single Traits Inherited?

Mendel started as simply as possible. He raised varieties of peas that were true-breeding for different forms of a single trait and cross-fertilized these, saving the resulting hybrid seeds and growing them the following year to determine their characteristics. The inheritance of dominant and recessive alleles on homologous chromosomes can explain the results of Mendel's crosses. Example: purple flowers (*PP*) crossed with white flowers (*pp*) produced purple flowered offspring (*Pp*). These, in turn, produced offspring, 75% of which had purple flowers (*PP*) and (*Pp*) and 25% of which had white flowers (*pp*), a 3:1 ratio.

The inheritance of dominant and recessive alleles on homologous chromosomes can explain the results of Mendel's crosses. His hypothesis in modern terms is: (1) each trait is determined by a pair of genes, found at corresponding loci in homologous chromosomes; (2) the members of a gene pair separate from each other during meiosis so that each gamete gets one copy (**law of segregation**); (3) which member of a pair gets into which gamete, relative to other pairs, is a matter of chance; (4) in heterozygous individuals, the **dominant** allele may mask the expression of the **recessive** allele without changing its structure. The actual combination of alleles carried by an organism is its **genotype** (*PP*, *Pp*, and *pp*), while the organism's traits (things that can be seen or measured) are its **phenotype** (purple or white flower color); and, (5) true-breeding organisms are homozygous (*PP* and *pp*), and hybrids are heterozygous (*Pp*). The possible genotypes and phenotypes of the offspring can be predicted by using the **Punnett square method**. This method lets you predict the probability that a certain outcome will occur.

Mendel's hypothesis can be used to predict the outcome of new types of single-trait crosses. For instance, cross-fertilization of an individual with dominant phenotype but unknown genotype (purple flower with either *PP* or *Pp* genotype) with a homozygous recessive individual (white flower = *pp*) is called a **test cross**, and can be used to determine the exact genotype of the dominant parent: *Pp* if recessive offspring occur, *PP* if no recessive offspring occur.

3) How Are Multiple Traits on Different Chromosomes Inherited?

Multiple traits may be controlled by genes on different chromosomes. Such traits are **inherited** independently of each other (the **law of independent assortment**). Example: smooth yellow seeds (*SSYY*) x wrinkled green seeds (*ssyy*) produce all smooth yellow offspring (*SsYy*). These, in turn, produce four types of offspring: 9/16 have smooth yellow seeds (*S-Y-*), 3/16 have smooth green (*S-yy*) seeds, 3/16 have wrinkled yellow (*ssY-*) seeds, and 1/16 have wrinkled green (*ssyy*) seeds, a 9:3:3:1 ratio. Genes for seed color and seed shape are inherited independently of each other (move independently of each other during meiosis) since they are located in different (non-homologous) chromosomes, which align themselves randomly during meiotic metaphase I. Punnett squares are used to determine all the random fertilizations possible from a particular mating.

4) How Are Genes Located on the Same Chromosome Inherited?

Mendel know nothing about the physical nature of genes or chromosomes. Later, it was discovered that chromosomes contain many hundreds of gene loci. Genes on different chromosomes show independent assortment. Genes on the same chromosome tend to be inherited together, a situation called **linkage**. However, **crossing over** may separate linked genes, since exchange of corresponding segments of DNA forms new gene combinations on both homologous chromosomes. This **genetic recombination** is one way that genetic variability occurs in gametes, since it creates new combinations of linked alleles.

5) How Is Sex Determined, and How Are Sex-Linked Genes Inherited?

In mammals and most insects, females have two identical **sex chromosomes**, called X chromosomes, whereas males have one X and one Y chromosome. All other chromosomes are identical in males and females and are called **autosomes**. For organisms in which males are XY and females are XX, the sex chromosomes carried by the sperm determines the sex of the offspring: X from egg + X from sperm = daughter, while X from egg + Y from sperm = son. **Sex-linked** genes are found only on the X or only on the Y chromosome. Males express any genes found on the X chromosome, and each male inherits his X chromosome from his mother. So, half of the sons of a mother heterozygous for a sex-linked recessive condition (like hemophilia or colorblindness) will be affected.

6) What Are Some Variations on the Mendelian Theme?

Alleles may display **incomplete dominance** in which the phenotype of heterozygotes is intermediate between the phenotypes of the homozygotes. In snapdragons, red flowers (*RR*) crossed with white flowers (*R'R'*) yield pink heterozygotes (*RR'*); which, when self-fertilized, yield 1/4 red + 2/4 pink + 1/4 white. Incomplete dominance occurs when the degree of color depends on the number of active genes present. In snapdragons, *R* makes 50 units of red pigment and *R'* makes 0 units, so *RR* makes 100 units of red in the homozygote, *RR'* makes 50 units of red pigment in the pink heterozygote, and *R'R'* flowers are white.

A single gene may have **multiple alleles**. There may be multiple (more than two) alleles of a gene present in a population of individuals. An individual diploid can have at most two different alleles for a given gene, even though more than two allelic forms of the gene (multiple alleles) are present in the population. For example, in the ABO blood type alleles in humans there are three alleles (A and B are **codominant** [both phenotypes are expressed in heterozygotes] and *o* is recessive) and six possible genotypes (type A = *AA* and *Ao*, type B =*BB* and *Bo*, type AB = *AB*, and type O = *oo*).

Many traits (like height, skin color, and eye color) are influenced by several genes. Some traits show a continuous variation of phenotypes in a population and cannot be split up into convenient, easily defined categories. These show **polygenic inheritance** since the effects of two or more pairs of functionally similar genes add up to produce a single phenotype. For instance, as the amount of melanin pigment increases in irises, eye color darkens from blue to green to brown to black. The more genes that contribute to a trait, the greater the number of phenotypes and the finer the distinction between them.

Single genes typically have multiple effects on phenotype, a situation called **pleiotropy**. For example, the *SRY* (Sex-determining Region of the Y chromosome) gene produces the many anatomical and hormonal differences between males and females.

The environment influences the expression of genes. The phenotype of an organism is influenced by both its genotype and its environment. For example, the cooler body areas of a Himalayan rabbit produce hair pigment while the warmer regions do not; human height and weight are products of genes and nutrition; human skin color is a product of genes and degree of sun exposure; and human IQ is a product of genes and educational environment. Both heredity and environment play major roles in the development of various mental abilities and almost certainly other personality traits as well.

7) How Are Human Disorders Investigated?

Because experimental crosses with humans are unethical, geneticists search medical, historical, and family records to study past crosses. Records including several generations can be arranged in the form of family **pedigrees**, diagrams that show the genetic relationships among a group of related individuals. Analysis of pedigrees, combined with molecular technology, has resulted in our understanding human genetic diseases.

8) How Are Human Disorders Caused by Single Genes Inherited?

Many human genetic disorders are caused by homozygous recessive alleles since the conditions occur because the recessive genes do not generate enough enzymes. Therefore, for many genes, a normal allele, which encodes a functional protein, is dominant over a mutant allele, which encodes a non-functional protein. Heterozygotes are **carriers** of the recessive genetic trait; they are phenotypically normal but can pass on their defective recessive allele to their offspring. Each of us carries a recessive allele for 5–15 harmful genes. An unrelated man and woman are unlikely to possess a defective allele in the same gene. However, first cousins and closer relatives are much more likely to carry a defective allele in the same gene, due to recent common relatives.

Albinism results from a defect in melanin production due to homozygosity for recessive alleles. **Sickle-cell anemia** is caused by homozygosity for a defective hemoglobin synthesis allele. Many human genetic disorders, like **Huntington disease**, are caused by dominant alleles. For dominant disorders to be passed on to offspring, at least one parent must have a copy of the disease gene, or the dominant allele may result from a new mutation. Some dominant alleles produce an abnormal protein that interferes with the function of the normal one. Other dominant alleles may encode proteins that carry out new, toxic reactions. Still other dominant alleles may encode a protein that is overactive, performing its normal function at inappropriate times and places. In Huntington disease, the abnormal protein forms large aggregates in nerve cells and the aggregates appear to result, ultimately, in nerve cell death.

Some human disorders, like red-green color blindness and **hemophilia**, are sex-linked. They have a unique pattern of inheritance: the disorders appear far more frequently in males and typically skip generations.

9) How Do Errors in Chromosome Number Affect Humans?

Nondisjunction, defined as errors involving chromosome distribution during meiosis, produces gametes with one (or a few) extra or missing sex chromosomes or autosomes. Most embryos formed from such gametes are miscarried, but a few types survive. Some genetic disorders among liveborn humans are caused by abnormal numbers of sex chromosomes. These include: (1) **Turner syndrome** (XO females): about 1 in 3000 female babies (47,000 in the US) has only one X. Hormonal deficiencies prevent XO females from menstruating or developing secondary sexual characteristics. XO females also have short stature, skin folds around the neck, increased risk of cardiovascular and kidney defects, and hearing loss. (2) **Trisomy X** (XXX females): about 1 in 1000 females (140,000 in the US) has three X's, with no major defects. (3) **Klinefelter syndrome** (XXY males): about 1 in 1000 males (134,000 in the US) is born with two X's and a Y. At puberty, some show partial breast development, hip broadening, and small testes. All XXY males are sterile but not impotent. (4) XYY males: about 1 in 1000 males (134,000 in the US) have XYY chromosomes. They have high testosterone levels, severe acne, tall stature, and slightly lower IQ scores. They may be genetically predisposed to violence, but research has not demonstrated this to be true.

Some genetic disorders among liveborn humans are caused by abnormal numbers of autosomes, particularly chromosomes 13, 18, and 21; the most common of which is **trisomy 21 (Down syndrome)**. About 1 in 900 babies (304,000 in the US) has trisomy 21 and displays weak muscle tone, a small mouth, distinctly shaped eyelids, heart malformations, and some degree of mental retardation. The frequency of nondisjunction increases with the age of the mother, and nondisjunction in sperm accounts for about 25% of the cases of Down syndrome.

Case study revisited. Analysis of narcolepsy in dogs results from a recessive allele of a single gene, located on chromosome 12. The gene is called *Hcrt2*, and it encodes a protein receptor on the cell surfaces of some cells in the brain hypothalamus. The *Hcrt2* protein normally binds to signaling molecules called hypocretins. In narcoleptic dogs, the receptor is defective, making the brain cells ignore the molecular signal delivered by the hypocretins. Narcoleptic mice have normal *Hcrt2* genes, but have a different mutation that prevents them from making hypocretins at all. In a study of humans with narcolepsy, seven of nine patients did not produce hypocretins (like narcoleptic mice) and two patients did produce hypocretins (like narcoleptic dogs).

KEY TERMS AND CONCEPTS

Fill-In: From the following list of key terms, fill in the blanks in the following statements.

allele	Down syndrome	genotype	locus	phenotype
carrier	dominant	heterozygote	nondisjunction	recessive
crossing over	gene	homozygote		

1. A unit of heredity containing the information for a particular characteristic is called a _____. One of several alterative forms of these units is an _____.

2. The physical location of a gene on a chromosome is its _____.

3. A _____ allele determines the phenotype of heterozygotes completely, whereas a _____ allele is expressed only in homozygotes.

4. An organism carrying two copies of the same allele of a gene is a _____, while an organism carrying two different alleles of a gene is a _____.

5. The genetic composition of an organism is its _____, while the physical properties of an organism are called its _____.

6. The exchange of corresponding segments of the chromatids of two homologous chromosomes during meiosis is called _____.

7. A _____ is an individual heterozygous for a recessive trait.

8. _____ is a genetic disorder caused by the presence of three copies of chromosome 21. This is caused by _____, an error in meiosis in which chromosomes fail to segregate properly into the daughter cells.

Key Terms and Definitions

allele (al-ēl´): one of several alternative forms of a particular gene.

autosome (aw´-tō-sōm): a chromosome that occurs in homologous pairs in both males and females and that does not bear the genes determining sex.

carrier: an individual who is heterozygous for a recessive condition; displays the dominant phenotype but can pass on the recessive allele to offspring.

codominance: the relation between two alleles of a gene, such that both alleles are phenotypically expressed in heterozygous individuals.

cross-fertilization: the union of sperm and egg from two individuals of the same species.

crossing over: the exchange of corresponding segments of the chromatids of two homologous chromosomes during meiosis.

dominant: an allele that can determine the phenotype of heterozygotes completely, such that they are indistinguishable from individuals homozygous for the allele; in the heterozygotes, the expression of the other (recessive) allele is completely masked.

Down syndrome: a genetic disorder caused by the presence of three copies of chromosome 21; common characteristics include mental retardation, distinctively shaped eyelids, a small mouth with protruding tongue, heart defects, and low resistance to infectious diseases; also called *trisomy 21*.

gene: a unit of heredity that encodes the information needed to specify the amino acid sequence of proteins and hence particular traits; a functional segment of DNA located at a particular place on a chromosome.

genetic recombination: the generation of new combinations of alleles on homologous chromosomes due to the exchange of DNA during crossing over.

genotype (jēn´-ō-tī p): the genetic composition of an organism; the actual alleles of each gene carried by the organism.

hemophilia: a recessive, sex-linked disease in which the blood fails to clot normally.

heterozygous (het-er-ō-zī´-gus): carrying two different alleles of a given gene; also called *hybrid*.

homozygous (hō-mō-zī´-gus): carrying two copies of the same allele of a given gene; also called *true-breeding*.

Huntington disease: an incurable genetic disorder, caused by a dominant allele, that produces progressive brain deterioration, resulting in the loss of motor coordination, flailing movements, personality disturbances, and eventual death.

hybrid: an organism that is the offspring of parents differing in at least one genetically determined characteristic; also used to refer to the offspring of parents of different species.

incomplete dominance: a pattern of inheritance in which the heterozygous phenotype is intermediate between the two homozygous phenotypes.

independent assortment: see *law of independent assortment*.

inheritance: the genetic transmission of characteristics from parent to offspring.

Klinefelter syndrome: a set of characteristics typically found in individuals who have two X chromosomes and one Y chromosome; these individuals are phenotypically males but are sterile and have several female-like traits, including broad hips and partial breast development.

law of independent assortment: the independent inheritance of two or more distinct traits; states that the alleles for one trait may be distributed to the gametes independently of the alleles for other traits.

law of segregation: Gregor Mendel's conclusion that each gamete receives only one of each parent's pair of genes for each trait.

linkage: the inheritance of certain genes as a group because they are parts of the same chromosome. Linked genes do not show independent assortment.

locus: the physical location of a gene on a chromosome.

multiple alleles: as many as dozens of alleles produced for every gene as a result of different mutations.

nondisjunction: an error in meiosis in which chromosomes fail to segregate properly into the daughter cells.

pedigree: a diagram showing genetic relationships among a set of individuals, normally with respect to a specific genetic trait.

phenotype (fēn´-ō-tī p): the physical characteristics of an organism; can be defined as outward appearance (such as flower color), as behavior, or in molecular terms (such as glycoproteins on red blood cells).

pleiotropy (plē´-ō-trō-pē): a situation in which a single gene influences more than one phenotypic characteristic.

polygenic inheritance: a pattern of inheritance in which the interactions of two or more functionally similar genes determine phenotype.

Punnett square method: an intuitive way to predict the genotypes and phenotypes of offspring in specific crosses.

recessive: an allele that is expressed only in homozygotes and is completely masked in heterozygotes.

segregation: see *law of segregation*.

self-fertilization: the union of sperm and egg from the same individual.

sex chromosomes: the pair of chromosomes that usually determines the sex of an organism; for example, the X and Y chromosomes in mammals.

sex-linked: referring to a pattern of inheritance characteristic of genes located on one type of sex chromosome (for example, X) and not found on the other type (for example, Y); also called X-linked. In sex-linked inheritance, traits are controlled by genes carried on the X chromosome; females show the dominant trait unless they are homozygous recessive, whereas males express whichever allele is on their single X chromosome.

sickle-cell anemia: a recessive disease caused by a single amino acid substitution in the hemoglobin molecule. Sickle-cell hemoglobin molecules tend to cluster together, distorting the shape of red blood cell shape and causing them to break and clog capillaries.

test cross: a breeding experiment in which an individual showing the dominant phenotype is mated with an individual that is homozygous recessive for the same gene. The ratio of offspring with dominant versus recessive phenotypes can be used to determine the genotype of the phenotypically dominant individual.

trisomy 21: see *Down syndrome*.

trisomy X: a condition of females who have three X chromosomes instead of the normal two; most such women are phenotypically normal and are fertile.

true-breeding: pertaining to an individual all of whose offspring produced through self-fertilization are identical to the parental type. True-breeding individuals are homozygous for a given trait.

Turner syndrome: a set of characteristics typical of a woman with only one X chromosome: sterile, with a tendency to be very short and to lack normal female secondary sexual characteristics.

THINKING THROUGH THE CONCEPTS

True or False: Determine if the statement given is true or false. If it is false, change the <u>underlined</u> word(s) so that the statement reads true.

9. _____ Each trait is determined by a pair of discrete units called <u>chromosomes</u>.

10. _____ Alternate forms of a gene are called <u>chromatids</u>.

11. _____ In a heterozygote, the gene that is not expressed is <u>recessive</u>.

12. _____ An *AabbDdEeGg* individual will produce <u>16</u> different types of gametes.

13. _____ Linkage modifies Mendel's law of <u>segregation</u>.

14. _____ New combinations of traits controlled by linked genes occur due to <u>crossing over</u>.

15. _____ When red snapdragons are crossed with white snapdragons to produce pink snapdragons, this is an example of <u>polygenic inheritance</u>.

16. _____ Human <u>females</u> determine the sex of their children.

17. _____ Males are <u>haploid</u> for sex-linked genes.

18. _____ A son inherits sex-linked traits from his <u>father</u>.

19. _____ A person with Turner syndrome is <u>XXY</u>.

20. _____ A person with Klinefelter syndrome is a <u>male</u>.

Matching: Chromosome anomalies in humans.

21. _____ have an abnormal number of autosomes

22. _____ sterile males with some breast development

23. _____ females with three X chromosomes per nucleus

24. _____ may be male or female

25. _____ fertile females with normal phenotypes

26. _____ short sterile females

27. _____ have a normal number of sex chromosomes

28. _____ much more common among the babies of older mothers

29. _____ males with more than one X chromosome

30. _____ sterile females with less than two X chromosomes

31. _____ may have a possible predisposition towards violence

32. _____ body cells have 45 chromosomes

33. _____ trisomy 21

34. _____ have 46 chromosomes in their body cells

Choices:

a. Turner syndrome

b. Klinefelter syndrome

c. trisomy X syndrome

d. XYY syndrome

e. Down syndrome

f. all of the above

g. none of the above

Multiple Choice: Pick the most correct choice for each question.

35. Traits controlled by sex-linked recessive genes are expressed more often in males because
    a. Males inherit these genes from their fathers.
    b. Males are always homozygous.
    c. All male offspring of a female carrier get the gene.
    d. The male has only one gene for the trait.

36. Each normal human possesses in his or her body cells
    a. 2 pairs of sex chromosomes and 46 pairs of autosomes
    b. 2 pairs of sex chromosomes and 23 pairs of autosomes
    c. 1 pair of sex chromosomes and 46 pairs of autosomes
    d. 1 pair of sex chromosomes and 23 pairs of autosomes
    e. 1 pair of sex chromosomes and 22 pairs of autosomes

37. A colorblind woman has children with a noncolorblind man. Which of the following is true of their children?
    a. All will be colorblind.
    b. All daughters will be normal and all sons will be carriers.
    c. All daughters will be colorblind and all sons will be normal.
    d. All daughters will be heterozygous and all sons will be colorblind.
    e. It is impossible to predict with any reasonable degree of certainty

38. Which of the following could be detected by counting up the number of chromosomes in a cell of the affected person?
    a. hemophilia
    b. albinism
    c. Huntington disease
    d. color-blindness
    e. trisomy 21

39. Which of these is not a genetic disorder?
    a. sickle-cell anemia
    b. hemophilia
    c. albinism
    d. Huntington disease
    e. malaria

40. A recessive allele on the X chromosome causes colorblindness. A noncolorblind woman (whose father is colorblind) has children with a colorblind man. What is the chance their son will be colorblind?
    a. 0
    b. 25%
    c. 50%
    d. 75%
    e. 100%

41. A colorblind boy has a noncolorblind mother and a colorblind father. From which parent did he get the colorblind gene?
    a. father
    b. mother
    c. either parent could have given him the gene

42. Hemophilia is an X-linked recessive gene causing a blood disorder. What are the chances that the daughter of a normal man and a heterozygous woman will have hemophilia?
    a. 0
    b. 25%
    c. 50%
    d. 75%
    e. 100%

43. A man who carries a harmful X-linked gene will pass the gene on to
    a. all of his daughters
    b. half of his daughters
    c. half of his sons
    d. all of his sons
    e. all of his children

APPLYING THE CONCEPTS

Genetics Problems. These practice problems and questions are intended to sharpen your ability to apply critical thinking and analysis to biological concepts covered in this chapter.

Crosses Involving One Trait:

44. When two plants with red flowers are mated together, the offspring always are red, but if two purple-flowered plants are mated together, sometimes some of the offspring have red flowers. Which flower color is dominant?

_____

45. In sheep, white (B) is dominant to black (b). Give the F_2 phenotypic and genotypic ratios resulting from the cross of a pure-breeding white ram with a pure-breeding black ewe.

_____

_____

46. If you found a white sheep and wanted to determine its genotype, what color animal would you cross it to?

_____

47. Squash may be either white or yellow. However, for a squash to be white, at least one of its parents must also be white. Which color is dominant?

_____

48. In peas, yellow seed color is dominant to green. Give the expected proportion of each color in the offspring of the following crosses: (a) a heterozygous yellow with a heterozygous yellow; (b) a heterozygous yellow with a green; and (c) a green with a green.

_____

_____

_____

49. If tall (D) is dominant to dwarf (d), give the genotypes of the parents that produce 3/4 tall plants and 1/4 dwarf plants among their progeny.

_____

Crosses Involving Two Traits:

50. In pigs, mule hoof (fused hoof) is dominant (C) while cloven foot is recessive (c). Belted coat pattern (S) is dominant to solid color (s). Give the F_2 genotype and phenotype ratios expected from the cross $CCSS$ x $ccss$.

_____

_____

_____

51. In the F_2 generation of the previous question, what proportion of the cloven-hoofed, belted pigs would be homozygous?

_____

_____

52. Flat tail (F) is dominant to fuzzy tail (f), and toothed (T) is dominant to toothless (t). Give the results of a cross between two completely heterozygous parents.

_____

_____

_____

53. In rabbits, black (B) is dominant to brown (b), and spotted coat (S) is dominant to solid coat (s). Give the genotypes of the parents if a black, spotted male is crossed with a brown, solid female and all the offspring are black and spotted.

_____

54. In the preceding problem, give the genotypes of the parents if some of the offspring were brown and spotted.

_____

55. In cattle, having horns (p) is recessive to hornless or polled (P). Coat color is controlled by incompletely dominant genes RR for red, rr for white, and RR' for roan. If two heterozygous roan-polled cattle are mated, what kinds of offspring are expected?

_____

_____

_____

56. If a yellow guinea pig is crossed with a white one, the offspring are cream-colored.

a. What is the simplest explanation for this result?

_____

b. What kinds of offspring are expected if two cream-colored guinea pigs mate?

_____

57. In carnations, red or white phenotypes are dependent on homozygous genotypes, while the heterozygotes are pink. Give the F_1 and F_2 genotypic and genotypic ratios expected from a cross: red x white.

_____

_____

Crosses Involving Sex-Linked Traits:

58. A normal woman whose father was a hemophiliac has children with a normal man. What are the chances of hemophilia occurring in their children?

_____

_____

59. Another woman with no history of hemophilia in her family has children with a normal man whose father was a hemophiliac. What are the chances of hemophilia occurring in their children?

_____

60. Colorblindness (*c*) is a sex-linked recessive trait, while normal color vision (*C*) is dominant.

a. If two normal-visioned parents have a colorblind son, what are the parents' genotypes?

_____

b. What are the chances that their daughter will be colorblind?

_____

61. In cats, yellow is due to gene *B*, and black to its allele *b*. These genes are sex-linked. The heterozygous condition results in tortoiseshell. What kinds of offspring (sex and color) are expected from a cross of a black male with tortoise-shell female?

_____

_____

62. In fruit flies, normal long wings are dominant (*V*) and vestigial (shortened) wings are recessive (*v*). These genes are autosomal. The sex-linked gene controlling red eye color (*W*) is dominant to white eyes (*w*). A male with red eyes and normal wings mates a white-eyed vestigial-winged female. Give the expected ratio of phenotypes in the F_2 generation.

_____

_____

Crosses Involving Gene Interactions;

63. In poultry, there are two independently assorting gene loci, each with two alleles that affect the shape of a chicken's comb. One locus has a dominant allele (*R*) for rose comb while its recessive allele (*r*) produces single comb. The other locus has a dominant gene (*P*) for pea comb while its recessive allele (*p*) also produces single combs. When the two dominant genes occur together (*R-P-*), a walnut comb is produced. So, *R-P-* = walnut, *R-pp* = rose, *rrP-* = pea, and *rrpp* = single. Give the expected phenotypic ratios of offspring from the following matings:

a. *RRPP* x *rrpp* _____

b. *RrPp* x *rrpp* _____

c. *Rrpp* x *rrPp* _____

d. *RrPP* x *RrPp* _____

e. *rrPp* x *RrPP* _____

64. In humans, deafness can be the result of a recessive allele affecting the middle ear (*dd* = deaf), or another recessive allele (*ee* = deaf) that affects the inner ear. Suppose two deaf parents have a child that can hear. Give the genotypes of all three individuals.

_____

65. If two hearing people, heterozygous at both loci (*DdEe*) for deafness have children, what are the chances that their first child would be normal hearing? What is the chance of deafness in this child?

_____

_____

_____

Crosses Involving Multiple Alleles:

66. Mallard ducks show a multiple allele pattern of inheritance in which M^R produces "restricted mallard" coloring and is dominant over *M* for mallard coloring, and both of these alleles are dominant over *m* for "dusky mallard" coloring. Give the phenotypic ratios expected among offspring from the following crosses:

a. $M^R M$ x $M^R m$ _____

b. $M^R M$ x *Mm* _____

c. *Mm* x *mm* _____

Crosses Involving Multiple Genes:

67. If there are two pairs of genes involved in producing skin color in black x white crosses, and if the five phenotypic classes are black, dark, medium, light, and white, give the expected F_2 results of a white x black mating.

_____

68. Give the darkest phenotype possible among the offspring of the following matings:

a. black x dark _____

b. black x medium _____

c. black x white_____

d. dark x medium _____

e. dark x white _____

f. medium x light _____

g. light x light _____

h. light x white _____

Short Answer:

69. Occasionally, a family occurs in which both parents have recessive albinism but all of their children have normal amounts of skin pigmentation. Propose a genetic explanation for the inheritance of albinism in these families.

_____

_____

_____

_____

_____

_____

_____

70. Referring to blood transfusions in humans, briefly explain why blood type O is the "universal donor" (which means that type O red blood cells can be given to anyone in an emergency situation) and blood type AB is the "universal recipient" (which means that a type AB person can receive anyone's red blood cells in an emergency without harm).

_____

_____

_____

_____

_____

_____

_____

_____

71. In some families where one parent is an albino (a recessive condition lacking in melanin pigment in the hair, skin, and eyes) and the other parent has normal pigmentation, none of the children have albinism, even if large numbers of children are produced. In other families where one parent is albino and the other parent is normal, half of their children have albinism. Explain the difference in the types of children produced in each type of family.

_____

_____

_____

_____

_____

_____

72. Understanding the role of chance in determining the outcome of genetics crosses is similar to understanding the role of chance in the outcome of coin tosses. If you were to flip two coins enough times, you would find that both coins land heads-up 25% of the time, and that one coin lands heads-up and the other lands tails-up 50% of the time. Can you explain these outcomes?

_____

_____

_____

_____

_____

_____

_____

73. Mendel used peas in his experiments to gain an understanding of the rules of inheritance. Why are peas better subjects for genetics studies than humans?

_____

_____

_____

_____

_____

_____

74. Narcolepsy occurs due to different genetic defects in dogs and mice. How does that relate to how narcolepsy occurs genetically in humans?

_____

_____

_____

_____

Use the Case Study and the Web sites for this chapter to answer the following questions.

75. Huntington Disease is a neurologic disorder, first described in the 1870's. What are the clinical symptoms of the disease? Is there a treatment for Huntington Disease?

_____

_____

_____

_____

_____

_____

Chapter 12: Patterns of Inheritance   143

76. In 1993 researchers identified both the genetic defect and the corresponding protein product associated with Huntington Disease. Describe the location of the gene and what the protein product is like.

_____

_____

_____

_____

_____

_____

_____

_____

_____

77. Describe the latest experimental therapies for Huntington Disease. How effective are they?

_____

_____

_____

_____

_____

_____

_____

_____

_____

_____

_____

ANSWERS TO EXERCISES

1. gene	9. false, genes	20. true	32. a
allele	10. false, alleles	21. e	33. e
2. locus	11. true	22. b	34. g
3. dominant	12. true	23. c	35. d
recessive	13. false, independent	24. e	36. e
4. homozygote	assortment	25. c	37. d
heterozygote	14. true	26. a	38. e
5. genotype	15. false, incomplete	27. e	39. e
phenotype	dominance	28. e	40. c
6. crossing over	16. false, males	29. b	41. b
7. carrier	17. true	30. a	42. a
8. Down syndrome	18. false, mother	31. d	43. a
nondisjunction	19. false, XO		

44. Purple is dominant.

45. Genotypic ratio: 1 *BB*: 2 *Bb*: 1 *bb*;
Phenotypic ratio: 3 white: 1 black

46. A test cross with a black (*bb*) sheep

47. White is dominant.

48. a. 3/4 yellow: 1/4 green;
b. 1/2 yellow: 1/2 green;
c. all green

49. *Dd* x *Dd*

50. 9 *C-S-* mule-foot, belted pigs:
3 *C-ss* mule-foot, solid pigs:
3 *ccS-* cloven-foot, belted pigs:
1 *ccss* cloven-foot, solid pig

51. There are 3 cloven-foot, belted pigs, but only
one is homozygous (*ccSS*); the other two are
heterozygous (*ccSs*).

52. 9 *F-T-* flat and toothed:
3 *F-tt* flat and toothless:
3 *ffT-* fuzzy and toothed:
1 *fftt* fuzzy and toothless

53. black spotted male (*BBSS*) mated with brown
solid female (*bbss*)

54. *BbSS* male mated with *bbss* female

55. 3 *P-RR* polled red:
6 *P-Rr* polled roan:
3 *P-rr* polled white:
1 *ppRR* horned red:
2 *ppRr* horned roan:
1 *pprr* horned white

56. a. Incomplete dominance with cream being
heterozygous
b. 1 yellow: 2 cream: 1 white

57. F$_1$: *Rr* pink;
F$_2$: 1/4 *rr* red: 2/4 *Rr* pink: 1/4 *rr* white

58. Half of the sons will inherit the defective gene
from mom and get Y from dad. Half of the
daughters will also inherit the defective gene
from mom but they will be heterozygous since
they also inherit a normal gene from dad.

59. None is expected to have hemophilia or inherit
the gene.

60. a. *Cc* (normal but carrier woman) x *CY*
(normal man) [X^{C}X^c x X^{C}Y]
b. No chance: she inherits a normal *C* gene
(X^C) from dad as well as either a *C* gene
(X^C) or *c* gene (X^c) from mom.

61. Black female, tortoiseshell female, black male,
yellow male

62. F$_2$ for both sexes: 3/8 red long, 3/8 white long,
1/8 red vestigial, 1/8 white vestigial

63. a. all walnut
b. 1/4 walnut: 1/4 rose: 1/4 pea: 1/4 single
c. 1/4 walnut: 1/4 rose: 1/4 pea: 1/4 single
d. 3/4 walnut: 1/4 pea
e. 1/2 walnut: 1/2 pea

64. *DDee* x *ddEE* → *DdEe*

65. 9/16 *D-E-* normal: 3/16 *D-ee* deaf:
3/16 *ddE-* deaf: 1/16 *ddee* deaf;
so, 9/16 normal: 7/16 deaf

66. a. 3/4 restricted mallard: 1/4 mallard
b. 1/2 restricted mallard: 1/2 mallard
c. 1/2 mallard: 1/2 dusky mallard.

67. F$_2$: 1/16 white: 4/16 light: 6/16 medium:
4/16 dark: 1/16 black

68. a. black
b. black
c. medium
d. black
e. dark
f. dark
g. medium
h. light

69. Like recessive deafness, there are several different genetic situations leading to albinism. In one type
of albinism, the gene necessary for making an enzyme needed to convert a substrate into melanin
pigment is missing (genotype *aa*), while in another type of albinism the gene necessary for transporting
the substrate for melanin into the pigment-producing cells is lacking (genotype *bb*). So, both *aaBB* and
AAbb people would have albinism. However, the children produced by individual *aaBB* and individual
AAbb would be *AaBb* and would have normal pigmentation because the children would make the
enzyme necessary to produce melanin and the enzyme necessary to transport the substrate for melanin
into the pigment cells.

70. In the human blood types, a type A person has red blood cells (RBCs) with type A antigen proteins, and blood plasma with anti-B antibodies. A type B person has RBCs with type B antigen and plasma with anti-A antibodies. A type AB person has RBCs with type A and type B antigens and plasma lacking anti-A and anti-B antibodies. A type O person has RBCs lacking type A and type B antigens and plasma with anti-A and anti-B antibodies. Since type O red blood cells do not contain either the A or the B antigen proteins, they can be given to anyone in an emergency situation, since they will not be attacked by either anti-A or anti-B antibodies in the blood plasma of the recipient. Since type AB people do not have anti-A or anti-B antibodies in their blood plasma, they can receive any type of RBC in a transfusion, since they lack the antibodies to attack the donated cells.

71. All albino individuals are recessive for the albinism gene (*aa*). There are two possible genotypes for the normal parents, however. Some normal parents may be homozygotes (*AA*) and others may be heterozygotes (*Aa*). In those families with an albino parent and a heterozygous normal parent, we expect the following outcome: *aa* x *Aa*, 50% with *aa* (albinism) and 50% with *Aa* (normal). In those families with an albino parent and a homozygous normal parent, we expect the following outcome: *aa* x *AA*, all children with Aa (normal).

72. Let's say we are flipping a quarter and a dime. There is only one way that both coins land heads-up: a 1/2 (50%) chance that the quarter lands heads-up and a 1/2 (50%) chance that the dime lands heads-up. Since these outcomes are independent events, we multiply their results together: 1/2 x 1/2 = 1/4 (25%). If one coin lands heads-up and the other lands tails-up, there are two ways for that to happen: quarter-heads (1/2 chance) and dime-tails (1/2 chance) = 1/4 (25%) total chance, or quarter-tails (1/2 chance) and dime-heads (1/2 chance) = 1/4 (25%) total chance. Adding these two chances together, we get a total chance of 1/2 (50%) that one coin lands heads-up and the other coin lands tails-up.

73. There are several reasons why peas are better organisms than humans to study genetic principles. Peas have a short life cycle and produce many offspring per plant, so that meaningful statistical analysis can be performed from the data obtained. Peas are inexpensive to raise and take up little space. Peas have many easy to recognize characteristics, and it is easy to control the types of matings desired. In addition, there are no ethical problems associated with the genetic manipulation of matings using pea plants.

74. In dogs, narcolepsy is caused by defects in a recessive gene called *Hcrt-2*, which normally makes a protein receptor (that normally binds to hypocretin) present on the surfaces of certain brain cells. Because the receptor is defective in narcoleptic dogs, their brains cannot respond to the molecular signals delivered by the hypocretin molecules. In mice, narcolepsy is caused by a lack of the hypocretin molecules, caused by a defective recessive gene. Narcoleptic mice have normal *Hcrt-2* genes and make normal receptor molecules for hypocretin. Studies on humans with narcolepsy indicate that some patients lack hypocretin molecules but have normal receptors (like the mice) and others make hypocretin molecules but may have defective receptors (like the dogs).

75. According to the Huntington's Disease Society of America, Huntington's Disease (HD) is a devastating, degenerative brain disorder for which there is, at present, no effective treatment or cure. HD slowly diminishes the affected individual's ability to walk, think, talk and reason. Eventually, the person with HD becomes totally dependent upon others for his or her care. Huntington's Disease profoundly affects the lives of entire families: emotionally, socially and economically. More than a quarter of a million Americans have HD or are "at risk" of inheriting the disease from an affected parent. HD affects as many people as Hemophilia, Cystic Fibrosis or muscular dystrophy. Early symptoms of Huntington's Disease may affect cognitive ability or mobility and include depression, mood swings, forgetfulness, clumsiness, involuntary twitching and lack of coordination. As the disease progresses, concentration and short-term memory diminish and involuntary movements of the head, trunk and limbs increase. Walking, speaking and swallowing abilities deteriorate. Death eventually follows from complications such as choking, infection or heart failure.

According to the National Institute of Neurological Disorders and Stroke, physicians may prescribe a number of medications to help control emotional and movement problems associated with HD. It is important to remember however, that while medicines may help keep these clinical symptoms under control, there is no treatment to stop or reverse the course of the disease.

76. According to the National Institute of Neurological Disorders and Stroke, in the study reporting the discovery of the gene for HD in 1993, over 50 investigators from nine different institutions, working together as The Huntington Disease Collaborative Research Group, isolated an expandable, unstable DNA segment on chromosome 4 that can change from generation to generation and even among affected siblings. The genetic mutation is a multiple repeat phenomenon similar to the defects responsible for fragile-X syndrome, myotonic dystrophy, and spino-bulbar muscular atrophy. The immediate effect of the discovery was to improve the accuracy of diagnostic testing, and the availability of a test almost immediately. A. T. Hoogeveen et. al. 1993, reported the identification of the protein made by the HD gene. They stated that the recent identification of the Huntington Disease (HD) gene, enabled them to synthesize peptides corresponding with the carboxy-terminal end [COOH end] of the predicted HD-gene. They used this peptide to create antibodies against the entire protein. Use of this antibody revealed the presence of a protein (huntingtin) with a molecular mass of approximately 330 kDa in both normal individuals and those with HD. Immunocytochemical studies showed a cytoplasmic localization of huntingtin in various cell types including neurons. In most of the neuronal cells the protein was also present in the nucleus. No difference in molecular mass or intracellular localization was found between normal and mutant cells.

77. In an article published in the St. Petersburg Times on December 1, 2000, it was reported that researchers at the University of South Florida (USF) have found that fetal tissue can survive being transplanted into the brains of patients with Huntington Disease, and that the transplanted tissue remains disease-free. Patients who received the transplants also showed marked improvements in their conditions, suggesting that transplantation of fetal neurons or other cells may one day offer life-changing treatment for people who suffer from Huntington, a fatal degenerative brain disease that has no cure. It could have applications for other brain and spinal cord diseases as well. "From this, we can conclude that these grafts will survive, connect with the brain, not be rejected," said the USF lead investigator. "Most importantly, we have concrete evidence that a disease that is otherwise fatal… finally has the opportunity to be treated therapeutically." The results were bolstered by nearly identical findings of French researchers that are being released in the British medical journal Lancet. The chairman of the medical and scientific advisory committee for the Huntington Disease Society of America said it's still "highly experimental," but it was encouraging that the genes that cause Huntington did not invade the transplanted neurons, too. The neurons are obtained from fetuses aborted in the first trimester. They're taken from the striatum, a central relay station in the brain and then transplanted in the damaged area of the patient's brain. "The reason this study is important is because it's helping to establish transplants as at least a potentially effective treatment method in degenerative diseases," said a professor of psychiatry and neurology at Johns Hopkins University in Baltimore.

Chapter 13: Biotechnology

OVERVIEW

In this chapter, you will learn about naturally occurring examples of genetic recombination, and several techniques used in recombinant DNA technology. The authors describe building DNA libraries, identifying and making copies of genes, and using the genes to modify living organisms. They also discuss some methods used to locate and sequence genes. Rice is the major food for about two-thirds of the Earth's humans, but rice is a poor source of most vitamins, especially vitamin A. Thus, millions of people suffer from vitamin A deficiency, a leading cause of death and blindness. In 1999, biotechnologists created genetically engineered rice, which is golden-yellow in color because it contains elevated levels of beta-carotene, a vitamin A precursor. How did scientists create this genetically modified plant? What are the potential risks and benefits of biotechnology?

1) What Is Biotechnology?

Biotechnology is any use or alteration of organisms, cells, or molecules for practical purposes. Common examples are the domestication of plants and animals that humans have been producing by selective breeding for 10,000 years. Modern technology also uses **genetic engineering** (modification of DNA) to achieve specific goals to: (1) understand more about cellular processes, including inheritance and gene expression; (2) better understand and treat various genetic disorders; and (3) generate economic and social benefits, including better agricultural organisms and valuable biological molecules. One important tool is **recombinant DNA**, DNA that is altered by the incorporation of genes from other organisms, often from different species. Recombinant DNA is transferred into animals or plants using vectors or carriers such as bacteria or viruses to make **transgenic** organisms expressing DNA that has been modified or derived from another species.

2) How Does DNA Recombination Occur in Nature?

DNA recombination occurs naturally through processes such as sexual reproduction, bacterial **transformation** (gene transfer between different bacteria as they pick up either free DNA from the environment or tiny circular DNA plasmids), and viral infections, which may transfer DNA between bacteria (via viruses called bacteriophages) or between eukaryotic species. Bacteria, fungi, algae, and protists contain **plasmids**, which are about 1,000-100,000 nucleotides long. A single bacterium may contain dozens or hundreds of copies of a plasmid. Plasmids may contain genes for metabolism of new energy sources (petroleum, for example), for diseases in organisms the bacteria infect (like diarrhea), for antibiotic resistance (such as resistance to penicillin).

Viruses may transfer DNA between bacteria and between eukaryotic species. Within infected cells, viral genes replicate and direct the synthesis of viral proteins, which assemble within the cells, forming new viruses that are released to infect new cells. During many viral infections, viral DNA sequences become incorporated into one of the host cell's chromosomes. In some cases, viruses can mistakenly incorporate human genes into the virus genome, creating a recombinant virus which can infect new cells and transfer the previous host's DNA sequence. On occasion, viruses cross species barriers to infect new species. In such cases, the new host may acquire some genes that originally belonged to an unrelated species.

3) How Does DNA Recombination Occur in Genetic Engineering Laboratories?

One way to increase the amount of Vitamin A in rice is to find the genes needed to make Vitamin A from one organism (a daffodil) and transfer these genes to the rice plant. **Restriction enzymes** from bacteria cut up large DNA molecules at specific nucleotide sequences to produce smaller pieces. For instance, the restriction enzyme EcoR1 cuts DNA at the sequence GAATTC (between the G and A), producing a staggered cut that leaves a small region of single-stranded DNA at the cut ends. Restriction enzymes naturally defend bacteria against viral infections by cutting apart invading viral DNA. Bacteria protect their own DNA by attaching methyl ($-CH_3$) groups to some of their nucleotides, blocking the action of restriction enzymes. Incubating DNA with a restriction enzyme produces many fragments which can be used to generate a collection of recombinant DNA called a "library."

Inserting foreign DNA into a **vector** can produce a recombinant DNA library. For example, DNA can be isolated from a daffodil and cut with EcoR1. Each daffodil DNA fragment has the same nucleotide sequence at its ends: TTAA or AATT. The vector is also cut with EcoR1 so its ends are also TTAA and AATT. Next, the vector and daffodil DNA fragments are combined, and DNA ligase is added to join the vector and daffodil DNA fragments together, creating recombinant DNA molecules. These daffodil/bacteria molecules are added to a culture of bacteria. By transformation, each bacterium can take up a different recombinant DNA molecule, so that each bacterium contains a different portion of the daffodil genome carried in a vector. Such a collection is called a **DNA library**. Isolation of a particular bacterial strain that carries a particular foreign gene from a DNA library is what scientists refer to as "cloning a gene."

4) How Can Researchers Identify Specific Genes?

Restriction fragment length polymorphisms (RFLPs) can be used to identify genes. Some of the differences in DNA sequences between people create differences in the lengths of DNA fragments produced by digestion with restriction enzymes. Differences in DNA sequences that alter the size of restriction fragments are called RFLPs. RFLPs can be detected by separating restriction fragments by **gel electrophoresis**. In gel electrophoresis, a mixture of restriction fragments is loaded into a well in a slab of agar (the "gel"). When an electrical field is applied to the gel, the negatively charged DNA fragments move forward towards the positively charged electrode with the smaller fragments moving faster than the larger fragments. Eventually, the fragments are separated by size, forming distinct bands on the gel.

If an RFLP is inherited along with the gene for a particular trait, the RFLP and the gene are close together if not identical. All that needs to be done is to identify the bacteria in the DNA library that contain the RFLP DNA. Researchers find genes of interest in a library by using **DNA probes** (short, often radioactive DNA sequences complementary to the gene sought) to identify bacteria in the library carrying vectors with the gene of interest. Genes from one organism can be identified based on similarity to related genes (with similar base sequences) in other organisms. Genes also can be identified based on their protein products.

5) What Are Some of the Applications of Biotechnology?

Once the gene of interest has been cloned (isolated), the cloned material can be put to a variety of uses:

A.   Cloned genes provide enough DNA for DNA sequencing. This can be done by either culturing the bacteria containing the gene to create multiple copies or using the **polymerase chain reaction** (PCR) that makes billions of copies of the gene using heat-stable DNA polymerase enzymes. The steps in PCR include: (1) separating double-stranded DNA into single strands by heating to about 90° C; (2) lowering the temperature to 50° C and using specific primers (short sequences of DNA) to form base pairs with the original DNA molecule and start replication at appropriate places; (3) using heat-resistant DNA polymerase (isolated from bacteria living in hot springs) to construct new DNA copies; (4) cooling the mixture to stop the reaction; repeating step (1); etc. Each cycle takes a few minutes, so that billions of genes can be made in a single afternoon starting with a single copy.

A variation of the standard PCR can be used to determine the nucleotide sequence of DNA. Only one primer is added to the PCR mixture. A small fraction of each nucleotide added to the reaction mixture is labeled with a fluorescent molecule or tag. Each type of nucleotide (A, T, G, and C) is labeled with a different colored tag. During PCR, if DNA polymerase adds an unlabeled nucleotide to the DNA mixture it is synthesizing, it continues along and adds the next nucleotide. But, when DNA polymerase adds a fluorescent nucleotide, it stops synthesis of the new strand. As a result, the modified PCR reaction produces a series of DNA fragments each one nucleotide longer than the next. For each fragment, whether the terminal nucleotide is A, T, G, or C is determined by its color. When the fragments are separated by gel electrophoresis, a sensor records the color of the terminal nucleotide of each increasingly longer DNA fragment, thus determining the base sequence. From the nucleotide sequence of the gene, the amino acid sequence of the protein it makes can be determined, which can lead to the determination of the protein's normal function in cells. This was done using the gene for cystic fibrosis. Information about a gene's base sequence can lead to rapid tests to determine the presence or absence of defective genes in fetuses (prenatal diagnosis), newborns, and adults.

   B.   **DNA fingerprinting** is a type of RFLP analysis for genetic inquiry in many different contexts. Each person's DNA produces a unique set of restriction fragments, creating a unique "DNA fingerprint" used to identify an individual. DNA fingerprinting can establish the innocence of suspected criminals before or after trials. It also can be used to test potential organ donors for compatibility with potential recipients, to monitor pedigree claims for livestock, to examine relationships among ancient human populations, and to track the fate of endangered species.

   C.   Genetic engineering is revolutionizing agriculture, replacing traditional selective breeding techniques to produce improved strains of crops. One technique is to insert recombinant DNA via a *Ti* plasmid from soil bacteria that infect plants. Genes can be directly implanted into plant cells using a "gene gun" that blasts DNA-coated particles into cells. Production of transgenic crops or animals requires FDA, EPA, and Dept. of Agricultural approval. Genetic engineering can make plants resistant to diseases, insects, and weeds. To create resistance to insects, scientists incorporate genes from a bacterium (*Bacillus thuringiensis*, or *Bt*). Genetic engineering can produce plants with therapeutic benefits, such as cheap sources of vaccines against, for instance, diarrhea caused by dangerous strains of *E. coli*. Genetic engineering may improve domesticated animals. To create transgenic animals, the cloned DNA is injected into a fertilized egg and, in the case of mammals, returned to a surrogate mother to allow it to develop. Growth hormone genes from rainbow trout have been introduced into carp and catfish, causing them to grow 40% faster than normal. Modern techniques for DNA analysis are yielding information from the DNA of extinct, even fossilized animals and plants, particularly those frozen in arctic ice or embedded in amber. Even if dinosaur DNA were recovered, the DNA probably would be too fragmented to allow for the reconstruction of a complete set of dinosaur genes.

6) What Are Some Medical Uses of Biotechnology?

   Knockout mice (with mutated genes similar to those in humans) provide models of human genetic diseases. Studies of such mice and their defects help researchers clarify the roles of human genes (like that for cystic fibrosis, sickle cell anemia, Alzheimer disease, and mad cow disease) and allow scientists to study the effects of different treatments for the disorder without risking human patients.

   Genetic engineering allows the production of therapeutic proteins like insulin and human growth hormone (both grown in *E. coli* bacteria). Using transgenic "pharm" animals (into which human genes have been introduced using viral vectors or DNA injected into fertilized eggs), products such as alpha-1-antitrypsin (to treat an inherited form of emphysema) can be harvested from milk.

   Human gene therapy is just beginning. More than 4,000 diseases in humans result from defects in single genes, and many more diseases involve impaired function of multiple genes. One challenge is to deliver the functional gene to a large number of a patient's cells. Another challenge is that the newly incorporated gene must be efficiently expressed so that the gene will produce enough of its protein to help the patient overcome the disease. The two potential types of human gene therapy are (1) replacing defective genes in body cells (being tested for cystic fibrosis and for SCID, using vectors such as liposomes and retroviruses); and (2) altering genes in fertilized eggs, thus permanently repairing the genetic defect. The Human Genome Project aims to sequence the entire human genome by 2003 or earlier.

7) What Are Some Ethical Implications of Human Biotechnology?

Technology exists that can simultaneously analyze DNA from hundreds of patients, looking for mutations in several genes at once. Genetic tests for cystic fibrosis and inherited forms of breast cancer illustrate potential problems. Since 850 babies are born each year in the US with cystic fibrosis (CF), scientists have recommended that all couples be tested to find carriers of CF. Will unbiased and affordable genetic counseling be available for all those couples who are both heterozygous for CF? Should society bear the expenses of treating a child born with CF to parents who knew they were carriers? Should insurance companies be allowed to deny coverage to such couples? The increasing availability and use of tests for genetic diseases raise concerns about genetic discrimination by employers and insurance companies.

Further dilemmas are posed by genetic tests that can indicate that a person has an elevated risk, but not a certainty, of developing Alzheimer Disease, breast cancer, or ovarian cancer. Insurance companies have denied or terminated health insurance based on required genetic tests, or family history of known diseases such as Huntington Disease. Many but not all states have passed laws prohibiting genetic discrimination. **Sickle cell anemia** and **Tay-Sachs Disease** illustrate the hazards and benefits of genetic screening programs. In the 1970s, a major screening program for sickle cell anemia, prevalent among African-Americans, was launched without adequate education in place, leading to misuse due to ignorance and prejudice. Testing for Tay-Sachs Disease, prevalent among Jews of Eastern European descent, is a success story: testing is voluntary, coupled with prenatal testing and careful genetic counseling.

The potential to clone humans raises ethical issues. Are there valid reasons to clone or not to clone humans? Would cloning lead to a "Brave New World" type of society? Is biological uniqueness a fundamental human right and intrinsic to human dignity? Human cloning is illegal in England and Norway, but is legal in the US.

Case study revisited. Scientists inserted three genes into the rice genome (two from daffodils and one from a bacterium), along with regulatory DNA sequences. The result is rice producing enough beta-carotene to prevent Vitamin A deficiency if three bowls are eaten daily. Seed of this rice is being supplied free to agricultural research centers in developing countries. Some argue that this crop will decrease the distrust of some about the motives of biotechnology, which for the most part have focused on the profit needs of farmers and the farming industry.

KEY TERMS AND CONCEPTS

Fill-In: From the following list of key terms, fill in the blanks in the following statements.

amniocentesis	DNA probe	PCR	restriction enzyme
chorionic villus sampling	gel electrophoresis	plasmid	RFLPs
cloning	DNA library	recombinant DNA	vector
DNA fingerprinting			

1. _____ is a procedure for sampling the fluid surrounding a fetus, and _____ is a procedure used to sample cells from the fetal chorionic villi.

2. _____ is a method of producing large numbers of copies of a specific piece of DNA.

3. _____ is a sequence of complementary nucleotides that can be used to identify a DNA segment that carries a gene.

4. _____ is used to separate DNA fragments on the basis of size.

5. _____ are the differences in the lengths of fragments produced by cutting DNA from different individuals of the same species with the same restriction enzymes. These differences can be used in _____ to identify an individual.

6. _____ is the process of producing a new organism genetically identical to an existing organism.

7. _____ is a bacterium, plasmid, or virus that carries DNA between different organisms. Specifically, a _____ is a small, circular piece of DNA found in the cytoplasm of many bacteria.

8. _____ is a protein, normally isolated from bacteria, that cuts double-stranded DNA at a specific nucleotide sequence.

9. A _____ is a readily accessible, easily duplicable, collection of all the DNA of a particular organism, normally cloned into bacterial plasmids.

10. _____ is genetic material altered by the incorporation of genes from a different organism.

Key Terms and Definitions

amniocentesis (am-nē-ō-sen-tē´-sis): a procedure for sampling the amniotic fluid surrounding a fetus: A sterile needle is inserted through the abdominal wall, uterus, and amniotic sac of a pregnant woman; 10 to 20 milliliters of amniotic fluid is withdrawn. Various tests may be performed on the fluid and the fetal cells suspended in it to provide information on the developmental and genetic state of the fetus.

biotechnology: any industrial or commercial use or alteration of organisms, cells, or biological molecules to achieve specific practical goals.

chorionic villus sampling (CVS): a procedure for sampling cells from the chorionic villi produced by a fetus: A tube is inserted into the uterus of a pregnant woman, and a small sample of villi are suctioned off for genetic and biochemical analyses.

DNA fingerprinting: the use of restriction enzymes to cut DNA segments into a unique set of restriction fragments from one individual that can be distinguished from the restriction fragments of other individuals by gel electrophoresis.

DNA library: a readily accessible, easily duplicable complete set of all the DNA of a particular organism, normally cloned into bacterial plasmids.

DNA probe: a sequence of nucleotides that is complementary to the nucleotide sequence in a gene under study; used to locate a given gene within a DNA library.

gel electrophoresis: a technique in which molecules (such as DNA fragments) are placed on restricted tracks in a thin sheet of gelatinous material and exposed to an electric field; the molecules then migrate at a rate determined by certain characteristics, such as length.

genetic engineering: the modification of genetic material to achieve specific goals.

plasmid (plaz´-mid): a small, circular piece of DNA located in the cytoplasm of many bacteria; normally does not carry genes required for the normal functioning of the bacterium but may carry genes that assist bacterial survival in certain environments, such as a gene for antibiotic resistance.

polymerase chain reaction (PCR): a method of producing virtually unlimited numbers of copies of a specific piece of DNA, starting with as little as one copy of the desired DNA.

recombinant DNA: DNA that has been altered by the recombination of genes from a different organism, typically from a different species.

restriction enzyme: an enzyme, normally isolated from bacteria, that cuts double-stranded DNA at a specific nucleotide sequence; the nucleotide sequence that is cut differs for different restriction enzymes.

restriction fragment: a piece of DNA that has been isolated by cleaving a larger piece of DNA with restriction enzymes.

restriction fragment length polymorphism (RFLP): a difference in the length of restriction fragments, produced by cutting samples of DNA from different individuals of the same species with the same set of restriction enzymes; the result of differences in nucleotide sequences among individuals of the same species.

sickle-cell anemia: a recessive disease caused by a single amino acid substitution in the hemoglobin molecule. Sickle-cell hemoglobin molecules tend to cluster together, distorting the shape of red blood cell shape and causing them to break and clog capillaries.

Tay-Sachs disease: a recessive disease caused by a deficiency in enzymes that regulate lipid breakdown in the brain.

transformation: a method of acquiring new genes, whereby DNA from one bacterium (normally released after the death of the bacterium) becomes incorporated into the DNA of another, living, bacterium.

transgenic: referring to an animal or a plant that expresses DNA derived from another species.

vector: a carrier that introduces foreign genes into cells.

THINKING THROUGH THE CONCEPTS

True or False: Determine if the statement given is true or false. If it is false, change the underlined word(s) so that the statement reads true.

11. _____   DNA recombination <u>does not</u> occur in nature.

12. _____   Sexual reproduction between humans <u>is an example</u> of DNA recombination in nature.

13. _____   Bacteria pick up free DNA and incorporate it into their chromosomes during <u>sexual reproduction</u>.

14. _____   A recessive condition caused by inability to break down fatty materials in nerve cells is called <u>sickle cell anemia</u>.

15. _____   A readily accessible, easy to duplicate collection of all the DNA of a particular organism is a <u>DNA library</u>.

16. _____   An enzyme that cuts DNA open to form sticky ends is called <u>ligase enzyme</u>.

17. _____   Bacterial DNA is protected from the action of restriction enzymes by <u>methylation</u>.

18. _____   A method for making many copies of a small amount of DNA is the <u>restriction fragment length polymorphism reaction</u>.

19. _____   RFLPs are helpful in gene <u>mapping</u>.

Fill-In: Using the figure below, fill in the answers to the following questions.

20. Describe what is happening in step 1.

21. Describe what is happening in step 2.

22. Describe what is happening in step 3.

23. Describe what is happening in step 4.

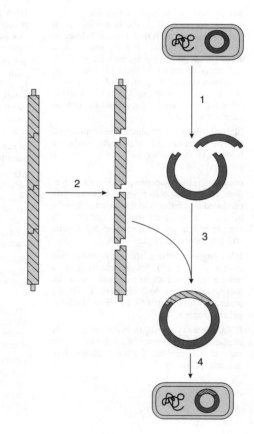

Matching: Recombinant DNA.

24. _____ accessory chromosomes in bacteria

25. _____ defends bacteria against viral infection by cutting apart the viral DNA

26. _____ self-replicating tiny loops of DNA

27. _____ used to identify marker genes in chromosomes

28. _____ readily accessible, easy to duplicate collection of all the DNA of a particular organism

29. _____ bacteria protect themselves from this by methylating their DNA

30. _____ a bacterium may contain hundreds of these

31. _____ can cut apart DNA to create single-stranded ends

32. _____ restriction fragment length polymorphisms

33. _____ cut at specific DNA sequences

34. _____ often contains genes for antibiotic-digesting enzymes

35. _____ cuts up DNA into fragments of various sizes that can be separated by gel electrophoresis

Choices:

a. DNA library

b. restriction enzymes

c. plasmids

d. RFLPs

Multiple Choice: Pick the most correct choice for each question.

36. Small accessory chromosomes found in bacteria and useful in recombinant DNA procedures are called
   a. plasmids
   b. palindromes
   c. centrioles

37. Which of the following is not a goal of biotechnology?
   a. generating economic benefits
   b. efficiently producing biologically important molecules
   c. improving agriculturally important food plants
   d. more effectively treating disease
   e. creating humans with higher intelligence levels

38. In biotechnology research, DNA fragments created by restriction enzyme action are separated from one another by
   a. crossing over
   b. gel electrophoresis
   c. centrifugation
   d. filtering
   e. the polymerase chain reaction

39. The enzymes used to cut genes in recombinant DNA research are called
   a. DNA polymerases
   b. RNA polymerases
   c. spliceosomes
   d. replicases
   e. restriction enzymes

40. DNA recombinations controlled by scientists in the laboratory
    a. are random and undirected
    b. involve specific pieces of DNA moved between deliberately chosen organisms
    c. use natural selection to determine their usefulness
    d. are of little practical use to humans
    e. usually cause harmful mutations

41. The polymerase chain reaction (PCR) is useful in
    a. analyzing a person's fingerprints
    b. cutting DNA into many small pieces
    c. allowing restriction enzymes to cut DNA
    d. creating recombinant plasmids
    e. making many copies of a small amount of DNA

42. Which vitamin deficiency was the golden-yellow rice engineered to combat?
    a. vitamin A
    b. vitamin B_{12}
    c. vitamin C
    d. vitamin D
    e. vitamin K

43. Which health condition was the golden-yellow rice engineered to combat?
    a. hearing loss
    b. anemia
    c. blindness
    d. soft bones
    e. dysentery

APPLYING THE CONCEPTS

These practice questions are intended to sharpen your ability to apply critical thinking and analysis to biological concepts covered in this chapter.

44. What do you think are some of the potential risks of releasing genetically engineered organisms into the environment?

_____

_____

_____

_____

_____

_____

45. What are some of the ethical issues raised by recombinant DNA technology?

_____

_____

_____

_____

_____

_____

_____

46. Explain how the use of restriction enzymes has allowed genetic engineers to easily splice together genes from different species into hybrid DNA molecules. How do the individual genes stick together?

_____

_____

_____

_____

_____

_____

_____

47. What are some of the reasons we should worry about the accidental or intentional release of genetically engineered organisms into the environment?

_____

_____

_____

_____

_____

48. During an RFLP analysis, describe how the DNA involved is cut into pieces and then separated to produce "DNA fingerprints." Would you expect different unrelated people to have identical DNA fingerprints? Explain your answer.

_____

_____

_____

_____

_____

49. State a few reasons why the people in underdeveloped countries might resist adding the genetically improved "golden" rice to their diets.

_____

_____

_____

_____

_____

_____

_____

50. Why do you think that most philosophers and ethicists say that human germinal cells (cells in the ovaries and testes that will produce eggs and sperm) should be off-limits to changes caused by genetic engineering?

_____

_____

_____

_____

_____

_____

_____

Use the Case Study and the Web sites for this chapter to answer the following questions.

51. Golden rice was genetically engineered to produce beta-carotene (vitamin A precursor) in its endosperm. Insufficient vitamin A leads to blindness but excess vitamin A is a teratogen (causes birth defects). What are the effects of excess beta-carotene?

_____

_____

_____

_____

_____

_____

52. Why was iron also added to the golden rice genome?

_____

_____

_____

_____

_____

53. The Institute of Science in Society site provides many details about the construction of golden rice. According to them, what are the major scientific and social flaws of this project?

_____

_____

_____

_____

_____

_____

ANSWERS TO EXERCISES

1. Amniocentesis
   chorionic villus sampling
2. PCR
3. DNA probe
4. Gel electrophoresis
5. RFLPs
   DNA fingerprinting
6. Cloning

7. Vector
   plasmid
8. Restriction enzyme
9. DNA library
10. Recombinant DNA
11. false, does
12. true
13. false, transformation

14. false, Tay-Sachs disease
15. true
16. false, restriction enzyme
17. true
18. false, polymerase chain
    reaction
19. true

20. A plasmid vector is removed from a bacterial cell and cut open with a restriction enzyme.

21. DNA from another organism is cut with the same restriction enzyme.

22. The plasmid DNA and a piece of DNA from the other organism are joined by complementary base pairing to form a recombinant molecule. The two pieces of DNA are joined by DNA ligase to form a permanent recombinant DNA molecule.

23. The recombinant plasmid is taken up by a host cell.

24. c	29. b	34. c	39. e
25. b	30. c	35. b	40. b
26. c	31. b	36. a	41. e
27. d	32. d	37. e	42. a
28. a	33. b	38. b	43. c

44. Recombinant organisms could compete with existing organisms and possibly replace them in nature, leading to undesirable consequences. Genetic engineering could produce organisms that act as hazardous pathogens, infecting plants, animals, and/or humans to cause hard-to-treat diseases. Newly engineered organisms also could transfer genes to other species, harming them or causing them to become harmful.

45. Some ethical questions that should be considered: Are the potential benefits worth the potential risks? Who decides what kinds of alterations in a species are acceptable? Should ecosystems be altered by introducing altered species? Do scientists or governments have the right to stop or forbid recombinant DNA research if there is a potential for it to solve medical or environmental problems? Who has the right to decide whether human genes should be altered and who should decide whose genes and which genes should be changed? What uses should be made of genetic information and who should decide what these are?

46. The restriction enzyme cuts DNA molecules at particular base sequences. The enzyme cuts the double stranded DNA molecule open unevenly, leaving sections of unpaired bases with complementary base sequences at either side of the cut region. This results in DNA fragments with "sticky ends" because the unpaired bases seek to base pair with complementary sequences. For instance:

$$\overset{\downarrow}{\underset{\underset{\uparrow}{CCCGGG}}{GGGCCC}} \quad makes \quad \frac{G}{CCCGG} \quad \frac{GGCCC}{G}$$

If different fragments of DNA are cut with the same restriction enzyme and then mixed together, genes from different organisms can base pair with each other at their sticky ends, and then be permanently joined together with DNA ligase.

47. Genetically engineered organisms could be harmful to humans, plants, or animals. Suppose a scientist splices a cancer gene into a bacterium in order to study its action, and the recombinant organism is accidentally ingested by a worker in the lab. This might give rise to a communicable form of cancer. Genetically engineered organisms could perhaps transfer some genes into other species, causing them harm or causing them to become harmful. Suppose a genetically engineered corn plant carrying a spliced gene for herbicide resistance transfers that gene to some weeds. This could result in weeds that are resistant to the herbicides normally used to kill them. Suppose that such corn carries a gene that kills invading bacteria, but the gene produces its toxin in pollen that is picked up by butterflies, killing them.

48. To make a DNA fingerprint, first you must extract all the DNA from some of the person's cells. Then you should add a restriction enzyme to cut the DNA up into fragments of many sizes ("restriction fragments"). Next, you should place all the restriction fragments into a well on a gel and perform gel electrophoresis by running an electrical current through the gel. This will cause the restriction fragments to migrate towards the positive electrode. There will be a negative correlation between the size of the fragments and the rate of speed they will travel through the gel (smaller fragments will move faster than larger fragments). After electrophoresis separates the fragments by size on the gel, they will be lined up in a manner resembling the bar codes seen on many packages. Since different people have DNA with slightly different DNA sequences due to the presence of different alleles for many genes, the fingerprints of unrelated people are expected to be different. Only identical twins, or clones, should have identical DNA fingerprints.

49. Fear of the unknown or of something that is different would be one reason to resist accepting the golden rice. There would be distrust of some "gift" from the rich foreigners for fear that the rice is actually something harmful. Also, since the rice looks different (is yellow) than what they expect (white rice), there would be reluctance to eat it, just as there would be reluctance for Americans to eat purple tomatoes containing more nutrients because "tomatoes are supposed to be red." In addition, if the scientific basis for blindness due to vitamin A deficiency is not understood by the people (if they believe that blindness is a sign punishment caused by gods who disfavor them, for instance), they will not accept the notion that eating the modified rice will prevent blindness.

50. Ethicists generally have no problem with gene therapy involving human somatic cells, since these changes affect only the individual receiving the cells, and the changes cannot be passed on to future generations. Ethicists recommend banning genetic engineering on germinal tissues because such manipulation could potentially affect unknown numbers of individuals in future generations, without their permission, and with unknown (and perhaps harmful) consequences on the genetic history of the human species. For instance, deliberately eliminating the gene for sickle cell anemia (an often fatal disorder of hemoglobin in red blood cells) in the human species sounds like a good idea, but doing so would eliminate one natural way to resist the effects of malaria, since humans heterozygous for the condition naturally filter the parasite out of their bloodstreams.

51. According to the Merck manual, Sec. 1, Ch. 3, although carotene is metabolized in the body to vitamin A at a slow rate, excessive ingestion of carotene does not cause vitamin A toxicity but produces carotenemia (high carotene blood levels). This condition is usually asymptomatic but may lead to carotenosis, in which the skin (but not the white of the eyes) becomes deep yellow, especially on the palms and soles. Carotenosis may also occur in diabetes mellitus, myxedema, and anorexia nervosa, possibly from a further reduction in the rate of conversion of carotene to vitamin A.

52. According to the Biotechnology Industry Organization, Swiss researchers have genetically enhanced rice to contain enough beta-carotene to satisfy daily requirements for vitamin A in as little as 300 grams of cooked rice per day. In addition, they significantly increased iron levels in the same rice strain. This extraordinary increase in micro-nutrient levels (rice contains virtually no beta-carotene or usable iron) was accomplished by incorporating seven new genes into a single strain of rice. The Swiss researchers gave the rice strain four genes that produce the enzymes that make beta-carotene from a precursor molecule. Two of these genes came from daffodils and were provided by researchers at the University of Freiburg in Germany; the other two were genes from bacteria. Other researchers were adding three new genes to increase the iron levels in a different strain of rice. The first gene produces a protein that breaks down the molecule in rice that prevents iron absorption. They then added a second gene from the French bean that is responsible for the production of the iron-storage protein, ferritin. This doubled the iron levels in the rice strain. Finally, to improve absorption they provided a third gene. Once both strains contained the new gene packets, the scientists cross bred the two strains to produce a hybrid containing all of the improvements.

53. According to the Institute of Science in Society, "Many have commented on the absurdity of offering 'golden rice' as the cure for vitamin A deficiency when there are plenty of alternative, infinitely cheaper sources of vitamin A or pro-vitamin A, such as green vegetables and unpolished rice, which would be rich in other essential vitamins and minerals besides. To offer the poor and malnourished a high-tech 'golden rice' tied up in multiple patents, that has cost US$100 million to produce and may cost as much to develop, is worse than telling them to eat cake." They also state: "It is clear that vitamin A deficiency is accompanied by deficiencies in iron, iodine and a host of micronutrients, all of which comes from the substitution of a traditionally varied diet with one based on monoculture crops of the Green Revolution. The real cure is to re-introduce agricultural biodiversity in the many forms of sustainable agriculture already being practiced successfully by tens of millions of farmers all over the world." They state that "Golden rice" exhibits all the undesirable, hazardous characteristics of existing GM plants, and in added measure on account of the increased complexity of the constructs and the sources of genetic material used. The hazards are highlighted below. (1) It is made with a combination of genes and genetic material from viruses and bacteria, associated with diseases in plants, and from other non-food species. (2) The gene constructs are new, and have never existed in billions of years of evolution. (3) Unpredictable by-products have been generated due to random gene insertion and functional interaction with host genes, which will differ from one plant to another. (4) The transgenic DNA is structurally unstable, leading to instability of the GM plants in subsequent generations, multiplying unintended, random effects. (5) Structural instability of transgenic DNA increases the likelihood of horizontal gene transfer and recombination.

Chapter 14: Principles of Evolution

OVERVIEW

In this chapter, you will learn about the concept of **evolution**. The authors present an historical account of pre-Darwinian thought, followed by the Darwin-Wallace theory of evolution by natural selection. In the final part of this chapter, you will learn about various proofs of evolution. Recently in China, the preserved remains of some previously undiscovered types of dinosaurs with feathers were found. This provided evidence that non-flying dinosaurs are the ancestors of today's birds. How does the discovery of feathered dinosaur fossils support the proposition that modern organisms arose by descent with modification from earlier organisms?

1) How Did Evolutionary Thought Evolve?

Pre-Darwinian science, heavily influenced by theology, held that all organisms were simultaneously created by God and that each distinct life form remained unchanged from the moment of creation. Plato said that each object on Earth is an imperfect and temporary reflection of an "ideal form" and Aristotle categorized all organisms into a linear hierarchy (the "ladder of Nature"). These ideas went unchallenged for nearly 2,000 years. However, as European explorers noted in the 1700s, the numbers of **species** (different kinds of organisms) in newly discovered lands was greater than expected and led to thoughts that similar species might have developed from a common ancestor.

Fossils are the preserved remains of organisms that lived long ago. Fossils found in rocks resembled parts of living organisms. The organization of fossils is consistent: (1) older fossils are found in rock layers beneath younger fossils; (2) the resemblance to modern forms of life gradually increased as increasingly younger fossils are examined, like a ladder of nature stretching back in time; and (3) many fossils are of species now extinct. Scientists concluded that different types of organisms have lived at various times in the past.

Geology provided evidence that the Earth is exceedingly old. Biblical calculations suggest the Earth is 4,000 to 6,000 years old. Georges Cuvier's theory of **catastrophism** claims that successive catastrophes on Earth, like the biblical Great Flood, produced layers of rock and caused many species to become extinct in short time periods, perhaps (as proposed by Louis Agassiz) with the creation of more species after each event. Actually, Earth is very old, having been formed from the forces of wind, water, earthquakes, and volcanoes in much the same way then as now. This is the theory of **uniformitarianism**, proposed by James Hutton and Charles Lyell. Modern geologists estimate the Earth to be about 4.5 billion years old.

The French biologist Jean Baptiste Lamarck proposed in 1801 that organisms evolved through the **inheritance of acquired characteristics**: through an innate drive for perfection (never scientifically demonstrated), living organisms can modify their bodies through the use or disuse of parts (actually true) and these modifications can be inherited by their offspring (actually false). Though Lamarck's theory was abandoned, by the mid-1800s biologists began to realize that the fossil record suggested that present-day species had evolved from preexisting ones. But how?

Both Charles Darwin and Alfred Russel Wallace proposed, in 1858, that evolution occurs by natural selection. In 1859, Darwin published his book *On the Origin of Species by Means of Natural Selection*, which attracted much attention to his new theory. The box below summarizes the Darwin-Wallace theory of evolution by means of natural selection.

SUMMARY OF THE DARWIN-WALLACE THEORY OF EVOLUTION

Observation 1: All natural **populations** have the potential to increase geometrically in size due to reproductive abilities. A population consists of all individuals of one species in a particular area

Observation 2: Most natural populations maintain a relatively constant size.

    Conclusion 1: Thus, many organisms must die young, producing few or no offspring each generation.

Observation 3: Individuals in a population differ in many abilities that affect survival and reproduction (some are "better adapted").

    Conclusion 2: The most well adapted organisms probably reproduce the most, since they survive the best. This differential reproduction is due to **natural selection**, the process in which individuals whose traits best adapt them to their environment leave a larger number of offspring.

Observation 4: Some of the variation in adaptiveness among individuals is genetic and is passed on to the offspring.

    Conclusion 3: Over many generations, differential reproduction among individuals with different genotypes changes the overall frequencies of genes in populations, resulting in evolution.

NOTES:

(1) Darwin did not know the mechanism of heredity (to explain Observation 4) and could not prove Conclusion 3.

(2) In desperation, Darwin resorted to a version of Lamarck's inheritance of acquired characteristics and this nearly destroyed his entire theory.

2) How Do We Know That Evolution Has Occurred?

Although scientists may still debate the relative importance of different mechanisms of evolutionary change, exceedingly few biologists dispute that evolution occurs. The fossil record provides evidence of evolutionary change over time. Giraffes, elephants, horses, and other types of organisms show a progressive series of fossils leading from ancient primitive organisms, through several intermediary stages, to the modern forms.

Comparative anatomy provides structural evidence of evolution. Through **convergent evolution**, unrelated species in similar environments evolve similar body functions from dissimilar underlying structures, called **analogous structures** (for example, wings of birds and butterflies). Also, closely related species in dissimilar environments evolve dissimilar body functions from similar underlying structures, called **homologous structures** (for example, among mammals, the forelimbs of apes, seals, dogs, and bats). Some species of organisms have **vestigial structures** (structures with no apparent purpose), which are homologous to functional structures in other species.

Embryological stages of animals can provide evidence of common ancestry. All vertebrate embryos look similar to one another early in their development (all have gill slits and tails, even humans), indicating that all vertebrate species have similar genes.

Modern biochemical and genetic analyses reveal relatedness among diverse organisms. All cells have DNA, RNA, ribosomes, similar genetic codes, similar amino acids in proteins, and similar chromosome structures.

3) What Is the Evidence That Populations Evolve by Natural Selection?

Artificial selection, the breeding of domestic organisms such as dogs to produce specific desired features, demonstrates that organisms may be modified by controlled breeding. Also, evolution by natural selection occurs today, as illustrated by the evolution of populations of roaches in Florida for which the poison roach bait called "Combat®" is ineffective.

Studies of evolution in natural and domestic populations of plants and animals have demonstrated that: (1) the variations on which natural selection works are produced by chance mutations, and (2) evolution by natural selection selects for organisms that are best adapted to a particular environment. Natural selection does not select for the "best" in any absolute sense, but only in the context of a particular environment. A trait that is advantageous under one set of circumstances may become disadvantageous if conditions change.

Case study revisited. In 2000, a new fossil of a feathered dinosaur turned out to be a fake. Skeptics of the original findings supporting feathered dinosaurs evolving into birds point out that the Chinese fossils of feathered dinosaurs are younger (120 million years old) than the fossils of actual flying birds (150 million years old). Perhaps the feathered "dinosaur" fossils are really ancient birds. But the Chinese fossils have dinosaurian skeletons that lack any birdlike feathers. Perhaps some lines of feathered dinosaurs persisted after the line leading to birds had branched off.

KEY TERMS AND CONCEPTS

Fill-In: From the following list of key terms, fill in the blanks in the following statements.

analogous	convergent	inheritance of acquired characteristics
artificial	fossils	uniformitarianism
catastrophism	homologous	vestigial

1. ___Fossils___, the preserved remains of organisms that lived long ago, are often found in rocks and typically resemble parts of living organisms.

2. The theory of ___CATASTROPHISM___ claims that successive worldwide events, like the biblical Great Flood, produced layers of rock and caused many species to become extinct in short periods of time.

3. According to the theory of ___CONVERGENT___ (UNIFORMATARIAN), Earth is very old, having been formed from the forces of wind, water, earthquakes, and volcanoes in much the same way then as now.

4. In 1801, the French biologist Lamarck proposed that organisms evolved through the ___CONVERGENT UNIFORMITARIAN___

5. Through ___CONVERGENT___ evolution, unrelated species in similar environments evolve similar body functions from dissimilar underlying structures, called ___analogous___ structures. Conversely, closely related species living in dissimilar environments, evolve dissimilar body functions from similar underlying structures, called ___Homologous___ structures.

6. Some species of organisms have ___VESTIGIAL___ structures with no apparent purpose, which are homologous to functional structures in other species.

7. The wing of a butterfly and the wing of a bird are ___ANALogous___ structures, while the forelimb of a human and the forelimb of a whale are ___Homologous___ structures.

8. ___*Artificial*___ selection, the breeding of domestic organisms such as dogs to produce specific desired features, demonstrates that organisms may be modified by controlled breeding.

Key Terms and Definitions

analogous structures: structures that have similar functions and superficially similar appearance but very different anatomies, such as the wings of insects and birds. The similarities are due to similar environmental pressures rather than to common ancestry.

artificial selection: a selective breeding procedure in which only those individuals with particular traits are chosen as breeders; used mainly to enhance desirable traits in domestic plants and animals; may also be used in evolutionary biology experiments.

catastrophism: the hypothesis that Earth has experienced a series of geological catastrophes, probably imposed by a supernatural being that accounts for the multitude of species, both extinct and modern, and preserves creationism.

convergent evolution: the independent evolution of similar structures among unrelated organisms as a result of similar environmental pressures; see *analogous structures*.

evolution: the descent of modern organisms with modification from preexisting life-forms; strictly speaking, any change in the proportions of different genotypes in a population from one generation to the next.

fossil: the remains of a dead organism, normally preserved in rock; may be petrified bones or wood; shells; impressions of body forms, such as feathers, skin, or leaves; or markings made by organisms, such as footprints.

homologous structures: structures that may differ in function but that have similar anatomy, presumably because the organisms that possess them have descended from common ancestors.

inheritance of acquired characteristics: the hypothesis that organisms' bodies change during their lifetimes by use and disuse and that these changes are inherited by their offspring.

natural selection: the unequal survival and reproduction of organisms due to environmental forces, resulting in the preservation of favorable adaptations. Usually, natural selection refers specifically to differential survival and reproduction on the basis of genetic differences among individuals.

population: all the members of a particular species within an ecosystem, found in the same time and place and actually or potentially interbreeding.

species (spē´-sēs): the basic unit of taxonomic classification, consisting of a population or series of populations of closely related and similar organisms. In sexually reproducing organisms, a species can be defined as a population or series of populations of organisms that interbreed freely with one another under natural conditions but that do not interbreed with members of other species.

uniformitarianism: the hypothesis that Earth developed gradually through natural processes, similar to those at work today, that occur over long periods of time.

vestigial structure (ves-tij´-ē-ul): a structure that serves no apparent purpose but is homologous to functional structures in related organisms and provides evidence of evolution.

THINKING THROUGH THE CONCEPTS

True or False: Determine if the statement given is true or false. If it is false, change the <u>underlined</u> word(s) so that the statement reads true.

9. __F__   Before Darwin, most people thought species were <u>capable of change</u>.

10. __T__   The idea that God created some new species after every catastrophe was proposed by <u>Agassiz</u>.

11. __F__   The remains of organisms preserved in rock are called <u>vestigial structures</u>.

12. __T__   <u>Lamarck</u> proposed that an internal drive toward complexity within cells is the driving force in evolution.

13. __F__   The similarity in the bones making up a bird's wing and a horse's foot are due to <u>convergent</u> evolution.

14. __T__   Amino acid sequences in proteins of different animals tend to <u>support</u> evolution.

15. __T__   Aristotle's "ladder of Nature" was considered <u>immutable</u>.

16. __F__   In convergent evolution, the two forms being modified are <u>closely related</u>.

17. __F__   The many different varieties of dogs are the result of <u>natural</u> selection.

18. __T__   Analogous structures arise due to <u>convergent</u> evolution.

Matching: Theories about life.

19. _____ uniformitarianism

20. _____ "ladder of Nature"

21. _____ "ideal forms"

22. _____ inheritance of acquired characteristics

23. _____ catastrophism

24. _____ multiple creations

25. _____ natural selection

Choices:

a. Lamarck   f. Agassiz

b. Aristotle   g. Cuvier

c. Wallace   h. Lyell

d. Plato   i. Darwin

e. Malthus

Multiple Choice: Pick the most correct choice for each question.

26. Which of the following proposes that living organisms inherited body parts modified through use or disuse?
    a. natural selection
    b. catastrophism
    c. inheritance of acquired characteristics
    d. evolution

27. Fossils resulted from a successive series of geological upheavals according to the theory of
    a. natural selection
    b. catastrophism
    c. uniformitarianism
    d. independent assortment

28. The evolution of adaptations between different species as a result of extensive interactions with each other is termed
    a. analogous evolution
    b. divergent evolution
    c. convergent evolution
    d. coevolution

29. Which of the following is homologous to the human arm?
    a. wing of insect
    b. wing of bird
    c. body of snake
    d. fin of fish
    e. tail of salamander

30. Supportive evidence for evolution is found in studies of
    a. biochemistry
    b. embryos
    c. comparative anatomy
    d. domestication of plants and animals
    e. all of the answers are correct

31. Fossils provide direct evidence for
    a. behavioral adaptations
    b. physiological characteristics
    c. habitat preference
    d. structural similarities and differences
    e. catastrophism

32. The fossil evidence seems to infer that
    a. birds gave rise to dinosaurs
    b. birds and dinosaurs have independent origins
    c. dinosaurs gave rise to birds
    d. birds gave rise to mammals
    e. mammals gave rise to birds

APPLYING THE CONCEPTS

These practice questions are intended to sharpen your ability to apply critical thinking and analysis to biological concepts covered in this chapter.

33. A species of moth has a very long proboscis (tubular mouth part) that is used to suck nectar from the inner base of a particular type of long, trumpet-shaped flower. Closely related moth species, however, have much shorter mouth parts and feed off nectar from plants with shorter tubular flowers. How would Darwin and Lamarck each explain the evolution of the species of moth with the exceptionally long proboscis?

_____

_____

_____

_____

_____

_____

_____

_____

34. The fossil record indicates that at progressively more recent levels of sediment (those progressively closer to the surface), fossils more closely resembling modern forms of plants and animals are found. How would you explain the changes in the fossil record if you believed in catastrophism or if you believed in uniformitarianism?

_____

_____

_____

_____

_____

_____

_____

35. Fossils from China of feathered dinosaurs with distinct reptilian features are about 120 million years old, while the oldest fossils of birds are about 150 million years old. Are these fossils compatible with the theory that dinosaurs evolved into birds? What additional evidence would be needed to support this theory?

_____

_____

_____

36. Over a 50 year period in the 1800s in England, in areas where industrial pollution covered the trees with soot, peppered moths evolved from having light colored wings to having dark colored wings. How would Lamarck (inheritance of acquired characteristics) and Darwin (evolution by natural selection) have explained the change?

_____

_____

_____

_____

_____

_____

_____

_____

_____

_____

_____

_____

Use the Case Study and the Web sites for this chapter to answer the following questions.

37. *Archaeopteryx* is considered the oldest known fossil bird. What features does it share with dinosaurs? Why is it considered a bird?

_____

_____

_____

_____

_____

_____

_____

_____

38. *Confuciusornis* is another fossil bird. It is almost as old as *Archaeopteryx* but shares one feature with birds that *Archaeopteryx* lacks. What is it?

_____

_____

_____

_____

39. Several years ago, scientists were excited to discover *Archaeoraptor* "the first flying dinosaur." *Archaeoraptor* is now more frequently referred to as the "Piltdown chicken." Why?

_____

_____

_____

_____

_____

_____

_____

_____

ANSWERS TO EXERCISES

1. Fossils
2. catastrophism
3. uniformitarianism
4. inheritance of
   acquired characteristics
5. convergent
   analogous
   homologous
6. vestigial
7. analogous
   homologous
8. Artificial

9. false, incapable of change
10. true
11. false, fossils
12. true
13. false, common ancestry
14. true
15. true
16. false, unrelated
17. false, artificial
18. true
19. h
20. b

21. d
22. a
23. g
24. f
25. c, i
26. c
27. b
28. d
29. b
30. e
31. d
32. c

33. Lamarck would explain the long proboscis by the "inheritance of acquired characteristics." The ancestral species was uniform for a short proboscis, but the moths began to habitually stretch their probosci deep into the long trumpet flowers to get nectar, stretching their probosci in the process. These slightly stretched probosci were in turn passed down to the next generation who repeated the process of stretching their probosci into the flowers and passing the longer probosci on. After many generations of this, all the moths of this species were born with very long probosci. Darwin would explain the long probosci by natural selection. Individuals in the ancestral species displayed genetic variation in proboscis length and those with longer probosci could get more nectar from the flowers, allowing them to live longer and out-reproduce those with shorter probosci. This produced the next generation with a higher percentage of moths with longer probosci. After many generations of natural selection favoring moths with longer probosci, the entire species showed this trait.

34. Catastrophism is the idea that the Earth underwent a series of periodic worldwide catastrophes resulting in significant physical changes as well as the extinction of life. After each of these catastrophes, new life forms more similar to modern forms could have been created. Thus, evolution did not occur. Uniformitarianism is the idea that physical changes that have occurred during the history of the Earth occur gradually and over long periods of time, just like they do today. Life forms would also change gradually in response to changes in their environments (in other words, evolve), resulting in the formation of fossils, the more recent of which more closely resemble modern forms.

35. It is possible that the group of feathered dinosaurs existed much earlier than 120 million years ago, and might have existed before the appearance of the first true birds 150 million years ago. If that is true, then the birds could have evolved from the ancient feathered dinosaurs. Evidence to support this hypothesis would be finding fossils of feathered dinosaurs more than 150 million years old.

36. Lamarck would say that because of the environmental change (the tree trunks became darker due to pollution), moths during their lifetimes gradually began to produce more dark pigment in their wings due to a desire to become less conspicuous. Each generation, these slightly darker wings were passed on to their offspring, who made them darker still. By the time 50 generations passed, all the moths had very dark wings. Darwin would say that initially, there was some genetic variation among the moths and while most of them had light wings, a few had dark wings. Before pollution occurred, the darker moths were conspicuous and were eaten quickly by birds, but the lighter moths lived longer and produced more offspring. As the tree trunks became darker due to pollution, however, the darker moths became less conspicuous than the lighter ones, so that birds ate the lighter moths quickly, more often not seeing the darker ones. This allowed the darker moths to begin out-reproducing the lighter moths, and the frequency of darker moths increased while the frequency of the lighter moths decreased. Eventually, the population would consist of many dark moths and few light moths due to the natural selection by birds of preferentially finding and eating the more conspicuous color of moth in the changing environment.

37. According to the American Museum of Natural History, *Archaeopteryx* is a very primitive bird, about the size of a pigeon, that lived around 150 million years ago, in what is today southern Germany. It had teeth like dinosaurs. It had claws in its hands, like dinosaurs (modern birds do not have claws in their wings). *Archaeopteryx* had a relatively small sternum, or breastbone, and also, behind this, it had abdominal ribs. This structure is similar to the one present in dinosaurs, but very different from the one present in modern birds, which do not have abdominal ribs, and have a very large sternum. *Archaeopteryx* had a long tail that was similar to the long tails of its dinosaurian relatives, in contrast to the short tails of modern birds. It is considered a bird since it possessed feathers arranged in a pattern similar to that of modern birds.

38. According to the Carnegie Museum of Natural History, scientists were amazed by *Confuciusornis'* similarity to *Archaeopteryx*, the oldest known true bird, which was unearthed in 1861. Both animals have long feathers and huge curved claws. However, where *Archaeopteryx* had a mouth filled with teeth, *Confuciusornis* — like modern birds — had a toothless beak.

39. *US News and World Report* (2/14/00) noted: "Did dinos soar? Imaginations certainly took flight over *Archaeoraptor liaoningensis*, a birdlike fossil with a meat-eater's tail that was spirited out of northeastern China, "discovered" at a Tucson, Ariz., gem-and-mineral show last year [1999], and displayed at the National Geographic Society in Washington, D.C. Some 110,000 visitors saw the exhibit, which closed January 17, 2000; millions more read about the find in November's National Geographic. Now, paleontologists are eating crow. Instead of "a true missing link" connecting dinosaurs to birds, the specimen appears to be a composite, its unusual appendage likely tacked on by a Chinese farmer, not evolution." "*Archaeoraptor* is not the first "missing link" to snap under scrutiny. In 1912, fossil remains of an ancient hominid were found in England's Piltdown quarries and quickly dubbed man's apelike ancestor. It took decades to reveal the hoax. But just as Piltdown man didn't derail scientists' views on human evolution, one "Piltdown chicken" doesn't overturn the "compelling" evidence that birds evolved from dinosaurs."

Chapter 15: How Organisms Evolve

OVERVIEW

In this chapter, you will learn about the basic principles of population genetics, including how to calculate gene and genotype frequencies and the Hardy-Weinberg principle. The authors examine the major forces causing evolution, which are mutation, migration, small population size, nonrandom mating, and natural selection. The authors also discuss the three ways that natural selection acts on populations, which are called stabilizing, disruptive, and directional selection. The authors also enumerate the results of natural selection.

In 1999, four children died of pneumonia and blood infections caused by an antibiotic-resistant "supergerm" called *Staphylococcus aureus* (or staph) contracted in the children's homes and communities. Antibiotics are losing their effectiveness against many bacterial diseases, raising the specter of incurable diseases, resistant to all known treatments. Ironically, the spread of resistance among bacteria is a response to humans using antibiotics to fight these bacteria, obviously not the outcome doctors wished to occur.

1) How Are Populations, Genes, and Evolution Related?

Evolutionary changes occur from generation to generation so that descendants are different from their ancestors. Evolution is a property not of individuals but of **populations** (all individuals of a species living in a given area). Inheritance provides the link between the lives of individuals and evolution of populations. **Population genetics** is the study of the frequency, distribution, and inheritance of alleles in populations. The **gene pool** is the sum of all the genes in a population. The relative frequencies of various alleles in a population are the **allele frequencies**. Evolution is changes in gene frequencies that occur in a gene pool over time.

The **Hardy-Weinberg principle** states that under certain conditions, allele and genotype frequencies in a population will remain constant over time (evolution will not occur). Such an evolution-free population is an **equilibrium population** that will remain in **genetic equilibrium** as long as: (1) there is no mutation; (2) there is no **gene flow** (migration) between populations; (3) the population is extremely large; (4) all mating is random; and (5) there is no natural selection (all genotypes reproduce equally well). Few natural populations are in genetic equilibrium.

2) What Causes Evolution?

There are five major causes of evolutionary change: mutation, gene flow, small population size, nonrandom mating, and natural selection. Mutations (changes in DNA sequence) are the ultimate source of genetic variability.

Mutations occur at rates between 1 in 100,000 and 1 in 1,000,000 genes per individual per generation. Although mutation by itself is not a major evolutionary force, without mutations there would be no evolution and no diversity among life forms. Mutations are not goal-directed, but random. Mutations provide potential for change; other forces, especially natural selection, act on that potential and may favor the spread or the elimination of a mutation within a population.

Gene flow, or migration between populations, changes allele frequencies. It spreads advantageous alleles throughout a species, and helps maintain all the organisms over a large area as one species since the populations cannot become very different in allele frequencies as long as gene flow occurs.

Small populations are subject to random changes in allele frequencies, a process called **genetic drift**. In large populations, chance events are unlikely to significantly alter the overall gene frequencies.

However, in small populations, chance events could reduce or eliminate alleles, greatly altering in a random way its genetic makeup. Genetic drift tends to reduce genetic variability within small populations, and genetic drift tends to increase genetic variability between or among populations. A **population bottleneck**, which is a drastic reduction in numbers followed by expansion in numbers from the few survivors, may cause both changes in allele frequencies and reduction in genetic variability, not allowing the population to evolve in response to environmental changes. Examples of this include the northern elephant seal and the cheetah. In the **founder effect**, some isolated populations are founded by a small number of individuals who may have different allele frequencies than the larger parent population due to chance, and this may lead to a sizable new population that differs greatly from the original. Populations occasionally become very small, and these may contribute significantly to major evolutionary changes.

Random mating rarely occurs within populations, and often individuals will seek mates with similar traits to themselves, a behavior known as assortative mating. All genotypes are not equally adaptive. Any time an allele confers a slight advantage to some individuals, natural selection will favor the enhanced reproduction of those individuals. Four important points about natural selection and evolution are: (1) natural selection does not cause genetic changes in individuals; (2) natural selection acts on individuals (causing unequal reproduction) but evolution (changes in gene frequencies) occurs in populations; (3) evolution is a change in the allele frequencies of a population, owing to unequal reproduction among organisms bearing different alleles (the **fitness** of an organism is a measure of its reproductive success); and (4) evolutionary changes are not "good" or "progressive" in an absolute sense, just relative to the environmental circumstances present at any particular time and place.

3) How Does Natural Selection Work?

Natural selection is primarily an issue of **differential reproduction**: organisms with favorable alleles leave more offspring (who inherit those alleles) than do other individuals with less favorable alleles. Natural selection acts on the phenotype, which reflects the underlying genotype. Natural selection can influence populations in three major ways.

Directional selection shifts character traits in a specific direction: it favors individuals at one end of a distribution range for a trait and selects against average individuals and those at the opposite extreme of the distribution. An example might be that directional selection favors small size against both average and large individuals in a population. If the environment gets colder, then a species may evolve in a constant direction for thicker fur.

Stabilizing selection acts against individuals who deviate too far from the average: it favors individuals having an average value for a trait and selects against individuals with extreme values due to opposing environmental pressures. An example might be that small lizards have a hard time defending territories, but large lizards are more likely eaten by owls. Opposing environmental pressures may produce **balanced polymorphisms**, in which two or more alleles are maintained in a population because each is favored by a separate environmental force. For example, in Africa, malaria is a health problem. People homozygous for the sickle cell anemia allele are unhealthy. People homozygous for the normal allele are susceptible to severe malaria. Heterozygous people suffer only mild anemia and are more resistant to malaria.

Disruptive selection adapts individuals within a population to different habitats: it favors individuals at both ends of the distribution of a trait and selects against average individuals. For example, in African birds called black-bellied seedcrackers, some have large beaks for cracking hard seeds, and others have small beaks for cracking soft seeds, but there are few birds with medium beaks.

A variety of processes can cause natural selection. Organisms with reproductively successful phenotypes have the best **adaptations** (characteristics that help an individual survive and reproduce) to their particular environments. **Competition** for scarce resources favors the best-adapted individuals. When two species interact extensively, as seen with predators and their prey, each exerts strong selection pressure on the other. When one evolves a new feature or modifies an old one, the other typically evolves new adaptations in response, a constant mutual feedback situation called **coevolution**. **Predation** includes any situation where one organism (the predator) eats another (the prey). Predation often leads to coevolution between predator and prey species. **Symbiosis** (individuals of different species live in intimate contact for long periods) leads to the most intricate coevolutionary adaptations. Sexual selection favors traits that help an organism mate.

   Kin selection favors altruistic behaviors. **Altruism** is any behavior that endangers an individual or reduces its reproductive success but benefits other members of the species. If the altruistic individual helps relatives who possess the same alleles, this is called kin selection. Altruistic behaviors make sense in evolution when you consider the relationship of the organisms involved and you consider **inclusive fitness**. Inclusive fitness is determined by an individual's success at contributing its genes to the next generation. If an individual is not reproducing and directly adding its own genes, they can at least ensure that the genes of close relatives are passed on.

   Case study revisited. The evolution of antibiotic resistance in bacterial populations is a direct consequence of natural selection applied by widespread use of antibiotic drugs, killing the sensitive bacteria and allowing the resistant ones to gain a reproductive advantage, producing disproportionally more offspring with resistant genes in the next generation. Physicians prescribe more antibiotics than is necessary, antibiotics are fed to farm animals, and antibiotic soaps and cleansers are used in households. How can further evolution of antibiotic resistance be prevented?

KEY TERMS AND CONCEPTS

Fill-In: From the following list of key terms, fill in the blanks in the following statements.

adaptation	disruptive selection	genetic drift
allele frequency	evolution	Hardy-Weinberg principle
altruism	fitness	kin selection
balanced polymorphism	founder effect	population
coevolution	gene flow	predation
Darwin	gene pool	prey

1. __ADAPTATION__ is a trait that helps an organism survive and reproduce in a particular environment.

2. The prolonged maintenance of two or more alleles in a population is called __Balance polymorphism__

3. When __gene flow__ occurs, alleles move from one population to another due to the migration of organisms.

4. __Genetic Drift__ is the change in the allele frequencies of a small population purely by chance.

5. __Kin selection__ favors a certain allele because it increases the survival or reproductive success of relatives bearing the same allele. The behavior that endangers an individual but benefits other members of its species is called __Altruism__.

6. __Evolution__ represents significant changes in allele frequencies over time in a population and __coevolution__ is the evolution of adaptations in different species due to extensive interactions with each other.

7. When ___DISRUPTIVE SELECTION___ occurs, both extremes are favored over the average phenotype.

8. ___DARWIN___ wrote *On the Origin of Species* in 1859.

9. The ___FOUNDER EFFECT___ occurs when an isolated population is founded by a small number of individuals, this population may develop allele frequencies that are very different from those of the parent population.

10. Under certain conditions, allele and genotype frequencies in a population will remain constant over time. This is called the ___HARDY-WEINBERG___

11. A ___POPULATION___ is a group of individuals of the same species found in the same time and place.

12. The ___ALLELE FREQUENCY___ is defined as the relative proportion of each allele of a gene found in a population.

13. The total of all alleles of all genes in the population is the ___GENE POOL___.

14. ___FITNESS___ is the measure of the reproductive success of an organism.

15. Any situation involving one organism eating another is ___PREDATION___. The organisms that are eaten by predators are called ___PREY___.

Key Terms and Definitions

adaptation: a trait that increases the ability of an individual to survive and reproduce compared to individuals without the trait.

allele frequency: for any given gene, the relative proportion of each allele of that gene in a population.

altruism: a type of behavior that may decrease the reproductive success of the individual performing it but benefits that of other individuals.

balanced polymorphism: the prolonged maintenance of two or more alleles in a population, normally because each allele is favored by a separate environmental pressure.

coevolution: the evolution of adaptations in two species due to their extensive interactions with one another, such that each species acts as a major force of natural selection on the other.

competition: interaction among individuals who attempt to utilize a resource (for example, food or space) that is limited relative to the demand for it.

differential reproduction: differences in reproductive output among individuals of a population, normally as a result of genetic differences.

directional selection: a type of natural selection in which one extreme phenotype is favored over all others.

disruptive selection: a type of natural selection in which both extreme phenotypes are favored over the average phenotype.

equilibrium population: a population in which allele frequencies and the distribution of genotypes do not change from generation to generation.

fitness: the reproductive success of an organism, usually expressed in relation to the average reproductive success of all individuals in the same population.

founder effect: a type of genetic drift in which an isolated population founded by a small number of individuals may develop allele frequencies that are very different from those of the parent population as a result of chance inclusion of disproportionate numbers of certain alleles in the founders.

gene flow: the movement of alleles from one population to another owing to the migration of individual organisms.

gene pool: the total of all alleles of all genes in a population; for a single gene, the total of all the alleles of that gene that occur in a population.

genetic drift: a change in the allele frequencies of a small population purely by chance.

genetic equilibrium: a state in which the allele frequencies and the distribution of genotypes of a population do not change from generation to generation.

Hardy-Weinberg principle: a mathematical model proposing that, under certain conditions, the allele frequencies and genotype frequencies in a sexually reproducing population will remain constant over generations.

inclusive fitness: the reproductive success of all organisms that bear a given allele, normally expressed in relation to the average reproductive success of all individuals in the same population; compare with *fitness*.

kin selection: a type of natural selection that favors a certain allele because it increases the survival or reproductive success of relatives that bear the same allele.

natural selection: the unequal survival and reproduction of organisms due to environmental forces, resulting in the preservation of favorable adaptations. Usually, natural selection refers specifically to differential survival and reproduction on the basis of genetic differences among individuals.

population: all the members of a particular species within an ecosystem, found in the same time and place and actually or potentially interbreeding.

population bottleneck: a form of genetic drift in which a population becomes extremely small; may lead to differences in allele frequencies as compared with other populations of the species and to a loss in genetic variability.

population genetics: the study of the frequency, distribution, and inheritance of alleles in a population.

predation (pre-dā´-shun): the act of killing and eating another living organism.

sexual selection: a type of natural selection in which the choice of mates by one sex is the selective agent.

stabilizing selection: a type of natural selection in which those organisms that display extreme phenotypes are selected against.

symbiosis (sim´-bī-ō´sis): a close interaction between organisms of different species over an extended period. Either or both species may benefit from the association, or (in the case of parasitism) one of the participants is harmed. Symbiosis includes parasitism, mutualism, and commensalism.

THINKING THROUGH THE CONCEPTS

True or False: Determine if the statement given is true or false. If it is false, change the underlined word(s) so that the statement reads true.

16. __F__ Individual plants or animals change in response to selection.

17. __T__ For a population to remain at equilibrium, it must be large.

18. __F__ Mutation is the factor that controls the direction of evolution.

19. __F__ Genetic drift is a characteristic of large populations.

20. __F__ Natural selection acts on genotypes directly.

21. __F__ Stabilizing selection results in change.

22. __T__ Gene flow tends to decrease differences between populations.

23. __T__ Both large-beaked and small-beaked varieties of a single species of birds can be maintained in an area by disruptive selection.

24. __F__ Behavior that endangers an organism but benefits its close relatives is symbiosis.

25. __F__ Most species eventually give rise to new species.

Short Answer:

A population of 600 plants contains 294 *AA*, 252 *Aa*, and 54 *aa* individuals. The *AA* and *Aa* plants produce purple flowers while the *aa* plants produce white flowers. What are the frequencies of the following?

26. _____ allele *A*

27. _____ allele *a*

28. _____ genotype *AA*

29. _____ genotype *Aa*

30. _____ genotype *aa*

31. _____ purple-flowered plants

32. _____ white-flowered plants

Matching: Evolutionary mechanisms.

33. _____   selection for giraffes with longer necks

34. _____   the frequencies of white and brown guinea pigs in a population change because a large number of white animals enters the population

35. _____   a small number of organisms begins a new colony that becomes large and has gene frequencies very different from the parent population and other neighboring populations

36. _____   provides "genetic potential" to a population, upon which natural selection acts

37. _____   chance loss of genetic variation from a population due to its small size

38. _____   temporary restriction in population size due to short term unfavorable environmental conditions

39. _____   causes the evolution in males of elaborate structures and behaviors related to reproduction

40. _____   average individuals survive and reproduce best since the population is well-adapted to a stable environment

41. _____   selection involving sickle cell anemia in Africans

42. _____   selecting a mate with traits similar to your own

43. _____   selection favoring both extreme phenotypes and not favoring average individuals

44. _____   a change in a prey species forces a change in a predator species

45. _____   selection of one extreme phenotype in populations living in a rapidly changing environment

46. _____   selection for sickle cell heterozygotes in environments with malaria

47. _____   behavior that endangers an organism or reduces its reproductive success but benefits others in the population

48. _____   selection within a bird species encountering only large and small seeds

49. _____   antibiotic resistance evolves in a population

50. _____   the total of all populations of organisms that interbreed under natural conditions

Choices:

a. sexual selection

b. population bottleneck

c. gene flow

d. mutation

e. altruism

f. assortative mating

g. founders effect

h. species

i. genetic drift

j. coevolution

k. disruptive selective

l. stabilizing selection

m. directional selection

Multiple Choice: Pick the most correct choice for each question.

51. All of the following are true except
    a. differential reproduction leads to a change in allele frequency
    b. evolution occurs in populations
    c. evolution is a change in allele frequency
    d. natural selection causes genetic changes in individuals
    e. with stabilizing selection, population averages do not change.

52. All of the following meet the Hardy-Weinberg requirements for equilibrium in a population except
    a. no random mating may occur
    b. no mutations may occur
    c. no migrations may occur
    d. no natural selection may occur
    e. populations must be very large

53. A population that meets the Hardy-Weinberg requirements
    a. evolves
    b. is small and usually isolated
    c. has allele frequency in equilibrium
    d. changes genotypic distribution from generation to generation
    e. always has 75% A and 25% a allele frequencies

54. In a hypothetical population, the frequency of the dominant A allele is 80% and the frequency of the recessive allele a is 20%. What percentage of the population would you expect to be heterozygous (Aa) in genotype?
    a. 4%
    b. 16%
    c. 32%
    d. 50%
    e. 25%

55. Which of the following is more likely to occur in a small population than in a large population?
    a. gene flow
    b. immigration
    c. genetic drift
    d. nonrandom mating
    e. natural selection

56. A characteristic that better enables an organism to survive and reproduce is
    a. a mutation
    b. an adaptation
    c. a bottleneck gene
    d. a stabilizing factor

APPLYING THE CONCEPTS

These practice questions are intended to sharpen your ability to apply critical thinking and analysis to biological concepts covered in this chapter.

57. In certain parts of Africa, the frequency of sickle cell anemia among newborn African infants is about 2% and holding steady, while in the United States, the frequency of sickle cell anemia among newborn African-Americans is about 0.2% and slowly declining. Why are these populations behaving differently in evolutionary terms regarding sickle cell anemia?

_____

_____

_____

_____

58. What are the five conditions that Hardy and Weinberg stated were necessary to keep a population in genetic equilibrium?

_____

_____

_____

_____

_____

_____

59. Explain why mutation by itself is not a major force in evolution but that mutation is necessary for evolution to occur.

_____

_____

_____

_____

_____

60. Explain why population size greatly influences the potential for chance events to change allele frequencies.

_____

_____

_____

_____

_____

_____

61. In a particular apple orchard, a large population of fruit flies exists in which 36% of the flies have dark bodies and 64% have light bodies. Birds have been observed feeding on the flies, and studies have shown that a typical bird will consume 60 dark flies and 40 light flies a day. What do you think is likely to happen to the fruit fly population over time?

_____

_____

_____

_____

_____

62. In each of the following situations, one particular force that may cause evolutionary change in a population is acting. Determine which force is in effect in each case, and the probable consequence for the population:

(a) An avalanche killed all but a random few members of a mouse population;

_____

(b) Because of the planting of trees on the Great Plains by human settlers, an increase in the frequency of interbreeding between birds from two different populations of orioles occurred;

_____

(c) Because of the leakage of radioactive material from a nuclear power plant, the incidence of genetic changes in a local population of mice increased dramatically;

_____

(d) Because of the use of DDT as an insecticide in a certain country, there was an dramatic increase in the frequency of grasshoppers that showed genetic resistance to DDT.

_____

63. Name three common practices by humans that collectively have contributed to the increase in the frequency of antibiotic resistance displayed by bacteria during the past 20 years.

_____

_____

_____

_____

_____

_____

_____

_____

Use the Case Study and the Web sites for this chapter to answer the following questions.

64. Antibiotic resistance is reaching crisis proportions. Diseases which were all but eradicated only a few years ago are back and resistant to all known treatments. There are two main categories of antibiotics and four mechanisms of antibiotic resistance. What are they?

_____

_____

_____

_____

_____

_____

_____

65. Initially, a DNA mutation may lead to a functional antibiotic resistance gene. Then the gene is passed from bacterial strain to bacterial strain. Name four methods for movement of resistance genes among bacteria. Which has the greatest clinical impact?

_____

_____

_____

_____

_____

_____

_____

66. What are the two biggest contributors to the development of resistant bacterial strains? How does the use of antibiotics in livestock affect humans?

_____

_____

_____

_____

_____

_____

_____

_____

_____

_____

67. What steps can individuals take to combat the "bacterial resistance epidemic"?

_____

_____

_____

_____

_____

_____

_____

_____

_____

_____

_____

ANSWERS TO EXERCISES

1. Adaptation
2. balanced polymorphism
3. gene flow
4. Genetic drift
5. Kin selection
   altruism
6. Evolution
   coevolution
7. disruptive selection
8. Darwin

9. founder effect
10. Hardy-Weinberg principle
11. population
12. allele frequency
13. gene pool
14. Fitness
15. predation
    prey
16. false, do not change
17. true

18. false, selection
19. false, small
20. false, phenotypes
21. false, lack of change
22. true
23. true
24. false, altruism
25. false, become extinct

26. 588 (294 *AA* x 2) + 252 (252 *Aa* x 1) = 840 ÷ 1200 (600 plants x 2 alleles each) = 0.70 = 70%.

27. 108 (54 *aa* x 2) + 252 (252 *Aa* x 1) = 360 ÷ 1200 = 0.30 = 30%.

28. 294 *AA* ÷ 600 total plants = 0.49 = 49%.

29. 252 ÷ 600 = 0.42 = 42%.

30. 54 ÷ 600 = 0.09 = 9%.

31. 294 *AA* + 252 *Aa* ÷ 600 total plants = 546÷600 = 0.91 = 91%.

32. 54 *aa* ÷ 600 = 0.09 = 9%.

33. m	39. a	45. m	51. d
34. c	40. l	46. l	52. a
35. g	41. l	47. e	53. c
36. d	42. f	48. k	54. c
37. i	43. k	49. m	55. c
38. b	44. j	50. h	56. b

57. In Africa, stabilizing selection is at work with sickle cell anemia: normal individuals (*HH*) are at a disadvantage due to susceptibility to malaria and those with sickle cell anemia (*hh*) are at a disadvantage as well. Heterozygotes (*Hh*) have the highest fitness since they are somewhat resistant to malaria and do not suffer from sickle cell disease. Thus, both alleles are maintained at relatively high frequencies in those areas of Africa. In the United States, normal individuals (*HH*) have the highest fitness since there is no malaria and their children will not have sickle cell disease. Heterozygotes (*Hh*) are healthy but can produce offspring with sickle cell disease (*hh*) if two heterozygotes mate; hence, their fitness is slightly reduced. Thus, in the United States, the *H* allele is slowly rising and the *h* allele is slowly declining due to directional selection favoring normal homozygotes.

58. A population will remain in genetic equilibrium as long as: (1) there is no mutation to change the frequency of genes; (2) there is no gene flow (migration) between populations to alter the gene frequencies in the populations; (3) the population is extremely large so that no chance deviations occur in gene frequencies; (4) all mating is random so that each individual has an equal chance of mating with any other individual; and (5) there is no natural selection so that all genotypes in the population will reproduce equally well.

59. Because the rate of mutation is very low (less than 1 in 100,000 for most genes), mutation alone will not change gene frequencies quickly enough to account for observed evolutionary changes. However, mutation is important in the process of evolution by natural selection because mutation is the ultimate source of the genetic variation upon which natural selection acts in natural populations.

60. Small populations are subject to relatively large random changes in allele frequencies (genetic drift). In large populations, chance events are unlikely to significantly alter the overall gene frequencies, but in small populations chance events could reduce or eliminate alleles, greatly altering in a random way its genetic makeup. Genetic drift tends to reduce genetic variability within small populations, and genetic drift tends to increase genetic variability between or among small populations.

61. In this population of flies, the dark ones are being eaten at a higher percentage (60%) than their frequency in the population (36%) would predict. Perhaps this is due to the dark flies being more easily visible to the birds than the light flies. If body color is an inherited trait in flies, the light flies are more likely than the dark flies to live longer and produce more offspring with light bodies. So, in each generation, the frequency of light flies should increase and the frequency of flies with dark bodies should decrease. This is an example of natural selection.

62. (a) Genetic drift is acting on this mouse population, usually with the consequence of loss of genetic variation.
(b) Gene flow or migration is occurring between the two bird populations, with the result that any genetic differences between the populations will be decreased over time.
(c) The rate of mutation is increasing in this population, which will increase genetic variation among the mice but probably decrease their fitness, since mutations are usually harmful.
(d) The directional selection form of natural selection is occurring in this population, which allows the population to become better adapted to the presence of DDT in the environment.

63. Physicians are in the habit of prescribing antibiotics for practically any illness, even when they will not have a therapeutic effect, as in the cases of viral infections. Patients simply insist upon it. Also, farmers have gotten into the habit of mixing antibiotics in the food given to their livestock animals, as a way of preventing infections in their herds. They feel that this is cheaper and easier than keeping the habitats of the animals clean and free from infectious microbes. In addition, people have been convinced by advertising that it is healthier to use soaps and household cleansers that contain antibiotics, even though this contributes to the environmental pressure leading to the evolution of microbes resistant to the antibiotics, which makes this practice self-defeating and dangerous.

64 According to K. Bush at NBCi.com, there are two main categories of antibiotics, based on their effect on cells. Some antibiotics do not allow bacteria to make cell wall components, and other antibiotics cause inhibition of synthesis of nucleic acids, proteins, or other cytoplasmic molecules. The main mechanisms of bacterial resistance to these antibiotics are: the bacteria modify the antibiotic's target of activity, the bacteria pump out the antibiotic, the bacteria enzymatically destroy the antibiotic, and the bacteria do not allow the penetration of the antibiotic into the bacteria.

65. In a report for the Food Safety Network, Guelph, Ont., W. J. Powell states that in response to environmental changes, bacteria quickly duplicate and exchange genes that confer a selective advantage. The prevalent mechanisms of genetic exchange are: plasmids that are involved in the exchange of genes, by conjugation, through a direct connection formed by a sex pilus; transposons, or "jumping genes" which facilitate the movement of genes in both directions, between chromosomes and plasmids and between bacteria; transduction, which is the transfer of genes by a virus, and is considered to be clinically important, particularly among the Gram-positive bacteria; and transformation, which is the movement of small pieces of DNA from the environment into the bacterial chromosome and is no longer considered clinically significant.

66. G. G. Khachatourians from the Univ. of Saskatchewan, reported in the Canadian Medical Association Journal that antibacterial agents may be used inappropriately in both human medicine (e.g., in response to demands from patients rather than according to medical indications) and agriculture (e.g., as growth-promoting and prophylactic agents in animals). In addition to medical misuse, inappropriate use of antibiotics in the agricultural setting is a major contributor to the emergence of antibiotic-resistant bacteria. Recently, the public has focused on the use of antibiotics as growth promoters in the livestock industry. Use as a growth promoter means feeding the antibiotics to livestock at low levels for long periods. High levels of antibiotics used for short periods will kill more bacteria, including the slightly resistant ones. Low levels of antibiotics are believed to promote the survival of resistant bacteria, especially when fed to animals for long periods. Resistant bacteria can be transferred to humans through direct contact with livestock or through consuming contaminated meat or vegetable products (contaminated by manure spread as fertilizer). Some bacteria found in livestock that do not cause disease in humans can pass their resistant genes onto bacteria that do which allows the problem to spread more rapidly. Some people believe food passes resistant bacteria to humans.

67. S. B. Levy reported in Scientific American that there are a number of things that can be done. Farmers should be helped to find inexpensive alternatives to antibiotics for encouraging animal growth and protecting fruit trees. Improved hygiene could enhance livestock development. The public can wash raw fruit and vegetables thoroughly to clear off both resistant bacteria and possible antibiotic residues. When they receive prescriptions for antibiotics, they should complete the full course of therapy (to ensure that all the pathogenic bacteria die) and should not "save" any pills for later use. Consumers also should refrain from demanding antibiotics for colds and other viral infections and might consider seeking non-antibiotic therapies for minor conditions. They can continue to put antibiotic ointments on small cuts, but they should think twice about routinely using hand lotions and a proliferation of other products now imbued with antibacterial agents. New laboratory findings indicate that certain of the bacteria-fighting chemicals being incorporated into consumer products can select for bacteria resistant both to the antibacterial preparations and to antibiotic drugs. Physicians can take some immediate steps to minimize any resistance ensuing from required uses of antibiotics. When possible, they should try to identify the causative pathogen before beginning therapy, so they can prescribe an antibiotic targeted specifically to that microbe instead of having to choose a broad-spectrum product. Washing hands after seeing each patient is a major and obvious, but too often overlooked, precaution. To avoid spreading multi-drug-resistant infections between hospitalized patients, hospitals place the affected patients in separate rooms, where they are seen by gloved and gowned health workers and visitors. This practice should continue.

Chapter 16: The Origin of Species

OVERVIEW

In this chapter, you will learn about species and speciation. After defining biological species, the authors describe the general mechanisms of allopatric and sympatric speciation, and premating and postmating mechanisms for maintaining reproductive isolation. Within the last ten years, three new species of mammals have been found in the isolated forested mountains of Vietnam: the saola, a large (200 pound) hoofed, horned quadruped, called the giant muntjac (the "barking deer") and a rabbit with short ears and a brown-striped coat. How do species like these arise?

1) What Is a Species?

Biologists define **species** (Latin for "appearance") as groups of actually or potentially interbreeding natural populations, which are reproductively isolated from other such groups. So, two organisms are the same species if they interbreed in nature and have normal, vigorous, fertile offspring. This definition does not apply to asexually reproducing species, however.

2) How Do New Species Form?

Speciation is the process by which new species form. Speciation depends on two factors: (1) isolation (gene flow between diverging populations must be small or nonexistent); and (2) genetic divergence of two populations (they must evolve large genetic differences so that they cannot interbreed or produce normal offspring).

Hypothetical mechanisms of speciation are **allopatric speciation**, in which the populations are geographically separated from each other and thus isolated from gene flow, and **sympatric speciation**, in which the populations share the same area but are isolated from gene flow. There is some debate as to whether genetic drift or natural selection normally plays the major role in allopatric speciation, but geographic isolation is involved in most cases of speciation in animals.

Allopatric speciation can occur in populations that are physically separated by an impassible barrier. If two or more populations become geographically isolated for any reason, little or no migration (gene flow) can occur between them, and the populations may accumulate genetic differences. Alternately, genetic differences may arise if one or more of the separated populations is small enough for genetic drift to occur (the founder effect). In either case, genetic differences between the separated populations may eventually become large enough to make interbreeding impossible.

Sympatric speciation can occur in populations that live in the same area. With sympatric speciation, two likely mechanisms can reduce gene flow. Ecological isolation occurs when a geographical area contains two distinct types of habitat and different members of the same species begin to specialize in one habitat or the other. This reduces gene flow. Chromosomal aberrations, specifically changes in chromosome number, can cause immediate reproductive isolation of a population. A common speciation mechanism in plants is **polyploidy**, the acquisition of multiple copies of each chromosome. If a fertilized egg duplicates its chromosomes but does not divide into daughter cells, the chromosome number can rise from diploid ($2n$, or pairs of chromosomes) to become tetraploid ($4n$, or 4 doses of each chromosome). Healthy $4n$ plants produce $2n$ gametes. Tetraploids breed successfully with other tetraploids. But if $4n$ breeds with $2n$, sterile $3n$ (triploid) offspring result. So, tetraploid plants and their diploid parents form distinct reproductive communities that cannot interbreed successfully. Speciation by polyploidy is common in plants (which can reproduce by self-fertilization and asexually), but rare in animals.

Change over time within a species can cause apparent "speciation" in the fossil record. Speciation events lead to forking branches in the evolutionary tree of life, as one species splits into two species. However, changes within a species over time also occur. Over time, the members of a species may come to be very different from their distant ancestors, even if no speciation occurs. Since fossils cannot breed, it is difficult to determine whether they were reproductively isolated from other fossils. Consequently, paleontologists typically assign extinct organisms to species without reference to the biological-species concept. During **adaptive radiation**, one species gives rise to many in a relatively short time. This occurs when populations of a single species invade a variety of new habitats and evolve in response to the differing environmental pressures.

3) How Is Reproductive Isolation Between Species Maintained?

Speciation occurs through the evolution of mechanisms that prevent interbreeding. Genetic divergence during a period of isolation is necessary for new species to arise, but speciation will occur only if mechanisms ensuring **reproductive isolation** also develop. **Isolating mechanisms** are structural and/or behavioral modifications that prevent interbreeding.

Premating isolating mechanisms include geographical isolation, ecological isolation, temporal isolation, behavioral isolation, and mechanical incompatibility. **Geographical isolation** prevents members of different species from meeting each other and usually is considered to be a mechanism that *allows* new species to form. **Ecological isolation** involves populations having different resource requirements involving the use of different local habitats within the same general area; thus, they are not likely to meet during the mating season. **Temporal isolation** occurs between species that breed at different times of the year. Even if two species occupy similar habitats, they cannot interbreed if they have different breeding seasons. **Behavioral isolation** involves species with different courtship rituals. Courtship rituals involve recognition and evaluation signals between males and females and also aid in distinguishing among species. Colors and songs of birds, frog croaks, cricket chirping patterns, and firefly flashing colors and frequencies are examples of signals used in courtship rituals. **Mechanical incompatibility** occurs when physical barriers between species prevent fertilization. For example, male and female sex organs of different species may not fit together properly for sperm transfer in animals, or different flower sizes or structures of different species may prevent pollen transfer in plants.

Postmating isolating mechanisms prevent production of vigorous, fertile offspring, and include gametic incompatibility, hybrid inviability, and hybrid inferiority. **Gametic incompatibility** occurs when sperm from one species are unable to fertilize eggs of another. **Hybrid inviability** occurs if hybrid offspring survive poorly. Hybrids may die during development or display behaviors that are mixtures of the two parental types and be unable to attract mates. **Hybrid infertility** occurs if hybrid offspring are unable to produce normal sperm or eggs. Most animal hybrids such as mules (horse mating with donkey) or ligers (lion mating with tiger) are sterile since their chromosomes do not pair properly during meiosis. Crosses between tetraploid ($4n$) and diploid ($2n$) plant species usually result in sterile triploid ($3n$) offspring.

4) What Causes Extinction?

Natural selection may lead to **extinction**, which is the death of all members of a species. Two characteristics predispose a species to extinction when the environment changes: extremely limited ranges (localized distribution), and very narrow structural or behavioral requirements (overspecialization). Wide-ranging species do not succumb to local environmental catastrophes, and species which feed on a variety of foods do not die off if one food supply vanishes. In addition, interactions with other organisms may drive a species to extinction, as happened when the Panama land bridge allowed North American species to migrate into South America, causing the extinction of most native South American species due to competition. Finally, habitat change and habitat destruction are the leading causes of extinction. Mass extinctions are disappearances of many varied species in a short time over a large area and may be caused by traumatic environmental events such as the effects of the impact of a large meteorite.

Case study revisited. In the Vietnamese mountains, the area covered by forests shrunk dramatically during the ice ages, becoming tiny "islands" of forest isolated from each other. In these isolated forests, allopatric speciation may have produced the saola, giant muntjac, striped rabbit, and other unique species. Heavy deforestation by the Vietnamese for economic development may possibly endanger these already rare animals, but the Vietnamese government has established a number of national parks and nature preserves in key areas. What benefit for humans is the search and discovery of new species?

KEY TERMS AND CONCEPTS

Fill-In: From the following list of key terms, fill in the blanks in the following story.

adaptive radiation	hybrid infertility	reproductive isolation
allopatric	mechanical incompatibility	speciation
behavioral isolation	polyploidy	species
ecological isolation	postmating isolating mechanisms	temporal isolation
geographic isolation	premating isolating mechanisms	

1. Michael took his son Zachary to the zoo. Zachary is taking a high school biology class and was loaded with questions for his dad. "How do we know that African lions and Asian tigers are different _SPECIES_?" he asked. Mike replied: "Zach, they have different physical characteristics, but more importantly they show _REPRODUCTIVE ISOLATION_ since they don't try to interbreed in nature or even in zoos."

2. This led to another question: "Dad, since both lions and tigers are mammals and cats, how could _SPECIATION_ have occurred in such closely related organisms?" "Well, Zach, when a single group of organisms, like mammals, gives rise to many closely related species, it's called _ADAPTIVE RADIATION_. In the case of lions and tigers, first there was _GEOGRAPHIC_ isolation since they evolved in different parts of the world, leading to _ALLOPATRIC_ speciation. And even if they lived for a time in the same general area, there would have been _ECOLOGICAL_ isolation since they would occupy different habitats there. Zach, these are examples of _PREMATING_ isolating mechanisms since they prevent lions and tigers from mating."

3. They entered the amphibian and reptile house, where many different species of frogs were living. "Dad, the wood frogs and green frogs breed in the same areas each year, but are different species. How can this be?" "Well Zach, they don't interbreed because wood frogs mate during early spring and green frogs mate during late spring, a situation called _TEMPORAL_ isolation. Also, the males of the two species have different sounding croaks and females are attracted only to males of their own kind. This is an example of _BEHAVIORAL_ isolation because they have different courtship rituals."

4. When they looked at the lizards, Zach said: "Dad, I just thought of something weird. Suppose a large lizard of one species tried to mate with a small lizard of another species. Could they succeed?" "Probably not," replied Mike as he chuckled at the thought of such an unlikely liaison, "since their sex organs couldn't fit together properly. This is a situation called _MECHANICAL INCOMPATIBILITY_ isolation."

5. Moving to the building housing hoofed animals, Zach noticed the horses, donkeys and mules. "Dad, since horses and donkeys produce mules when farmers force them to mate, aren't they all members of the same species?" "Son, horses and donkeys would never interbreed under natural conditions, and besides, the mules they produce are sterile, an example of _HYBRID INFERTILITY_. This is an example of a _POSTMATING_ isolating mechanism since it acts after mating between different species takes place."

6. "Dad, this is all way cool, but now I have a really important question. When do we eat?" "Soon, just eat your banana right now. By the way, Zach, bananas don't have seeds and are sterile because they have three copies of each chromosome, a condition called _polyploidy_." "Banana, smanana!" exclaimed Zach. "I want a burger and fries!"

Key Terms and Definitions

adaptive radiation: the rise of many new species in a relatively short time as a result of a single species that invades different habitats and evolves under different environmental pressures in those habitats.

allopatric speciation (al-ō-pat´-rik): speciation that occurs when two populations are separated by a physical barrier that prevents gene flow between them (geographical isolation).

behavioral isolation: the lack of mating between species of animals that differ substantially in courtship and mating rituals.

ecological isolation: the lack of mating between organisms belonging to different populations that occupy distinct habitats within the same general area.

extinction: the death of all members of a species.

gametic incompatibility: the inability of sperm from one species to fertilize eggs of another species.

geographical isolation: the separation of two populations by a physical barrier.

hybrid infertility: reduced fertility (typically, complete sterility) in the hybrid offspring of two species.

hybrid inviability: the failure of a hybrid offspring of two species to survive to maturity.

isolating mechanism: a morphological, physiological, behavioral, or ecological difference that prevents members of two species from interbreeding.

mechanical incompatibility: the inability of male and female organisms to exchange gametes, normally because their reproductive structures are incompatible.

polyploidy (pahl´-ē-ploid-ē): having more than two homologous chromosomes of each type.

postmating isolating mechanism: any structure, physiological function, or developmental abnormality that prevents organisms of two different populations, once mating has occurred, from producing vigorous, fertile offspring.

premating isolating mechanism: any structure, physiological function, or behavior that prevents organisms of two different populations from exchanging gametes.

reproductive isolation: the failure of organisms of one population to breed successfully with members of another; may be due to premating or postmating isolating mechanisms.

speciation: the process of species formation, in which a single species splits into two or more species.

species (spē´-sēs): the basic unit of taxonomic classification, consisting of a population or series of populations of closely related and similar organisms. In sexually reproducing organisms, a species can be defined as a population or series of populations of organisms that interbreed freely with one another under natural conditions but that do not interbreed with members of other species.

sympatric speciation (sim-pat´-rik): speciation that occurs in populations that are not physically divided; normally due to ecological isolation or chromosomal aberrations (such as polyploidy).

temporal isolation: the inability of organisms to mate if they have significantly different breeding seasons.

THINKING THROUGH THE CONCEPTS

True or False: Determine if the statement given is true or false. If it is false, change the <u>underlined</u> word(s) so that the statement reads true.

7. ___F___   Temporal isolation is isolation by <u>distance</u>.

8. ___T___   Polyploidy is most common in <u>plants</u>.

9. ___F___   Mechanical incompatibility is a <u>postmating</u> isolating mechanism.

10. ___T___   The most valid way to determine whether two organisms belong to different species is to look at <u>mating behavior</u>.

11. ___F___   Speciation depends on <u>lack of isolation</u> between populations.

12. __T__   Populations physically separated are <u>allopatric</u>.

13. __T__   Sympatric speciation can occur if <u>chromosome aberrations</u> occur.

14. __T__   Geographic isolation is a <u>premating</u> isolating mechanism.

15. __T__   Hybrid infertility is a <u>postmating</u> isolating mechanism.

16. __T__   <u>Adaptive radiation</u> may occur when a species encounters a wide variety of unoccupied habitats.

Matching: Speciation.

17. __C__   causes "instant speciation"

18. __B__   causes most animal speciation

19. __B__   populations are separated by a physical barrier

20. __C__   causes much plant speciation but little animal speciation

21. __A__   reduces genetic differences between populations, retarding animal speciation

22. __D__   one species gives rise to many new species in a short time

23. __C__   acquisition of more than two copies of each chromosome in the nucleus of cells

Choices:

a. gene flow
b. geographical isolation
c. polyploidy
d. adaptive radiation

Matching: Maintaining reproductive isolation.

24. __A__   Interbreeding does not occur in nature between British peppered moths and Canadian peppered moths.

25. __D__   Interbreeding does not occur between closely related species of fruit flies with slightly different courtship rituals.

26. __H__   When horses and donkeys are forced to interbreed, sterile mules are produced.

27. __B__   Leopard frogs and pickerel frogs that live in the same area with similar mating seasons do not interbreed because one species breeds in swamps and the other breeds in clear lakes.

28. __E__   Closely related species of katydid insects do not interbreed because the male and female sex organs cannot fit together properly to allow sperm transfer to occur.

29. __C__   Wood frogs and green frogs breed in the same lakes but do not interbreed because one species breeds in April while the other species breeds in May.

30. __F__   Two closely related species of fruit flies sometimes mate but the female's immune system kills the male's sperm as though it were a foreign invading microbe.

Choices:

a. geographical isolation
b. ecological isolation
c. temporal isolation
d. behavioral isolation
e. mechanical isolation
f. gamete incompatibility
g. hybrid inviability
h. hybrid infertility

Multiple Choice: Pick the most correct choice for each question.

31. Among animals in particular, which of the following is the most common event necessary for speciation to happen?
    a. geographic isolation
    b. adaptive radiation
    c. reproductive isolation
    d. ecological isolation
    e. migration

32. In plants, a common method of sympatric speciation is
    a. ecological isolation
    b. geographical isolation
    c. adaptive radiation
    d. polyploidy
    e. nondisjunction

33. Two species of pines releasing pollen at separate times in the same habitat is an example of
    a. geographical isolation
    b. ecological isolation
    c. temporal isolation
    d. behavioral isolation
    e. mechanical incompatibility

34. The Everglades kite is almost extinct because of
    a. fire in the Everglades
    b. saltwater intrusion in the Everglades
    c. invasion of the walking catfish
    d. disappearance of the apple snail
    e. industrial melanism

APPLYING THE CONCEPTS

These practice questions are intended to sharpen your ability to apply critical thinking and analysis to biological concepts covered in this chapter.

35. Name and describe two genetic models for speciation.

_____

_____

_____

_____

_____

_____

_____

36. Name the two general conditions necessary for speciation in animals.

_____

_____

_____

_____

_____

_____

_____

37. Briefly describe three postmating reproductive isolating mechanisms.

_____

_____

_____

_____

_____

_____

Answer the following question, based on the figure below.

(a)   (b)

38. Compare parts (a) and (b) in the figure and comment on which evolutionary pattern is the result of speciation and which is not. Please briefly explain your answer.

_____

_____

_____

_____

_____

39. Dogs (*Canis familiaris*) and coyotes (*Canis latrans*) are given different species names by biologists. Interestingly, dogs and coyotes will eagerly mate with each other in captivity and produce perfectly healthy, fertile offspring. What sorts of criteria do you think biologists have used to determine that dogs and coyotes are different species? Do you personally think that dogs and coyotes should be considered different species? Do you think that dogs and coyotes are closely related types of animals?

_____

_____

_____

_____

_____

40. The drug colchicine disrupts formation of the mitotic spindle and prevents cell division after the chromosomes have doubled before the start of meiosis. Describe how you might use colchicine to produce a new polyploid species of plant.

_____

_____

_____

_____

_____

_____

41. Triploid varieties of fruits and vegetables produce larger fruits that are seedless. Explain why these varieties are seedless, and what is the consequence of not producing seeds?

_____

_____

_____

_____

_____

_____

42. Design a study to determine whether the Vietnam muntjac ("barking deer") is a species different than the typical deer in the US.

_____

_____

_____

_____

_____

Use the Case Study and the Web sites for this chapter to answer the following questions.

43. The Annanite mountains of Vietnam/Laos are the home of a recently discovered species: the "saola." How was it discovered? What is its normal habitat? What is its survival status?

_____

_____

_____

_____

_____

_____

44. The Annanite mountains of Vietnam/Laos are the home of a recently discovered species: the "giant muntjac." How was it discovered? What is its normal habitat? What is its survival status?

_____

_____

_____

_____

_____

_____

45. The Annanite mountains of Vietnam/Laos are also the home of a recently discovered species: a strange striped rabbit. How was it discovered? What is its normal habitat? What is its survival status?

_____

_____

_____

_____

_____

_____

_____

_____

_____

_____

_____

ANSWERS TO EXERCISES

1. species
   reproductive
2. speciation
   adaptive radiation
   geographic
   allopatric
   ecological
   premating
3. temporal
   behavioral
4. mechanical incompatibility
5. hybrid infertility
   postmating
6. polyploidy

7. false, time
8. true
9. false, premating
10. true
11. false, isolation
12. true
13. true
14. true
15. true
16. true
17. c
18. b
19. b
20. c

21. a
22. d
23. c
24. a
25. d
26. h
27. b
28. e
29. c
30. f
31. a
32. d
33. c
34. d

35. Two hypothetical mechanisms of speciation are allopatric speciation, in which the populations are geographically separated from each other and thus isolated from gene flow, and sympatric speciation, in which the populations share the same area but are isolated from gene flow due to different behaviors.

36. Speciation is the process by which new species form. Speciation in animals generally depends on two factors: isolation, since gene flow between diverging populations must be small or non-existent; and the genetic divergence of two populations, since they must evolve large genetic differences so that they cannot interbreed or produce normal offspring.

37. Postmating isolating mechanisms prevent production of vigorous, fertile offspring, and include gametic incompatibility, hybrid inviability, and hybrid inferiority. Gametic incompatibility occurs when sperm from one species are unable to fertilize eggs of another. Hybrid inviability occurs if hybrid offspring survive poorly. Hybrids may die during development, or display behaviors that are mixtures of the two parental types and be unable to attract mates. Hybrid infertility occurs if hybrid offspring are unable to produce normal sperm or eggs due to problems in chromosome pairing during meiosis.

38. Speciation has occurred twice in part (a) to produce three contemporary species at the top of the figure. The branched arrows indicate the speciation events which resulted in reproductive isolation between the new species formed. In part (b), a large amount of phenotypic change has occurred over an extended period of time, but no reproductive isolation has occurred so that the entire lineage has remained as a single species.

39. Since dogs and coyotes successfully interbreed in captivity, they must be closely related, having diverged from common ancestors in the not too distant past. Biologists consider them different species because they hardly ever interbreed under natural conditions in the wild. Also, recent molecular genetics studies have shown that dogs are much more closely related to wolves than they are to coyotes.

40. Let's say that the plant species has 20 pairs of chromosomes in its diploid cells and 20 chromosomes in its haploid cells. If colchicine is used on cells undergoing meiosis, instead of the resulting spores having 20 chromosomes (haploid), they will have 40 chromosomes (diploid). If these diploid spores develop into diploid gametes (20 pairs of chromosomes), they could mate with normal haploid gametes (20 chromosomes) and produce triploid organisms with each of the 20 chromosomes being present three times in each cell of the offspring. If the triploid organism thus produced can be propagated asexually (by budding or by planting cuttings that could take root and grow into complete plants), a new polyploid species of plant will have been produced.

41. Cultivated bananas are triploid organisms lacking seeds. Their wild diploid relatives have hard seeds within the fruits. The triploid bananas are seedless because meiosis cannot occur normally in triploid organisms, meaning that no normal gametes can be produced. During normal meiosis, pairs of homologous chromosomes pair up and segregate during the first meiotic division, so that each gamete receives an exact haploid number of chromosomes. In triploid organisms, there are three copies of each chromosome type so that during meiosis two of them pair up and the other remains unpaired. When segregation occurs, the extra chromosomes move randomly, some to one of the poles and some to the other pole. This leads to genetically unbalanced gametes, each with a random combination of one and two doses of various chromosomes. Such gametes are usually non-functional, so that no seeds are produced.

42. The most reliable criterion to determine whether a male and a female belong to the same species or not is to give them an opportunity to mate under normal conditions and see whether they produce normal, healthy, fertile offspring. Such matings between horses and donkeys produce sterile hybrids called mules, a sign that horses and donkeys belong to different species. Determining the chromosome numbers of the male and female also can provide a clue: if the numbers are different, chances are they belong to different species. So, in the case of a Vietnam muntjac and a typical U.S. deer, observation

of their mating behaviors towards each other, whether they can together produce normal, healthy, fertile offspring, and whether their chromosome numbers are identical or not, are valid ways to determine whether they represent one or two species.

43. According to P. Massicot of animalinfo.org, the saola is a forest-dwelling ox weighing about 100 kg (220 lb). All known locations for the species are mountainous with steep river valleys, covered by evergreen or semi-deciduous forests between 300–1800 meters (1000–6000 feet), with low human disturbance. Current knowledge indicates that the saola prefers the edge areas of wet lowland evergreen forest habitats and evergreen montane forests. Villagers say that the ox eats the leaves of fig trees and other bushes along riverbanks. The saola is said to travel in small groups of 2–3 animals, rarely up to 6–7 animals. The most serious threats to the saola are hunting and loss of forest habitat due to logging and conversion to farmland. In May 1992, the discovery of three pairs of horns in the only remaining area of pristine forest in northern Vietnam led to the first documentation by Western scientists of a new species of ox. Saola are shot for their meat. Because of their scarcity, local people place a higher value on saola than on more common species. Hunters were also aware of the intense interest from the world's scientific community, increasing their motivation to capture live specimens. Among local communities, there was some awareness that forest resources were declining, but the general perception was that resources were still plentiful. Moreover, the hunters showed little or no understanding of the principles of resource management of the species that they hunted. Everything that was encountered during hunting trips was shot or captured, if possible. Although the saola was valued by hunters, there seemed to be little concern about its decline or local extinction. There are an estimated several hundred saola left.

44. According to an Agence France-Presse article reprinted by forests.org, this mammal is a small stag belonging to the muntjac (or barking deer) family. It has black fur and weighs about 15 kilos (33 pounds), only half the weight of the other muntjac species in Vietnam. The scientists were not able to observe the animal directly and made the discovery from skulls found in villages in the area. They established it was a new species after analyzing the animal's tissue while further information was provided by Vietnamese hunters. This muntjac species lives in forests between 400 and 1,000 meters (1320 and 3,300 feet) in altitude. Its small size allows the animal to move easily through the dense vegetation.

45. Researchers at the University of East Anglia have identified a new and highly distinctive species of rabbit — striped and with a red rump — found in the Annamite Mountains in Laos and Vietnam. Three of the striped rabbits were found, freshly hunted, in a meat market in Laos by a British biologist. Samples from these animals were sent for identification to one of the world experts on rabbits. Researchers extracted DNA from the Laos rabbits and also from 100-year-old museum specimens of the only other known species of striped rabbit, the critically endangered Sumatran rabbit, endemic to mountain forests in Sumatra. The Annamite rabbit, which has since been seen in a nature reserve in neighboring Vietnam, closely resembles the Sumatran rabbit, both possessing black/dark brown stripes on the face and back, a red rump and short tails and ears. However, despite the striking external similarity, genetic analysis reveals significant differences between the two. The genetic data suggests that these two species may have diverged about eight million years ago. The Sumatran rabbit had been feared extinct as there had only been one sighting since 1916. But, in early 1998, a team from Fauna and Flora International captured automatic camera-trap photos of the species in Mt. Kerinci National Park, in Sumatra. Only 15 specimens of the Sumatran rabbit had ever been collected, all at the turn of the century, and the species remains critically endangered due to the destruction of its mountain forest habitat.

Chapter 17: The History of Life on Earth

OVERVIEW

In this chapter, you will learn about the history of life on this planet. The authors trace the evolution of life from the beginning of the universe through prebiotic evolution, spontaneous generation of the first living prokaryotic cells, metabolic evolution, the rise of eukaryotes and multicellularity, the invasion of land, and finally human evolution. Europa, one of the 16 moons of Jupiter, is among the celestial objects deemed most likely to hold extraterrestrial life. Its surface is frozen ice littered with huge cracks, humps, and crevices, suggesting that the surface ice is floating on liquid water, warmed perhaps by heat from the moon's rocky core. Water is a key component of the conditions under which life can arise, as it did on Earth, and might have on Europa.

1) How Did Life Begin?

In the 1600s, biologists thought life arose through **spontaneous generation** from nonliving matter and unrelated life forms (like trees giving rise to fish and birds). This was not substantially refuted until the mid-1800s by Louis Pasteur. In the 1930s, Oparin and Haldane proposed **prebiotic** (or chemical) **evolution**: evolution before life began, when the atmosphere of the Earth contained hydrogen, methane, carbon dioxide, nitrogen, hydrogen sulfide, hydrochloric acid, water vapor, and ammonia gases but no free oxygen gas. In 1953, Miller and Urey did lab experiments to show that under prebiotic atmospheric conditions, organic molecules like amino acids, nucleotides, and ATP could be produced. Since no organisms or oxygen gas existed in prebiotic times, organic molecules could accumulate in shaded pools not subjected to ultraviolet solar radiation, forming a "primordial soup." So, organic molecules can be synthesized spontaneously under prebiotic conditions, and prebiotic conditions would allow organic molecules to accumulate.

RNA may have been the first self-reproducing molecule. In the 1980s, Cech and Altman discovered that certain small RNA molecules called **ribozymes** act as enzymes to make more RNA. Possibly, molecular evolution began in the primordial soup when ribozymes began to copy themselves and make other molecules. Cellular life requires self-replicating molecules enclosed within membranes. If water containing proteins and lipids is agitated, hollow membrane-like balls called **microspheres** form. If microspheres formed around ribozymes, nonliving **protocells** formed, possibly evolving into living cells.

None of this has been proven beyond doubt, but it is suggestive of what probably happened.

2) What Were the Earliest Organisms Like?

Life began about 3.5 billion years ago, according to fossils and chemical data. The first cells were prokaryotic and obtained nutrients and energy by absorbing organic molecules from the environment and breaking down these molecules anaerobically (no oxygen gas was present) and gaining a little energy. These cells were primitive anaerobic prokaryotes (bacteria). Eventually, some cells evolved photo-synthesis, the ability to use solar energy to make their own complex, energy-rich molecules from water and carbon dioxide. The cyanobacteria evolved, and through photosynthesis, released oxygen gas into the atmosphere. About 2.2 billion years ago, free oxygen gas began to accumulate, and atmospheric levels of oxygen gas increased steadily until reaching a stable level about 1.5 billion years ago. This, in turn, allowed the evolution of microbes capable of aerobic respiration, breaking down organic molecules completely using oxygen gas to release significant amounts of chemical energy.

Eukaryotes developed membrane-enclosed organelles and a nucleus. Predation soon evolved, with larger prokaryotic cells engulfing bacteria. About 1.7 billion years ago, eukaryotic cells having membrane-bound nuclei and cytoplasmic organelles evolved from predatory bacteria. According to Lynn Margulis' **endosymbiont hypothesis**, primitive cells acquired the precursors of mitochondria and chloroplasts by engulfing certain types of bacteria and forming a symbiotic (mutually supportive) relationship with them. The fact that these organelles retain their own bacterial-like DNA supports the hypothesis. Cilia, flagella, centrioles, and microtubules may have evolved from a symbiosis between spiral-shaped bacteria and a primitive eukaryotic cell. The origin of the nucleus is more obscure.

3) How Did Multicellularity Arise?

Increased size was an advantage, but large unicellular organisms could not survive due to the slowness of diffusion. Multicellular organisms evolved about one billion years ago. Multicellular algae developed specialized rootlike and leaflike structures that facilitated their invasion of diverse habitats.

Multicellular animals developed specializations to help them capture prey and feed. They evolved muscular movement so that predators could chase prey and prey could escape. Hydrostatic skeletons (water-filled tubes) for locomotion evolved in some worms and then external skeletons evolved in the arthropods. Finally, internal skeletons in the vertebrates developed. Additionally, greater sensory capabilities and more-sophisticated nervous systems evolved.

4) How Did Life Invade the Land?

Terrestrial organisms must find adequate water, protect their gametes from drying out, and resist the effects of gravity without a buoyant watery environment. The plants that first colonized the land, however, had ample sunlight, rich nutrient sources in the soil, and no predators. Some plants developed specialized structures that adapted them to dry land. Waterproof coatings on the aboveground parts reduced water loss, rootlike structures were anchored in the soil, mining water and nutrients, and extra-thick cell walls enabled stems to stand erect.

Primitive land plants (mosses and ferns) retained swimming sperm and required water to reproduce, but the **conifers** (cone-bearing plants) retained their eggs internally and encased sperm within pollen grains blown around by the wind, allowing the conifers to flourish in dry habitats. Landing on a female cone near the egg, the pollen released sperm cells directly into living tissue, eliminating the need for a surface film of water. As the moist climate dried up, conifers flourished. Flowering plants enticed animals (mainly insects) to carry pollen from flower to flower, thus wasting much less pollen than conifers. Flowering plants also reproduced more rapidly and grew more quickly than conifers.

Some animals evolved specialized structures that adapted them to life on dry land. Some animals were **preadapted** for land life: they already had structures suitable for life on land, such as **exoskeletons** in the arthropods. Amphibians evolved from lobefin fishes. Lobefins had two preadaptations for land: stout fleshy fins for crawling and a pouch off their digestive tract that acted as a primitive lung. With these improvements, lobefins evolved into amphibians. But amphibians still depended on water for egg laying and to keep their skin moist for gas exchange.

Reptiles, which evolved from amphibians, developed several adaptations to dry land: internal fertilization; shelled, waterproof eggs containing a supply of water and food; scaly, waterproof skin; and improved lungs. Two groups of smaller reptiles developed insulation to retain body heat. Reptiles gave rise to both birds (with feathers) and mammals (with hair). Unlike birds which lay eggs, mammals evolved live birth and mammary (milk-producing) glands to feed the young. When the reptilian dinosaurs became extinct 65 million years ago, the mammals adaptively radiated out into the vast array of modern forms.

5) What Role Has Extinction Played in the History of Life?

In the history of life, nothing lasts forever. The upward trend in species diversity and the slow, steady turnover of species have been interrupted by episodes of **mass extinction**. These episodes are characterized by the disappearances of many varied species in a relatively short time over a large area. The worst episode occurred about 245 million years ago at the end of the Permian period, when 90% of the world's species became extinct.

Mass extinctions may be caused by climate changes. Organisms that are adapted for survival under one set of environmental conditions may be unable to survive under a drastically different set of conditions. One cause of climate change is plate tectonics (movement of the Earth's plates resulting in changes in latitude). More sudden events also play a role in mass extinctions, such as massive volcanic eruptions and the effects of the impact of a large meteorite, causing an "impact winter" that severely lowered the temperature by blocking out sunlight for a period of years.

6) How Did Humans Evolve?

Regarding human evolution, paleontologists disagree about the interpretation of the skimpy fossil evidence and many ideas may have to be revised as new fossils are found. **Primates** are lemurs, monkeys, apes, and humans. The most likely primate ancestors were probably insect-eating tree shrews, small nocturnal mammals whose fossils are 80 million years old. Early primates ate fruits and leaves and evolved several adaptations for life in the trees. Primate evolution has been linked to: (1) grasping hands for powerful (club swinging) and precise (writing, sewing) manipulations; (2) binocular vision (forward-looking eyes with overlapping fields of vision) for accurate depth perception and color vision for finding ripe fruit; and (3) a large brain with high intelligence that facilitated hand-eye coordination and complex social interactions.

Between 20-30 million years ago, in the tropical forests of Africa, the *dryopithecine* primates diverged from the monkey line. *Dryopithecines* evolved into the **hominids** (humans) and pongids (great apes) some 5 to 8 million years ago, perhaps from a hominid with ape-like traits like *Ardipithecus ramidus*. Early hominids, called the *australopithecines* (southern apes of Africa) which lived about 4 million years ago, could stand and walk upright. Upright posture freed the hands to carry weapons and manipulate tools.

The genus *Homo* diverged from the australopithecines about 2.5 million years ago. The evolution of *Homo* was accompanied by advances in tool technology. *Homo neanderthalensis* (Neanderthals) appeared about 150,000 years ago and had large brains and ritualistic behaviors but did not lead to modern humans based on DNA analysis. Modern humans (*Homo sapiens*) evolved about 150,000 years ago in Africa, perhaps from *Homo heidelbergensis*, which had developed earlier from *Homo ergaster*. *Homo sapiens* spread into the Near East, Europe, and Asia, supplanting all other hominids. The evolution of Homo was accompanied by advances in tool technology. Cro-Magnons (humans from Europe and the Middle East, beginning about 90,000 years ago) had domed heads, smooth brows, and prominent chins, just like us. Humans and Neanderthals coexisted in Europe until humans overran and displaced the Neanderthals.

The evolution of human behavior is highly speculative. Human **cultural evolution** (learned behavior passed down from previous generations) now far out-paces biological evolution. There have been three major surges of human population growth, each associated with a cultural revolution: (1) development of tools (ending 10,000 years ago, with 5 million humans worldwide); (2) agricultural revolution (the past 8000 years, with 750 million humans worldwide in 1750); and (3) industrial revolution (the past 250 years, with over 6 billion humans worldwide at the present time).

Case study revisited. If liquid water does exist under the frozen ice of Europa, what sort of life might exist there? On Earth, a similar situation exists with Lake Vostok, a huge lake the size of Lake Ontario buried more than two miles beneath the Antarctic ice pack. Efforts are underway to devise methods to drill into Lake Vostok without the risk of cross contamination of life forms. In addition to water, what other conditions would be necessary for life to have arisen on Europa?

KEY TERMS AND CONCEPTS

Fill-In: From the following list of terms, fill in the following statements.

aquatic	endosymbiont hypothesis	preadapted
conifers	hominids	ribozymes
Cro-Magnon	microspheres	spontaneous generation
cultural evolution	Neanderthals	terrestrial

1. In the 1600s, biologists thought life arose through _____ from nonliving matter.

2. Certain small RNA molecules called _____ act as enzymes to make more RNA. Possibly, molecular evolution began in the primordial soup when they began to copy themselves and make other molecules.

3. If water containing proteins and lipids is agitated, hollow membrane-like balls called _____ form.

4. According to Lynn Margulis' _____, primitive cells acquired the precursors of mitochondria and chloroplasts by engulfing certain types of bacteria and forming a symbiotic (mutually supportive) relationship with them.

5. _____ organisms must find adequate water, protect their gametes from drying out, and resist the effects of gravity without a buoyant watery environment. Primitive land plants (mosses and ferns) have retained swimming sperm and require water to reproduce, but the _____ retain their eggs internally within cones and encased sperm within pollen grains blown around by the wind, allowing them to flourish in dry habitats. However, some animals were _____ for land life since they already had structures suitable for life on land, such as exoskeletons in the arthropods.

6. Dryopithecine primates evolved into the _____ (humans) and the pongids (great apes) about 5 to 8 million years ago. _____ appeared about 150,000 years ago and had large brains and ritualistic behaviors, but did not lead to modern humans based on DNA analysis.

7. Human _____ (learned behavior passed down from previous generations) now far out-paces biological evolution.

Key Terms and Definitions

conifer (kon´-eh-fer): a member of a class of tracheophytes (Coniferophyta) that reproduces by means of seeds formed inside cones and that retains its leaves throughout the year.

cultural evolution: changes in the behavior of a population of animals, especially humans, by learning behaviors acquired by members of previous generations.

endosymbiont hypothesis: the hypothesis that certain organelles, especially chloroplasts and mitochondria, arose as mutually beneficial associations between the ancestors of eukaryotic cells and captured bacteria that lived within the cytoplasm of the pre-eukaryotic cell.

exoskeleton (ex´-ō-skel´-uh-tun): a rigid external skeleton that supports the body, protects the internal organs, and has flexible joints that allow for movement.

hominid: a human or a prehistoric relative of humans, beginning with the Australopithecines, whose fossils date back at least 4.4 million years.

mass extinction: the extinction of an extraordinarily large number of species in a short period of geologic time. Mass extinctions have recurred periodically throughout the history of life.

microsphere: a small, hollow sphere formed from proteins or proteins complexed with other compounds.

preadaptation: a feature evolved under one set of environmental conditions that, purely by chance, helps an organism adapt to new environmental conditions.

prebiotic evolution: evolution before life existed; especially, the abiotic synthesis of organic molecules.

primate: a mammal characterized by the presence of an opposable thumb, forward-facing eyes, and a well-developed cerebral cortex; includes lemurs, monkeys, apes, and humans.

protocell: the hypothetical evolutionary precursor of living cells, consisting of a mixture of organic molecules within a membrane.

ribozyme: an RNA molecule that can catalyze certain chemical reactions, especially those involved in the synthesis and processing of RNA itself.

spontaneous generation: the proposal that living organisms can arise from nonliving matter.

THINKING THROUGH THE CONCEPTS

True or False: Determine if the statement given is true or false. If it is false, change the underlined word(s) so that the statement reads true.

8. _____ Primitive Earth was characterized by an abundance of free oxygen.

9. _____ The first living organisms were most likely prokaryotic.

10. _____ Mitochondria, chloroplasts, and centrioles have their own DNA.

11. _____ Diffusion occurs quickly and efficiently in large cells.

12. _____ Coal, mined today, is made of fossilized plants.

13. _____ Conifers, as a rule, produce more pollen than flowering plants.

14. _____ Animals with exoskeletons were preadapted to life in the water.

15. _____ Amphibians are fully adapted to life on land.

16. _____ Feathers evolved for insulation.

17. _____ The most severe mass extinction wiped out 50% of the species on Earth.

18. Arrange the following events into the sequence that scientists assert occurred during the history of Earth.
    a. evolution of terrestrial organisms
    b. oxygen gas begins to accumulate in the atmosphere
    c. spontaneous formation of simple organic molecules which, in the absence of O_2, accumulated in the seas
    d. evolution of anaerobic prokaryotic cells
    e. evolution of mitochondria and chloroplasts (the Endosymbiont hypothesis)
    f. chance formation of ribozymes with the ability to make accurate and inaccurate copies of itself
    g. evolution of aerobic prokaryotic cells
    h. evolution of multicellular eukaryotic organisms
    i. evolution of primitive photosynthetic anaerobic cells
    j. by chance, primitive microspheres surround the proper mix of organic molecules and form primitive living cells

Identify: Determine whether the following statements refer to **club mosses and tree ferns**, **conifers**, or **flowering plants**.

19. _____   better adapted to colder climates

20. _____   use insects to transport pollen

21. _____   need water for sexual reproduction since sperm must swim to the eggs

22. _____   primarily use wind to passively transport pollen

23. _____   evolved from conifer-like ancestors

24. _____   dominant plants today

25. _____   dominant plants 250 million years ago

26. _____   dominant plants 325 million years ago

Matching: Terrestrial animals. Some questions may have more than one correct answer.

27. _____   evolved from reptilian ancestors

28. _____   evolved from lobefins whose fins evolved into limbs for
crawling

29. _____   dinosaurs

30. _____   use their lungs and moist skin to exchange gases
with the air

31. _____   first land animals to evolve waterproof eggs

32. _____   humans

33. _____   first land animals

34. _____   evolved with feathers for insulation

35. _____   land animals that shed their eggs and sperm into the water

36. _____   were preadapted to life on land due to their exoskeletons

37. _____   directly evolved from amphibian-like ancestors

Choices:

a. reptiles

b. amphibians

c. mammals

d. arthropods

e. birds

Multiple Choice: Pick the most correct choice for each question.

38. It is proposed that the primitive atmosphere contained all of the following except
    a. CO_2
    b. O_2
    c. NH_2
    d. H_2O

39. The first living "cells" probably were hollow, ball-shaped structures called
    a. microspheres
    b. protenoids
    c. polypeptides
    d. ribozymes
    e. bacteria

40. The first organisms probably were primitive
    a. photosynthetic bacteria
    b. cyanobacteria
    c. anaerobic bacteria
    d. aerobic microbes
    e. viruses

41. The first cells probably
    a. produced their own food
    b. absorbed food from the environment
    c. engulfed food from the environment
    d. did not require food
    e. underwent sexual reproduction

42. Reptiles are more advanced than amphibians because of
    a. internal fertilization
    b. eggs with shells
    c. scaly skin
    d. improved lungs
    e. all of the choices are correct

43. Why are scientists so excited about studying one of the moons of Jupiter, named Europa?
    a. it is about the size of Europe
    b. it is about the same size of Earth's moon
    c. it may have water beneath its frozen surface
    d. it may actually be a small planet
    e. it appears to be younger in origin than Jupiter itself

44. Lake Vostok, a huge body of water buried two miles beneath the Antarctic ice pack, is being studied by scientists because
    a. it is similar to how water may exist on Jupiter's moon Europa
    b. it may contain a large oil deposit
    c. it could solve the world's water shortage
    d. it appears to be contaminated
    e. its temperature is much lower than the ice above it.

45. What is the main problem keeping scientists from examining the contents of Lake Vostok?
    a. fear of a cave-in
    b. fear of contaminating the water with surface bacteria
    c. the lake is too deep to reach with known drilling technology
    d. the region is too cold to work in
    e. the weather is so unpredictable that workers may be stranded there

Short Answer: Based on the following figure, answer the questions below.

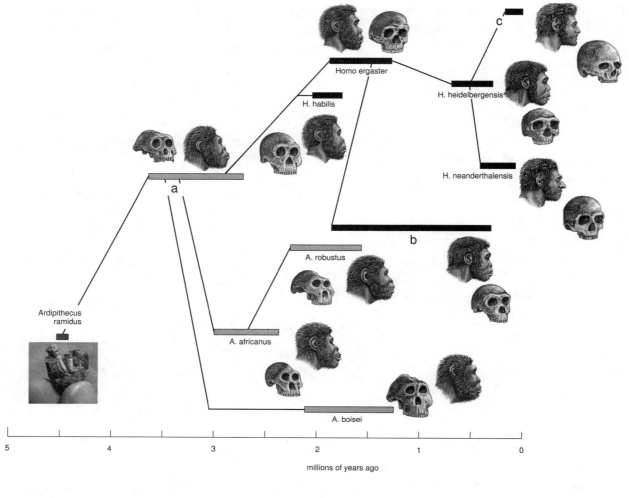

46. Identify member "a" of the human family tree: _____

47. Identify member "b" of the human family tree: _____

48. Identify member "c" of the human family tree: _____

APPLYING THE CONCEPTS

These practice questions are intended to sharpen your ability to apply critical thinking and analysis to biological concepts covered in this chapter.

49. Scientists have evidence that life developed from non-living chemicals over 3 billion years ago (spontaneous generation). Define spontaneous generation and explain why scientists assert that it cannot happen today. Also, explain why scientists say that it probably happened many years ago.

_____

_____

_____

50. List four adaptations reptiles have that help them cope successfully with a fully terrestrial lifestyle.

_____

_____

51. Which of the following discoveries, if true, would force scientists to revise drastically their thinking about the origin of life on Earth? (1) The Earth is 5 billion years old, not 4.6 billion years old. (2) Lipids can spontaneously form selectively permeable membranes without the presence of proteins. Or, (3) the atmosphere of the Earth 4.5 billion years ago contained about the same amount of oxygen gas as it does today. Explain your choice.

_____

_____

_____

_____

_____

52. Which two physical changes are the most important ones that occurred during the evolution of humans from their primate relatives?

_____

_____

_____

53. Scientists think that there may be liquid water deep beneath the frozen surface of Europa, a moon of Jupiter. Why does this excite scientists?

_____

_____

_____

_____

54. Explain the differences between the "Single African origin" and the "Multiregional" theories about the origins of humans. What type of evidence would provide convincing evidence regarding which theory is more accurate?

_____

_____

_____

_____

_____

_____

_____

_____

55. What was the most significant event in the history of life on Earth?

_____

_____

_____

_____

_____

_____

_____

Use the Case Study and the Web sites for this chapter to answer the following questions.

56. Since water is one of the basic elements of life as we know it, many astrobiologists believe that the best way to detect life in space is to look for liquid water. Recent observations suggest that there may be liquid water on Europa, one of the moons of Jupiter. Define "astrobiology."

_____

_____

_____

_____

57. What is the evidence that Europa has a liquid water layer under its icy crust?

_____

_____

_____

_____

_____

_____

_____

_____

_____

_____

_____

_____

_____

_____

_____

58. Does Europa have an atmosphere likely to support life?

_____

_____

_____

_____

_____

_____

_____

_____

_____

_____

_____

_____

_____

_____

_____

_____

_____

ANSWERS TO EXERCISES

1. spontaneous generation
2. ribozymes
3. microspheres
4. endosymbiotic hypothesis
5. Terrestrial
   conifers
   preadapted
6. hominids
   Neanderthals
7. cultural evolution
8. false, absence
9. true
10. true
11. false, slow and inefficient
12. true
13. true
14. false, on land

15. false, are not fully adapted
16. true
17. false, over 90%
18. c, f, j, d, i, b, g, e, h, a
19. conifers
20. flowering plants
21. club mosses and tree ferns
22. conifers
23. flowering plants
24. flowering plants
25. conifers
26. club mosses and tree ferns
27. c and e
28. b
29. a
30. b
31. a

32. c
33. d
34. e
35. b
36. d
37. a
38. b
39. a
40. c
41. b
42. e
43. c
44. a
45. b
46. *Australopithecus afarensis*
47. *Home erectus*
48. *Homo sapiens*

49. Spontaneous generation is the generation of living cells from nonliving matter. The reason that spontaneous generation cannot happen today is that present-day conditions are quite different than conditions on Earth before life. The main difference is that Earth's present atmosphere is rich in oxygen, which reacts with chemical bonds and breaks them down, so that lots of complex molecules cannot accumulate to provide the raw material for new life to begin. Early Earth lacked free oxygen, so that raw materials for life formed and accumulated in the seas, making spontaneous generation possible.

50. Reptiles developed several adaptations to dry land: (1) internal fertilization; (2) shelled, waterproof eggs containing a supply of water and food; (3) scaly, waterproof skin; and (4) improved lungs.

51. The atmosphere of Earth 4.5 billion years ago contained about the same amount of oxygen gas as it does today. If this were true, the thinking of scientists regarding the possibility of spontaneous generation in prebiotic times would be erroneous since large molecules would be broken down by free oxygen gas, not allowing large quantities of such molecules to accumulate and serve as the raw materials for the chance development of living cells from nonliving materials.

52. One important change was the development of the ability to walk upright on the hind legs, freeing the forelimbs to develop other uses. The other important change was the development of a larger brain, allowing for the development of greater degrees of intelligence.

53. The possible presence of water on Europa may mean that living cells or organisms might have developed there and may still be living in the waters. Since life on Earth initially developed in a watery environment, and seems to have been a necessary part of the environment leading to the origin of life, there is hope that life might have developed on a watery Europa some time in the past as well.

54. The "Multiregional" theory proposes that human ancestors migrated many times from Africa during the past 2 million years and these different groups of migrants traveled to various areas in Europe and Asia. In those regions, each group independently evolved into separate branches of the present human species, more or less simultaneously. In the "Single African origin" theory, the human species evolved once in Africa and dispersed from there during the past 150,000 years, spreading into the near East, Europe, and Asia, displacing all other hominids. The degree of similarity and difference between the base sequences of genes, using molecular genetics techniques, could provide evidence favoring one or the other of these theories. Human DNA from people living in all parts of the world are quite similar to each other, indicating a recent common origin for all humans.

55. The buildup of free oxygen gas in the atmosphere caused the most significant changes in the history of life on Earth. Before the evolution of photosynthesis, the atmosphere contained little or no oxygen gas, and spontaneous generation continued to be possible in the seas. After oxygen gas accumulated in the atmosphere due to widespread photosynthesis, spontaneous generation was no longer possible. Also, the buildup of ozone (molecules containing three atoms of oxygen) in the upper atmosphere filtered out much ultraviolet radiation, making life on land possible. In addition, the presence of oxygen gas allowed cellular respiration to evolve in cells, using mitochondria to completely break down sugar to release more energy.

56. According to L.D. Harper and G. Schmidt at nasa.com, astrobiology is the study of the origin, evolution, distribution, and destiny of life in the universe. It uses multiple scientific disciplines and space technologies to address fundamental questions: How does life begin and develop? Does life exist elsewhere in the universe? What is life's future on Earth and beyond?

57. According to the report from spaceref.com, NASA researchers have the strongest evidence yet that one of Jupiter's most mysterious moons hides a fermenting ocean of water underneath its icy coat. This evidence comes from magnetic readings by NASA's Galileo spacecraft, reported in the Friday, Aug. 25, 2000 edition of the journal *Science*. Europa, the fourth largest satellite of Jupiter, has long been suspected of harboring vast quantities of water. Since life as we know it requires water, this makes the moon a prime target for the search of exobiology — life beyond Earth. "The direction that a magnetic compass on Europa would point to flips around in a way that's best explained by the presence of a layer of electrically conducting liquid, such as saltwater, beneath the ice," explained Dr. Margaret Kivelson, one of five co-authors at the University of California, Los Angeles (UCLA).

Researchers at The Johns Hopkins University Applied Physics Laboratory (APL) and Brown University may have solved a 20-year-old geological mystery surrounding Jupiter's icy moon Europa. In the August 11, 2000 issue of *Science*, Louise Prockter of APL and Robert Pappalardo of Brown report evidence of "folds" on the moon's frozen surface. The researchers say the mountain-like features — found in three regions — are the first indication of compression on the fractured Europan crust. "We learned from Voyager images in the late 1970s that there was a lot of extension on Europa — that the surface was pulling apart and a slushy material was moving up through the gaps — but no one could find out how this new material was being accommodated," Prockter says. "Now, we have finally found folds where the icy surface material compresses, and this will help us start to understand how Europa evolved and how it resurfaces." Prockter and Pappalardo first noticed the folds in high-resolution images of Europa's Astypalaea Linea fracture region, taken by the Galileo spacecraft. Near the large fracture zone they spotted fine-scale features that typically occur in fold structures (such as the Appalachian Mountains) on Earth: regional patterns of fractures and small ridges which mark adjacent crests and valleys. The folds' direction and location along Astypalaea Linea coincide with models of tidal stress, the gravitational pull from Jupiter that scientists believe creates the pattern of large, canyon-like cracks on Europa's rotating surface. The size and nature of the folds — crests possibly tens to hundreds of meters high and spaced about 25 kilometers (16 miles) apart — also tell the researchers about the surface itself. They indicate warping of a thin brittle lithosphere covering a thicker region, or asthenosphere, of "warmer" and mobile glacier-like ice.

58. NASA's Galileo spacecraft has found an ionosphere on Jupiter's moon Europa, an indication that the icy moon also has an atmosphere. "While this discovery does not relate to the question of possible life on Europa, it does show us there is a surface process occurring there, and Europa is not just some dead hunk of material," said lead investigator Dr. Arvydas Kliore of NASA's Jet Propulsion Laboratory, Pasadena, CA. Kliore reports his findings in the July 18, 1997 issue of *Science* magazine. The ionosphere was detected through a series of six occultation experiments performed during Galileo's encounters with Europa in December 1996 and February 1997. During occultation, Europa was positioned between the spacecraft and Earth, causing interruption in the radio signal. Measurements of the Galileo radio signal received at the Deep Space Network stations in Goldstone, CA, and Canberra, Australia, showed that the radio beam was refracted by a layer of electrons, or charged particles, in Europa's ionosphere. An ionosphere is a layer of charged particles (ions and electrons) found in the upper levels of an atmosphere, created when gas molecules in the atmosphere are ionized. On Europa, this ionized layer can be caused either by the Sun's ultraviolet radiation or by energetic particles trapped in Jupiter's magnetic field, known as the magnetosphere. Europa and the other Jovian satellites are immersed in this magnetosphere. "Most likely the charged particles in Jupiter's magnetosphere are hitting Europa's icy surface with great energy, knocking atoms of water molecules off the moon's surface," Kliore said. Europa's ionosphere has a maximum density of 10,000 electrons per cubic centimeter, which is significantly lower than the average density of 20,000 to 250,000 electrons per cubic centimeter found in Jupiter's ionosphere. This indicates that Europa's ionosphere is tenuous; nonetheless, it is strong enough for scientists to infer the presence of an atmosphere.

Chapter 18: Systematics: Seeking Order Amidst Diversity

OVERVIEW

In this chapter, you will learn about the basic principles used to classify organisms into discrete groups for the purpose of systematic study. Acquired immune deficiency syndrome (AIDS) appeared seemingly from nowhere in the early 1980s; no one knew what caused it or where it came from. Soon, scientists found that human immunodeficiency virus (HIV) is the infectious agent that causes AIDS. Systematists, biologists who strive to categorize organisms according to their evolutionary history, asked what kind of virus was the ancestor of AIDS? When a systematist concludes that two species are closely related, it means that the two species share a recent common ancestor from which both species evolved. The closest relative of HIV-1 (the most common virus causing AIDS) is a virus strain that infects a particular chimpanzee subspecies that inhabits a limited range in West Africa. So, HIV-1 must have somehow "jumped" from West African chimpanzees to humans!

1) How Are Organisms Named and Classified?

Systematics is the science of reconstructing **phylogeny** (evolutionary history). **Taxonomy** is the science of naming organisms and placing them into categories based on their evolutionary relationships. The more characteristics two organisms share, the closer their evolutionary relationship. The eight major taxonomic categories include (1) **domain**, (2) **kingdom**, (3) **phylum**, (4) **class**, (5) **order**, (6) **family**, (7) **genus**, and (8) **species**. These form a nested hierarchy in which each level includes all the levels below it. As we move down the hierarchy, fewer and fewer groups are included. Each category is increasingly narrow and specifies a group whose common ancestor is increasingly recent. The **scientific name** (always underlined or italicized) of an organism is formed from its genus (closely related species) and species (populations of organisms that can interbreed under natural conditions). This system was devised by Carl von Linne' (Carolus Linnaeus). These names are recognized by biologists worldwide.

Modern systematists use numerous criteria for classification. Past evolutionary relationships are inferred based on features shared by living organisms who inherited them from common ancestors. Historically, the most important distinguishing characteristics have been anatomical in nature (bones and teeth, for instance). Developmental stages also provide clues to common ancestry. Ultimately, evolutionary relationships among species must reflect genetic similarities, namely DNA nucleotide sequences and chromosome structures.

2) What Are the Kingdoms of Life?

Before 1970, taxonomists classified all life into two kingdoms (plants and animals), an oversimplification. In 1969, Robert Whittaker proposed five kingdoms: (1) Monera (generally unicellular prokaryotes); (2) Protista (generally unicellular eukaryotes); (3) Plantae (mostly multicellular eukaryotes that obtain nutrients by photosynthesis); (4) Fungi (mostly multicellular eukaryotes that absorb externally digested nutrients); and (5) Animalia (mostly multicellular eukaryotes that ingest food and digest it internally).

Recent evidence based on rRNA nucleotide sequences indicates the Monera are two distinct groups. Carl Woese has proposed classifying life into three broad categories or domains: (1) Bacteria (or eubacteria); (2) Archaea (or archaebacteria); and (3) Eukarya (eukaryotic organisms). Each domain could contain a number of kingdoms, but kingdom-level classification remains unsettled. The two domains, Bacteria and

Archaea, include the organisms from the kingdom Monera, and the domain Eukarya would contain the four kingdoms Protista, Fungi, Plantae, and Animalia. A three-domain system more accurately reflects life's history and the composition of kingdoms may change in the future.

3) Why Do Taxonomies Change?

Taxonomic categories are revised as taxonomists learn more about evolutionary relationships, particularly using molecular biology techniques. For instance, red wolves are no longer considered to be a distinct species because mitochondrial DNA analysis indicates that red wolves are actually hybrids formed by matings between gray wolves and coyotes. The goal of modern systematics is to devise classifications that accurately reflect evolutionary history. Each designated group should contain all of the living descendants of a common ancestor; such groups are said to be **monophyletic**.

Asexually reproducing organisms pose a particular challenge to taxonomists since the criteria of interbreeding cannot be used to distinguish among species. Most bacteria, archaea, and protists reproduce asexually most of the time. The phylogenetic species concept offers an alternative definition of species: the smallest diagnosable group that contains all the descendants of a single common ancestor.

4) Exploring Biodiversity: How Many Species Exist?

Each year, between 7,000 and 10,000 new species are named, most of them insects, many from the tropical rain forests. Today, about 1.4 million species have been identified (5% are prokaryotes and protists, 22% are plants and fungi, and 73% are animals). However, scientists think that 7 to 10 million species may exist on Earth. This total range of species diversity, and the interrelationships among species, is known as **biodiversity**. The destruction of tropical rain forests is a true tragedy, since even though these forests cover only 6% of Earth's land mass, they probably are home to two-thirds of the world's existing species, most of which have never been studied or named.

Case study revisited. Comparison of DNA nucleotide sequences among different viruses reveals that one strain of HIV-1 is more closely related to a chimpanzee virus than to the other strain of HIV-2, and both HIV-1 and HIV-2 are more closely related to ape or monkey viruses than to one another. Conclusion? The HIV-1 and HIV-2 viruses jumped from ape and monkey hosts into human hosts, that HIV-1 and HIV-2 viruses are not directly related to each other, and that HIV did not evolve strictly within human hosts. Cross-species infections have occurred, most probably through human consumption of monkeys (HIV-2) and chimpanzees (HIV-1).

KEY TERMS AND CONCEPTS

Fill-In: From the following list of terms, fill in the blanks below.

Animalia	Eukarya	genus	Plantae	species
class	family	kingdom	Protista	systematics
domain	Fungi	phylum		

1. _____ is the science of reconstructing the evolutionary histories of life forms on Earth.

2. A _____ is a new taxonomic category based on molecular analysis indicating that two groups of bacteria exist. _____ is the domain containing all eukaryotic organisms.

3. _____ is the broadest taxonomic category within the domain Eukarya, consisting of phyla. Within this category, _____ are unicellular, eukaryotic organisms, _____ are multicellular, eukaryotic organisms that absorb externally digested nutrients, _____ are multicellular, eukaryotic, photosynthetic organisms, and _____ are multicellular, eukaryotic organisms that ingest food and digest it internally.

4. A _____ is the taxonomic category of animals contained within a kingdom and consisting of related classes; a _____ is the taxonomic category composed of related orders; a _____ is the taxonomic category contained within an order and consisting of related genera; a _____ is the taxonomic group containing closely related species; and a _____ is a group of organisms within a genus that interbreed under natural conditions.

Key Terms and Definitions

biodiversity: the total number of species within an ecosystem and the resulting complexity of interactions among them.

class: the taxonomic category composed of related genera. Closely related classes form a division or phylum.

domain: the broadest category for classifying organisms; organisms are classified into three domains: Bacteria, Archaea, and Eukarya.

DNA sequencing: the process of determining the chemical composition of a DNA molecule (in particular, the order in which the molecule's constituent nucleic acids are arranged).

family: the taxonomic category contained within an order and consisting of related genera.

genus (jē-nus): the taxonomic category contained within a family and consisting of very closely related species.

kingdom: the second broadest taxonomic category, contained within a domain and consisting of related phyla or divisions. This textbook recognizes four kingdoms within the domain Eukarya: Protista, Fungi, Plantae, and Animalia.

monophyletic: referring to a group of species that contains all the known descendents of an ancestral species.

order: the taxonomic category contained within a class and consisting of related families.

phylogeny (fī-lah´-jen-ē): the evolutionary history of a group of species.

phylum (fī-lum): the taxonomic category of animals and animal-like protists that is contained within a kingdom and consists of related classes.

scientific name: the name of an organism formed from the two smallest major taxonomic categories—the genus and the species.

species (spē´-sēs): the basic unit of taxonomic classification, consisting of a population or series of populations of closely related and similar organisms. In sexually reproducing organisms, a species can be defined as a population or series of populations of organisms that interbreed freely with one another under natural conditions but that do not interbreed with members of other species.

systematics: the branch of biology concerned with reconstructing phylogenies and with naming and classifying species.

taxonomy (tax-on´-uh-mē): the science by which organisms are classified into hierarchically arranged categories that reflect their evolutionary relationships.

THINKING THROUGH THE CONCEPTS

True or False: Determine if the statement given is true or false. If it is false, change the <u>underlined</u> word(s) so that the statement reads true.

5. _____   The science that places organisms into categories based on their evolutionary relationships is <u>classification</u>.

6. _____   A genus contains several <u>classes</u>.

7. _____   A kingdom is a more <u>general</u> category than a phylum.

8. _____   The two-part name for a type of organism is its <u>genus and species</u>.

9. _____   Organisms within the same <u>genus</u> interbreed in nature.

10. _____   Eastern and mountain bluebirds have the same <u>genus</u> name.

11. _____   <u>Darwin</u> developed the classification system we use today.

12. _____   One important criterion used to classify organisms is <u>geographical location</u>.

13. _____   Most bacterial and protistan organisms are <u>prokaryotic</u>.

14. _____   Plants are, but fungi are not, <u>multicellular</u>.

Multiple Choice: Pick the most correct choice for each question.

15. The classification of organisms is called

    a. taxonomy

    b. morphology

    c. ecology

    d. nomenclature

    e. phylogeny

16. Which hierarchical order goes from more specific to more general?

    a. kingdom division class

    b. genus family order

    c. order family genus

    d. order kingdom species

    e. family order species

17. In the scientific name, *Canis lupis*, "*Canis*" is the name of the

    a. genus

    b. species

    c. family

    d. wolf group only

    e. dog group only

18. Which of the following is true?

    a. the human immunodeficiency virus (HIV) originated in humans

    b. HIV-1 and HIV-2 are identical viruses

    c. the way HIV initially got into humans is known beyond doubt

    d. the progenitor of HIV is found in chimpanzees

    e. HIV-1 gave rise to HIV-2

19. Which of the following is false?

    a. HIV-1 is more closely related to HIV-2 than to a virus in chimpanzees

    b. HIV-1 is a direct relative of a chimpanzee virus

    c. HIV-2 is a direct relative of a chimpanzee virus

    d. HIV is the infectious agent that causes AIDS

    e. HIV did not evolve strictly within human hosts

Matching: The kingdoms.

20. _____   unicellular, eukaryotic organisms

21. _____   multicellular, non photosynthetic organisms with extracellular digestion

22. _____   all organisms are photosynthetic

23. _____   multicellular, non-photosynthetic organisms with digestion within the body

24. _____   organisms are unicellular and have nuclei

Choices:

    a. Kingdom Protista

    b. Kingdom Fungi

    c. Kingdom Plantae

    d. Kingdom Animalia

APPLYING THE CONCEPTS

These practice questions are intended to sharpen your ability to apply critical thinking and analysis to biological concepts covered in this chapter.

25. In what way did Darwin reinterpret the significance of Linnaeus' system of classification?

_____

_____

_____

_____

_____

26. How can understanding the origin of HIV help scientists devise better ways to treat and control the spread of AIDS?

_____

_____

_____

_____

_____

27. Arrange the following classification of the wolf into a sequence of categories from the most encompassing to the least encompassing. Class Mammalia, Family Canidae, Domain Eukarya, Order Carnivora, Genus *Canis*, Kingdom Animalia, species *lupus*, Phylum Chordata.

_____

_____

_____

Use the Case Study and the Web sites for this chapter to answer the following questions.

28. Most researchers agree that both HIV viruses were originally transmitted from primates to humans. What they disagree about is how and when the transfer took place. What primate species were the "original" hosts?

_____

_____

_____

_____

_____

_____

_____

29. Edward Hooper has popularized the iatrogenic (medically-caused) oral polio vaccine theory, i.e., HIV was passed to humans when HIV-1 contaminated chimpanzee cells were used to prepare vaccines. According to this theory, when should the first cases of HIV-1 have occurred?

_____

_____

_____

_____

_____

_____

30. After extensive testing, researchers identified an African HIV-1 infected plasma sample originally collected in 1959. However, these researchers think that HIV-1 evolved much earlier. Why?

_____

_____

_____

_____

_____

_____

ANSWERS TO EXERCISES

1. Systematics	genus	14. false, photosynthetic
2. domain	species	15. a
Eukarya	5. false, taxonomy	16. b
3. Kingdom	6. false, species	17. a
Protista	7. true	18. d
Fungi	8. true	19. a
Plantae	9. false, species	20. a
animalia	10. true	21. b
4. phylum	11. false, Linnaeus	22. c
class	12. false, anatomical similarities	23. d
family	13. false, unicellular	24. a

25. Linnaeus placed organisms into a series of hierarchically arranged categories based on resemblances to other life forms, a system he thought was related to the creation of these organisms by God. Darwin, however, proposed that Linnaeus' categories reflected the evolutionary relatedness of organisms, reasoning that the more physically similar organisms are to each other, the more closely related evolutionarily they are as well.

26. Knowing the origin of any illness can help in the control of its spread, since health officials can tell people how to avoid being exposed. Knowing that HIV originated in apes and monkeys can help in treating the condition, since the virus apparently is benign in its normal hosts. By studying natural molecular defense mechanisms in apes and monkeys, strategies for treating AIDS in humans may be developed.

27. Domain Eukarya (all eukaryotic organisms), Kingdom Animalia (all animals), Phylum Chordata (all animals with notochords), Class Mammalia, Order Carnivora, Family Canidae, Genus *Canis*, species *lupus* (all wolves)

28. According to A. Kanabus and S. Allen from AVERT, debate around the origin of AIDS has sparked considerable interest and controversy since the beginning of the epidemic. However, in trying to identify where AIDS originated, there is a danger that people may try and use the debate to attribute blame for the disease to particular groups of individuals or certain lifestyles. The first cases of AIDS occurred in the USA in 1981, but they provide little information about the source of the disease. There is now clear evidence that the disease AIDS is caused by the virus HIV. So to find the source of AIDS we need to look for the origin of HIV. It is now generally accepted that HIV is a descendant of simian (monkey) immunodeficiency virus (SIV). Certain simian immunodeficiency viruses bear a very close resemblance to HIV-1 and HIV-2, the two types of HIV. For example, HIV-2 corresponds to a simian immunodeficiency virus found in the sooty mangabey monkey (SIVsm), sometimes known as the green monkey, which is indigenous to western Africa. The more virulent strain of HIV, namely HIV-1, was until very recently more difficult to place. In February 1999 it was announced that a group of researchers from the University of Alabama had studied frozen tissue from a chimpanzee and found that the simian virus it carried (SIVcpz) was almost identical to HIV-1. The chimpanzee came from a sub-group of chimpanzees known as *Pan troglodytes troglodytes*, which were once common in west-central Africa. It is claimed by the researchers that this shows that these chimpanzees were the source of HIV-1, and that the virus at some point crossed species from chimpanzees to humans. However, it is not necessarily clear that chimpanzees are the original reservoir for HIV-1 because chimpanzees are only rarely infected with SIVcpz. It is therefore possible that both chimpanzees and humans have been infected from a third, as yet unidentified, primate species. In either case at least two separate transfers into the human population would have been required.

29. According to M. Ridley of Prospect Magazine, most scientists believe that AIDS was "naturally" transferred from primates to human beings via a hunter who ate a chimpanzee. But a competing theory claims that AIDS was caused in the 1950s when thousands of Africans were given a live polio vaccine derived from chimp kidneys. According to Hooper, a particular type of live polio vaccine called Chat may have been grown in the 1950s in cells derived from chimpanzee kidneys. Chimpanzees are the probable animal source of the AIDS virus; live vaccines could have been contaminated if an infected animal was used. Chat was tested on more than one million Africans in 1957–60, in the very areas where AIDS subsequently became epidemic for the first time. Two other less serious forms of AIDS developed in parts of west Africa at about the same time, each epidemic closely associated with an area in which similar live polio vaccines may have been tested. As for HIV itself, the oldest positive test consists of a blood sample taken from an unknown African man in Kinshasa (then Leopoldville) in 1959 in unrecorded circumstances. Since there was a Chat vaccination trial in the city at the same time, it is possible that this sample was taken in a post-vaccination follow-up. So it cannot be ruled out that this man had contracted his HIV from polio vaccine a few weeks before. So, it appears that human AIDS dates from the same years during which live polio vaccines were tested.

30. In 1998, Dr. Tuofu Zhu of the Aaron Diamond AIDS Research Center in New York and colleagues reported the characterization of viral DNA sequences of an HIV-1 plasma sample taken from an adult Bantu male in 1959 living in what is now the Democratic Republic of Congo. The researchers studied 1,213 plasma samples obtained from Africa between 1959 and 1982; the 1959 sample tested positive for HIV-1 through various testing methods. The plasma sample is the oldest confirmed case of HIV-1. The research team notes that "given the large genetic differences between HIV-1 and HIV-2, the divergence of these viruses could not have occurred in the late 1940s; that branching point must have come considerably earlier."

Chapter 19: The Hidden World of Microbes

OVERVIEW

In this chapter, you will learn about the three major groups of microorganisms: viruses, prokaryotic unicellular microbes in the bacterial and archaean domains, and eukaryotic unicellular microbes in the Kingdom Protista. In April, 1979, an explosion in a top-secret biological warfare factory in Sverdlovsk, USSR, released a cloud of disease-causing bacteria called *Bacillus anthracis*, which causes anthrax in animals and humans. Within days, thousands of citizens of Sverdlovsk became ill and died of respiratory infection. Anthrax bacteria are easily isolated from infected animals, cheap and easy to culture, can be dried into a powder that remains viable for years, and is easily weaponized by packing the powder into a missile warhead.

1) What Are Viruses, Viroids, and Prions?

Viruses have no cells, membranes, ribosomes, cytoplasm, or energy source, cannot grow or move, and can reproduce only within **host** (virus-infected) cells. A virus consists of a molecule of DNA or RNA surrounded by a protein coat and is too small to see under the light microscope. The protein coat is specialized to allow a virus to enter a specific host cell. Once inside, the viral genetic material takes command, forcing the host to make more viral protein and genetic material that assemble to form new viruses to burst out of the host cell and invade new host cells. Viral infections cause diseases that are difficult to treat. Viruses that infect bacteria are called **bacteriophages**. Within a particular organism, viruses specialize in attacking particular cell types. Antiviral agents may destroy host cells as well as virus. Antibiotics are useless against viruses.

Some infectious agents are even simpler than viruses. Some plant diseases are caused by **viroids**, pieces of RNA without protein coats and only one-tenth the size of viruses. **Prions**, infectious "replicating" particles made of protein alone, cause degenerative brain diseases in humans (**kuru** and *Creutzfeldt-Jacob disease*) and livestock (*scrapie*). Of current concern is the possibility that humans can become infected with *bovine spongiform encephalopathy* ("mad cow disease"), a prion disease that affects cattle, by eating beef from infected animals. Apparently, these proteins fold abnormally, inducing normal proteins to refold abnormally as well, causing nerve cell damage and degeneration. If a small mutation in the gene that codes for the normal prion protein increases the likelihood that the protein will fold into its abnormal form, a tendency to develop a prion disease could be inherited. The origin of viruses, viroids and prions is obscure.

2) Which Organisms Make Up the Prokaryotic Domains — Bacteria and Archaea?

Prokaryotes lack nuclei, chloroplasts, and mitochondria and are at least 10 times smaller than eukaryotic cells. The **Archaea** diverged from the **Bacteria** very early in the history of life and are only distantly related. The archaea are more closely related to eukaryotes than are bacteria. Thus, eukaryotes evolved from the archaea. Bacterial **cell walls** contain **peptidoglycan** (sugar chains attached by short chains of amino acids), but archaean cell walls lack this substance. The structure and composition of plasma membranes, ribosomes, and RNA polymerases differ between the two domains, as do the processes of transcription and translation.

Lack of sexual reproduction and fossil evidence make prokaryotes hard to classify. Taxonomists use criteria such as shape, means of movement, pigments, staining properties, nutrient requirements, DNA and RNA sequences, and the morphology of colonies to classify them. Bacterial cell walls afford protection from osmotic rupture in watery environments, and give bacteria their shapes (rodlike, corkscrew-shaped, and spherical). The **Gram staining** of cell walls reveals two types of bacteria: (1) *gram-positive* bacteria have peptidoglycan walls only and are often sensitive to penicillin; and (2) *gram-negative* bacteria have an additional outer membrane-like coating often toxic to mammals.

Outside the cell wall, sticky polysaccharide or protein **capsules** (help bacteria escape detection by host immune systems) and **slime layers** (help bacteria adhere) aid in bacterial survival. Protein **pili** (hairlike projections) help bacteria attach to other cells. Some bacteria and archaea can move by using simple **flagella**, orienting toward various stimuli, a behavior called **taxis**: (1) **chemotactic** bacteria move toward food or away from toxins; (2) **phototactic** bacteria move to or from light; and (3) **magnetotactic** bacteria contain iron crystals and orient to the Earth's magnetic field. In hostile environments, many rod-shaped bacteria form internal **endospores** (thick-walled coating around the chromosome), which often are used for bacterial dispersal.

Bacteria reproduce rapidly by asexual cell division called **binary fission** (see Chapter 11). However, some transfer genetic material in **plasmids** (small circular DNA molecules often containing drug-resistance genes) from donor to recipient bacteria during **bacterial conjugation** using hollow *sex pili*. Conjugation produces new genetic combinations in the recipient cell. Prokaryotes are specialized for specific habitats.

Prokaryotes exhibit diverse metabolism. The **cyanobacteria** engage in plantlike photosynthesis. The **chemosynthetic** bacteria and archaeans release *sulfates* (sulfur) and *nitrates* (nitrogen) into the soil. Some archaeans are **methanogens** that convert carbon dioxide to methane. The **anaerobes** (use no oxygen) release sulfur into the atmosphere during a special type of photosynthesis. Prokaryotes perform many functions that are important to other life forms. Some **symbiotic** bacteria live in the digestive tracts of animals and help them break down food molecules, making nutrients like vitamins K and B$_{12}$ in humans, for instance. Ruminants (cow, sheep, and deer) eat leaves and rely on symbiotic bacteria in their digestive tract to break down the cellulose. **Nitrogen-fixing bacteria** live in the roots of **legume** plants (beans and their relatives), converting nitrogen gas into ammonium for the plants. Prokaryotes also play a crucial role in recycling wastes. The term *biodegradable* refers largely to the work of prokaryotes. Some **pathogenic** bacteria cause disease and pose a threat to human health. Tetanus, botulism, the plague, Lyme disease, tuberculosis, gonorrhea, syphilis, strep throat, and pneumonia are some diseases caused by bacteria.

3) Which Organisms Make Up the Kingdom Protista?

Protists are a diverse group including funguslike water molds and slime molds, plantlike unicellular algae, and animal-like protozoa, among others (See Table 19-1 in the textbook).

Funguslike protists absorb nutrients from the environment and typically decompose dead organisms. **Water molds**, or *oomycetes*, cause downy mildew of grapes and late blight of potatoes. *Slime molds* have a mobile feeding stage and a stationary reproductive stage (**fruiting body**) producing spores. The **acellular slime molds** consist of a mass of cytoplasm (**plasmodium**) containing thousands of nuclei. The **cellular slime molds** consist of masses of independent cells that move and feed by producing extensions (**pseudopods**). These cells release a chemical signal that causes the cells to form a dense aggregation (**pseudoplasmodium**) that forms a fruiting body.

The algae are plantlike protists. Unicellular **algae**, or **phytoplankton**, gain energy by photosynthesis. **Dinoflagellates** have two flagella for movement and have red pigment, sometimes causing "red tides" when the water is warm and rich in nutrients. **Diatoms** are called "pastures of the sea" since they are important as food in marine food webs. They encase themselves in silica (glassy) pillbox-like shells that accumulate as "diatomaceous earth." **Euglenoids** each have a flagellum for movement and an eyespot for light detection. Red algae are mostly marine and strictly multicellular. The brown algae dominate in cool coastal waters and are strictly multicellular. The green algae include both unicellular and multicellular species. They live mostly in ponds and lakes and probably gave rise to land plants.

Animal-like **protozoa** ingest food and digest it internally. **Zooflagellates** have at least one flagellum for movement or food gathering. One parasitic type (*Trypanosoma*) causes African sleeping sickness while another (*Giardia*) causes dysentery in the United States.

Sarcodines are the freshwater **amoebas** (shell-less, one form causes amoebic dysentery) and **heliozoans** ("sun animals" with glassy shells). Other sarcodines include the marine **foraminiferans** (with chalky limestone shells) and **radiolarians** (with glassy shells). Sarcodines move and capture food by forming pseudopods.

Sporozoans have no means of movement and live as parasites. The sporozoan *Plasmodium* causes malaria. **Ciliates** (Ciliphora), which move by cilia, are the most complex protozoan in cell structure; they behave as if they have well-developed nervous systems and they have mouth openings, anal openings, and contractile vacuoles to excrete water. A *Paramecium* is a common ciliate.

Case study revisited. Some biological weapons under development include anthrax, smallpox, Ebola hemorrhagic fever, and the bacteria that cause plague. Genetic engineering may be used to make the pathogens antibiotic resistant. Defending a civilian population against biological attacks is difficult. Even treating victims after an attack would probably be ineffective, and would overwhelm our public health system.

KEY TERMS AND CONCEPTS

Fill-In: Using the terms below, fill in the boxes in the following concept map.

bacilli	dinoflagellate	protistans	spirilla
bacteria and archaea	DNA and RNA	sarcodines	viruses
ciliates	euglenoids	slime molds	water molds
cocci	protein coats	sporozoans	zooflagellates
diatoms			

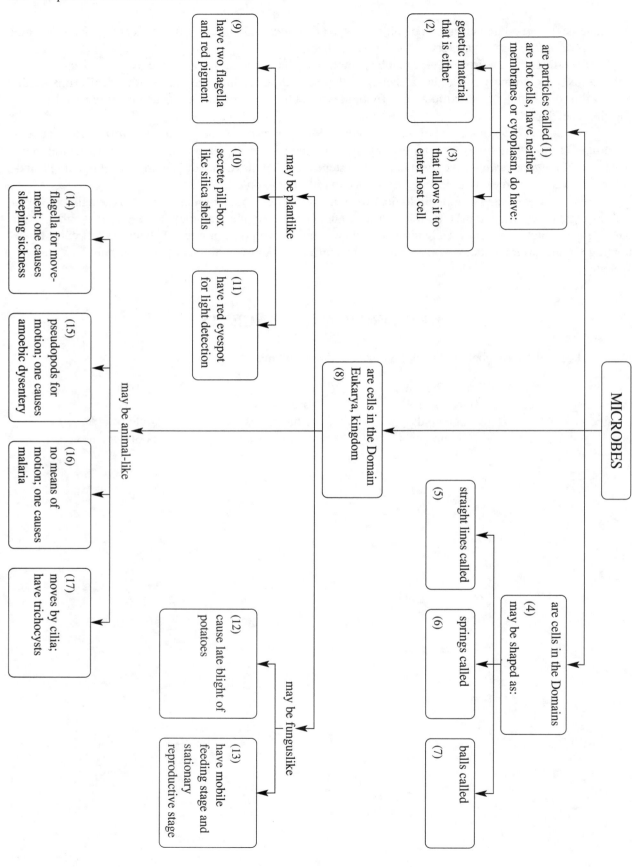

Key Terms and Definitions

acellular slime mold: a type of funguslike protist that forms a multinucleate structure that crawls in amoeboid fashion and ingests decaying organic matter; also called *plasmodial slime mold.*

alga (al´-ga; pl., **algae, al´-jē)**: any photosynthetic member of the eukaryotic Kingdom Protista.

amoeba: a type of animal-like protist that uses a characteristic streaming mode of locomotion by extending a cellular projection called a *pseudopod.*

anaerobe: an organism whose respiration does not require oxygen.

Archaea: one of life's three domains; consists of prokaryotes that are only distantly related to members of the domain Bacteria.

Bacteria: one of life's three domains; consists of prokaryotes that are only distantly related to members of the domain Archaea.

bacterial conjugation: the exchange of genetic material between two bacteria.

bacteriophage (bak-tir´-ē-ō-fāj): a virus specialized to attack bacteria.

binary fission: the process by which a single bacterium divides in half, producing two identical offspring.

capsule: a polysaccharide or protein coating that some disease-causing bacteria secrete outside their cell wall.

cellular slime mold: a funguslike protist consisting of individual amoeboid cells that can aggregate to form a sluglike mass, which in turn forms a fruiting body.

cell wall: a layer of material, normally made up of cellulose or cellulose-like materials, that is outside the plasma membrane of plants, fungi, bacteria, and some protists.

chemosynthetic (kēm´-ō-sin-the-tik): capable of oxidizing inorganic molecules to obtain energy.

chemotactic (kēm-ō-tak´-tik): moving toward chemicals given off by food or away from toxic chemicals.

ciliate (sil´-ē-et): a protozoan characterized by cilia and by a complex unicellular structure, including harpoonlike organelles called *trichocysts.* Members of the genus *Paramecium* are well-known ciliates.

cilium (sil´-ē-um; pl., **cilia)**: a short, hairlike projection from the surface of certain eukaryotic cells that contains microtubules in a $9 + 2$ arrangement. The movement of cilia may propel cells through a fluid medium or move fluids over a stationary surface layer of cells.

cyanobacterium: a photosynthetic prokaryotic cell that utilizes chlorophyll and releases oxygen as a photosynthetic byproduct; sometimes called *blue-green algae.*

diatom (dī´-uh-tom): a protist that includes photosynthetic forms with two-part glassy outer coverings; important photosynthetic organisms in fresh water and salt water.

dinoflagellate (dī-nō-fla´-jel-et): a protist that includes photosynthetic forms in which two flagella project through armorlike plates; abundant in oceans; can reproduce rapidly, causing "red tides."

endospore: a protective resting structure of some rod-shaped bacteria that withstands unfavorable environmental conditions.

euglenoid (ū´-gle-noid): a protist characterized by one or more whiplike flagella that are used for locomotion and by a photoreceptor that detects light. Euglenoids are photosynthetic, but if deprived of chlorophyll, some are capable of heterotrophic nutrition.

flagellum (fla-jel´-um; pl., **flagella)**: a long, hairlike extension of the plasma membrane; in eukaryotic cells, it contains microtubules arranged in a $9 + 2$ pattern. The movement of flagella propel some cells through fluids.

foraminiferan (for-am-i-nif´-er-un): an aquatic (largely marine) protist characterized by a typically elaborate calcium carbonate shell.

fruiting body: a spore-forming reproductive structure of certain protists, bacteria, and fungi.

Gram stain: a stain that is selectively taken up by the cell walls of certain types of bacteria (gram-positive bacteria) and rejected by the cell walls of others (gram-negative bacteria); used to distinguish bacteria on the basis of their cell wall construction.

heliozoan (hē-lē-ō-zō´-un): an aquatic (largely freshwater) animal-like protist; some have elaborate silica-based shells.

host: the prey organism on or in which a parasite lives; is harmed by the relationship.

kuru: a degenerative brain disease, first discovered in the cannibalistic Fore tribe of New Guinea, that is caused by a prion.

legume (leg´-ūm): a member of a family of plants characterized by root swellings in which nitrogen-fixing bacteria are housed; includes soybeans, lupines, alfalfa, and clover.

magnetotactic: able to detect and respond to Earth's magnetic field.

methanogen (me-than´-ō-jen): a type of anaerobic archaean capable of converting carbon dioxide to methane.

nitrogen-fixing bacterium: a bacterium that possesses the ability to remove nitrogen (N_2) from the atmosphere and combine it with hydrogen to produce ammonium (NH_4^+).

pathogenic (path´-ō-jen-ik): capable of producing disease; refers to an organism with such a capability (a pathogen).

peptidoglycan (pep-tid-ō-glī´-kan): a component of prokaryotic cell walls that consists of chains of sugars cross-linked by short chains of amino acids called *peptides.*

phototactic: capable of detecting and responding to light.

phytoplankton (fī´-tō-plank-ten): photosynthetic protists that are abundant in marine and freshwater environments.

pilus (pil´-us; pl., **pili)**: a hairlike projection that is made of protein, located on the surface of certain bacteria, and is typically used to attach a bacterium to another cell.

plasmid (plaz´-mid): a small, circular piece of DNA located in the cytoplasm of many bacteria; normally does not carry genes required for the normal functioning of the bacterium but may carry genes that assist bacterial survival in certain environments, such as a gene for antibiotic resistance.

plasmodial slime mold: see *acellular slime mold.*

plasmodium (plaz-mō´-dē-um): a sluglike mass of cytoplasm containing thousands of nuclei that are not confined within individual cells.

prion (prē´-on): a protein that, in mutated form, acts as an infectious agent that causes certain neurodegenerative diseases, including kuru and scrapie.

Protista (prō-tis´-tuh): a taxonomic kingdom including unicellular, eukaryotic organisms.

protozoan (prō-tuh-zō´-an; pl., **protozoa)**: a nonphotosynthetic or animal-like protist.

pseudoplasmodium (soo´-dō-plaz-mō´-dē-um): an aggregation of individual amoeboid cells that form a sluglike mass.

pseudopod (sood´-ō-pod): an extension of the plasma membrane by which certain cells, such as amoebae, locomote and engulf prey.

radiolarian (rā-dē-ō-lar´-ē-un): an aquatic protist (largely marine) characterized by typically elaborate silica shells.

sarcodine (sar-kō´-dīn): a nonphotosynthetic protist (protozoan) characterized by the ability to form pseudopodia; some sarcodines, such as amoebae, are naked, whereas others have elaborate shells.

slime layer: a sticky polysaccharide or protein coating that some disease-causing bacteria secrete outside their cell wall; helps the cells aggregate and stick to smooth surfaces.

sporozoan (spor-ō-zō´-un): a parasitic protist with a complex life cycle, typically involving more than one host; named for their ability to form infectious spores. A well-known sporozoan (genus *Plasmodium*) causes malaria.

symbiotic: referring to an ecological relationship based on symbiosis.

taxis (taks´-is; pl., **taxes)**: an innate behavior that is a directed movement of an organism toward or away from a stimulus such as heat, light, or gravity.

viroid (vī´-roid): a particle of RNA that is capable of infecting a cell and of directing the production of more viroids; responsible for certain plant diseases.

virus (vī´-rus): a noncellular parasitic particle that consists of a protein coat surrounding a strand of genetic material; multiplies only within a cell of a living organism (the host).

water mold: a funguslike protist that includes some pathogens, such as the downy mildew, which attacks grapes.

zooflagellate (zō-ō-fla´-jel-et): a nonphotosynthetic protist that moves by using flagella.

THINKING THROUGH THE CONCEPTS

True or False: Determine if the statement given is true or false. If it is false, change the underlined word(s) so that the statement reads true.

18. _____   Protistan organisms are underlined eukaryotic.

19. _____   Viruses are cellular.

20. _____   Antibiotics are a better defense against viral infection than vaccines.

21. _____   Bacteria reproduce asexually by mitosis.

22. _____   Cyanobacteria are heterotrophic.

23. _____   Unicellular algae are often called phytoplankton.

24. _____   Dinoflagellates are unicellular, eukaryotic, feeding protists.

25. _____   The organism causing African sleeping sickness is a sporozoan.

26. _____   The organism causing malaria is a zooflagellate.

Identify: Determine whether the following statements refer to **euglenoids**, **dinoflagellates**, **diatoms** or to **all** of these algae.

27. _____   can move by a flagellum or by wriggling

28. _____   produce glassy shells that fit together like shoe boxes

29. _____   have pairs of flagella

30. _____   lack a rigid outer covering

31. _____   many are bioluminescent

32. _____   have single flagella for locomotion

33. _____   most live in fresh water

34. _____   cause "red tides" that kill fish

35. _____   have simple light-sensitive organelles

36. _____   shells have been used in toothpaste

37. _____   produce a toxic nerve poison

38. _____   make "food" by photosynthesis

Matching: Protozoans.

39. _____ the "white cliffs of Dover" are made from their calcium shells

40. _____ all are parasites

41. _____ move by using flagella

42. _____ use cilia for locomotion

43. _____ some are symbiotic

44. _____ use pseudopods for locomotion

45. _____ have no means of locomotion

46. _____ prey items enter their oral groove

47. _____ one type digests cellulose in the guts of termites

48. _____ one type causes malaria

49. _____ one type causes amoebic dysentery

50. _____ some have contractile vacuoles to excrete excess water

51. _____ one type causes African sleeping sickness

52. _____ some types form glassy (silica) shells

Choices:

a. sarcodines

b. zooflagellates

c. ciliates

d. sporozoans

Multiple Choice: Pick the most correct choice for each question.

53. Of the following, which are eukaryotic?
    a. bacteria
    b. blue-green algae
    c. protists
    d. viruses
    e. viroids

54. All of the following are true except
    a. viruses are prokaryotic
    b. viruses require a host cell
    c. viruses are intracellular parasites
    d. viruses possess proteins and hereditary material
    e. viruses may have DNA or RNA

55. Viral infections do <u>not</u> cause
    a. AIDS
    b. herpes
    c. common cold
    d. flu
    e. malaria

56. Nitrogen-fixing bacteria
    a. convert nitrates into nitrogen gas
    b. convert nitrogen gas into a usable form
    c. remove nitrates from the soil
    d. remove nitrogen gas from plants
    e. are found on the roots of corn and wheat plants

57. All protists are characterized by
    a. single cell
    b. cell wall
    c. chlorophyll
    d. filamentous body form
    e. all of the choices are correct

58. The malarial parasite, *Plasmodium*, is a
    a. sporozoan
    b. ciliate
    c. zooflagellate
    d. sarcodine
    e. blue-green alga

59. Protozoans include all of the following except
    a. dinoflagellates
    b. sporozoans
    c. zooflagellates
    d. sarcodines
    e. ciliates

60. "Red tides" are caused by a population explosion of
    a. diatoms
    b. zooflagellates
    c. dinoflagellates
    d. euglenoids
    e. bacteria

APPLYING THE CONCEPTS

These questions are intended to sharpen your ability to apply critical thinking and analysis to biological concepts covered in this chapter.

61. The common intestinal bacterium called *E. coli* is a typical bacterium in that it can reproduce rapidly under favorable growth conditions. Suppose you placed 10 bacteria in a very large flask containing the ideal culture medium for bacterial growth and reproduction. Under these conditions, the bacteria will divide every 20 minutes. How long would it take until the flask contained about 1 billion bacteria? How many bacteria would be present after 12 hours? If bacteria reproduce so rapidly, why aren't we neck-deep in bacteria all over the world?

_____

_____

_____

_____

_____

_____

_____

62. Name several differences between the two large domains of prokaryotic organisms: the bacteria and the archaea.

_____

_____

_____

63. In what ways are prokaryotic organisms important to the healthy existence of humans on Earth?

_____

_____

_____

_____

_____

64. For each of the following groups of organisms, can you think of two characteristics that makes them different from members of the other groups: viruses, methanogens, cyanobacteria, nitrogen-fixing bacteria?

_____

_____

_____

_____

65. Penicillin interferes with the formation of new cell walls when cells divide. Explain why taking penicillin for a bacterial infection is not effective in the case of a viral infection.

_____

_____

_____

_____

66. What are some reasons that terrorist groups are developing biological warfare capabilities involving pathogenic bacteria instead of developing nuclear weapons?

_____

_____

_____

_____

_____

_____

Use the Case Study and Web sites for this chapter to answer the following questions.

67. While biological weapons are not new, the bacterial strains, delivery mechanisms, and toxins have become much more sophisticated. Most pre-modern biological weapons were bacteria. How were they dispersed? Did they work?

_____

_____

_____

_____

_____

_____

_____

_____

_____

68. What are the characteristics of a "Perfect Biological Weapon"?

_____

_____

_____

_____

_____

_____

_____

69. In recent years, many biological weapons programs have focused on anthrax caused by *Bacillus anthracis*. Why is the disease such a threat?

_____

_____

_____

_____

_____

_____

_____

_____

ANSWERS TO EXERCISES

1. viruses	21. false, binary fission	41. b
2. DNA and RNA	22. false, autotrophic	42. c
3. protein coats	23. true	43. b
4. bacteria and Archaea	24. false, protozoa	44. a
5. rodlike	25. false, zooflagellate	45. d
6. corkscrew shaped	26. false, sporozoan	46. c
7. spherical shaped	27. euglenoids	47. b
8. protistans	28. diatoms	48. d
9. water molds	29. dinoflagellates	49. a
10. slime molds	30. euglenoids	50. c
11. dinoflagellates	31. dinoflagellates	51. b
12. diatoms	32. euglenoids	52. a
13. euglenoids	33. euglenoids	53. c
14. zooflagellates	34. dinoflagellates	54. a
15. sarcodines	35. euglenoids	55. e
16. sporozoans	36. diatoms	56. b
17. ciliates	37. dinoflagellates	57. a
18. true	38. all	58. a
19. false, non-cellular	39. a	59. a
20. false, an ineffective	40. d	60. c

61. Since bacteria double in number every 20 minutes, after one hour there would be 10 x 2 x 2 x 2 = 80 bacteria (an eight-fold increase every hour). At that rate of increase, it would take about 9 hours to accumulate 1.342 billion bacteria. After 12 hours, there would be approximately 700 billion bacteria in the flask (about 1.342 billion after 9 hours x 8 = about 10.74 billion after 10 hours x 8 = about 85.92 billion after 11 hours x 8 = about 687.36 billion after 12 hours). There are estimates that the total weight of all bacteria on Earth is much higher than the total weight of all other organisms combined. But we are not swamped with bacteria because seldom do they find the ideal conditions necessary to reproduce as rapidly as they can.

62. Bacterial cell walls contain peptidoglycans (molecules with sugar and amino acid groups), while archaeans lack peptidoglycans. The genes of archaeans are more closely related to eukaryotic organisms than they are to bacteria. The structure and composition of plasma membranes, ribosomes, and RNA polymerases also differ between bacteria and archaeans.

63. Prokaryotic organisms are important to the well-being of humans (and other living organisms) because they aid in the cycling of chemical elements between living organisms and the non-living environment. The cyanobacteria, through the process of photosynthesis, produce free oxygen gas to the atmosphere. The chemosynthetic prokaryotes release sulfates and nitrates into the soil. Some methanogens convert carbon dioxide into methane gas. The nitrogen-fixing bacteria convert nitrogen gas into ammonium nitrates in certain plants. Symbiotic bacteria in the digestive system of animals help them break down food molecules and produce nutrients like vitamins K and B_{12} in humans. Prokaryotes are responsible for most biodegradable recycling.

64. The viruses have only one type of nucleic acid and cannot make their own proteins or ATP for energy. The methanogens produce methane from carbon dioxide and lack peptidoglycans in their cell walls. The cyanobacteria use chlorophyll in photosynthesis and produce oxygen gas. The nitrogen-fixing bacteria live in the roots of legume plants and convert nitrogen gas into ammonium for plants.

65. Penicillin works by interfering with the synthesis of cell wall components, and thus is effective against bacteria, which have cell walls. Since viruses do not have cell walls, penicillin has no effect against viral infections.

66. Developing biological warfare capability is much less expensive than developing nuclear weapons, and involves less sophisticated techniques and equipment. Also, defending a civilian population against biological attacks is more difficult. Even treating victims after a biological warfare attack would be a major problem since public health facilities are incapable of handling an unusually large number of critically ill people all at once.

67. According to G. W. Christopher et. al. and the whyfiles.com, in the 14th century, during an attack on Kaffa (now Feodossia, Ukraine), Tatar armies catapulted comrades who died of bubonic plague into the city. The defenders were infected, and they fled, apparently carrying the plague to Europe. Thus the second outbreak of "black death" in Europe may be partly blamed on biological warfare. In 1763, during the French and Indian Wars, Sir Jeffrey Amherst, commander of British forces in North America, directed that smallpox-bearing blankets be given to enemy tribes in the Ohio River Valley. Nobody knows whether a devastating smallpox epidemic that happened almost simultaneously was due to the infamous blankets — or natural transmission. During World War I, Germany aimed an ambitious biological weapons project at its enemies' livestock. Anthrax and glanders (both bacterial diseases) were used to infect sheep that were shipped to Russia. Germany also tried to infect American horses that would be shipped to the Western front. From 1932–1945, Japan employed more than 3,000 scientists and support staff in its biological weapons project in occupied China. Prisoners were deliberately infected with several biological agents, and at least 10,000 died. Up to 11 Chinese cities were attacked with anthrax, cholera, salmonella and other agents. A 1941 attack on Changteh killed at least 1,700 Japanese troops, demonstrating that biological weapons are tricky to use. In 1979, escaped

bacteria from the ongoing Soviet bio-war effort caused an anthrax epidemic that killed at least 66 people in and near the city of Sverdlovsk. In 1984, the Rajneeshee religious cult intentionally contaminated salad bars in Oregon restaurants with Salmonella, causing 751 cases of enteritis — gut infections. Forty-five of these people needed hospitalization. In 1995, the Aum Shinrikyo cult attacked Tokyo subways with the nerve gas sarin, killing 12 and injuring 5,000. Investigators learned the cult was also working with several biological warfare agents, including anthrax and botulism toxin.

68. R. E. Hurlbert reports that the perfect biological organism for use as a weapon should have the following characteristics: (1) It should be highly infectious; it should require only a few organisms to cause the desired effect (e.g. smallpox); it should require a small quantity of material to cause the desired effect (e.g. botox). (2) It should be efficiently dispersible, usually in the air; contagious or effective on contact. (3) It should be readily grown and produced in large quantities. (4) It should be stable in storage; preferably in a ready-to-deliver state. (5) It should be resistant enough to environmental conditions so as to remain infectious or operational long enough to affect the majority of the target, but not so persistent as to affect the occupying army. (6) It should be resistant to treatment by antibiotics, antibodies, pharmaceutical drugs, etc.

69. According to the Center for Disease Control, human anthrax has three major clinical forms: cutaneous, inhalation, and gastrointestinal. Cutaneous anthrax is a result of introduction of the spore through the skin; inhalation anthrax, through the respiratory tract; and gastrointestinal anthrax, by ingestion. If untreated, anthrax in all forms can lead to septicemia and death. Early treatment of cutaneous anthrax is usually curative, and early treatment of all forms is important for recovery. Patients with gastrointestinal anthrax have reported-case fatality rates ranging from 25% to 75%.

According to the Center for Civilian Biodefense Studies at John Hopkins Univ., given appropriate weather and wind conditions, 50 kilograms of anthrax released from an aircraft along a 2 kilometer line could create a lethal cloud of anthrax spores that would extend beyond 20 kilometers downwind. The aerosol cloud would be colorless, odorless and invisible following its release. Given the small size of the spores, people indoors would receive the same amount of exposure as people on the street. There are currently no atmospheric warning systems to detect an aerosol cloud of anthrax spores. The first sign of a bioterrorist attack would most likely be patients presenting with symptoms of inhalation anthrax.

According to the Department of Defense's Anthrax Vaccine Immunization Program, anthrax is effective as a biological weapon because: (1) Anthrax is almost always deadly if not treated early. (2) Spores can be produced in large quantities using basic knowledge of biology. (3) Spores can be stored for decades without losing potency. (4) Spores can be easily spread in the air by missiles, rockets, artillery, aerial bombs and sprayers.

Chapter 20: The Fungi

OVERVIEW

In this chapter, you will learn about the general features and ecology of fungi, including a brief description of the major fungal divisions. Fungi are a very diverse group. Some fungi are edible mushrooms, some are yeast used in the baking industry (because they produce carbon dioxide gas which helps bread "rise" by forming gas bubbles in the dough) and the brewing industry (because they produce alcohol under certain conditions), and others are used to make cheese. Some fungi are poisonous (like the "death cap mushroom"), and others produce substances that are hallucinogenic to humans. Fungi affect human lives as gourmet treats, everyday foods, deadly poisons, and mind-altering drugs.

1) What Are the Main Adaptations of Fungi?

Most fungi have a filamentous body called a **mycelium**, an interwoven mass of threadlike filaments called **hyphae**. The hyphae are either made of one cell with many nuclei, or made of many cells separated by porous partitions called **septa** (plural of **septum**). Fungal cell walls contain chitin. Fungi do not move about, but hyphae can grow rapidly in any direction within suitable environments.

Fungi obtain their nutrients from other organisms. Some are **saprobes**, digesting the bodies of dead organisms, secreting enzymes outside their bodies to digest complex molecules, and then absorbing the smaller subunits. Others are parasitic, feeding on living organisms and causing disease. Still others live in mutually beneficial relationships (are **symbiotic**) with other organisms. A few are predatory.

Most fungi can reproduce both sexually and asexually. In simple asexual reproduction, a mycelium breaks into pieces, each growing into a new mycelium. Some fungi produce sexual and asexual **spores**. Spores are small, spherical, thick-walled single cells that are dispersed by wind or water and develop into new mycelia. Mitotic division of haploid mycelium cells produces asexual spores. Fusion of haploid cells produces diploid zygotes, which undergo meiosis to make haploid sexual spores.

2) How Are Fungi Classified?

Nearly 100,000 species of fungi are known, with about 1000 new species described each year. The chytrid fungi (Phylum Chytridiomycota) live in water and have swimming spores, each with a flagellum, that need water for dispersal. The chytrids are an ancient group of fungi that gave rise to the other groups of modern fungi. Most feed on dead aquatic plants or other detritus in watery environments, but some types are parasites, especially of frogs.

The **zygote fungi** (Phylum Zygomycota) cause soft fruit rot and black bread mold (*Rhizopus*). Asexual reproduction in zygote fungi is initiated by the formation of haploid spores in black spore cases called **sporangia**. These spores disperse through the air. Haploid cells from hyphae of different mating types fuse sexually to produce diploid **zygospores** that disperse through the air, then undergo meiosis to produce new hyphae.

The **sac fungi** (Phylum Ascomycota) reproduce by forming sexual spores within sacs called **asci** (plural of **ascus**). Sac fungi are colorful molds, morels, and truffles; others cause Dutch elm disease and chestnut blight. Some yeasts cause vaginal infections, while others are used in the baking and brewing industries. One species produces penicillin.

The **club fungi** (Phylum Basidiomycota) produce club-shaped reproductive structures (**basidia**, plural of **basidium**) containing sexual spores (**basidiospores**). Germinating basidiospores form hyphae of two different mating types, which fuse to form underground mycelia. The filaments grow outward below ground from the original spore in a roughly circular pattern, occasionally sending up numerous mushrooms in a ringlike pattern called a **fairy ring**. One mycelium covers 38 acres in northern Michigan state and is estimated to be 1500 years old. Common club fungi are mushrooms, puffballs, shelf fungi (monkey-stools), as well as parasitic rusts and smuts of grain crops.

The **imperfect fungi** (Phylum Deuteromycota) are a group in which the means of sexual reproduction (if any) has not been observed by scientists. Therefore, they cannot be classified into one of the four main phyla.

Some fungi form symbiotic associations. **Lichens** are associations of fungi (mostly ascomycetes) with photosynthetic green algae or cyanobacteria. The fungus provides support and water, and the algal or bacterial partner provides food. They grow on bare rock, acting as original colonizers of new environments. Some lichens in the Arctic are 4000 years old.

Mycorrhizae (plural of **mycorrhiza**) are fungi associated symbiotically with the roots of about 80% of all rooted plants. The hyphae invade the root cells, digest organic nutrients in the soil, absorb the digested nutrients and water, and pass them directly into the root cells. The plant provides sugar to the fungal cells.

3) How Do Fungi Affect Humans?

Some fungi, such as mushrooms and truffles, serve as food. Some are decomposers of other organisms, releasing nutrients with carbon, nitrogen, and phosphorous, and minerals used by plants. Cycling of nutrients within ecosystems would cease without fungal and bacterial decomposers. Parasitic fungi can cause skin diseases such as ringworm and athlete's foot, lung diseases such as valley fever and histoplasmosis, and vaginal yeast infections. Molds of the genus *Aspergillus* produce aflatoxins, highly toxic, carcinogenic compounds. The ascomycete *Claviceps* infects rye plants and causes a disease called ergot poisoning. The symptoms of ergot poisoning include gangrene, burning sensations, vomiting, convulsive twitching, and vivid hallucinations. LSD is derived from a component of the ergot toxins. Fungi cause the majority of plant diseases. Fungi cause plant diseases such as chestnut blight, Dutch elm disease, and corn smut. Farmers use "fungal pesticides" to kill crop predators such as termites, tent caterpillars, aphids, citrus mites, and rice weevils. Fungi are used to make bread rise, produce wine from grapes, flavor cheeses, and make beer. An ascomycete mold produces the antibiotic medicine penicillin.

Fungi play a crucial ecological role. The fungi are Earth's undertakers, consuming the dead of all species and returning their component substances to the ecosystems from which they came.

Case study revisited. Some of the deadliest known poisons are made by mushrooms. *Amanita* mushrooms (the death cap) make amatoxins that inhibit the action of an RNA polymerase in human cells, resulting in no mRNA production, causing such severe liver damage that only a liver transplant will prolong life. Mushrooms of the genus *Psilocybe* produce the mind-altering alkaloid chemicals psilocybin and psilocin, causing intoxication, euphoria, and hallucinations.

KEY WORDS AND CONCEPTS

Fill-In: From the following list of terms, fill in the blanks below.

ascus	fairy ring	lichen	sac fungi	symbiosis
basidium	hypha	mycelium	septum	zygote fungi
club fungi	imperfect fungi	mycorrhizae		

1. A _____ is a close relationship between two types of organisms. This occurs when _____ form an association with plant roots. It also occurs when an association between an alga and a fungus forms a new entity called a _____.

2. _____ are molds of fruit and bread. Morels, truffles, and yeasts belong to the _____. Mushrooms, puffballs, shelf fungi, rusts, and smuts belong to the _____. And, _____ have no known mechanism of sexual reproduction.

3. The saclike case in which sexual spores are formed is called an _____. A diploid cell, often club-shaped, which produces spores by meiosis is called a _____.

4. A _____ is a threadlike structure, many of which make up the fungal body. The body of a fungus, consisting of a mass of threadlike structures is called a _____. The threadlike structures may be separated into individual cells by a _____.

5. A outward underground growth of mycelia from club fungi which periodically emerge as numerous fruiting bodies in a ringlike pattern is called a _____.

Key Terms and Definitions

ascus (as´-kus): a saclike case in which sexual spores are formed by members of the fungal division Ascomycota.

basidiospore (ba-sid´-ē-ō-spor): a sexual spore formed by members of the fungal division Basidiomycota.

basidium (bas-id´-ē-um): a diploid cell, typically club-shaped, formed by members of the fungal division Basidiomycota; produces basidiospores by meiosis.

club fungus: a fungus of the division Basidiomycota, whose members (which include mushrooms, puffballs, and shelf fungi) reproduce by means of basidiospores.

fairy ring: a circular pattern of mushrooms formed when reproductive structures erupt from the underground hyphae of a club fungus that has been growing outward in all directions from its original location.

hypha (hī´-fuh; pl., hyphae): a threadlike structure that consists of elongated cells, typically with many haploid nuclei; many hyphae make up the fungal body.

imperfect fungus: a fungus of the division Deuteromycota; no species in this division has been observed to form sexual reproductive structures.

lichen (lī´-ken): a symbiotic association between an alga or cyanobacterium and a fungus, resulting in a composite organism.

mycelium (mī-sēl´-ē-um): the body of a fungus, consisting of a mass of hyphae.

mycorrhiza (mī-kō-rī´zuh; pl., mycorrhizae): a symbiotic relationship between a fungus and the roots of a land plant that facilitates mineral extraction and absorption.

sac fungus: a fungus of the division Ascomycota, whose members form spores in a saclike case called an *ascus*.

saprobe (sap´-rōb): an organism that derives its nutrients from the bodies of dead organisms.

septum (pl., septa): a partition that separates the fungal hypha into individual cells; pores in septa allow the transfer of materials between cells.

sporangium (spor-an´-jē-um; pl., sporangia): a structure in which spores are produced.

spore: a haploid reproductive cell capable of developing into an adult without fusing with another cell; in the alternation-of-generation life cycle of plants, a haploid cell that is produced by meiosis and then undergoes repeated mitotic divisions and differentiation of daughter cells to produce the gametophyte, a multicellular, haploid organism.

symbiotic: referring to an ecological relationship based on symbiosis.

zygospore (zī´-gō-spor): a fungal spore, produced by the division Zygomycota, that is surrounded by a thick, resistant wall and forms from a diploid zygote.

zygote fungus: a fungus of the division Zygomycota, which includes the species that cause fruit rot and bread mold.

THINKING THROUGH THE CONCEPTS

True or False: Determine if the statement given is true or false. If it is false, change the underlined word(s) so that the statement reads true.

6. _____   The single threadlike filaments that make up the body of a fungus are called mycelia.

7. _____   The cell walls of most fungi are composed of chitin.

8. _____   Fungi are autotrophic.

9. _____   Most fungal nuclei are diploid.

10. _____   Fungi digest food particles outside their bodies.

11. _____   In fungi, haploid sexual spores are produced from diploid zygotes.

12. _____   Fungi and protozoans are decomposers.

13. _____   A fungus causes Dutch elm disease.

14. _____   A fungus found growing beneficially with plant roots is called a lichen.

15. _____   Yeast belong to the fungal phylum Zygomycota.

Identify: Determine whether the following statements refer to **lichens**, **mycorrhizae**, **both**, or **neither**.

16. _____   hyphae often invade plant root cells

17. _____   association between ascomycete fungi and cyanobacteria

18. _____   harm the non-fungal partner

19. _____   early colonizers of bare rocky volcanic islands

20. _____   association between basidiomycetes fungi and the roots of vascular plants

21. _____   association between ascomycete fungi and animals

22. _____   digest organic compounds in the soil

23. _____   represent symbiotic relationships

Matching: Fungi.

24. _____   mushrooms

25. _____   sac fungi

26. _____   causes ringworm

27. _____   produces zygospores

28. _____   no known sexual reproduction

29. _____   yeasts

30. _____   bread mold

31. _____   makes penicillin

Choices

a. zygomycetes

b. ascomycetes

c. deuteromycetes

d. basidiomycetes

32. _____ causes Dutch-elm disease

33. _____ puffballs

34. _____ truffles

35. _____ club fungi

36. _____ causes athlete's foot

Multiple Choice: Pick the most correct choice for each question.

37. With few exceptions, the fungal plant body is composed of
    a. hyphae
    b. chitin
    c. mycorrhizae
    d. vascular tissue
    e. cellulose

38. Fungal cell walls are characterized by the presence of
    a. algin
    b. cellulose
    c. glucose
    d. chitin
    e. calcium carbonate

39. Ecologically, fungi as well as bacteria are important because they are
    a. predators
    b. the basis of food chain
    c. producers of organic materials
    d. decomposers of organic materials
    e. capable of nitrogen-fixation

40. Spore formation occurs in sac-like structures within the
    a. Zygomycetes
    b. Basidiomycetes
    c. the imperfect fungi
    d. Ascomycetes
    e. Deuteromycetes

41. Lichens are a symbiotic relationship between
    a. two different fungi
    b. a protozoan and an alga
    c. mold and mildew
    d. a protozoan and a fungus
    e. an alga and a fungus

42. Which of the following is false?
    a. Fungi are used in the wine-making process
    b. Fungi are used in the bread-making process
    c. Fungi are used to make some cheeses
    d. All mushrooms are safe to eat
    e. Some mushrooms make substances that cause hallucinations in humans

43. Which of the following is true?
    a. The alcohol that yeast produces is important in bread making
    b. The gas that yeast produces is important in making alcoholic drinks
    c. Yeast is a fungus
    d. Bleu cheese is ruined if fungus is present while it is being made
    e. All mushrooms are safe to eat.

APPLYING THE CONCEPTS

These practice questions are intended to sharpen your ability to apply critical thinking and analysis to biological concepts covered in this chapter.

44. Be careful what you wish for, in case it comes true. The owner of a peach orchard was worried because his trees were attacked by a parasitic fungus. He wished to get rid of all fungi in his orchard, so he sprayed his trees and the ground with a powerful fungicide. The next season, no fungi were found in the orchard, but the trees grew even more poorly and produced even less fruit than the year before, when they were infected. Why did the orchard owner experience such poor results?

_____

_____

_____

_____

_____

45. Name several reasons why a fungal infection (like athlete's foot) might be more difficult to get rid of than a bacterial infection.

_____

_____

_____

_____

_____

Use the Case Study and the Web sites for this chapter to answer the following questions.

46. Wild mushrooms can be delicious, debilitating or deadly. Even the most common wild mushrooms can have toxic look-a-likes. How does the King boletes (aka *Boletus edulis*) differ from its look-a-likes listed on the "A Few Good Mushrooms" site?

_____

_____

_____

_____

_____

_____

_____

_____

_____

_____

_____

_____

47. The "Death Cap mushroom" (*Amanita phalloides*) is considered to be "The World's Most Dangerous Mushroom." Why?

_____

_____

_____

_____

_____

_____

48. The "Death Cap mushroom" (*Amanita phalloides*) actually produces two types of toxic compounds. What are they?

_____

_____

_____

_____

_____

_____

_____

_____

_____

_____

ANSWERS TO EXERCISES

1. symbiosis
   mycorrhizae
   lichen
2. zygote fungi
   sac fungi
   club fungi
   imperfect fungi
3. ascus
   basidium
4. hypha
   mycelium
   septum
5. fairy ring
6. false, hyphae
7. true
8. false, heterotrophic
9. false, haploid

10. true
11. true
12. false, bacteria
13. true
14. false, mycorrhiza
15. false, Ascomycota
16. mycorrhizae
17. lichens
18. neither
19. lichens
20. mycorrhizae
21. neither
22. mycorrhizae
23. both
24. d
25. b
26. b

27. a
28. c
29. b
30. a
31. b
32. b
33. d
34. b
35. d
36. b
37. a
38. d
39. d
40. d
41. e
42. d
43. c

44. The orchard owner should not have wished for all the fungi to die, for his over-zealous spraying of fungicide killed not only the harmful parasites but also the beneficial mycorrhizae fungi. Without the mycorrhiza, the peach trees could not absorb water and minerals from the soil as efficiently and, therefore, grew poorly and yielded less fruit.

45. One reason fungi are difficult to treat is that they often grow into their host cells, making it difficult to kill them without also harming the host cells as well. Fungi also grow rapidly, so that infections can reoccur quickly if a few cells remain after treatment. In addition, since fungi are eukaryotic cells, their metabolism is more like ours than bacteria, making it more difficult to find treatments that will kill the fungus without also harming the host.

46. According to matkurja.com, Jesenski goban (*Boletus edulis*), also called "jurcek" ("George"). Jesenski goban grows in conifer forests, mostly under firs and pines; loves the Alpine area, due to its popularity it has become quite rare in lowlands near larger cities. Unfortunately, we are not the only ones who love its taste — so do snails and worms, therefore mature mushrooms may frequently be "inhabited." If you are lucky enough to find them, there is a wide choice of dishes you can make with them. The flesh of Jesenski goban is white and fragrant. Small ones are usually pickled; Jesenski goban is also ideal for sun-drying. It is perfect for soups or risotto, sauteed, grilled — gourmets particularly enjoy them fresh, as a salad. Jesenski goban is one of the safest mushrooms, it can be easily distinguished from poisonous ones. The most dangerous look-alikes are Vrazji goban and Leponogi goban: both have reddish and yellowish stems (while the stipe of Jesenski goban is white), but more importantly, both will turn first dark blue then black where cut or bruised. Another similar mushroom is Zolcasti goban (*Tylopilus felleus*), which is not poisonous (merely inedible) and has a distinctly bitter taste.

47. According to D. W. Fischer, this single, widespread species of mushroom is solely responsible for the majority of fatal and otherwise serious mushroom poisoning cases, worldwide as well as in North America. In North America, Death Cap poisonings have been reported from California, Oregon, and New York. In New York, the only known victims to date were natives of Laos. In California and in Oregon, most reported Death Cap poisonings have also involved Southeast Asian immigrants. The poisoning cases typically involve several victims — often including children — who "enjoyed" the mushrooms as a group. One or two deaths per case are common.

48. According to J. Parmentier of microscopy UK, the main poisons of the Death Cap are phallotoxins and amanitins. These compounds are all bicyclic peptides; amanitins are octapeptides (eight amino acids), phallotoxins are heptapeptides (seven amino acids), so the latter have one less amino acid. Alpha-amanitin is extraordinarily toxic for humans: 5–10 milligrams can kill the average person and the average cap of *Amanita phalloides* contains 30–90 milligrams. The poison inhibits the workings of RNA polymerase with the consequence of inhibiting all protein synthesis. Cell death ensues. Because the liver is responsible for a major portion of protein synthesis in the body, the failure of the liver is the most prominent feature of a poisoning by the Death Cap.

Again, D. W. Fischer states that amatoxins contained in the Death Cap are responsible for many of the symptoms suffered by its victims. They are present in all the tissues of the mushroom, in sufficient concentration that two or three grams of mushroom are considered a potentially lethal dose. Amatoxins are cyclopeptides composed of a ring of amino acids that inhibit the production of specific proteins within liver and kidney cells. Without these proteins, cells cease to function. Following ingestion, five to twenty-four hours (average, twelve hours) pass before nausea, vomiting, abdominal pain, and diarrhea begin. These initial symptoms are followed by a brief period of apparent improvement, but without treatment, severe liver damage and kidney failure often result in coma and death.

Chapter 21: The Plant Kingdom

OVERVIEW

In this chapter, you will learn about plants. After a discussion of evolutionary trends, the authors describe the bryophytes, ferns, conifers, and flowering plants. Taxol is especially effective in treating early-diagnosed cases of ovarian cancer. Taxol occurs naturally in the bark of the American yew tree. A single 100 year old yew tree, however, produces only one dose of taxol. The search for medically useful drugs that occur naturally in plants is called *bioprospecting*. Can we find ways to garner the benefits of bioprospecting without sacrificing the very ecosystems that we will need for future bioprospecting?

1) What Are the Key Features of Plants?

Plants are multicellular and use photosynthesis to convert carbon dioxide and water into sugar and oxygen gas. They have an **alternation of generations** life cycle; they produce separate diploid and haploid generations that alternate with each other. The diploid **sporophyte** generation grows by mitosis from a **zygote**. The sporophyte produces haploid spores by meiosis. The haploid **gametophyte** generation grows by mitosis from a spore. The gametophyte produces, by mitosis, gametes (eggs and sperm) that fuse to produce the diploid zygote. As plants become more highly evolved, the sporophyte generation becomes more obvious and long lived, and the gametophyte generation decreases in size and duration.

2) What Is the Evolutionary Origin of Plants?

Plant ancestors were most likely aquatic, photosynthetic protists, similar to present-day green algae. The green algae are mostly multicellular and colonial forms living in fresh water ponds and lakes. Evidence that land plants evolved from green algal ancestors includes similar DNA sequences, and: (1) similar types of chlorophyll and accessory pigments between them, and (2) they both store food as starch and have cell walls made of cellulose. Green algae live in fresh water and their ancestors probably did as well. This suggests that the green algae experienced conditions, such as fluctuating temperatures and periods of dryness, that led to the evolution of adaptations for life on land.

3) How Did Plants Invade and Flourish on Land?

The plant body increased in complexity as plants made the evolutionary transition from water to dry land. Plants adapted to dry land by: (1) anchoring roots to absorb water and nutrients; (2) developing a waxy waterproof **cuticle** on leaf and stem surfaces to reduce water loss; (3) developing **stomata** (pores) in leaves and stems that open for gas exchange and close when water is scarce to reduce evaporation; (4) developing conducting vessels to transport water and minerals upwards from roots to leaves, and move sugars from leaves to other body parts; and (5) producing the stiffening polymer **lignin** in the conducting vessels to support the plant body.

The invasion of land required protection and a means of dispersal for sex cells and developing plants. Instead of flagellated gametes and spores that algae release into the water, plants evolved pollen, seeds, and later flowers and fruits to protect spores, gametes, and young embryos from dessication, attract pollinators, and aid in dispersion of offspring. Plants initially developed *pollen* and *seeds*, and later, *flowers* and *fruits*.

Two major groups of land plants arose from the ancient algal ancestors. The *non-vascular plants*, or **bryophytes**, require a moist environment to reproduce. The *vascular plants*, or **tracheophytes**, have colonized drier habitats. The bryophytes (mosses and liverworts) are nonvascular plants lacking true roots, stems, and leaves. Anchoring rootlike structures, called **rhizoids**, bring water and nutrients into the plant body; they then diffuse throughout the body which must remain small (less than 1 inch tall). Enclosed reproductive structures are present to prevent dessication: **archegonia** (plural of **archegonium**) produce eggs and **antheridia** (plural of **antheridium**) produce sperm. Sperm must swim to eggs through a film of water; thus, most bryophytes are confined to moist areas. In bryophytes, the leafy gametophyte is larger than the leafless sporophyte. In fact, sporophytes grow out of the archegonia located on the gametophyte.

Adaptations that allowed plants to grow taller than the bryophytes included support structures for the body and vessels to conduct water and nutrients. Evolution of rigid conducting cells ("**vessels**") in the **vascular plants** allowed plants to live on dryer land. In vascular plants, and especially the seed plants, the diploid sporophyte is dominant over the smaller, shorter-lived gametophyte.

The seedless vascular plants include the smaller divisions of club mosses (*Lycopodium*) and horsetails (*Equisetum*), and a large division of ferns. Ferns are the only seedless vascular plants with broad leaves. Ferns, and their relatives, share two features with the bryophytes; the small gametophytes lack conducting vessels, and the sperm must swim through water to reach the egg. However, the haploid spores made by the sporophyte are produced in a specialized structure, the *sporangium*.

The seed plants dominate the land due to the evolution of **pollen** (also called "pollen grains") to allow for sperm to reach eggs without swimming through water, and **seeds** to allow embryos to develop without being immersed in water. The pollen, containing sperm-producing cells, is the male gametophyte generation in seed plants. The female gametophyte plant is a small group of haploid cells that produces an egg. Pollen is dispersed by wind or by animal pollinators like bees. Analogous to the eggs of birds and reptiles, seeds consist of: (1) an embryonic plant; (2) a supply of food for the embryo; and (3) a protective outer coat. Seed plants are grouped into two types: *gymnosperms*, which lack flowers, and *angiosperms*, the flowering plants.

Gymnosperms ("naked seed") evolved earlier than the flowering plants and include the conifers (500 species) and two smaller groups, the cycads and gingkos. **Conifers** (500 species of pines, firs, spruces, hemlocks, and cypresses) are most abundant in the far north and at high elevations. They are adapted to dry, cold conditions due to: (1) retention of green leaves throughout the year (**evergreens**), allowing them to photosynthesize and grow all year long; (2) leaves that are thin needles covered with a thick cuticle to minimize evaporation; and (3) production of "antifreeze" in their sap to allow transport of nutrients in sub-freezing temperatures.

A pine tree is a diploid sporophyte. It makes smaller, male and larger, female cones. Male cones release pollen (the male gametophyte) that is carried by wind to female cones. At the base of each scale on the female cone are two **ovules** (immature seeds). In the ovules, diploid spores form and undergo meiosis to produce haploid female gametophytes, which develop and produce eggs. The pollen grains that land on a female cone send out pollen tubes that burrow into the female cone. After 14 months, they each may reach an egg cell, release sperm and fertilization occurs. A fertilized egg is enclosed in a seed and develops into an embryo. The seed is released when the cone matures and its scales separate.

Angiosperms (over 230,000 species) evolved from gymnosperm ancestors that formed an association with insects. The insects ate some pollen for food and carried pollen from plant to plant, Thus, these plants wasted less pollen than gymnosperms. Angiosperms have been successful due to (1) the evolution of **flowers**. Flowers are sporophyte structures containing the male and female gametophytes. Fertilizations occur within the flower after a pollen tube grows to the eggs in the ovary. (2) The evolution of **fruits**. Fruits are ripened flower ovaries containing seeds, housing the embryo plants, that attract animals and entice them to disperse the seeds. And (3), the evolution of broad leaves to increase the amount of sunlight trapped for photosynthesis in warm, moist climates. However, in most temperate climates, angiosperms drop their leaves annually, during fall and winter, and become dormant to reduce water loss.

To discourage animals from eating tender leaves, angiosperms have evolved many defenses. These include thorns, spines, resins, and chemical defenses now harvested by humans to be used in such substances as taxol, aspirin, nicotine, caffeine, mustard, and peppermint.

Case study revisited. Commercial success of a plant-derived drug may threaten the source species' survival, and the best solution may be to find alternative means of producing the drug. Researchers have been working hard to find ways to synthesize taxol without using American yew trees. A great deal of taxol is now made by modifying a similar molecule made by European yews, a species that is easily cultivated and grows relatively quickly, and produces the molecule in its wood and leaves as well as in the bark (the American yew produces taxol only in its bark). Ideally, drug manufacturers seek to make completely synthetic taxol that requires no plant input.

KEY TERMS AND CONCEPTS

From the following list of key terms, fill in the blanks in the following statements.

algae	ferns	seeds
alternation of generations	gametophyte	sporophyte
angiosperms	gymnosperms	stomata
bryophytes	lignin	vessels
cuticle	roots	

1. Most plants have an _____ life cycle consisting of both a diploid _____ generation (that develops from a zygote and produces haploid spores by meiosis) and a haploid _____ generation (that develops from a spore and produces, by mitosis, gametes that fuse to produce a diploid zygote).

2. Plant ancestors were most likely aquatic, photosynthetic protists, similar to present-day _____.

3. Plant adaptations to dry land include: _____ to absorb water and nutrients; _____ to transport water and minerals upward from roots to leaves, and move sugars from leaves to other body parts; _____ in the conducting vessels to support the plant body; waterproof _____ on leaf and stem surfaces to limit water evaporation; _____ in leaves and stems that open for gas exchange and close when water is scarce to reduce evaporation; and _____ to protect young embryos from dessication.

4. The _____ are nonvascular plants lacking true roots, stems, and leaves. They are represented by mosses and liverworts.

5. _____ are seedless vascular plants with broad leaves. They dominated Earth during the Carboniferous period.

6. The _____ include the conifers, the cycads, and the gingkos.

7. The _____ produce seeds, flowers, and fruits. They dominate Earth today.

Key Terms and Definitions

alternation of generations: a life cycle, typical of plants, in which a diploid sporophyte (spore-producing) generation alternates with a haploid gametophyte (gamete-producing) generation.

angiosperm (an´-jē-ō-sperm): a flowering vascular plant.

antheridium (an-ther-id´-ē-um): a structure in which male sex cells are produced, found in the bryophytes and certain seedless vascular plants.

archegonium (ar-ke-gō´-nē-um): a structure in which female sex cells are produced; found in the bryophytes and certain seedless vascular plants.

bryophyte (brī´-ō-fīt): a simple nonvascular plant of the division Bryophyta, including mosses and liverworts.

conifer (kon´-eh-fer): a member of a class of tracheophytes (Coniferophyta) that reproduces by means of seeds formed inside cones and that retains its leaves throughout the year.

cuticle (kū´-ti-kul): a waxy or fatty coating on the exposed surfaces of epidermal cells of many land plants, which aids in the retention of water.

evergreen: a plant that retains green leaves throughout the year.

flower: the reproductive structure of an angiosperm plant.

fruit: in flowering plants, the ripened ovary (plus, in some cases, other parts of the flower), which contains the seeds.

gametophyte (ga-mēt´-ō-fīt): the multicellular haploid stage in the life cycle of plants.

gymnosperm (jim´-nō-sperm): a nonflowering seed plant, such as a conifer, cycad, or gingko.

lignin: a hard material that is embedded in the cell walls of vascular plants and provides support in terrestrial species; an early and important adaptation to terrestrial life.

ovule: a structure within the ovary of a flower, inside which the female gametophyte develops; after fertilization, develops into the seed.

pollen/pollen grain: the male gametophyte of a seed plant.

rhizoid (rī´-zoid): a rootlike structure found in bryophytes that anchors the plant and absorbs water and nutrients from the soil.

seed: the reproductive structure of a seed plant; protected by a seed coat; contains an embryonic plant and a supply of food for it.

sporophyte (spor´-ō-fīt): the diploid form of a plant that produces haploid, asexual spores through meiosis.

stoma (stō´-muh; pl., stomata): an adjustable opening in the epidermis of a leaf, surrounded by a pair of guard cells, that regulates the diffusion of carbon dioxide and water into and out of the leaf.

tracheophyte (trā´-kē-ō-fīt): a plant that has conducting vessels; a vascular plant.

vascular: describing tissues that contain vessels for transporting liquids.

vessel: a tube of xylem composed of vertically stacked vessel elements with heavily perforated or missing end walls, leaving a continuous, uninterrupted hollow cylinder.

zygote (zī´-gōt): in sexual reproduction, a diploid cell (the fertilized egg) formed by the fusion of two haploid gametes.

THINKING THROUGH THE CONCEPTS

True or False: Determine if the statement given is true or false. If it is false, change the underlined word(s) so that the statement reads true.

8. _____ The gametophyte plant generation is diploid.

9. _____ Spores develop into the gametophyte generation.

10. _____ The gametophyte plant produces gametes.

11. _____ The archegonium is a male reproductive organ.

12. _____ The dominant generation of the bryophytes is the gametophyte.

13. _____ A liverwort is an example of a vascular plant.

14. _____ The gametophyte generation of ferns is the large, visible generation.

15. _____ The angiosperms are the flowering plants.

16. _____ Flowers generally contain both male and female gametophytes.

Identify: Determine whether the following statements refer to the **sporophyte** generation, the **gametophyte** generation, **both**, or **neither**.

17. _____ haploid

18. _____ begins as a zygote

19. _____ multicellular

20. _____ produces sex cells by meiosis

21. _____ begins as a spore

22. _____ produces spores by meiosis

23. _____ grows by mitosis

24. _____ dominant form in flowering plants

Matching: Land plants. Some questions may have more than one correct answer.

25. _____ do not produce seeds

26. _____ make pollen grains

27. _____ lack true vascular tissues

28. _____ some rely on insect pollinators

29. _____ sperm must swim to eggs

30. _____ produce fruits

31. _____ seedless vascular plants

32. _____ gametophytes larger than sporophytes

33. _____ produce flowers

34. _____ makes cones

35. _____ most successful land plants today

36. _____ vascular plants

37. _____ mosses

38. _____ seed plants

Choices:
   a. angiosperms
   b. gymnosperms
   c. ferns
   d. bryophytes

Multiple Choice: Pick the most correct choice for each question.

39. In the alternation of generations life cycle, the haploid stage is called the
    a. zygote
    b. gametophyte
    c. sporophyte
    d. antheridium
    e. archegonium

40. The haploid stage of an alternation of generations life cycle produces
    a. seeds
    b. spores
    c. gametes
    d. zygotes
    e. cones

41. During alternation of generations, meiosis results in the production of
    a. egg cells
    b. sperm cells
    c. spores
    d. answers a and b are correct
    e. answers a, b and c are correct

42. The ancestors of the terrestrial plants are considered to be
    a. bryophytes
    b. green algae
    c. red algae
    d. brown algae

43. Which of the following is seen as a basic trend in plant evolution?
    a. gametophyte generation increases in size
    b. gametophyte generation decreases in size
    c. sporophyte generation produces gametes
    d. sporophyte generation decreases in size
    e. water becomes more necessary for reproduction

44. Gymnosperms are characterized by all of the following except
    a. cones
    b. seeds
    c. needle-shaped leaves
    d. evergreen
    e. fruit

45. The term "bioprospecting" refers to
    a. looking for gold
    b. looking for plants that produce oil
    c. looking for medically useful plants
    d. looking for plants with pretty flowers
    e. looking for plants that smell good

46. The American yew tree is a natural source of
    a. wood for making fine furniture
    b. taxol, a cancer fighting agent
    c. antibiotics
    d. drugs to fight baldness
    e. a natural insecticide

APPLYING THE CONCEPTS

These practice questions are intended to sharpen your ability to apply critical thinking and analysis to biological concepts covered in this chapter.

From the figure below, answer the following questions concerning the alternation of generations life cycle.

47. The type of plant body depicted by (1) is commonly called a _____.

48. The type of cell division (2) producing spores is called _____.

49. The type of plant body depicted by (3) is commonly called _____.

50. The fusion of gametes (4) is commonly referred to as _____.

51. Are zygotes typically haploid or diploid cells? _____

52. Are spores typically haploid or diploid cells? _____

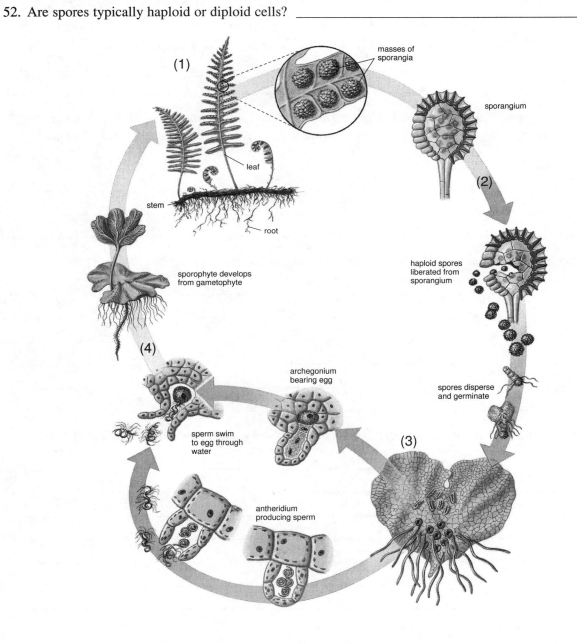

53. In recent years, bees domestic to the United States have decreased in numbers for several reasons: the rise in bee parasites, unfavorable changes in climate, and the spread of so-called killer bees from South America. The decline of domestic bees has worried many farmers, especially those who own orchards. Why should farmers be concerned with bees when they grow apples and peaches?

_____

_____

_____

_____

54. Ferns seem to be the evolutionary link between the mosses below them and the seed plants above them. Can you think of some characteristics of ferns that link them to mosses and to seed plants?

_____

_____

_____

_____

55. Suppose you were shown a piece of an organism that looked something like a pale green mushroom. What things would you look for or test for to determine whether the object is a fungus or a plant?

_____

_____

_____

_____

56. What does "alternation of generations" mean in plants, and how does each generation reproduce?

_____

_____

_____

_____

_____

57. What are three adaptations that terrestrial plants have evolved which allow them to live out of the water?

_____

_____

_____

_____

_____

_____

58. What are several reasons why scientists consider that protists similar to green algae were the ancestors of plants?

_____

_____

_____

_____

_____

59. The ironic thing about finding useful drugs in plants growing naturally, especially in tropical rain forests, is that in order to obtain enough of the drug, the plant species producing it, and the tropical forest in which it grows, may be seriously threatened by over-harvesting. Using taxol as an example, suggest ways to obtain taxol without threatening the American yew tree, the natural source of taxol.

_____

_____

_____

_____

_____

_____

Use the Case Study and the Web sites for this chapter to answer the following questions.

60. For every new plant-derived miracle drug, thousands of plants are tested that have no therapeutic function. *Cyberbotanica* lists a number of possible anti-cancer agents isolated from plants. Are any clinically useful?

_____

_____

_____

61. According to the National Cancer Institute (NCI) Fact Sheet: *Questions and Answers About NCI's Natural Products Branch*, 20,000 plant samples are tested a year, but only seven plant-derived anti-cancer drugs have received Food and Drug Administration approval since 1960. Non-cancer *Plant-Based Medicinal Drugs* have similar discovery rates. Is this an effective drug identification system?

_____

_____

_____

_____

_____

_____

62. Look at the history of five medicines that changed the world. What are they, and what led to their discoveries?

_____

_____

_____

_____

_____

_____

_____

_____

ANSWERS TO EXERCISES

1. alternation of generation
   sporophyte
   gametophyte
2. algae
3. roots
   vessels
   lignin
   cuticle
   stomata
   seeds
4. bryophytes
5. Ferns
6. gymnosperms
7. angiosperms
8. false, haploid
9. true
10. true
11. false, female
12. true
13. false, nonvascular

14. false, sporophyte
15. true
16. true
17. gametophyte
18. sporophyte
19. both
20. neither
21. gametophyte
22. sporophyte
23. both
24. sporophyte
25. c, d
26. a, b
27. d
28. a
29. c, d
30. a
31. c
32. d
33. a

34. b
35. a
36. a, b, c
37. d
38. a, b
39. b
40. c
41. c
42. b
43. b
44. e
45. c
46. b
47. sporophyte
48. meiosis
49. gametophytes
50. fertilization
51. diploid
52. haploid

53. Orchard owners are concerned about the decline in native bee populations because bees are the major organisms responsible for cross-pollination of the flowering trees in their orchards, carrying pollen from one tree's flowers to those of another tree. Without adequate numbers of bees, maximum pollination will not occur, leading to reductions in the amount of fruit produced.

54. Ferns are like seed plants in that they have true roots, stems, and leaves, as well as vascular tissue that transports water and nutrients through the plant, and a life cycle in which the diploid body stage predominates. Ferns are like mosses because they have flagellated sperm that must swim to the eggs, and they do not make seeds or cones or flowers.

55. You should test for whether chlorophyll is present (a plant trait) or not (a fungal trait). Also, test to see whether chitin is present (a fungal trait) or whether cellulose is present (a plant trait). Finally, look at the object under the microscope to see whether it is made of a number of individual filaments (a fungal trait) or not (a plant trait).

56. Alternation of generation refers to a pattern in plants in which a haploid, gamete-producing "gametophyte" generation alternates with a diploid, spore-producing "sporophyte" generation. Diploid sporophyte cells give rise to haploid spores through meiosis. Each spore grows by mitosis into a haploid gametophyte plant, the haploid cells of which give rise, by mitosis, into haploid gametes, which join by fertilization to produce the diploid sporophyte, which grows by mitosis.

57. Land plants have evolved protection from drying out for the gametes and the developing diploid sporophyte plant within specialized organs. Land plants have evolved decreased dependence on water as a medium for sperm transfer, especially the gymnosperms and angiosperms which have developed pollen grains. Land plants have also developed a vascular system to supply water to all areas of the plant, and have reduced the size and duration of the haploid gametophyte stage in the alternation of generations life cycle.

58. Evidence that land plants evolved from green algae ancestors includes similarities in their DNA sequences, and similar types of chlorophyll and accessory pigments in green algae and land plants. In addition, green algae and land plants both store food as starch, and have cell walls made of cellulose. Also, green plants live in fresh water, allowing them to adapt to environmental conditions found on land, such as fluctuating temperatures and periods of dryness.

59. Commercial success of a plant-derived drug may threaten the source species' survival, and the best solution may be to find alternative means of producing the drug. Researchers have been working hard to find ways to synthesize taxol without using American yew trees. A great deal of taxol is now made by modifying a similar molecule made by European yews, a species that is easily cultivated and grows relatively quickly, and produces the molecule in its wood and leaves as well as in the bark (the American yew produces taxol only in its bark). Drug manufacturers seek to make completely synthetic taxol that requires no plant input.

60. As reported by Cyberbotanica, there are over 70 "anti-cancer botanicals" or compounds derived from plants which display anti-cancer properties. Many of those compounds have been proven effective.

61. It depends on your point of view. At the rate of 7 out of 20,000, finding an effective drug is like looking for a needle in a haystack, but if you are a patient whose life was saved by one of these drugs, it surely was worth the effort to find the medication! The possibilities for developing new drugs from forest resources should figure heavily in any calculation of the forest's true worth. All 119 plant-derived drugs used worldwide in 1991 came from fewer than 90 of the 250,000 plant species that have been identified. Each such plant is a unique chemical factory, says Norman R. Farnsworth of the University of Illinois at Chicago, "capable of synthesizing unlimited numbers of highly complex and unusual chemical substances whose structures could [otherwise] escape the imagination… forever." In other words, scientists may be able to synthesize these plant compounds in the laboratory, but dreaming them up, rather than plucking them from the forest and then replicating them, is quite another matter. How much money is this effort worth? Since the mid-1960s, says Farnsworth, one-fourth of all prescription drugs dispensed from American pharmacies contained active ingredients derived from flowering plants. Commercially, these plant-derived medicines are worth about $14 billion a year in the United States and $40 billion worldwide. In 1985, Lilly Research Laboratories sold roughly $100 million worth of vincristine and vinblastine—the periwinkle derivatives used to treat childhood leukemia and Hodgkin's disease—and turned a stunning 8 percent profit. Surprisingly, U.S. pharmaceutical companies do very little research on developing new drugs from wild plants. Why? For one thing, the industry has come to rely more on synthesized chemicals than on natural compounds for drugs, so the

backlog of active natural substances still waiting to be tested is growing. For another, drug companies worry about whether they will be able to patent uses of natural products. On the other hand, in the early 1990s, the U.S. National Cancer Institute earmarked $8 million to screen 50,000 natural substances for activity against 100 cancer cell lines and the AIDS virus. China, Germany, India, and Japan, among others, are also screening wild species for new drugs. effective in clinical trials and others are undergoing clinical trials at the present time.

62. According to S. R. King of Shaman Pharmaceuticals, five medicines that changed the world include: (1) Quinine, from the bark of the cinchona tree, is used to treat malaria. (2) Coca, from the coca tree is used to treat ailments as diverse as toothache and altitude sickness. Cocaine is derived from coca. (3) Curare, a mixture of ingredients from several plants, is used as a muscle relaxant to ease the stiffened muscles caused by polio and to treat such diverse conditions as lockjaw, epilepsy, and chorea; the active ingredient of curare, d-tubocurarine, led to the skeletal muscle relaxant Intocostrin, used in surgery. (4) If you hold the leaf of the jaborandi tree (*Pilocarpus jaborandi*) up to the light, you see translucent droplets on its surface. Each droplet is a gland that secretes an alkaloid-rich oil. Several substances are extracted from this aromatic oil, including the alkaloid pilocarpine, a weapon against the blinding disease glaucoma. Applied to the scalp, it is said to prevent baldness. An infusion of the powdered leaves has been used as a stimulant and expectorant in diabetes and asthma. It has been incorporated into the treatment of a number of diseases including pleurisy (inflammation of the lung tissue) and rheumatism (muscle and joint pain). It is used in tablet form, under the name Salegen, to treat xerostoma, or dry mouth syndrome. (5) Natives used pineapple poultices to reduce inflammation in wounds and other skin injuries. Native people also drank the juice to aid digestion and to cure stomach ache. An enzyme that broke down proteins (bromelain) was isolated from the flesh of the pineapple, accounting for many of the pineapple's healing properties. It has been found that bromelain can also break down blood clots, which consist mainly of protein. Research continues. This enzyme may well play a major part in heart attack treatment in the near future, as well as in the treatment of burned tissue, abscesses, and ulcers.

All these medicinal properties were first noticed by native peoples. Companies from developed countries are now researching plants, some of which are known to have been used for medicinal purposes and others which offer potential. Some emphasize a respectful collaboration with native and indigenous peoples as their primary method of drug discovery.

Chapter 22: The Animal Kingdom

OVERVIEW

In this chapter, you will learn about the major groups of animals. The authors discuss characteristics of the invertebrates including the sponges, cnidarians, flatworms, roundworms, segmented worms, arthropods, mollusks, and echinoderms, as well as the chordate invertebrates; the lancelets and the tunicates. Vertebrate characteristics are discussed, including the chordate fish, amphibians, reptiles, birds, and mammals. **Table 22-1** is an excellent summary of the major characteristics of each phylum of animals discussed in this chapter. Learn as much of it as you can.

The giant squid is the world's largest invertebrate animal, growing 60 feet long, with eyes each the size of a human head, and having ten tentacles with suckers and hooks, and a powerful beak. Scientists know nothing of its habits, lifestyle, or behavior, because no one has seen this squid in its natural habitat, since they inhabit deep ocean waters, beyond the reach of human divers. All specimens have been dead or were dying.

1) What Characteristics Define an Animal?

The characteristics that collectively define an animal include: (1) multicellular bodies; (2) obtaining energy by eating the bodies of other organisms; (3) sexual reproduction (typically); (4) cells that lack cell walls; (5) motility; and (6) rapid response to external stimulation.

2) Which Anatomical Features Mark Branch Points on the Animal Evolutionary Tree?

Animal systematists have looked at features of animal anatomy and embryological development for clues about the evolutionary history of animals. Lack of tissues separates sponges from all other animals. One of the earliest innovations was the appearance of **tissues** — groups of similar cells integrated into a functional unit, such as a muscle. Only sponges lack tissues. An ancient common ancestor without tissues gave rise to both the sponges and the remaining tissue-containing phyla.

All animals with tissues also have symmetrical bodies. That is, their bodies can be bisected along a plane so that the resulting halves are mirror images of one another. Any symmetrical animal has an upper (**dorsal**) surface and a lower (**ventral**) surface. The symmetrical, tissue-bearing animals are divided into two groups: those with **radial symmetry** and those with **bilateral symmetry**. Animals with radial symmetry can be divided into roughly equal halves by <u>any</u> plane cutting through their central axis. Animals with bilateral symmetry can be divided into mirror-image halves only along <u>one</u> plane cutting through their central axis. So, another split in animal evolution separated the ancestors of radially symmetrical cnidarians and ctenophores from the rest of the phyla, which have bilateral symmetry.

Radially symmetrical animals have two embryonic tissue layers (**germ layers**), while bilaterally symmetrical animals have three. Radial animals have an inner layer of **endoderm** that forms the lining of most hollow organs, and an outer layer of **ectoderm** that forms the tissue covering the body and lining the inner cavities and forms the nerve cells. Between these layers, bilateral animals have a layer of **mesoderm** that forms muscles and circulatory and skeletal systems (when present). Echinoderms confuse this issue a bit. Adult echinoderms are radially symmetrical, but they have three tissue layers and their larvae are bilateral. Therefore, they are classified as bilateral animals.

Radial animals are either **sessile** (fixed to one spot) or float around. Bilateral animals move under their own power in a particular direction, typically in the direction of the head. Thus, as **cephalization**, the concentration of sense organs and brain in the head, evolved in the bilateral animals, they had an advantage.

Cephalization produces an **anterior** (head) end, with sense organs, and a **posterior** (tail) end.

After the origin of bilateral symmetry, a fluid-filled body cavity between the digestive tube (gut) and the outer body wall evolved. Some bilateral animals (like flatworms) lack this cavity. Those with the cavity had a "tube-within-a-tube" body plan. This cavity created a space where new organs could evolve. Based on the structure of the cavity, two groups of animals exist. Some phyla, like roundworms (nematodes), have a **pseudocoelom** (the *pseudocoelomates*). A pseudocoelom is a cavity that is not completely surrounded by mesoderm tissue. Phyla with a **coelom** are called *coelomates*. A coelom is a body cavity completely lined with mesoderm. Annelids, arthropods, mollusks, echinoderms, and chordates are coelomate phyla. Recent molecular data (DNA comparisons) indicate that the pseudocoelomate phyla are merely different branches within the coelomate group. Thus, the pseudocoelom may not be a precursor of the coelom, but rather a modification of it.

Coelomates include two distinct evolutionary lines. In **protostome** development, the coelom forms within the space between the body wall and the digestive cavity. In **deuterostome** development, the coelom forms as outgrowths of the digestive cavity. Annelids, arthropods, and mollusks are protostomes, while echinoderms and chordates are deuterostomes.

3) What Are the Major Animal Phyla?

There are two major categories of animals. **Invertebrates**, animals without backbones, are the earliest animals, comprising 97% of animals today. The invertebrate category contains many phyla. **Vertebrates**, animals with backbones (the fish, amphibians, reptiles, birds, and mammals), are all in the phylum Chordata.

The Sponges: Phylum Porifera. Sponges lack true tissues and organs. The sponge body is perforated with numerous tiny pores. Three major cell types are present: (1) **epithelial cells** on the body surface, including pore cells; (2) **collar cells** with flagella to control water flow; and (3) **amoeboid cells** to digest and distribute nutrients, make reproductive cells, and make spines called **spicules**. Sponges may reproduce asexually by **budding** or sexually through fusion of sperm and egg.

The Hydra, Anemones, and Jellyfish: Phylum Cnidaria. Cnidarian bodies are radially symmetrical with two germ layers (ectoderm and endoderm) and a jellylike **mesoglea** in between. They have true tissues and a **nerve net** to control contractile tissue and feeding, but lack organs and brains. They have two body plans. The sessile, tubular **polyp** is usually attached to rocks and possesses **tentacles** to attack and seize prey. The mobile, swimming **medusa** ("jellyfish") has trailing tentacles armed with **cnidocyte** cells that eject poisonous or sticky darts to capture prey. The prey is moved into the **gastrovascular cavity** that has one opening, the mouth/anus. Cnidarians can reproduce both asexually (by budding) and sexually. One group, the corals, with limestone shells, form reefs that are the basis for a diverse ecosystem.

The Flatworms: Phylum Platyhelminthes. Flatworms have bilateral symmetry, a gastrovascular (GV) digestive cavity, and cephalization with clusters of nerve cells called **ganglia** (plural of **ganglion**) in a simple anterior brain. They also have **nerve cords** that carry nerve signals to and from the ganglia. Flatworms are either **free-living** or **parasites**. *Planaria* is a free-living flatworm. Parasitic flatworms include intestinal tapeworms (ingested as encapsulated **cysts**) and liver and blood flukes. Flatworms lack circulatory and respiratory systems, relying on diffusion to move molecules. Flatworms reproduce asexually and sexually; most possess both male and female sex organs, so they are **hermaphroditic**.

The Roundworms: Phylum Nematoda. Nematodes have a tubular, one-way digestive tract (from mouth to intestine to anus), a fluid-filled pseudocoelom that acts as a **hydrostatic skeleton**, and a head with a simple brain. They lack circulatory and respiratory systems, relying on diffusion to move molecules. Most reproduce sexually and have separate male and female sexes. Billions thrive in each acre of topsoil. Parasitic roundworms include trichinella, hookworm, and heartworm.

The Segmented Worms: Phylum Annelida. *Annelid* bodies consist of a series of repeating segments, each with nerve ganglia, excretory structures (**nephridia**, plural of **nephridium**), and muscles. This **segmentation** allows for complex movement. Annelids have a true coelom that acts as a hydrostatic skeleton. Their blood is confined to the heart and blood vessels, so they have a **closed circulatory system**. The digestive system of earthworms consists of a mouth, **pharynx**, esophagus, *crop*, *gizzard*, intestine, and anus. Segmented worms include earthworms (the *oligochaetes*), marine tube worms (the *polychaetes*), and the freshwater *leeches* (both carnivorous and parasitic forms).

   The Insects, Arachnids, and Crustaceans: Phylum Arthropoda. *Arthropods* (more than 1,000,000 species) are the most successful group of animals on Earth due to the following adaptations: an **exoskeleton** of **chitin** with jointed appendages; segmentation; and efficient gas exchange mechanisms including **gills** (in crustaceans), **tracheae** (in insects), or **book lungs** (in spiders). Their blood flows through vessels and enclosed body cavities called **hemocoels**, so they have an **open circulatory system**. Arthropods have well-developed sensory and nervous systems, including **compound eyes**. The exoskeleton periodically must be shed (**molted**) and replaced with a larger one. *Insects* are the most diverse and abundant arthropods (850,000 species, class Insecta), with three pairs of legs and two pairs of wings, and body regions called the **head, thorax**, and **abdomen**. Some insects undergo complete **metamorphosis**: from the egg to the **larva,** which is adapted for feeding, to the **pupa**, a nonfeeding form in which physical changes occur, to the adult, which is adapted for reproduction. Spiders, scorpions, ticks, and mites are *arachnids* (class Arachnida) with four pairs of legs and simple eyes. Most are carnivorous predators. Crabs, shrimp, crayfish, lobsters, and barnacles are *crustaceans* (class Crustacea), the only primarily aquatic arthropods. Crustaceans have two pairs of antennae and many other appendages, compound eyes, and gills.

   The Snails, Clams, and Squid: Phylum Mollusca. *Mollusks* have a moist, muscular body with a hydrostatic skeleton, an open circulatory system (except for the cephalopods), and a **mantle**. The mantle is an extension of the body wall that forms a gill chamber and may secrete a calcium carbonate shell. Snails and slugs are *gastropod* mollusks (class Gastropoda) with a muscular foot and a rasping **radula** used to scrape algae from rocks for food. Scallops, mussels, clams, and oysters are *bivalve* mollusks (class Bivalvia) with two shells connected by a flexible hinge. Bivalves are filter feeders, using gills for respiration and feeding. Octopuses, squids, nautiluses, and cuttlefish are *cephalopod* mollusks (class Cephalopoda), that are mostly marine predators. In cephalopods, the foot has evolved into tentacles with suction discs. They move by jet propulsion caused by forceful expulsion of water from their mantle cavity.

   The Sea Stars, Sea Urchins, Sand Dollars, and Sea Cucumbers: Phylum Echinodermata. The *echinoderms* have free swimming embryos with bilateral symmetry, but the adults have radial symmetry, lack a head, have an internal **endoskeleton** of calcium carbonate plates, and move slowly by using tube feet (rows of suction cups). The tube feet are part of a **water vascular system**; water enters through a **sieve plate**, passes through a central canal and into radial canals. Each radial canal has many tube feet, each with a muscular **ampulla** or squeeze bulb. Echinoderms have no circulatory system. Many, especially sea stars, can regenerate lost parts.

   The Tunicates, Lancelets, and Vertebrates: Phylum Chordata. *Chordates* share certain characteristics. (1) They have a **notochord**, which is a stiff flexible anterior-posterior rod made of **cartilage**, for muscle attachment. The notochord is replaced by the bony backbone (**vertebral column**) in vertebrates. (2) They have a **dorsal, hollow nerve cord** with an anterior brain. They have (3) **pharyngeal gill slits** and (4) a **post-anal tail** early in development. The invertebrate chordates lack a backbone and belong to two groups: the small, fishlike *lancelets*, and the *tunicates* (including the sea squirts). Vertebrates have an endoskeleton of cartilage (sharks) or bone, paired appendages (fins, limbs, wings), and large complex brains. There are seven classes of vertebrates:

   The **jawless fishes** (class Myxani: marine *hagfish*, and class Petromyzontiformes: aquatic *lampreys*) have unpaired fins, cartilage skeletons, no scales, and circular gill slits. The **cartilaginous fish** (class Chondrichthyes: sharks, rays, skates) have skeletons made of cartilage, leathery skin with tiny scales, a two-chambered heart, and rows of razor-sharp teeth. The **bony fish** (class Osteichthyes) have bony skeletons, gills, and swim bladders (for buoyancy) and a two-chambered heart. **Amphibians** (class Amphibia: frogs and salamanders) have limbs, primitive lungs, and moist skin for gas exchange in adult forms. They have a three-chambered heart, and use external fertilization with juvenile gilled forms developing in water.

Reptiles and Birds (class Reptilia: lizards, snakes, turtles, alligators, and crocodiles) have tough, scaly, waterproof skin, and internal fertilization. They have shelled **amniote eggs** that prevent desiccation of the embryos due to the presence of an internal **amnion** membrane that encloses the embryo in a watery environment. Reptiles have efficient lungs, a three-chambered heart (alligators and crocodiles have four-chambered hearts), and limbs (except the snakes). Birds are considered a distinctive group of "reptiles" (often called class Aves). They have wings for flight, feathers for heat insulation, hollow bones, internal fertilization, amniote eggs, warm-bloodedness, a four-chambered heart, and a respiratory system with lungs and air-sacs.

Mammals (class Mammalia) have limbs modified for running, swimming, flying, or grasping, hair for heat insulation, warm-bloodedness, a four-chambered heart, and highly developed brains They have internal fertilization, **mammary glands** that produce milk for feeding young, and embryonic development within the uterus in females (except the egg-laying **monotreme** mammals: platypus and spiny anteater). Most mammals are placental. They retain developing young for long periods in a uterine **placenta** where gas, nutrient, and waste exchange occurs. However, the young of **marsupial** mammals (opossums, koalas, kangaroos) leave the uterus and crawl into a protective pouch to continue development.

MAJOR EVOLUTIONARY ADVANCES

Porifera:	multicellular
Cnidaria:	two tissue layers gastrovascular cavity
Platyhelminthes:	true organs bilateral symmetry three tissue layers
Nematodes:	separate mouth and anal openings to the digestive tract
Annelids:	segmentation circulatory system internal body cavity
Arthropods:	exoskeleton jointed appendages
Mollusca:	enlarged brain
Echinoderms:	internal skeleton
Chordates:	internal cartilaginous or bony skeleton complex nervous system

KEY TERMS AND CONCEPTS

1. Fill in the information requested in the following table.

Phylum	Number of germ (tissue) layers	Type of body symmetry present	Type of body cavity present	Is body segmentation present?
Porifera				
Cnidaria				
Platyhelminthes				
Nematoda				
Annelida				
Arthropoda				
Chordata				

Key Terms and Definitions

abdomen: the body segment at the posterior end of an animal with segmentation; contains most of the digestive structures.

amnion (am´-nē-on): one of the embryonic membranes of reptiles, birds, and mammals; encloses a fluid-filled cavity that envelops the embryo.

amniote egg (am-nē-ōt´): the egg of reptiles and birds; contains an amnion that encloses the embryo in a watery environment, allowing the egg to be laid on dry land.

amoeboid cell: a protist or animal cell that moves by extending a cellular projection called a pseudopod.

ampulla: a muscular bulb that is part of the water-vascular system of echinoderms; controls the movement of tube feet, which are used for locomotion.

anterior: the front, forward, or head end of an animal.

bilateral symmetry: a body plan in which only a single plane through the central axis will divide the body into mirror-image halves.

book lung: a structure composed of thin layers of tissue, resembling pages in a book, that are enclosed in a chamber and used as a respiratory organ by certain types of arachnids.

budding: asexual reproduction by the growth of a miniature copy, or bud, of the adult animal on the body of the parent. The bud breaks off to begin independent existence.

cartilage (kar´-teh-lij): a form of connective tissue that forms portions of the skeleton; consists of chondrocytes and their extracellular secretion of collagen; resembles flexible bone.

cephalization (sef-ul-ī-zā´-shun): the tendency of sensory organs and nervous tissue to become concentrated in the head region over evolutionary time.

chitin (kī´-tin): a compound found in the cell walls of fungi and the exoskeletons of insects and some other arthropods; composed of chains of nitrogen-containing, modified glucose molecules.

closed circulatory system: the type of circulatory system, found in certain worms and vertebrates, in which the blood is always confined within the heart and vessels.

cnidocyte (nīd´-ō-sīt): in members of the phylum Cnidaria, a specialized cell that houses a stinging apparatus.

coelom (sē´-lōm): a space or cavity that separates the body wall from the inner organs.

collar cell: a specialized cell lining the inside channels of sponges. Flagella extend from a sievelike collar, creating a water current that draws microscopic organisms through the collar and into the body, where they become trapped.

compound eye: a type of eye, found in arthropods, that is composed of numerous independent subunits called *ommatidia*. Each ommatidium apparently contributes a piece of a mosaiclike image perceived by the animal.

cyst (sist): an encapsulated resting stage in the life cycle of certain invertebrates, such as parasitic flatworms and roundworms.

deuterostome (doo´-ter-ō-stōm): an animal with a mode of embryonic development in which the coelom is derived from outpocketings of the gut; characteristic of echinoderms and chordates.

dorsal (dor´-sul): the top, back, or uppermost surface of an animal oriented with its head forward.

ectoderm (ek´-tō-derm): the outermost embryonic tissue layer, which gives rise to structures such as hair, the epidermis of the skin, and the nervous system.

endoderm (en´-dō-derm): the innermost embryonic tissue layer, which gives rise to structures such as the lining of the digestive and respiratory tracts.

endoskeleton (en´-dō-skel´-uh-tun): a rigid internal skeleton with flexible joints to allow for movement.

epithelial cell (eh-puh-thē´-lē-ul): a flattened cell that covers the outer body surfaces of a sponge.

exoskeleton (ex´-ō-skel´-uh-tun): a rigid external skeleton that supports the body, protects the internal organs, and has flexible joints that allow for movement.

free-living: not parasitic.

ganglion (gang´-lē-un): a cluster of neurons.

gastrovascular cavity: a saclike chamber with digestive functions, found in simple invertebrates; a single opening serves as both mouth and anus, and the chamber provides direct access of nutrients to the cells.

germ layer: a tissue layer formed during early embryonic development.

gill: in aquatic animals, a branched tissue richly supplied with capillaries around which water is circulated for gas exchange.

head: the anteriormost segment of an animal with segmentation.

hemocoel (hē´-mō-sēl): a blood cavity within the bodies of certain invertebrates in which blood bathes tissues directly; part of an open circulatory system.

hermaphroditic (her-maf´-ruh-dit´-ik): possessing both male and female sexual organs. Some hermaphroditic animals can fertilize themselves; others must exchange sex cells with a mate.

hydrostatic skeleton (hī-drō-stat´-ik): a body type that uses fluid contained in body compartments to provide support and mass against which muscles can contract.

invertebrate (in-vert´-uh-bret): an animal that never possesses a vertebral column.

larva (lar´-vuh): an immature form of an organism with indirect development prior to metamorphosis into its adult form; includes the caterpillars of moths and butterflies and the maggots of flies.

mammary gland (mam´-uh-rē): a milk-producing gland used by female mammals to nourish their young.

mantle (man´-tul): an extension of the body wall in certain invertebrates, such as mollusks; may secrete a shell, protect the gills, and, as in cephalopods, aid in locomotion.

marsupial (mar-soo´-pē-ul): a mammal whose young are born at an extremely immature stage and undergo further development in a pouch while they remain attached to a mammary gland; includes kangaroos, opossums, and koalas.

medusa (meh-doo´-suh): a bell-shaped, typically free-swimming stage in the life cycle of many cnidarians; includes jellyfish.

mesoderm (mēz´-ō-derm): the middle embryonic tissue layer, lying between the endoderm and ectoderm, and normally the last to develop; gives rise to structures such as muscle and skeleton.

mesoglea (mez-ō-glē´-uh): a middle, jellylike layer within the body wall of cnidarians.

metamorphosis (met-a-mor´-fō-sis): in animals with indirect development, a radical change in body form from larva to sexually mature adult, as seen in amphibians (tadpole to frog) and insects (caterpillar to butterfly).

molt: to shed an external body covering, such as an exoskeleton, skin, feathers, or fur.

monotreme: a mammal that lays eggs; for example, the platypus.

nephridium (nef-rid´-ē-um): an excretory organ found in earthworms, mollusks, and certain other invertebrates; somewhat resembles a single vertebrate nephron.

nerve cord: a paired neural structure in most animals that conducts nervous signals to and from the ganglia; in chordates, a nervous structure lying along the dorsal side of the body; also called spinal cord.

nerve net: a simple form of nervous system, consisting of a network of neurons that extend throughout the tissues of an organism such as a cnidarian.

notochord (nōt´-ō-kord): a stiff but somewhat flexible, supportive rod found in all members of the phylum Chordata at some stage of development.

open circulatory system: a type of circulatory system found in some invertebrates, such as arthropods and mollusks, that includes an open space (the hemocoel) in which blood directly bathes body tissues.

parasite (par´-uh-sīt): an organism that lives in or on a larger prey organism, called a *host*, weakening it.

pharyngeal gill slit (far-in´-jē-ul): an opening, located just posterior to the mouth, that connects the digestive tube to the outside environment; present (as some stage of life) in all chordates.

pharynx (far´-inks): in vertebrates, a chamber that is located at the back of the mouth and is shared by the digestive and respiratory systems; in some invertebrates, the portion of the digestive tube just posterior to the mouth.

placenta (pluh-sen´-tuh): in mammals, a structure formed by a complex interweaving of the uterine lining and the embryonic membranes, especially the chorion; functions in gas, nutrient, and waste exchange between embryonic and maternal circulatory systems and secretes hormones.

placental (pluh-sen´-tul): referring to a mammal, possessing a placenta (that is, species that are not marsupials or monotremes).

polyp (pah´-lip): the sedentary, vase-shaped stage in the life cycle of many cnidarians; includes hydra and sea anemones.

post-anal tail: a tail that extends beyond the anus; exhibited by all chordates at some stage of development.

posterior: the tail, hindmost, or rear end of an animal.

protostome (prō´-tō-stōm): an animal with a mode of embryonic development in which the coelom is derived from splits in the mesoderm; characteristic of arthropods, annelids, and mollusks.

pseudocoelom (soo´-dō-sēl´-ōm): "false coelom"; a body cavity that has a different embryological origin than a coelom but serves a similar function; found in roundworms.

pupa: a developmental stage in some insect species in which the organism stops moving and feeding and may be encased in a cocoon; occurs between the larval and the adult phases.

radial symmetry: a body plan in which any plane along a central axis will divide the body into approximately mirror-image halves. Cnidarians and many adult echinoderms have radial symmetry.

radula (ra´-dū-luh): a ribbon of tissue in the mouth of gastropod mollusks; bears numerous teeth on its outer surface and is used to scrape and drag food into the mouth.

segmentation (seg-men-tā´-shun): an animal body plan in which the body is divided into repeated, typically similar units.

sessile (ses´-ul): not free to move about, usually permanently attached to a surface.

sieve plate: in plants, a structure between two adjacent sieve tube elements in phloem, where holes formed in the primary cell walls interconnect the cytoplasm of the elements; in echinoderms, the opening through which water enters the water-vascular system.

spicule (spik´-ūl): a subunit of the endoskeleton of sponges that is made of protein, silica, or calcium carbonate.

tentacle (ten´-te-kul): an elongate, extensible projection of the body of cnidarians and cephalopod mollusks that may be used for grasping, stinging, and immobilizing prey, and for locomotion.

thorax: the segment between the head and abdomen in animals with segmentation; the segment to which structures used in locomotion are attached.

tissue: a group of (normally similar) cells that together carry out a specific function; for example, muscle; may include extracellular material produced by its cells.

trachea (trā´-kē-uh): in birds and mammals, a rigid but flexible tube, supported by rings of cartilage, that conducts air between the larynx and the bronchi; in insects, an elaborately branching tube that carries air from openings called *spiracles* near each body cell.

tube foot: a cylindrical extension of the water-vascular system of echinoderms; used for locomotion, grasping food, and respiration.

ventral (ven´-trul): the lower side or underside of an animal whose head is oriented forward.

vertebral column (ver-tē´-brul): a column of serially arranged skeletal units (the vertebrae) that enclose the nerve cord in vertebrates; the backbone.

vertebrate: an animal that possesses a vertebral column.

water-vascular system: a system in echinoderms that consists of a series of canals through which seawater is conducted and is used to inflate tube feet for locomotion, grasping food, and respiration.

THINKING THROUGH THE CONCEPTS

True or False: Determine if the statement given is true or false. If it is false, change the <u>underlined</u> word(s) so that the statement reads true.

2. _____   Large animals are likely to have <u>less</u> stable internal environments.

3. _____   Adult sponges generally are <u>motile</u>.

4. _____   Water <u>leaves</u> a sponge through the osculum.

5. _____   A free-swimming cnidarian is called a <u>medusa</u>.

6. _____   Flatworms have <u>radial</u> symmetry.

7. _____   Terrestrial arthropods have <u>open</u> circulatory systems.

8. _____   Birds have <u>three-chambered</u> hearts.

9. _____   Echinoderms have <u>endoskeletons</u>.

10. _____   The osteichthyes are <u>cartilaginous</u> fishes.

11. _____   Mammals evolved from <u>birds</u>.

Identify: Determine whether the following statements refer to **poriferans** (sponges), **cnidarians** (jellyfish), **both**, or **neither**.

12. _____   coral reefs

13. _____   some extracellular digestion

14. _____   resemble colonies of cells

15. _____   have polyp and medusae stages

16. _____   hydras

17. _____   simplest multicellular animals

18. _____   separate mouth and anal openings

19. _____   lack true tissues

20. _____   have radial symmetry

21. _____   predatory

22. _____   have stinging cells

23. _____   body wall has many pores

24. _____   have collar cells

25. _____   have bilateral symmetry

26. _____   have internal skeletons of spicules

Matching: Worms.

			Choices
27. _____	have bilateral symmetry		a. roundworms
28. _____	planarians		b. flatworms
29. _____	have nerve ganglia		c. annelid worms
30. _____	segmented		d. all worms
31. _____	hookworms		
32. _____	lacks a separate anus		
33. _____	have coelomic cavities		
34. _____	have true organs		
35. _____	heartworms		
36. _____	are multicellular		
37. _____	have circulatory systems		
38. _____	trichinosis		
39. _____	have nephridia		
40. _____	tapeworms		
41. _____	leeches		

Identify: Determine whether the following statements refer to **arthropods**, **mollusks**, **both**, or **neither**.

42. _____ octopuses

43. _____ have internal rigid skeletons

44. _____ largest Phylum of animals

45. _____ some have closed circulatory systems

46. _____ have open circulatory systems

47. _____ have exoskeletons

48. _____ clams

49. _____ have tracheae

50. _____ are segmented

51. _____ evolved from annelid worms

52. _____ metamorphosis during development

53. _____ snails

54. _____ crabs

55. _____ can learn and remember

56. _____ have water-vascular systems

57. _____ barnacles

Matching: Chordates. Some questions may have more than one correct answer.

58. _____ evolved amniote eggs

59. _____ have reduced segmentation

60. _____ produce placentas

61. _____ frogs

62. _____ have non-bony skeletons

63. _____ have 2-chambered hearts

64. _____ dinosaurs

65. _____ have notochords at some stage of development

66. _____ generally have four-chambered hearts

67. _____ on the boundary between aquatic and terrestrial existence

68. _____ have swim bladders

69. _____ feed milk to their young

70. _____ have open circulatory systems

71. _____ generally have three-chambered hearts

72. _____ have backbones

73. _____ are warmblooded

74. _____ evolved into amphibians

75. _____ have hollow bones and air sacs

76. _____ whales

77. _____ sharks

Choices

a. cartilage fish

b. bony fish

c. amphibians

d. reptiles

e. birds

f. mammals

g. all of these

h. none of these

Multiple Choice: Pick the most correct choice for each question.

78. The simplest multicellular animals are
   a. hydras
   b. anemones
   c. sponges
   d. jellyfishes
   e. flatworms

79. Radial symmetry is exhibited by all of the following except
   a. polyp
   b. medusa
   c. sponge
   d. jellyfish

80. Which characteristic do cnidarians and flatworms have in common?
   a. gastrovascular cavity
   b. bilateral symmetry
   c. flame cells
   d. hydrostatic skeleton
   e. an anus

81. Segmentation first appeared in the
   a. segmented worms
   b. insects
   c. roundworms
   d. mollusks
   e. flatworms

82. Oral and anal openings of the digestive tract first developed in
    a. annelids
    b. platyhelminths
    c. nematodes
    d. cnidarians
    e. sponges

83. The largest and smartest of the invertebrates belongs to
    a. phylum Arthropoda
    b. phylum Echinodermata
    c. phylum Mollusca
    d. phylum Chordata
    e. phylum Vertebrata

84. Which is the largest invertebrate animal on Earth?
    a. the elephant
    b. the giant squid
    c. the blue whale
    d. the brontosaurus
    e. the giant redwood tree

85. What is completely unknown about the giant squid?
    a. the size of its brain
    b. the number of tentacles it has
    c. the composition of its blood
    d. the size of its eyes
    e. its natural behavior

APPLYING THE CONCEPTS

These practice questions are intended to sharpen your ability to apply critical thinking and analysis to biological concepts covered in this chapter.

Based on the following figure, answer the questions below.

(a)

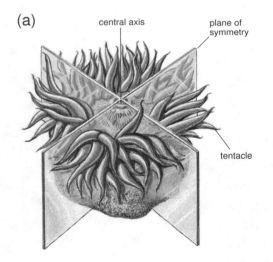

central axis   plane of symmetry   tentacle

(b)

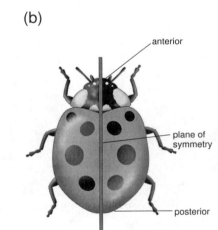

anterior   plane of symmetry   posterior

86. The type of symmetry depicted by (a) is called _____ symmetry.

87. The type of symmetry depicted by (b) is called _____ symmetry.

88. Which groups of land vertebrate animals are found in the Arctic and Antarctic regions, the coldest parts of Earth? Why are they able to live in these regions while other types of animals, especially the arthropods, which are tremendously successful elsewhere, are unable to survive there very well?

_____

_____

_____

_____

_____

_____

89. Complete the following animal classification key by filling in the appropriate phylum.

   A. Animals without tissues. _____

   B. Animals with tissues: go to C, D, or E.

      C. Animals with radial symmetry in all life stages. _____

      D. Animals with bilateral symmetry as larvae and radial symmetry as adults. _____

      E. Animals with bilateral symmetry as adults: go to F or G.

         F. Animals without a coelom (acoelomates). _____

         G. Animals with some kind of a coelom: go to H or I.

            H. Animals with a false body cavity (pseudocoelomates). _____

            I. Animals with a true body cavity (coelomates) and segmentation: go to J or K.

               J. Animals with coelom forming from outgrowths of digestive cavity (deuterostomates). _____

               K. Animals with coelom forming from cell masses within the space between the body wall and the digestive cavity (protostomates): go to L or M.

                  L. Animals with an external skeleton (exoskeleton). _____

                  M. Animals lacking an exoskeleton: go to N or O.

                     N. Animals with an open circulatory system. _____

                     O. Animals with a closed circulatory system. _____

90. Which characteristics do animals have that collectively make them different than the other domains or kingdoms of living things?

_____

_____

_____

91. Describe the type of symmetry possessed by each of the following: a pencil, a vase, a doughnut, a baseball, a spoon, an automobile, a compact disc, and a roast beef sandwich on toast.

_____

_____

Use the Case Study and the Web sites for this chapter to answer the following questions.

92. Many fascinating creatures can be found in the oceans' depths. The first whole giant squid specimen was discovered over 100 years ago. Why are they so difficult to study?

_____

_____

93. Probably the most famous deep sea resident is the coelacanth. Why is it called a "living fossil"?

_____

_____

_____

_____

_____

_____

94. The coelacanth has two biological features more characteristic of mammals than fish. What are they?

_____

_____

_____

_____

_____

_____

95. What is the most unusual feature of the gulper eel?

_____

_____

_____

_____

_____

_____

96. Like many deep water species, gulpers have distinctive markers on their tails. What are they?

_____

_____

ANSWERS TO EXERCISES

1.

Phylum	Number of germ (tissue) layers	Type of body symmetry present	Type of body cavity present	Is body segmentation present?
Porifera	no true tissues	N.A.	none	no
Cnidaria	two	radial	none	no
Platyhelminthes	three	bilateral	none	no
Nematoda	three	bilateral	pseudocoelom	no
Annelida	three	bilateral	coelom	yes
Arthropoda	three	bilateral	coelom	yes
Chordata	three	bilateral	coelom	yes

2. false, more
3. false, sessile
4. true
5. true
6. false, bilateral
7. true
8. false, four-chambered
9. true
10. false, bony
11. false, reptiles
12. cnidarians
13. cnidarians
14. poriferans
15. cnidarians
16. cnidarians
17. poriferans
18. neither
19. poriferans
20. cnidarians
21. cnidarians
22. cnidarians

23. poriferans
24. poriferans
25. neither
26. poriferans
27. d
28. b
29. d
30. c
31. a
32. b
33. c
34. d
35. a
36. d
37. c
38. a
39. c
40. b
41. c
42. mollusks
43. neither
44. arthropods

45. mollusks
46. both
47. arthropods
48. mollusks
49. arthropods
50. arthropods
51. both
52. arthropods
53. mollusks
54. arthropods
55. mollusks
56. neither
57. arthropods
58. d
59. g
60. f
61. c
62. a
63. a, b
64. d
65. g
66. e, f

67. c
68. b
69. f
70. h
71. c, d
72. g
73. e, f
74. b
75. e
76. f
77. a
78. c
79. c
80. a
81. a
82. c
83. c
84. b
85. e
86. radial
87. bilateral symmetry

88. Just about all the land vertebrates in the Arctic and Antarctic regions are birds (penguins, for instance) and mammals (polar bears, for instance). What sets these vertebrate groups apart from the rest is that they are warm blooded animals. They maintain a high constant body temperature that enables them to function well even in the constant coldness of the polar regions. Reptiles and Amphibians, the other types of land vertebrates, cannot remain active in the Arctic or Antarctic environments since their body temperature would be as cold as their surroundings. The same is true for all types of terrestrial invertebrates.

89.
A. Porifera	F. Platyhelminthes	L. Arthropoda
C. Cnidaria	H. Nematoda	N. Mollusca
D. Echinodermata	J. Chordata	O. Annelida

90. Animals have eukaryotic cells (thus, not bacteria or archaebacteria). Animals are multicellular (hence, not protistans). Animals cannot make their own food (thus, not plants). Animals ingest their food and usually digest it within a digestive tract (thus, not fungi).

91. The following show radial symmetry: a pencil, a vase, a doughnut, a baseball, and a compact disc. The following show bilateral symmetry: a spoon, an automobile, and a roast beef sandwich on toast.

92. According to the National Museum of Natural History, the giant squid lives deep in the ocean and spends little time near the surface when healthy. Therefore, they are extremely difficult to study.

93. According to J. D. Knight of Sea and Sky, the coelacanth fish is considered to be a living fossil. They were thought to have been extinct since the end of the Cretaceous period over 65 million years ago. Fossils of the fish have been found that date back over 410 million years. But in 1938, a local fisherman caught a live coelacanth off the coast of South Africa. A second specimen was discovered in 1952 off the coast of the Comoro islands. Since then, live coelacanths have been sighted and photographed in the wild. They have even been found in the waters off southern England. The coelacanth is an endangered species. Because they are so elusive, their exact numbers are not known. However, only 6 to 8 have ever been seen in the wild at one time, and their population off the coast of Africa seems to be on the decline.

94. J. D. Knight of Sea and Sky reports that the coelacanth fish is believed to be closely related to another animal which became the first fish to leave the sea and live on land. Coelacanths have long pectoral and pelvic fins with traces of bones in them. This is believed to be the first step in the evolution of fishes into land-dwelling amphibians. These fins have helped them to earn the nickname, "old fourlegs."

In 1975, a scientist in the American Museum of Natural History's Department of Ichthyology helped dissect a coelacanth. Inside were five fully formed babies, showing us that coelacanth eggs hatch internally. Because of this method of reproduction, the coelacanth cannot produce many eggs, and is able to brood only a few young. It is therefore extremely rare and vulnerable to overfishing.

95. J. D. Knight of Sea and Sky also states that the gulper eel is one of the most bizarre looking creatures in the deep ocean. Its most notable attribute is its large mouth. The eel's mouth is loosely-hinged, and can be opened wide enough to swallow an animal much larger than itself. The hapless prey is then deposited into a pouch-like lower jaw, which resembles that of a pelican. In fact, it is sometimes referred to as the pelican eel. This giant mouth gives the eel its other common name of umbrella-mouth gulper. The gulper's stomach can also stretch to accommodate its large meals. The eel also has a very long, whip-like tail. Some specimens that have been brought to the surface in fishing nets have been known to have their long tails twisted into several knots. The gulper eel grows to a length of about two to six feet and is found in all of the world's oceans at depths ranging from 3,000 to 6,000 feet.

96. According to the Monterey Bay Aquarium, a gulper eel has a bioluminescent organ on the tip of its tail.

Chapter 23: Plant Form and Function

OVERVIEW

In this chapter you will be introduced to the structure of flowering plants. You will also see how the individual structures function so that the plant can survive. You will also learn how plants have modified their structures over time in order to live in particular habitats. Additionally, plants have evolved important interactions with other organisms such as fungi and bacteria These interactions will be discussed.

1) How Are Plant Bodies Organized and How Do Plants Grow?

Plant bodies are organized into the **shoot system**, usually above-ground, and the **root system**, usually below-ground. The shoot system is composed of **stems**, leaves, buds, and when appropriate, flowers and fruits. The role of the above-ground portion is to absorb sunlight energy and conduct photosynthesis, to transport nutrients and other substances between structures, and to produce certain hormones. (Yes, plants have hormones.) Flowers and fruits are specialized for reproduction.

The root system is composed of **roots** and has many functions. It anchors the plant to a substrate (such as soil), absorbs water and minerals from the soil, and stores excess sugars (usually as starch). It also transports water, minerals, and other substances to the shoot, produces certain hormones, and it interacts with soil microorganisms.

The way in which plant structures are organized depends, in part, on their relationship to one of two evolutionary groups: **monocots** and **dicots**. Monocots and dicots differ in the organization of their leaves, flowers, vascular tissue, root systems, and seed structures.

Most plants grow throughout their lives. Growth occurs from undifferentiated cells found in the **meristem** regions. These cells divide by mitosis. Some of the daughter cells produced will mature and specialize according to form and function; these cells have **differentiated**. So, plants grow taller or longer from the tips of shoots and roots through division of **apical meristems**. This is referred to as **primary growth**. Plants grow and become larger in diameter due to divisions of **lateral meristems**, also called the **cambium**. Growth from lateral meristems is called **secondary growth**.

2) What Are the Tissues and Cell Types of Plants?

Plants, just like animals, are made up of different types of tissue. In plants, these include the **dermal tissue system**, which covers the outer surfaces of the plant body; the **ground tissue system**, which consists of all tissues that are not dermal or vascular; and the **vascular tissue system**, which transports water, nutrients, minerals, and hormones throughout the plant body.

Dermal tissue may be either **epidermal tissue** (the epidermis) or the periderm. The **epidermis** is a single cell layer covering leaves, stems, roots, flowers, seeds, and fruit. The epidermis of above-ground organs is protected by a waxy coating, the **cuticle**, which reduces water loss from the plant. The epidermis of below-ground structures does not have a cuticle. (Why not?) Some epidermal cells may elongate, forming hairs. Hairs located on the root system are called **root hairs** and they greatly increase the ability of the roots to absorb water and nutrients.

As woody stems and roots age, their epidermis is replaced by the **periderm**. The periderm is made up of specialized waterproof **cork cells**.

Ground tissue is made up of parenchyma, collenchyma, and sclerenchyma cells. **Parenchyma** cells are thin-walled and alive when mature. The parenchyma cells perform most of the metabolic functions for the plant; therefore, they are quite diverse. Parenchyma cells may also be storage cells. When you eat potatoes, you are eating parenchyma cells storing starch. **Collenchyma** cells are commonly described as having unevenly thickened cell walls. And, although they are alive when mature, they often do not divide. Collenchyma cells help provide support in plants. The strings of celery are made of collenchyma cells and vascular tissue. **Sclerenchyma** cells have thick cell walls that are hardened with a substance called lignin. The gritty texture of pears is because of a type of sclerenchyma cells. Even though these cells are not alive when mature, they continue to support the plant.

Vascular tissue is composed of the transport tissues xylem and phloem. **Xylem** transports water and minerals from the roots to the shoot. The conducting cells found here are either tracheids or vessel elements. **Tracheids** lie end to end, allowing water and minerals to flow through pits in the ends of the cells. **Vessel elements** are larger than tracheids but also lie end to end. The ends may have open pores or they may be completely open, forming a long tube. The **phloem** transports metabolic substances like sugars and amino acids. The phloem is made up of cells called **sieve-tube elements** that are connected end to end. They form long, continuous **sieve tubes**. Between each sieve-tube element are sieve plates, the ends of the sieve-tube elements with pores in them allowing substances to flow through the sieve tube. Although sieve-tube elements are alive when they are mature, they do not contain a nucleus. The functions of sieve-tube elements are directed, instead, by **companion cells**.

3) Roots: Anchorage, Absorption, and Storage

If you were to plant a garden from seed, the first visible sign that the seeds were germinating would be the appearance of the **primary root**. The primary root of a dicot (like carrots or tomatoes) will grow deep into the soil and form a **taproot system**. In taproot systems, the taproot is maintained and smaller, much less developed roots grow out from its sides. The primary root of a monocot (like grass or corn) will grow down into the soil and then die. New roots are produced from the base of the stem, forming a **fibrous root system**.

The growth and development of individual roots is similar between taproot and fibrous root systems. A **root cap** is produced at the tip of a root by the apical meristem. The root cap protects the root apical meristem from the soil as the root grows downward. The root apical meristem also produces a water-permeable epidermis, a cortex, and the vascular cylinder.

Most of the root is made up of parenchyma cells in the **cortex**. Specialized parenchyma cells form the **endodermis**, a ring inside the cortex, surrounding the vascular cylinder. The cell walls of the endodermis are embedded with wax, forming a water barrier called the **casparian strip**. The surfaces of the endodermal cells facing the parenchyma cells and the vascular cylinder do not contain the wax. Therefore, the presence of the casparian strip forces water into the endodermal cells and then into the vascular cylinder. Otherwise, water would flow around the cells and back out the other side of the root.

The **vascular cylinder** of the root contains the xylem and the phloem. Surrounding the xylem and the phloem is the **pericycle**. The pericycle comes from meristem tissue so it is able to divide. Therefore, the pericycle produces **branch roots** off the existing root.

4) Stems: Reaching for the Light

As the stem grows, it may differentiate into buds, leaves, or flowers. The apical meristem is at the tip of the stem in the **terminal bud**. Clusters of meristem cells along the stem form **leaf primordia**, which are the beginnings of new leaves. **Lateral buds** are also produced by the meristem cells. Lateral buds will later develop into branches. Leaves and lateral buds are located at **nodes** on the stem. The areas of the stem between the nodes are called **internodes**.

Stems are formed from dermal tissue, vascular tissue, and two types of ground tissue, cortex and pith. Since the waxy cuticle covers the epidermis of stems (and leaves) and retards water loss, CO_2 and O_2 gas exchange between the cells and the atmosphere is compromised. To let gas exchange occur, specialized cells form pores or **stomata** in the epidermis. As in roots, the cortex is located between the epidermis and the vascular tissue. Additionally, stems have **pith** which is located inside the vascular tissue. Cells of both the cortex and pith provide support to stems, store starch, and may conduct photosynthesis.

Vascular tissue in stems is similar to that of roots and is continuous from the root system, through the stem structures, to the leaves. The apical meristem produces vascular tissue in young stems. In dicots, the xylem that is produced is called **primary xylem** and the phloem that is produced is called **primary phloem**. In woody plants, meristematic tissue called the **vascular cambium** produces **secondary xylem** and **secondary phloem**, giving rise to lateral growth. The young, newly formed secondary xylem is called **sapwood**, while the older xylem is called **heartwood**. In the spring, the xylem that is produced tends to contain large cells since water is plentiful; as the season progresses and summer droughts occur, smaller xylem cells are produced. This pattern creates the typical **annual rings** you have probably seen in tree trunks. To the outside of the secondary phloem, epidermal cells are stimulated by hormones to produce the **cork cambium**. This meristematic tissue produces cork cells that protect the trunk and prevent it from drying out. The **bark** of a tree consists of phloem, the cork cambium, and cork cells.

(note: Only dicots produce woody tissue; monocots do not.)

5) Leaves: Nature's Solar Collectors

You've seen leaves and have probably noticed that most have a broad, flat area. That area is called the **blade** and it is specialized to capture sunlight. The stalk that attaches the blade to the stem of the plant is called the **petiole**. Vascular tissue runs through the petiole and branches out in the blade, forming **vascular bundles** or **veins**. The cuticle covers the leaf, retarding water loss and compromising gas exchange, just like in the stem. Specialized leaf epidermal cells, called **guard cells**, form openings called stomata allowing the exchange of CO_2 and O_2. Beneath the epidermis, parenchyma cells specialized for photosynthesis constitute the **mesophyll**. The mesophyll is often organized into two layers of cells, the **palisade cells** and the **spongy cells**. The mesophyll cells are packed with chloroplasts and they are the cells that conduct most of the photosynthesis for the plant.

6) How Do Plants Acquire Nutrients?

Nutrients, in the form of **minerals** from the soil, are taken into root hairs of roots by <u>active transport</u>. The nutrients then <u>diffuse</u> into pericycle cells through connections between cells called plasmodesmata. <u>Active transport</u> is again used to get the nutrients out of the pericycle and into the extracellular space around the xylem cells within the vascular cylinder. From the extracellular space, the nutrients <u>diffuse</u> into the tracheids of the xylem. (This is why the casparian strip is important.)

Symbiotic relationships between plants and fungi and between plants and soil bacteria increase the ability of plants to get nutrients from nutrient-poor soils. Fungi that form a symbiotic relationship with plant roots are called **mycorrhizae**. Mycorrhizae help plants acquire minerals, especially phosphorous. In return, the fungus receives carbohydrates and amino acids from the plant. Nitrogen is also often limited for many plants. Legumes (peas and beans) are able to get enough nitrogen because they have evolved an association (symbiosis) with **nitrogen-fixing bacteria**. These bacteria have the enzymes needed to convert N_2 from the atmosphere into ammonium (NH_4^+) or nitrate (NO_3^-). This process is called **nitrogen fixation**. The bacteria enter the root hairs of a legume and travel to the cortex of the root. The bacteria multiply and stimulate the cortex cells to multiply as well. All this cell division produces a root **nodule**. Within the nodule nitrogen fixation occurs, supplying the plant with a steady supply of usable nitrogen. In return, the bacteria receive carbohydrates for their energy needs.

7) How Do Plants Acquire Water and Transport Water and Minerals?

It is commonly known that roots take in water for the plant; but how? Water flows into root cells by <u>osmosis</u>. Root cells have a high concentration of minerals in them; this helps to maintain a concentration gradient needed for osmosis to occur. Water also flows freely through and around epidermal and cortex cell walls. When the water encounters the casparian strip around the endodermal cells, it goes into the endodermal cells by <u>osmosis</u>. From the endodermal cells, water flows into the vascular cylinder.

How does water get to the top of a tree 100 m tall? Think about the trees outside your window, or the redwood trees in the Pacific Northwest... Those leaves at the top need water; how does it get there?

Pressure is created by water moving into the root xylem by active transport, osmosis, and diffusion, but this is only sufficient to move water up a plant less than a meter tall. So, there must be more to it.

According to the **cohesion-tension theory**, the water moving through plants is actually pulled up from the top. The water moves by **bulk flow** (that's the movement of fluid molecules as one unit rather than as individual molecules). Minerals are carried passively along with it. The movement of water up a plant is best understood if you remember that water molecules are polar and have <u>cohesion</u> and <u>adhesion</u> properties. The <u>cohesive</u> nature of water molecules holds them together, since they are attracted to one another. The water molecules are also attracted to the walls of the xylem cells; they <u>adhere</u> to them and actually creep up the walls. These properties account for half of the reason water makes it to the top of a tree. The other half relates to the water that is evaporating from the surface of leaves, **transpiration**. As water molecules leave the plant through transpiration, a tension is created. The tension is created because the water molecules leaving the plant are, in essence, attached (hydrogen bonded) to the ones still in the plant. The tension created pulls water molecules up the plant.

However, plants can't always lose water freely. Water loss from the leaves must be controlled. This control occurs by epidermal guard cells closing and opening the stomata. The opening and closing of stomata is regulated by the concentration of potassium in the guard cells. When the potassium concentration is high, water flows into the guard cells by <u>osmosis</u>; the guard cells swell, creating the stomatal opening. When the potassium concentration is lowered in the guard cells, water flows out by <u>osmosis</u>, the guard cells shrink, closing the stomatal opening. This movement of potassium into or out of guard cells is regulated by light reception by the guard cells, CO_2 concentration, and water availability. When water is lost from the plant faster than it can be replaced from the soil, the mesophyll cells release the hormone **abscisic acid**. If abscisic acid has been stimulated, potassium is stopped from moving into the guard cells. Thus they remain closed, conserving water.

8) How Do Plants Transport Sugars?

The end product of photosynthesis is sugar. The areas of the plant that produce these carbohydrates are referred to as a **source** (usually a leaf). And, any structure that needs carbohydrates produced by a source is referred to as a **sink** (the rest of the plant, especially growing areas, roots, and fruits). Therefore, movement of sugars between sources and sinks must occur. The **pressure flow theory** explains how the movement occurs. Glucose is produced by photosynthesis, usually in leaves. The glucose is converted to sucrose and moved by <u>active transport</u> into companion cells of the phloem. The concentration gradient of sucrose increases in companion cells, but it is still low in sieve-tube elements. This allows the sucrose to <u>diffuse</u> into the sieve-tube elements. Water follows the sugar into the sieve-tube by <u>osmosis</u> from nearby xylem, increasing the hydrostatic pressure within the sieve tube. The reverse occurs to unload sucrose into a sink region. Sucrose is <u>actively transported</u> into the sink, such as a developing fruit. This lowers the sucrose concentration in the sieve tube. The lower sucrose concentration causes water to flow out of the sieve tube by <u>osmosis</u>, lowering the hydrostatic pressure again. The hydrostatic pressure difference between the sieve tube at the source and the sieve tube at the sink causes water to flow along the sieve tube, carrying sucrose with it.

Evolutionary Connections: What Are Some Special Adaptations of Roots, Stems, and Leaves?

Things are not always as they appear... Plants have evolved to fit their environment and with this has come modifications of their organs. Roots may be specialized to store a relatively large amount of food as starch. Other plants, such as the orchid, have roots that are capable of photosynthesis. Stems may be used by the plant as a way of starting miniature versions of themselves in a new area. **Runners** serve to propagate plantlets away from the "mother" plant. Stems may also serve as storage receptacles. The stems of cacti store water, while underground stems called **rhizomes** store carbohydrates. Modified branches such as **thorns** protect the plant from would-be predators, whereas **tendrils** grasp support structures, enabling the plant to climb. Modified leaves may store water if they are succulent, or reduce water loss if they are spiny. Other leaves function as tendrils in climbing up a support structure or in nitrogen gathering by capturing and digesting insects.

KEY TERMS AND CONCEPTS

Fill-In: From the following list of key terms, fill in the blanks of the following statements.

abscisic acid	differentiated	pericycle	secondary phloem
annual rings	epidermis	periderm	secondary xylem
apical meristems	endodermis	petiole	shoot system
bark	fibrous root system	pit	sieve plate
blade	ground tissue system	pith	stems
branch roots	heartwood	primary growth	stomata
bulk flow	internode	primary phloem	taproot system
casparian strip	lateral buds	primary root	theory
cambium	lateral meristem	primary xylem	transpiration
cohesion-tension theory	leaf primordia	root system	vascular bundle
cork cambium	leaves	rootcap	vascular cambium
cortex	minerals	sapwood	vascular cylinder
cuticle	node	secondary growth	vascular tissue system
dermal tissue system			

1. The above ground portion of a plant is called the _____ and is composed of _____, _____, buds, flowers, and fruits. The below ground portion of a plant is the _____ which functions to absorb water and _____ from the soil.

2. The area of the stem where leaves attach is called a _____, while the area of the stem between leaves is an _____. A leaf is composed of the _____, which functions to absorb sunlight, and the _____, which attaches to the stem.

3. Shoots grow taller and roots grow longer by cell division of the _____ which are responsible for _____. Cells that specialize in form and function have _____. Shoots grow in diameter by division of cells of the _____, also called the _____. The growth in diameter leads to _____.

4. The three tissue systems of a plant include the _____, _____, and the _____.

5. Epidermal tissue is also referred to as the _____ or, in older tissue, it is called the _____ made up of cork cells. A waxy layer, the _____ covers the shoot system.

6. The first plant structure to emerge from a seed is the _____. In dicots, this root matures into a _____, while in monocots, it matures into a _____. The tip of a growing root is protected by the _____, and water can permeate the cells of the root epidermis, the _____.

7. New leaves are formed by growth of _____ and branches mature from _____.

8. Water entering the vascular cylinder is first forced into cells by the presence of the waxy _____ around cells of the _____. Surrounding the vascular cylinder is a layer of meristematic cells, the _____ which may give rise to _____.

9. In young, developing stems, the vascular tissue is formed by the apical meristem. The xylem that is formed is _____ and the phloem that is formed is _____. As woody plants mature, the _____ produces _____ and _____.

10. Young, newly formed secondary xylem is called _____, while the older xylem is called _____. Variations in rainfall during the growing season produces alternating spring wood and summer wood, the recognizable _____ of tree trunks.

11. Outside the secondary phloem, hormones stimulate epidermal cells to produce the _____ which, in turn, produces cork cells.

12. Plants move water to the top of plants using the _____ which involves the movement of water as one unit or movement in _____ due to the pull from water leaving the plant through _____. The loss of water from the plant is regulated by the opening and closing of _____ often controlled by the hormone _____.

KEY TERMS AND DEFINITIONS

abscisic acid (ab-sis´-ik): a plant hormone that generally inhibits the action of other hormones, enforcing dormancy in seeds and buds and causing the closing of stomata.

annual ring: a pattern of alternating light (early) and dark (late) xylem of woody stems and roots, formed as a result of the unequal availability of water in different seasons of the year, normally spring and summer.

apical meristem (āp´-i-kul mer´-i-stem): the cluster of meristematic cells at the tip of a shoot or root (or one of their branches).

bark: the outer layer of a woody stem, consisting of phloem, cork cambium, and cork cells.

blade: the flat part of a leaf.

branch root: a root that arises as a branch of a preexisting root, through divisions of pericycle cells and subsequent differentiation of the daughter cells.

bulk flow: the movement of many molecules of a gas or fluid in unison from an area of higher pressure to an area of lower pressure.

cambium (kam´-bē-um; pl., cambia): a lateral meristem, parallel to the long axis of roots and stems, that causes secondary growth of woody plant stems and roots. See *cork cambium; vascular cambium.*

Casparian strip (kas-par´-ē-un): a waxy, waterproof band, located in the cell walls between endodermal cells in a root, that prevents the movement of water and minerals into and out of the vascular cylinder through the extracellular space.

cohesion−tension theory: a model for the transport of water in xylem, by which water is pulled up the xylem tubes, powered by the force of evaporation of water from the leaves (producing tension) and held together by hydrogen bonds between nearby water molecules (cohesion).

collenchyma (kōl-en´-ki-muh): an elongated, polygonal plant cell type with irregularly thickened primary cell walls that is alive at maturity and that supports the plant body.

companion cell: a cell adjacent to a sieve tube element in phloem, involved in the control and nutrition of the sieve tube element.

cork cambium: a lateral meristem in woody roots and stems that gives rise to cork cells.

cork cell: a protective cell of the bark of woody stems and roots; at maturity, cork cells are dead, with thick, waterproofed cell walls.

cortex: the part of a primary root or stem located between the epidermis and the vascular cylinder.

cuticle (kū´-ti-kul): a waxy or fatty coating on the exposed surfaces of epidermal cells of many land plants, which aids in the retention of water.

dermal tissue system: a plant tissue system that makes up the outer covering of the plant body.

dicot (dī´-kaht): short for dicotyledon; a type of flowering plant characterized by embryos with two cotyledons, or seed leaves, modified for food storage.

differentiated cell: a mature cell specialized for a specific function; in plants, differentiated cells normally do not divide.

endodermis (en-dō-der´-mis): the innermost layer of small, close-fitting cells of the cortex of a root that form a ring around the vascular cylinder.

epidermal tissue: dermal tissue in plants that forms the epidermis, the outermost cell layer that covers young plants.

epidermis (ep-uh-der´-mis): in plants, the outermost layer of cells of a leaf, young root, or young stem.

fibrous root system: a root system, commonly found in monocots, characterized by many roots of approximately the same size arising from the base of the stem.

ground tissue system: a plant tissue system consisting of parenchyma, collenchyma, and sclerenchyma cells that makes up the bulk of a leaf or young stem, excluding vascular or dermal tissues. Most ground tissue cells function in photosynthesis, support, or carbohydrate storage.

guard cell: one of a pair of specialized epidermal cells surrounding the central opening of a stoma of a leaf, which regulates the size of the opening.

heartwood: older xylem that contributes to the strength of a tree trunk.

internode: the part of a stem between two nodes.

lateral bud: a cluster of meristematic cells at the node of a stem; under appropriate conditions, it grows into a branch.

lateral meristem: a meristematic tissue that forms cylinders parallel to the long axis of roots and stems; normally located between the primary xylem and primary phloem (vascular cambium) and just outside the phloem (cork cambium); also called *cambium.*

leaf: an outgrowth of a stem, normally flattened and photosynthetic.

leaf primordium (pri-mor´-dē-um; pl., primordia): a cluster of meristem cells, located at the node of a stem, that develops into a leaf.

legume (leg´-ū m): a member of a family of plants characterized by root swellings in which nitrogen-fixing bacteria are housed; includes soybeans, lupines, alfalfa, and clover.

meristem cell (mer´-i-stem): an undifferentiated cell that remains capable of cell division throughout the life of a plant.

mesophyll (mez´-ō-fil): loosely packed parenchyma cells beneath the epidermis of a leaf.

mineral: an inorganic substance, especially one in rocks or soil.

monocot: short for monocotyledon; a type of flowering plant characterized by embryos with one seed leaf, or cotyledon.

mycorrhiza (mī-kō-rī´zuh; pl., mycorrhizae): a symbiotic relationship between a fungus and the roots of a land plant that facilitates mineral extraction and absorption.

nitrogen fixation: the process that combines atmospheric nitrogen with hydrogen to form ammonium (NH_4^+).

nitrogen-fixing bacterium: a bacterium that possesses the ability to remove nitrogen (N_2) from the atmosphere and combine it with hydrogen to produce ammonium (NH_4^+).

node: in plants, a region of a stem at which leaves and lateral buds are located.

nodule: a swelling on the root of a legume or other plant that consists of cortex cells inhabited by nitrogen-fixing bacteria.

nutrient: a substance acquired from the environment and needed for the survival, growth, and development of an organism.

palisade cell: a columnar mesophyll cell, containing chloroplasts, just beneath the upper epidermis of a leaf.

parenchyma (par-en´-ki-muh): a plant cell type that is alive at maturity, normally with thin primary cell walls, that carries out most of the metabolism of a plant. Most dividing meristem cells in a plant are parenchyma.

pericycle (per´-i-sī-kul): the outermost layer of cells of the vascular cylinder of a root.

periderm: the outer cell layers of roots and a stem that have undergone secondary growth, consisting primarily of cork cambium and cork cells.

petiole (pet´-ē-ōl): the stalk that connects the blade of a leaf to the stem.

phloem (flō´-um): a conducting tissue of vascular plants that transports a concentrated sugar solution up and down the plant.

pit: an area in the cell walls between two plant cells in which secondary walls did not form, such that the two cells are separated only by a relatively thin and porous primary cell wall.

pith: cells forming the center of a root or stem.

pressure-flow theory: a model for the transport of sugars in phloem, by which the movement of sugars into a phloem sieve tube causes water to enter the tube by osmosis, while the movement of sugars out of another part of the same sieve tube causes water to leave by osmosis; the resulting pressure gradient causes the bulk movement of water and dissolved sugars from the end of the tube into which sugar is transported toward the end of the tube from which sugar is removed.

primary growth: growth in length and development of the initial structures of plant roots and shoots, due to the cell division of apical meristems and differentiation of the daughter cells.

primary phloem: phloem in young stems produced from an apical meristem.

primary root: the first root that develops from a seed.

primary xylem: xylem in young stems produced from an apical meristem.

rhizome (rī´-zōm): an underground stem, usually horizontal, that stores food.

root: the part of the plant body, normally underground, that provides anchorage, absorbs water and dissolved nutrients and transports them to the stem, produces some hormones, and in some plants serves as a storage site for carbohydrates.

root cap: a cluster of cells at the tip of a growing root, derived from the apical meristem; protects the growing tip from damage as it burrows through the soil.

root hair: a fine projection from an epidermal cell of a young root that increases the absorptive surface area of the root.

root system: the part of a plant, normally below ground, that anchors the plant in the soil, absorbs water and minerals, stores food, transports water, minerals, sugars, and hormones, and produces certain hormones.

runner: a horizontally growing stem that may develop new plants at nodes that touch the soil.

sapwood: young xylem that transports water and minerals in a tree trunk.

sclerenchyma (skler-en´-ki-muh): a plant cell type with thick, hardened secondary cell walls that normally dies as the last stage of differentiation and both supports and protects the plant body.

secondary growth: growth in the diameter of a stem or root due to cell division in lateral meristems and differentiation of their daughter cells.

secondary phloem: phloem produced from the cells that arise toward the outside of the vascular cambium.

secondary xylem: xylem produced from cells that arise at the inside of the vascular cambium.

shoot system: all the parts of a vascular plant exclusive of the root; normally aboveground, consisting of stem, leaves, buds, and (in season) flowers and fruits; functions include photosynthesis, transport of materials, reproduction, and hormone synthesis.

sieve plate: in plants, a structure between two adjacent sieve tube elements in phloem, where holes formed in the primary cell walls interconnect the cytoplasm of the elements; in echinoderms, the opening through which water enters the water-vascular system.

sieve tube: in phloem, a single strand of sieve tube elements that transports sugar solutions.

sieve-tube element: one of the cells of a sieve tube, which form the phloem.

sink: in plants, any structure that uses up sugars or converts sugars to starch and toward which phloem fluids will flow.

source: in plants, any structure that actively synthesizes sugar and away from which phloem fluid will be transported.

spongy cell: an irregularly shaped mesophyll cell, containing chloroplasts, located just above the lower epidermis of a leaf.

stem: the portion of the plant body, normally located above ground, that bears leaves and reproductive structures such as flowers and fruit.

stoma (stō´-muh; pl., **stomata**): an adjustable opening in the epidermis of a leaf, surrounded by a pair of guard cells, that regulates the diffusion of carbon dioxide and water into and out of the leaf.

taproot system: a root system, commonly found in dicots, that consists of a long, thick main root and many smaller lateral roots, all of which grow from the primary root.

tendril: a slender outgrowth of a stem that coils about external objects and supports the stem; normally a modified leaf or branch.

terminal bud: meristem tissue and surrounding leaf primordia that are located at the tip of the plant shoot.

thorn: a hard, pointed outgrowth of a stem; normally a modified branch.

tracheid (trā´-kē-id): an elongated xylem cell with tapering ends that contains pits in the cell wall; forms tubes that transport water.

transpiration (trans´-per-ā-shun): the evaporation of water through the stomata of a leaf.

vascular bundle (vas´-kū-lar): a strand of xylem and phloem in leaves and stems; in leaves, commonly called a *vein*.

vascular cambium: a lateral meristem that is located between the xylem and phloem of a woody root or stem and that gives rise to secondary xylem and phloem.

vascular cylinder: the centrally located conducting tissue of a young root, consisting of primary xylem and phloem.

vascular tissue system: a plant tissue system consisting of xylem (which transports water and minerals from root to shoot) and phloem (which transports water and sugars throughout the plant).

vein: in vascular plants, a vascular bundle, or a strand of xylem and phloem in leaves.

vessel: a tube of xylem composed of vertically stacked vessel elements with heavily perforated or missing end walls, leaving a continuous, uninterrupted hollow cylinder.

vessel element: one of the cells of a xylem vessel; elongated, dead at maturity, with thick, lignified lateral cell walls for support but with end walls that are either heavily perforated or missing.

xylem (zī-lum): a conducting tissue of vascular plants that transports water and minerals from root to shoot.

THINKING THROUGH THE CONCEPTS

True or False: Determine if the statement given is true or false. If it is false, change the underlined word(s) so that the statement reads true.

13. _____   The epidermis of the <u>root system</u> is covered by the cuticle.

14. _____   The function of root systems is greatly enhanced by the presence of <u>root hairs</u>.

15. _____   <u>Parenchyma</u> cells store starch in white potatoes.

16. _____   <u>Sclerenchyma</u> cells create the "gritty" texture of pears.

17. _____   Tracheids and vessel elements transport water and minerals through <u>phloem</u> tissue.

18. _____   Sieve-tube elements <u>function independently</u> because they have no nucleus.

19. _____   A <u>monocot</u> such as a dandelion produces a taproot.

20. _____   The <u>rootcap</u> must be continuously replaced as the root pushes its way through the soil.

21. _____   When regulating stomatal openings, plants must balance <u>the need for CO$_2$ with the need to conserve water</u>.

22. _____   <u>Some stems</u> are the primary site for the conduction of photosynthesis.

23. _____   <u>Leaves</u> grow from lateral buds.

24. _____   Xylem and phloem cells are differentiated <u>sclerenchyma</u> cells.

25._____   The secondary phloem, cork cambium, and cork cells constitute the <u>pith</u> of a tree.

26. _____   Removing a <u>ring of bark</u> will ultimately kill a tree.

27. _____   Leaf mesophyll cells are differentiated from <u>parenchyma</u> cells.

28. _____ A <u>tendril</u> may be either a modified stem or a modified leaf.

29. _____ The <u>casparian strip</u> plays a vital role in getting water into the vascular cylinder.

30. _____ The cohesion-tension theory explains how plants <u>circulate sugar</u>.

31. _____ Water is <u>pushed</u> through a plant.

32. Complete the following table comparing features of monocots and dicots.

	Monocots	Dicots
flowers		
leaves		
vascular tissue		
root pattern		
embryo in seed		

Matching: Plant structures and their functions.

33. _____ leaf

34. _____ stem

35. _____ root system

36. _____ petiole

37. _____ epidermis

38. _____ internode

39. _____ node

40. _____ root hair

41. _____ phloem

42. _____ lateral bud

43. _____ cuticle

44. _____ stomata

45. _____ xylem

Choices:

a. absorbs water and minerals

b. covers and protects leaves, stems, roots

c. waxy layer; prevents water loss

d. transports water and minerals

e. regulate diffusion of O_2 and CO_2 and water loss

f. increases absorptive surface area

g. transports sugars, amino acids, and hormones

h. absorbs sunlight

i. develops into branches

j. stem area between leaves

k. site of leaf and lateral bud attachment

l. stalk of leaf, attaches blade to stem

m. supports plant

Matching: Cell types and their functions or characteristics.

46. _____   increase absorption surface area

47. _____   waterproof cells of stems and roots

48. _____   transports water and minerals

49. _____   transports carbohydrates

50. _____   unevenly thickened walls; provides support

51. _____   lie end to end; connect through pits

52. _____   hardened with lignin; not alive when mature

53. _____   conducts photosynthesis

54. _____   controls functions of sieve-tube element

55. _____   lie end to end; open ends

56. _____   controls stomatal openings

57. _____   active cellular division; allows plant growth

58. _____   makes up phloem; open tubes

59. _____   conduct most metabolic functions in a plant;
             are alive when mature

Choices:

a. collenchyma

b. parenchyma

c. sclerenchyma

d. guard cell

e. mesophyll cell

f. companion cell

g. cork cell

h. meristem cell

i. root hair

j. xylem

k. tracheids

l. phloem

m. vessel elements

n. sieve-tube elements

Short Answer.

60. Complete the following table of stem and leaf modifications.

Stem Modifications	Function	Plant Example
runners		
succulent stems		
storage		
rhizome		
thorns		
stem tendrils		
Leaf Modifications	**Function**	**Plant Example**
spines		
succulent leaves		
insectivorous		
bulb		
leaf tendrils		

Complete the blanks on the following leaf diagram using the terms below.

cuticle   lower epidermis   phloem   upper epidermis
guard cell   mesophyll   stoma   xylem

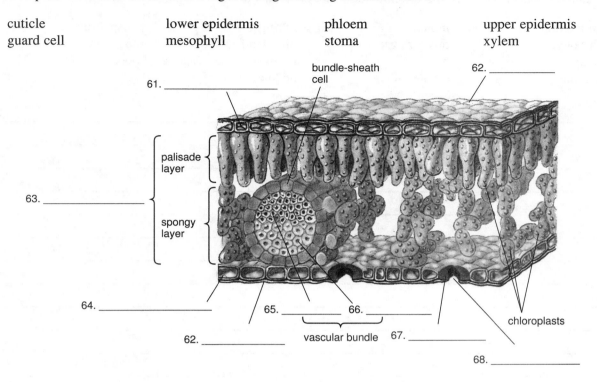

61. _____

62. _____

bundle-sheath cell

palisade layer

63. _____

spongy layer

64. _____

65. _____ 66. _____

62. _____

vascular bundle   67. _____

chloroplasts

68. _____

APPLYING THE CONCEPTS

These practice questions are intended to sharpen your ability to apply critical thinking and analysis to the biological concepts covered in this chapter.

69. Compare a taproot system and a fibrous root system.

_____

_____

_____

70. If you were to pull a carrot or a dandelion from the soil then pull out a handful of grass, which would pull out easier, which would bring the most soil up with it? Therefore,which root system do you think would best reduce erosion along a hillside? Why?

_____

_____

_____

_____

71. Provide two examples of legumes:

a. _____   b. _____

72. Using the following terms, discuss how fungi and bacteria aid roots in acquiring nutrients.

ammonium	mineral	nitrogen	phosphorous
carbohydrate	mycorrhizae	nitrogen fixation	root hairs
fungus	N$_2$	nitrogen-fixing bacteria	roots
legume	nitrate	nodule	

_____

_____

_____

_____

_____

_____

_____

_____

_____

_____

_____

73. Using the following terms, discuss how plants transport carbohydrates throughout the plant body.

active transport	hydrostatic pressure	pressure flow theory	sink
companion cell	leaf	sieve tube	source
diffusion	osmosis	sieve-tube element	xylem
fruit	phloem		

_____

_____

_____

_____

_____

_____

_____

_____

_____

_____

_____

_____

The sphagnum moss bogs are inhabited by many species of plants adapted to the cold, wet, acidic habitat found there. Because of the high acidity of the habitat, nutrients, especially nitrogen, are unavailable to the plants living there and they have evolved unique ways of getting this important nutrient. Use the Case Study and Web sites listed for this chapter to answer the following questions.

74. How do sundews, venus fly traps, and pitcher plants differ in their mechanism for trapping insects?

_____

_____

_____

_____

_____

_____

_____

_____

75. Why can't the insects simply fly off of the trap-leaf of a sundew? What causes the insect to become "trapped"?

_____

_____

76. The trap-leaves of the venus fly trap close because of water pressure changes that occur in the leaf. This causes the inward movement, trapping the fly. How do you think the leaf opens again to catch another fly once the first has been digested?

_____

_____

_____

77. Can the trap-leaves of venus fly traps catch flies indefinitely or are they limited in the number of times they can open and close? If they are limited, do they die once they have reached their limit?

_____

_____

_____

ANSWERS TO EXERCISES

1. shoot system
   leaves
   stems
   root system
   minerals
2. node
   internode
   blade

   petiole
3. apical meristem
   primary growth
   differentiated
   lateral meristem
   cambium
   secondary growth
4. dermal tissue system

   groundtissue system
   vascular tissue
   system
5. epidermis
   periderm
   cuticle
6. primary root
   taproot system

   fibrous root system
   rootcap
   cortex
7. casparian strip
   endodermis
   pericycle
   branch roots

8. leaf primordia
   lateral buds
9. primary xylem
   primary phloem
   vascular cambium
   secondary xylem
   secondary phloem
10. sapwood
    heartwood
    annual rings

11. cork cambium
    bark
12. cohesion-tension
    theory
    bulk flow
    transpiration
    stomata
    abscisic acid
13. false, shoot system
14. true

15. true
16. true
17. false, xylem
18. false, are dependent
    on companion cells
19. false, dicot
20. true
21. true
22. true
23. false, branches

24. false, parenchyma
25. false, bark
26. true
27. true
28. true
29. true
30. false, transport water
    and minerals
31. false, pulled

32.

	Monocots	Dicots
flowers	multiples of 3	multiples of 4 or 5
leaves	smooth, narrow, parallel veins	oval or palmate, net-like veins
vascular tissue	vascular bundles scattered	vascular bundles in ring
root pattern	fibrous root system	taproot system
embryo in seed	one cotyledon	two cotyledons

33. h
34. m
35. a
36. l
37. b
38. j
39. k
40. f
41. g
42. i
43. c
44. e
45. d
46. i
47. g
48. j
49. l
50. a
51. k
52. c
53. e
54. f
55. m
56. d
57. h
58. n
59. b

60.

Stem Modifications	Function	Plant Example
runners	sprout young plants	strawberry plants, spider plant
succulent stems	store water	cactus, baobab
storage	starch	white potato
rhizome	horizontal under ground stem	iris, ginger
thorns	protect	roses, honey locust
stem tendrils	grasping, support	ivy, grapes
Leaf Modifications	Function	Plant Example
spines	protection	cacti
succulent leaves	store water	some desert plants, jade plant
insectivorous	nitrogen acquisition	venus fly-traps, sundews, pitcher plant
bulb	store nutrients, to over-winter	onion, daffodils, tulips
leaf tendrils	grasping, support	garden peas

61. upper epidermis
62. cuticle
63. mesophyll
64. lower epidermis
65. xylem
66. phloem
67. guard cell
68. stoma

69. Taproot systems are composed of a persistent primary root with many smaller roots growing out from the sides of the taproot. In fibrous root systems, the primary root does not persist. New roots are produced from the base of the stem. These new roots are approximately equal in size.

70. The tap root system of the carrot or dandelion would most likely pull out easier than the fibrous root system of the grass. The root system of the grass would have considerably more soil attached to it than the tap root of the carrot or dandelion. This is because of the numerous small lateral roots formed in fibrous root systems. Since fibrous root systems tend to hold on to soil particles, they help tremendously in the prevention of soil erosion.

71. Peas, clover, soybeans, alfalfa

72. Symbiotic relationships between plants and <u>fungi</u> and between plants and soil bacteria increase nutrient acquisition in nutrient poor soils. <u>Fungi</u> that form a symbiotic relationship with plant <u>roots</u> are called <u>mycorrhizae</u>. <u>Mycorrhizae</u> help plants acquire <u>minerals</u>, especially <u>phosphorous</u>. In return, the <u>fungus</u> receives <u>carbohydrates</u> and amino acids from the plant. <u>Nitrogen</u> is also often in limiting quantities to many plants. <u>Legumes</u> have overcome this by forming an association with <u>nitrogen-fixing bacteria</u>. These bacteria possess the enzymes needed to convert N_2 from the atmosphere into <u>ammonium</u> or <u>nitrate</u>, the process of <u>nitrogen fixation</u>. The bacteria enter the <u>root hairs</u> of a legume and travel to the cortex of the <u>root</u>. The bacteria multiply and stimulate the cortex cells to multiply as well. A root <u>nodule</u> is produced. It is within the <u>nodule</u> that <u>nitrogen fixation</u> occurs, supplying the plant with a steady supply of usable <u>nitrogen</u>. In return, the bacteria receive <u>carbohydrates</u> for their energy needs.

73. An area of the plant that produces carbohydrates such as a <u>leaf</u> is referred to as a <u>source</u>. Any structure that needs carbohydrates produced by a <u>source</u> is referred to as a <u>sink</u>. Movement of sugars between <u>sources</u> and <u>sinks</u> must occur. It is believed that the <u>pressure flow theory</u> describes how this happens. Glucose is produced by photosynthesis, usually in <u>leaves</u>. The glucose is converted to sucrose and moved by <u>active transport</u> into <u>companion cells</u> of the <u>phloem</u>. As the concentration gradient of sucrose increases in <u>companion cells</u>, it is low in <u>sieve-tube elements</u>. Thus, sucrose <u>diffuses</u> into the <u>sieve-tube</u> <u>elements</u>. Water enters the <u>sieve-tube</u> by <u>osmosis</u> from nearby <u>xylem</u>, increasing the <u>hydrostatic pressure</u> within the <u>sieve tube</u>. The reverse sequence occurs, unloading sucrose at a <u>sink</u> region. Sucrose is <u>actively transported</u> into a <u>sink</u>, such as a developing <u>fruit</u>, thus lowering the sucrose concentration in the <u>sieve tube</u>. With a lower sucrose concentration, water flows out of the <u>sieve tube</u> by <u>osmosis</u>, lowering the <u>hydrostatic pressure</u> again. The <u>hydrostatic pressure</u> difference between the <u>sieve tube</u> at the <u>source</u> and the <u>sieve tube</u> at the <u>sink</u> causes water to flow down the <u>sieve tube</u>, carrying sucrose with it.

74. According to B. Meyers-Rice of the International Carnivorous Plant Society, the Venus flytrap has sensitive hairs on the inner surface of their trap-leaves. These hairs detect the minute, rapid movements of an insect landing on the leaf. Water pressure changes occur within the leaf causing it to snap shut, trapping the insect within. Sundews produce leaves with stalks or tentacles with rich nectar that attracts insects. The insects get stuck in the adhesive that is also produced by the stalk. The stalk then closes over the insect, trapping and digesting it. The pitcher plant leaves form a deep pitcher containing digestive enzymes and wetting agents (so the insect is too wet to fly away). Pitcher plants also produce downward sloping hairs on the inside of the pitcher so that an insect trying to climb out slips back in.

75. According to B. Meyers-Rice of the International Carnivorous Plant Society, the stalks on the leaf of a sundew contain mucilage and adhesives that the insect gets trapped in or even coated with.

76. Over time, the water pressure will reduce again. As this happens, the trap-leaves will open again, getting ready for a new victim. Or, see Evolutionary Connections, Chapter 25, page 529 in the text.

77. According to B. Meyers-Rice of the International Carnivorous Plant Society, the trap-leaves of a venus flytrap can only function as a trap a few times. After they can no longer trap insects, they do not die, instead they open wide to capture sunlight for photosynthesis.

Chapter 24: Plant Reproduction and Development

OVERVIEW

This chapter will present to you an overview of how flowering plants reproduce. It will review the organization of the flower and show you the importance of each part. Then, the ways that plants are pollinated and some of the animals involved, will be identified. These animals have **coevolved** with plants to develop pollinator-nutrient resource relationships. Once a flower has been pollinated, fertilization may take place, leading to the development of the embryo plant within a dormant seed. The chapter ends by showing you the importance of seed development, its dormancy, and its germination.

1) What Are the Features of Plant Life Cycles?

The plant life cycle is different from what you are used to thinking about (which is probably the "typical" animal development pattern). Plant life cycles are represented by <u>two distinct generations</u>, the diploid **sporophyte** and the haploid **gametophyte**. Because of this, the plant life cycle is referred to as the **alternation of generations**.

Ferns are most commonly used as an example of the alternation of generations because it is possible for you to see both generations. The fern sporophyte is what you may be familiar with and can see during a casual walk in the woods. The sporophyte produces haploid **spores** through meiosis (sporophyte means "spore producing plant"). The spores are dispersed to the soil by the wind. When conditions are favorable, the spore will **germinate** and grow into a mature, haploid gametophyte ("gamete producing plant"). The gametophyte is actually a very small plant that produces eggs and sperm (gametes). Gametophytes produce the eggs and sperm without another meiotic event because the gametophytes are already haploid. Sperm swim to the egg through water (either dew or rain drops). When egg and sperm fuse together, a diploid zygote is formed that will grow into another sporophyte. This alternating between diploid and haploid plants continues as long as conditions are favorable.

This life cycle occurs in all plants. In primitive land plants, such as mosses and ferns, the gametophyte is actually an independent plant that requires free water (dew or rain drops) to complete reproduction. However, more evolved land plants (evergreens and flowering plants) have microscopic gametophytes that are dependent on the sporophyte. In flowering plants the sporophyte produces the **flower**. Within the flower, meiosis produces the spores that develop into gametophytes. The gametophytes grow and mature within the flower. A megaspore divides by mitosis to produce the female gametophyte which will produce the egg. A microspore divides by mitosis to produce the male gametophyte, called a **pollen grain**, which will produce two sperm.

When pollen is distributed from one flower to another, the pollen grain germinates and grows a tube that will deliver the two sperm to the female gametophyte for fertilization. The zygote that is formed develops into an embryo sporophyte, dormant within a **seed** until the right conditions for growth exist.

2) How Did Flowers Evolve?

Flowers evolved as plants acquired characteristics that attracted pollinators. Plants that happened to carry genetic mutations to produce energy-rich nectar or pollen and a way to display their availability to pollinators passed those traits on to their offspring. Ultimately, the diverse assortment of flowers seen today arose. The parts of the flower, the **sepals**, **petals**, **stamens**, and **carpels** are really modified leaves. If the flower is a **complete flower**, then sepals, petals, stamens, and carpels are all present. If one or more of these flower parts is not present, then the flower is an **incomplete flower**. The flower parts develop in whorls. The sepals form the outermost whorl of flower parts and function to protect the flower bud. The petals are located just inside the sepals. They are often brightly colored to attract attention. If the stamens are present, they are located just inside the petals. The stamens include a **filament** that supports the **anther** that produces pollen. If the carpel is present, it is the innermost structure. The carpel is composed of one or more **ovaries**, each with a rather long **style** ending in a sticky **stigma**. Within each ovary is an **ovule** containing the female gametophyte. After fertilization of the egg, the ovule will mature into the seed and the ovary will develop into the **fruit**.

3) How Do Gametophytes Develop in Flowering Plants?

Inside the anthers and the ovaries meiosis produces the spores that develop into gametophytes. Development of the male gametophyte is rather straightforward. The anthers of stamens produce diploid, **microspore mother cells** inside structures called pollen sacs. Each microspore mother cell will divide by meiosis and produce four **microspores**. In turn, each microspore will divide by mitosis producing the male gametophyte, a pollen grain. In each pollen grain, there are two cells: a **tube cell** and a **generative cell** that is inside the tube cell's cytoplasm.

Development of the female gametophyte is a little more complex. Within the ovaries, ovules are produced and are made of layers of cells. The outer layers are called the **integuments**, and they enclose the diploid, **megaspore mother cell**. The megaspore mother cell divides by meiosis producing four haploid **megaspores**. Three of the megaspores deteriorate. The surviving megaspore divides by mitosis three times, producing eight haploid nuclei. From these eight nuclei, seven cells are produced. This seven-celled structure is the female gametophyte, called the **embryo sac**. The seven cells of the female gametophyte are distributed so that there are three cells at each end of the structure. The seventh cell has two nuclei called **polar nuclei**. The **egg** is located at the bottom of the female gametophyte, near an opening in the integuments.

4) How Does Pollination Lead to Fertilization?

When a compatible pollen grain lands on the stigma of a flower, **pollination** has occurred. The pollen grain absorbs water from the stigma; this allows it to germinate. Germination results in growth of a tube down through the style to the ovary. The generative cell divides by mitosis producing two sperm. As the tube cell elongates, it carries the two sperm with it. When the tip of the tube cell reaches the ovule, it penetrates the embryo sac and releases both of the sperm. One sperm **fertilizes** the egg forming the diploid zygote. The second sperm enters the seventh cell, fusing with both polar nuclei. This new cell is triploid since three haploid nuclei fused together. It will divide by mitosis, forming **endosperm**. Since there are two fertilization events, this process is called **double fertilization**. It is important that you distinguish between pollination, the pollen grain landing on the stigma, and fertilization, the fusion of sperm and egg, as two separate events!

5) How Do Seeds and Fruits Develop?

Once fertilization has occurred, the embryo sac and the integuments develop into the seed. The integuments specifically form the **seed coat.** The triploid endosperm will provide nutrients for the developing embryo. As the embryo forms from the zygote, it develops a primary root, the shoot, and **cotyledons** or seed leaves. The developing shoot is divided at the attachment site of the cotyledons. The region above the cotyledons is called the **epicotyl**, and the region below the cotyledons is the **hypocotyl**.

The wall of the ovary develops into the fruit. Fruits have evolved in many different ways; for example, they may be dry and papery, hard and spiked, or soft and fleshy. The purpose of all fruits is for seed dispersal.

The mature embryo within the seed enters a state of lowered metabolic activity; it becomes **dormant**. In this dormant state, the embryo can endure adverse environmental conditions, such as a harsh, cold winter, or a long period of drought. Often, the dormant state will not cease, and the seed will not germinate until an adaptive set of specific requirements has been met. As a protection against early germination, many seeds must actually dehydrate. Others must experience a prolonged cold period followed by a warm, moist period. This avoids germination during warm, moist autumn days that will be followed by freezing winter temperatures. The seed coat may function as a deterrent to germination. In this case, the seed coat must crack or split to allow water and or oxygen to enter before the seed will germinate.

6) How Do Seeds Germinate and Grow?

When a seed breaks dormancy and begins to germinate, it absorbs a great deal of water, causing it to swell. The swelling puts enormous pressure on the seed coat. Finally, it breaks open and the primary root emerges, followed by the shoot. As the shoot grows, it pushes upward, out of the soil. In order to protect the fragile apical meristem, monocot shoots are protected by a protective sheath, the **coleoptile**. Dicots do not produce a coleoptile. Instead, they protect their shoot apical meristem by forming a hook, either at the hypocotyl or the epicotyl. In other words, the shoot remains bent as it pushes upward. As the young plant is becoming established, it receives nutrients and energy stored in its cotyledons until newly formed leaves develop and can begin photosynthesizing.

KEY TERMS AND CONCEPTS

Fill-In: From the following list of key terms, fill in the blanks of the following statements.

alternation of generations	endosperm	incomplete	pollen grain
anthers	epicotyl	integuments	pollination
carpels	fertilize	megaspore	seed coat
coevolution	filament	megaspore mother cell	sepals
coleoptile	flower	microspore	spores
complete flower	fruit	microspore mother cell	sporophyte
cotyledons	gametophyte	polar nuclei	stamens
dormant	generative cell	ovary	stigma
double fertilization	germination	ovules	style
embryo sac	hypocotyl	petals	tube cell

1. The plant life cycle includes a haploid generation and a diploid generation. The cycling between the two is referred to as _____. During the plant life cycle, the diploid _____ produces _____ and the haploid _____ produces the gametes.

2. In flowering plants the male gametophyte is produced by a _____ that has divided by mitosis. The male gametophyte is actually the _____. The female gametophyte is produced by mitotic division of the _____. The zygote that is formed after fertilization occurs stays protected in a drought resistant _____.

3. _____ are composed of modified leaves. The _____ function to protect the bud, and the _____ function to attract pollinators.

4. The male reproductive structures are called the _____ and are composed of a long stalk, the _____, with the pollen producing _____ at its tip.

5. The female reproductive structures are called the _____ and are composed of a sticky _____ that is at the tip of an elongated _____. At the base is a bulbous _____ which contains _____. It is the ovary that will develop into the _____ which may be edible.

6. If a flower contains all flower parts, it is considered to be a _____. If, however, a flower is missing one or more of the structures, it is an _____ flower.

7. Flowering plants and their animal pollinators have adapted to one another; this is an example of _____.

8. Development of the male gametophyte begins with division of the diploid microspore mother cells producing four haploid _____. As these divide and develop into the pollen grain, two cells are produced within the pollen grain, the _____ and the _____.

9. When the pollen has matured, it will be distributed to the stigma of another flower. This is referred to as _____.

10. The female gametophyte is the _____. The process of its development occurs within an ovule. The outer layers of the ovule are the _____ which surround the diploid megaspore mother cell. Through meiosis, four haploid _____ will be produced. The embryo sac is a seven celled structure. One cell, in the center of the embryo sac contains two nuclei. These are the _____.

11. Once pollination has occurred, the pollen grain germinates producing the pollen tube with two sperm inside. One sperm will _____ the egg, while the second sperm will fuse with the two polar nuclei, producing the triploid _____. The use of two sperm in the fertilization process is termed _____.

12. After fertilization, the integuments of the ovule develop into the _____, and the endosperm nourishes the developing embryo. Most of the endosperm is absorbed into the seed leaves or _____. The seed will remain _____ until conditions are favorable for _____.

13. Once germination is initiated, the primary root will emerge from the seed followed by the young shoot. In monocots, the emerging shoot is protected by the _____. In dicots, the shoot is divided into regions with respect to the cotyledons. The shoot above the cotyledons is called the _____, and the shoot below them is called the _____.

KEY TERMS AND DEFINITIONS

alternation of generations: a life cycle, typical of plants, in which a diploid sporophyte (spore-producing) generation alternates with a haploid gametophyte (gamete-producing) generation.

anther (an´-ther): the uppermost part of the stamen, in which pollen develops.

carpel (kar´pel): the female reproductive structure of a flower, composed of stigma, style, and ovary.

coevolution: the evolution of adaptations in two species due to their extensive interactions with one another, such that each species acts as a major force of natural selection on the other.

coleoptile (kō-lē-op´-tīl): a protective sheath surrounding the shoot in monocot seeds, allowing the shoot to push aside soil particles as it grows.

complete flower: a flower that has all four floral parts (sepals, petals, stamens, and carpels).

cotyledon (kot-ul-ē´don): a leaflike structure within a seed that absorbs food molecules from the endosperm and transfers them to the growing embryo; also called *seed leaf*.

dormancy: a state in which an organism does not grow or develop; usually marked by lowered metabolic activity and resistance to adverse environmental conditions.

double fertilization: in flowering plants, the fusion of two sperm nuclei with the nuclei of two cells of the female gametophyte. One sperm nucleus fuses with the egg to form the zygote; the second sperm nucleus fuses with the two haploid nuclei of the primary endosperm cell, forming a triploid endosperm cell.

egg: the haploid female gamete, normally large and nonmotile, containing food reserves for the developing embryo.

embryo sac: the haploid female gametophyte of flowering plants.

endosperm: a triploid food storage tissue in the seeds of flowering plants that nourishes the developing plant embryo.

epicotyl (ep´-ē-kot-ul): the part of the embryonic shoot located above the cotyledons but below the tip of the shoot.

fertilization: the fusion of male and female haploid gametes, forming a zygote.

filament: in flowers, the stalk of a stamen, which bears an anther at its tip.

flower: the reproductive structure of an angiosperm plant.

fruit: in flowering plants, the ripened ovary (plus, in some cases, other parts of the flower), which contains the seeds.

gametophyte (ga-mēt´-ō-fīt): the multicellular haploid stage in the life cycle of plants.

generative cell: in flowering plants, one of the haploid cells of a pollen grain; undergoes mitosis to form two sperm cells.

germination: the growth and development of a seed, spore, or pollen grain.

hypocotyl (hī´-pō-kot-ul): the part of the embryonic shoot located below the cotyledons but above the root.

incomplete flower: a flower that is missing one of the four floral parts (sepals, petals, stamens, or carpels).

integument (in-teg´-ū-ment): in plants, the outer layers of cells of the ovule that surrounds the embryo sac; develops into the seed coat.

megaspore: a haploid cell formed by meiosis from a diploid megaspore mother cell; through mitosis and differentiation, develops into the female gametophyte.

megaspore mother cell: a diploid cell, within the ovule of a flowering plant, that undergoes meiosis to produce four haploid megaspores.

microspore: a haploid cell formed by meiosis from a microspore mother cell; through mitosis and differentiation, develops into the male gametophyte.

microspore mother cell: a diploid cell contained within an anther of a flowering plant, which undergoes meiosis to produce four haploid microspores.

ovary: in animals, the gonad of females; in flowering plants, a structure at the base of the carpel that contains one or more ovules and develops into the fruit.

ovule: a structure within the ovary of a flower, inside which the female gametophyte develops; after fertilization, develops into the seed.

petal: part of a flower, typically brightly colored and fragrant, that attracts potential animal pollinators.

polar nucleus: in flowering plants, one of two nuclei in the primary endosperm cell of the female gametophyte; formed by the mitotic division of a megaspore.

pollen/pollen grain: the male gametophyte of a seed plant.

pollination: in flowering plants, when pollen grains land on the stigma of a flower of the same species; in conifers, when pollen grains land within the pollen chamber of a female cone of the same species.

seed: the reproductive structure of a seed plant; protected by a seed coat; contains an embryonic plant and a supply of food for it.

seed coat: the thin, tough, and waterproof outermost covering of a seed, formed from the integuments of the ovule.

sepal (sē´-pul): the set of modified leaves that surround and protect a flower bud, typically opening into green, leaflike structures when the flower blooms.

spore: a haploid reproductive cell capable of developing into an adult without fusing with another cell; in the alternation-of-generation life cycle of plants, a haploid cell that is produced by meiosis and then undergoes repeated mitotic divisions and differentiation of daughter cells to produce the gametophyte, a multicellular, haploid organism.

sporophyte (spor´-ō-fīt): the diploid form of a plant that produces haploid, asexual spores through meiosis.

stamen (stā´-men): the male reproductive structure of a flower, consisting of a filament and an anther, in which pollen grains develop.

stigma (stig´-muh): the pollen-capturing tip of a carpel.

style: a stalk connecting the stigma of a carpel with the ovary at its base.

tube cell: the outermost cell of a pollen grain; digests a tube through the tissues of the carpel, ultimately penetrating into the female gametophyte.

THINKING THROUGH THE CONCEPTS

True or False: Determine if the statement given is true or false. If it is false, change the <u>underlined</u> word(s) so that the statement reads true.

14. _____ The <u>ovule</u> develops into the fruit after fertilization occurs.

15. _____ During <u>pollination</u>, pollen is attached to the stigma.

16. _____ Spores are produced by the <u>gametophyte</u>.

17. _____ The <u>pollen grain</u> is the male gametophyte.

18. _____   Conifers are wind pollinated, a very <u>efficient</u> process.

19._____   <u>Flower parts</u> are modified leaves.

20. _____   A complete flower has <u>only stamens</u>.

21. _____   Flower evolution occurred <u>separately from</u> the evolution of insects.

22. _____   Pollination and fertilization are <u>the same event</u>.

23. _____   Double fertilization occurs <u>only in flowering plants</u>.

24. _____   The <u>generative cell</u>, from the pollen grain, divides to produce two sperm.

25. _____   The megaspore mother cell is <u>haploid</u>.

26. _____   The <u>embryo</u> is uniquely triploid.

27. _____   The flesh of a peach develops from the <u>ovary</u>.

28. _____   The <u>seed coat</u> is mature integuments.

29. _____   Flour is made from wheat seed <u>endosperm</u>.

30. _____   The purpose of the fruit is to aid in <u>pollen</u> distribution.

31. _____   Newly developed seeds often <u>germinate immediately</u>.

Matching: Flower characteristics and animal pollinators. Some questions may have more than one correct answer.

32. _____   odors are sweet, "flowery"   Choices:

33. _____   mimics female   a. bee

34. _____   red flowers   b. moth

35. _____   <u>deep</u>, narrow nectar tubes   c. hummingbird

36. _____   bright yellow flowers   d. beetle

37. _____   odors are strong, musky   e. fly

38. _____   produce a <u>large</u> amount of nectar   f. wasp

39. _____   no odor   g. butterflies

40. _____   odors of rotting flesh

41. _____   tubular flowers

42. _____   white flowers

43. _____   reflect ultraviolet light

44. _____   produce sexual scent

Identify: Determine which of the following adaptations are characteristics of dispersal by **animal**, **wind**, **water**, "**shot gun**" or because they are **edible**.

45. _____ seeds with hairy tufts

46. _____ explosive fruits

47. _____ featherweight fruits

48. _____ round and buoyant fruits

49. _____ fruit with hooked spines

50. _____ juicy and tasty fruits

51. _____ winged fruit

52. _____ seeds have ready supply of fertilizer

Identify: Name the parts of the flower in the diagram below. Indicate the function of each.

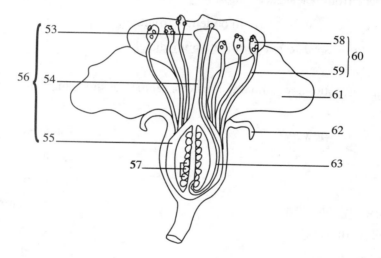

Flower Part   Function

53. _____   _____

54. _____   _____

55. _____   _____

56. _____   _____

57. _____   _____

58. _____   _____

59. _____   _____

60. _____   _____

61. _____   _____

62. _____   _____

63. _____   _____

APPLYING THE CONCEPTS

These practice questions are intended to sharpen your ability to apply critical thinking and analysis to the biological concepts covered in this chapter.

64. Using the following terms, explain "alternation of generations."

alternation of generations
diploid
egg(s)
gametes
gametophyte

germinate
haploid
life cycle(s)
meiosis
mitosis

sperm
spores
sporophyte
zygote

_____

_____

_____

_____

_____

_____

65. What is meant by the term "double fertilization"?

_____

_____

_____

66. The Case Study "A walk though a meadow" asks why flowers are so "strikingly and diversely colored." Below, explain why some flowers are so brightly colored and others are not; why some smell wonderful, others literally stink, and others have no smell at all; and why some flowers have such an odd (but beautiful) shape.

_____

_____

_____

_____

_____

_____

_____

_____

67. The Evolutionary Connections box discusses the dependence that has developed between the yucca plant and the yucca moth. After reading the Earth Watch box, think about the risk the yucca plant and the yucca moth each face. What if pesticide use killed the population of yucca moths? What if, instead, global climate change led to a wetter habitat for the yucca plant such that it could no longer survive? Project the consequences for these species.

_____

_____

_____

_____

_____

_____

_____

_____

Use the Web sites for this chapter to answer the following questions.

68. Discuss the differences between self-fertilization and cross fertilization. Which method is more likely to make use of pollinators? Which method is more likely to lead to evolution of the species (because of increased genetic variability)?

_____

_____

_____

_____

_____

69. How do some plants guarantee cross pollination or inhibit self-fertilization?

_____

_____

_____

_____

_____

_____

_____

_____

_____

_____

70. Why would it do little good to remove a plant causing pollen allergy from your backyard or even from your neighborhood?

_____

_____

_____

_____

_____

_____

_____

71. Why is it unlikely for you to be allergic to pollen produced by pine trees, roses, or lilies?

_____

_____

_____

_____

_____

_____

_____

_____

72. Comparing non-GM and GM soybeans and corn, which field of non-GM plants would be more likely to be contaminated by GM plants planted in a nearby field? Why?

_____

_____

_____

_____

_____

_____

_____

_____

73. Why would a farmer want to grow a genetically modified crop?

_____

_____

_____

_____

_____

Identify the seedling structures on the diagram below.

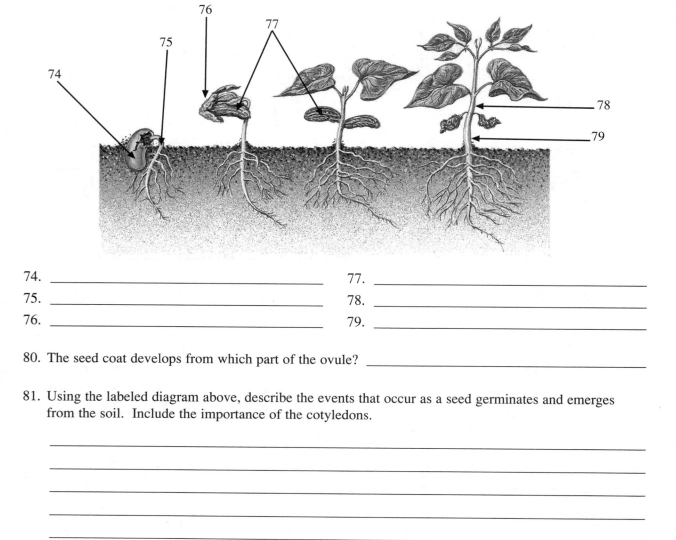

74. _____

75. _____

76. _____

77. _____

78. _____

79. _____

80. The seed coat develops from which part of the ovule? _____

81. Using the labeled diagram above, describe the events that occur as a seed germinates and emerges from the soil. Include the importance of the cotyledons.

_____

_____

_____

_____

_____

_____

_____

_____

_____

_____

_____

_____

_____

_____

_____

_____

_____

ANSWERS TO EXERCISES

1. alternation of generations
   sporophyte
   spores
   gametophyte
2. microspore mother cell
   pollen grain
   megaspore mother cell
   seed
3. Flowers
   sepals
   petals
4. stamens
   filament
   anther
5. carpel
   stigma
   style
   ovary
   ovules
   fruit
6. complete flower
   incomplete
7. coevolution
8. microspores
   tube cell
   generative cell
9. pollination
10. embryo sac
    integuments
    megaspores
    polar nuclei

11. fertilize
    endosperm
    double fertilization
12. seed coat
    cotyledons
    dormant
    germination
13. coleoptile
    epicotyl
    hypocotyl
14. ovary
15. true
16. false, sporophyte
17. true
18. false, very inefficient
19. true
20. false, all flower parts
21. false, along with
22. false, separate events
23. true
24. true
25. false, diploid
26. false, endosperm
27. true
28. true
29. true
30. false, seed
31. false, are dormant
32. a
33. f
34. c
35. c

36. a
37. b
38. c
39. c
40. d/e
41. b/g
42. a/b
43. a
44. f
45. wind
46. "shot gun"
47. wind
48. water
49. animal
50. edible
51. wind
52. edible
53. stigma, catches pollen
54. style, connects stigma to ovary
55. ovary, contains ovules
56. carpel, female reproductive structures
57. ovules, become seeds
58. anther, produces pollen
59. filament, supports anther
60. stamen, male reproductive structure
61. petal, attracts pollinators
62. sepal, protects flower bud
63. tube, carries sperm to ovary

64. <u>Alternation of generations</u> describes the <u>life cycle</u> of a plant. Plant <u>life cycles</u> are represented by two distinct generations, the <u>diploid sporophyte</u> and the haploid <u>gametophyte</u>. The sporophyte produces <u>haploid spores</u> through <u>meiosis</u>. The spores are dispersed to the soil by the wind. When conditions are favorable, the spore will germinate and grow into a mature, <u>haploid gametophyte</u>. The <u>gametophyte</u> produces <u>eggs</u> and <u>sperm</u>, the <u>gametes</u>. Sperm swim to the <u>egg</u> through water. When <u>egg</u> and <u>sperm</u> fuse together, a <u>diploid zygote</u> is formed that will grow into another <u>sporophyte</u>.

65. The term "double fertilization" is used to describe the fertilization events in flowering plants. One sperm fertilizes the egg, producing the zygote. A second sperm enters the primary endosperm cell and fuses with the two polar nuclei, producing the triploid endosperm.

66. The diversity in flower color, shape, and scent is due to the coevolution of plants and pollinators. As mutations occurred in plants that resulted in the attraction of pollinators, mutations also occurred in pollinators that resulted in traits that helped them pollinate certain plants. Plants and pollinators have acted as a significant force of natural selection in the evolution of each other. Thus, the two have coevolved. For example, bee-pollinated flowers attract the bees by emitting a sweet smell. At the same time, the flowers often have ultraviolet markings that direct the bee to the nectar it desires. When the bee feeds, pollen is brushed onto her back. With a visit to another flower, she deposits some of the pollen from the first flower on the stigma of the second flower. In contrast, flowers that have evolved without a sweet smell and that are red do not attract bees for pollination. Instead, these flowers attract hummingbirds. The long bill and tongue of the bird are perfect for reaching the copious nectar at the base of the long, tubular flower.

67. Some species have so perfectly coevolved that they rely solely on one another for survival. If the yucca moth population was wiped out due to pesticide use, the yucca plant would no longer have a pollinator and would not be able to reproduce. The species would eventually die-out as well. If the situation was reversed, and the yucca plant population died because of rapid climate changes, the yucca moth would have no place to lay her eggs, thus, its population would die out as well. Such dependencies are extremely risky for the long term survival of a species.

68. According to the Brooklyn Botanic Garden, self-pollination is the transfer of pollen from an anther to a stigma in the same flower or to a flower on the same plant. Cross-pollination is the transfer of pollen from one plant to a stigma on a different plant, but of the same species. Self-pollination can be an effective method of pollination, but it has the disadvantage of excluding genetic variability. Cross-pollination, on the other hand, increases genetic variability within species and is promoted by many plants in a variety of ways.

69. The Brooklyn Botanic Garden indicates that in some species, male and female flowers are produced on separate plants. This condition guarantees cross-pollination. Chemical reactions between the pollen and stigmas of the same plant may prevent self-pollination in many species, usually by inhibiting pollen grain germination or by preventing pollen tube growth. In some species, the anthers release pollen before the stigmas of flowers on the same plant are ready to receive it, or the mature pollen does not reach the stigma because its styles have not elongated. Since pollen is usually able to germinate for only a short period after being released from the anthers, these species are much more likely to be pollinated by another plant.

70. According to the U.S. Department of Health and Human Services and reported by Maryland Public Television, complete avoidance of allergenic pollen means moving to a place where the offending plant does not grow and where its pollen is not present in the air. But even this extreme solution may offer only temporary relief since a person who is sensitive to one specific weed, tree, or grass pollen may often develop allergies to others after repeated exposure. Thus, people allergic to ragweed could move to an area where ragweed does not grow, and develop allergies to other weeds or even to grasses and trees in their new surroundings. Based on this, removing a plant that causes pollen allergy from your backyard or neighborhood would not work either.

71. According to the U.S. Department of Health and Human Services reported by Maryland Public Television, the chemical makeup of pollen determines whether a particular type is likely to cause hay fever. Pine tree pollen is produced in large amounts but its chemical composition makes it less allergic than other types. Pine pollen also tends to fall straight downward and is not widely scattered. Therefore, it is rarely inhaled. Additionally, it is common to hear people say that they are allergic to colorful or scented flowers like roses (or lilies). Only florists, gardeners, and others who have close contact with flowers are likely to become sensitized to pollen from these plants. The large, heavy, waxy pollen grains of many flowering plants is not carried by wind so most people have little contact with this type of pollen.

72. According to E. Nafziger at the Univ. of Illinois, soybeans are a self-pollinated crop; there is little movement of genetic material between fields, so a soybean crop grown using non-GM seed and handled carefully by the producer should be GM-free. However, some of the corn in the United States labeled as "GMO-free" would probably not meet a "zero-tolerance" standard. Because corn is a cross-pollinated crop, and about one-third of the acreage (at least in Illinois) is GM, it is likely that most corn grown has at least a few grains that were pollinated with GM corn pollen. Even though a very small amount of pollen typically moves out of a field where it was shed, it is possible for corn pollen to move on the wind for more than a mile. Even under low wind conditions, some corn plants on the edge of a field can be pollinated by pollen from another field.

73. Biotech-info.net presents the President of the Iowa Corn Growers Association, D. Boettger's explanation. Farmers would be interested in growing genetically modified crops because it helps them use less pesticide and to grow better products more economically. Most GM crops are engineered to include a bacterium toxic to the corn borer, an insect that must otherwise be controlled with pesticides. For instance, 59% of Iowa's soybean fields were planted with soybean varieties genetically modified to permit spraying with herbicides so that weeds can be controlled without damaging the soybean plants.

74. seed

75. primary root

76. seed coat

77. cotyledons

78. hypocotyl

79. epicotyl

80. the integuments

81. The seed absorbs water, breaking open the seed coat. The primary root emerges and grows downward in the soil. The hypocotyl elongates, maintaining a hook to protect the apical meristem. The cotyledons emerge from the soil, breaking free of the seed coat. The cotyledons provide nutrients to the developing shoot and new leaves, shriveling as the nutrients become depleted. New leaves expand and begin photosynthesizing.

Chapter 25: Plant Responses to the Environment

OVERVIEW

This chapter will explain how chemicals produced by plants act as hormones and cause the plant to respond to environmental signals. An environmental signal might be gravity or the direction of the sun's rays, which induces a growth response. The signal may be the length of a day, indicating the appropriate time for flowering. Thus, how a plant will grow, flower, or initiate dormancy is due to stimuli from the environment that results in hormone production.

1) What Are Plant Hormones, and How Do They Act?

Hormones are chemicals that are produced by cells in one area of an organism (the plant) and are transported to another area where they cause a response. Five classes of **plant hormones** have been identified to date. **Auxins** promote elongation of cells in the shoot. They affect roots differently; in low concentrations, they stimulate elongation; however, slightly higher concentrations inhibit root growth. The distribution of auxin within the plant body is affected by light and gravity. Therefore, the hormone influencing **phototropism** and **gravitropism** is auxin. **Gibberellins** also stimulate stem elongation. But, they also influence flowering, fruit development, seed germination, and bud sprouting. **Cytokinins** tend to stimulate cell division. Thus, they influence events relating to cell division, including bud sprouting, fruit development, endosperm production, and embryo growth. Cytokinins also influence plant metabolism. **Ethylene** is a unique hormone because it is a gas at room temperature. Ethylene stimulates fruit ripening and induces leaf drop, as well as flower and fruit drop. **Abscisic acid** controls stomatal opening and closing and inhibits the effect of gibberellins so that seeds and buds remain dormant until conditions are favorable.

2) How Do Hormones Regulate the Plant Life Cycle?

Assume the plant life cycle begins with the dormant seed. Abscisic acid produced by the embryo reduces the metabolic rate of the embryo inside the seed so that its growth is suspended. The high concentration of abscisic acid in the seed must be reduced before the seed can germinate. This may occur by heavy rains rinsing the abscisic acid out of the seed or by freezing temperatures causing the abscisic acid to break down. As abscisic acid levels decrease, gibberellin levels increase. Gibberellin triggers the production of enzymes that will digest the endosperm, providing energy and nutrients for the embryo to grow.

When an embryo grows out of the seed, it must determine which way to grow its shoot and root systems. It must be able to tell which way is up and which way is down. It can do this because organelles in the stem detect gravity and cause auxin to accumulate on the lower side of the stem. The accumulation of auxin causes these lower cells to elongate, curving the stem upward. The effect of auxin on roots may cause opposite results. Again, organelles in the root detect gravity and cause auxin to accumulate on the lower side of the root. As auxin accumulates on the lower side of a horizontal root, the cells are inhibited from elongating, and the cells on the upper side grow more quickly, causing the root to curve downward. The organelles that detect gravity are starch-filled plastids.

The size and shape of a plant's shoot and root systems are determined by the interactions between auxin and cytokinin. For instance, in some plants, the shoot tip produces auxin which inhibits lateral buds from growing. This is called **apical dominance**. The auxin concentration decreases as you move down the stem, away from the shoot tip. Lateral buds become less inhibited and grow to produce branches. At the same time, cytokinins produced in the roots move up the stem, stimulating lateral bud growth. The interactions between the two hormones give the plant a characteristic shape. The interactions between auxin and cytokinin also influence the size of the shoot system relative to the size of the root system.

Once the plant is mature enough, it is able to flower. Some plants, such as snapdragons and tomatoes, flower as soon as the plant has matured. These plants are **day-neutral**. However, most plants use day length to regulate the timing of their flower production. If day length is important, a critical length of daylight must be reached or not exceeded. If a species is a **long-day plant**, then it must receive daylight for periods longer than a specific critical value. For example, spinach must experience days longer than 13 hours or it will not flower. If a species is a **short-day plant**, then it must not receive daylight for periods longer than a specific critical value. For example, cockleburs will not flower if the day length exceeds 15.5 hours. Day length stimulates the production of hormones called **florigens** to induce flowering. Florigens, however, have yet to be identified.

Day length must be detected and the **biological clock** of a plant must be set. This is done by the pigment **phytochrome**, which absorbs red light and far-red light. Phytochrome is a protein that has two different shapes, active and inactive. When phytochrome absorbs red light, it is changed to its P_{fr} form. In the P_{fr} form, phytochrome is active and may either stimulate or inhibit physiological processes. During the night and when phytochrome P_{fr} absorbs far-red light, it is converted back to the P_r, or the inactive, form. Phytochrome regulates flowering, leaf growth, chlorophyll synthesis, and epicotyl/hypocotyl hook straightening.

Once a flower is mature, pollination followed by fertilization is likely to occur and will lead to fruit and seed development. Hormones are also involved in regulating fruit and seed development. The pollen that lands on the stigma releases auxin or gibberellin, which causes the ovary to begin developing into a fruit. Fertilization further stimulates fruit development as the seeds mature. Ripening of fruit is caused by the production of ethylene. Ethylene is made by the fruit cells after they have been exposed to a high concentration of auxin made by the mature seeds.

As the plant reaches the end of a growing season, the leaves and fruits age rapidly. This is called **senescence**. An **abscission layer** forms at the point where the leaves or fruits attach to the stem. The formation of the abscission layer occurs after stimulation by ethylene and allows the plant to drop its leaves and its ripened fruit. Senescence and the abscission layer are induced by a decrease in the production of cytokinin by the roots and a decrease in production of auxin by the leaves and fruits. Lateral buds will become dormant due to the presence of abscisic acid. The plant has ended its growing season and will remain dormant until environmental stimuli activate hormones again in the spring.

KEY TERMS AND CONCEPTS

Fill-In: From the following list of key terms fill in the blanks in the following statements.

abscisic acid	biological clock	florigen	long-day plant	plastids
abscission layer	cytokinin	gibberellin	phytochrome	senescence
apical dominance	day-neutral plant	gravitropism	phototropism	short-day plant
auxin	ethylene	hormone	plant hormone	

1. _____ can be observed as a plant grows toward the light. This response to an environmental stimulus is induced by the action of a _____ called_____.

2. To date, five classes of _____ have been classified. As indicated above, auxins are involved in stem elongation. Auxins also influence root growth downward in response to gravity. This response is called _____. Organelles called _____ filled with starch are believed to be the gravity sensors in a plant.

3. Fruit development and seed germination are influenced by the hormone _____.
Meanwhile, growth responses involving cell division, such as bud sprouting, are controlled by
_____. One of the more economically important hormones is the gas
_____ since it regulates fruit ripening. Plants growing under drought conditions will
be able to regulate water loss by closing their stomata, which are controlled by _____.

4. Although a flowering hormone, per se, has not been identified, the term _____ is
used to refer to its presence. It is known, however, that plants have a _____ and
flowering, in most, is regulated by day length.

5. Plants that need short days (with long nights) are called _____. In turn, plants
that need long days are called _____. Some species, however, will flower
independent of day length if the plant is mature; these plants are _____. There is a
pigment, _____, that absorbs red and far-red light, regulating flowering.

6. At the close of the growing season for a plant, a rapid aging process occurs in leaves and fruit. This
event is known as _____ and is followed by leaf or fruit drop. Leaf or fruit drop
can only occur after the formation of the _____ at the point of stem attachment.

KEY TERMS AND DEFINITIONS

abscisic acid (ab-sis´-ik): a plant hormone that generally inhibits the action of other hormones, enforcing dormancy in seeds and buds and causing the closing of stomata.

abscission layer: a layer of thin-walled cells, located at the base of the petiole of a leaf, that produces an enzyme that digests the cell wall holding leaf to stem, allowing the leaf to fall off.

apical dominance: the phenomenon whereby a growing shoot tip inhibits the sprouting of lateral buds.

auxin (awk´-sin): a plant hormone that influences many plant functions, including phototropism, apical dominance, and root branching; generally stimulates cell elongation and, in some cases, cell division and differentiation.

biological clock: a metabolic timekeeping mechanism found in most organisms, whereby the organism measures the approximate length of a day (24 hours) even without external environmental cues such as light and darkness.

cytokinin (sī-tō-kī´-nin): a plant hormone that promotes cell division, fruit growth, and the sprouting of lateral buds and prevents the aging of plant parts, especially leaves.

day-neutral plant: a plant in which flowering occurs as soon as the plant has grown and developed, regardless of daylength.

ethylene: a plant hormone that promotes the ripening of fruits and the dropping of leaves and fruit.

florigen: one of a group of plant hormones that can both trigger and inhibit flowering; daylength is a stimulus.

gibberellin (jib-er-el´-in): a plant hormone that stimulates seed germination, fruit development, and cell division and elongation.

gravitropism: growth with respect to the direction of gravity.

hormone: a chemical that is synthesized by one group of cells, secreted, and then transported to other cells, whose activity is influenced by reception of the hormone.

long-day plant: a plant that will flower only if the length of daylight is greater than some species-specific duration.

phototropism: growth with respect to the direction of light.

phytochrome (fī´-tō-krōm): a light-sensitive plant pigment that mediates many plant responses to light, including flowering, stem elongation, and seed germination.

plant hormone: the plant-regulating chemicals auxin, gibberellins, cytokinins, ethylene, and abscisic acid; somewhat resemble animal hormones in that they are chemicals produced by cells in one location that influence the growth or metabolic activity of other cells, typically some distance away in the plant body.

senescence: in plants, a specific aging process, typically including deterioration and the dropping of leaves and flowers.

short-day plant: a plant that will flower only if the length of daylight is shorter than some species-specific duration.

THINKING THROUGH THE CONCEPTS

True or False: Determine if the statement given is true or false. If it is false, change the <u>underlined</u> word(s) so that the statement reads true.

7. _____   <u>Charles Darwin</u> conducted hormone experiments using grass seedlings.

8. _____   The high concentration of auxin produced in stems would <u>inhibit</u> the growth of roots.

9. _____   Abscisic acid maintains dormancy in seeds by <u>reducing</u> the metabolic rate of the embryo.

10. _____   Roots "know" to grow downward because of specialized <u>plastids</u> in cells.

11. _____   The branching pattern of a plant is determined by the interaction of <u>ethylene and water</u>.

12. _____   Pinching off the tip of a *Coleus* plant will create a <u>bushier plant</u> since the dominant influence of auxin has been removed.

13. _____   The environmental cue that induces flowering is <u>temperature</u>.

14. _____   Day-neutral plants flower <u>only under long-day conditions</u>.

15. _____   The <u>hormone</u> phytochrome sets the biological clock of a plant.

16. _____   Phytochrome in the active P_{fr} form absorbs <u>red light</u> and stimulates or inhibits physiological processes.

17. _____   As a shoot emerges from the soil and is exposed to red light, <u>cytokinin</u> changes form, causing the stem to straighten from its protective bend.

18. _____   When fruits are ripe, they are often <u>green</u> in color.

19. _____   Increased ethylene production by aging leaves causes <u>loosening</u> of cell walls of the abscission layer.

20. _____   Methylsalicylate production <u>decreases</u> the immune defenses of a plant under a virus attack.

21. _____   The chemical <u>volicitin</u> in caterpillar saliva causes maize plants to "signal" a parasitic wasp by releasing a mixture of chemicals.

Identify: Determine whether the following growth responses are related to **phototropism, gravitropism,** or **both**.

22. _____   growth toward light

23. _____   movement induced by starch filled plastids

24. _____   regulated by auxin

25. _____   coleoptile bending

26. _____   root growth into soil

27. _____   bending due to cell elongation

28. _____   growth in response to gravity

29. _____   bending due to stem inhibition

Matching: Responses to inducing hormones.

30. _____ seed germination

31. _____ phototropism

32. _____ maintains seed dormancy

33. _____ shoot elongation

34. _____ leaf drop

35. _____ stimulates lateral bud sprout

36. _____ influences flowering

37. _____ gravitropism

38. _____ closes stomata

39. _____ prevents lateral bud sprout (apical dominance)

40. _____ fruit development

41. _____ fruit ripening

42. _____ is a gas

43. _____ has not yet been found

Choices:

a. auxin

b. gibberellins

c. cytokinins

d. abscisic acid

e. ethylene

f. florigen

APPLYING THE CONCEPTS

These practice questions are intended to sharpen your ability to apply critical thinking and analysis to the biological concepts covered in this chapter.

44. Give a general definition of a hormone.

_____

_____

_____

45. "One bad apple spoils the bunch…" is a common phrase often referring to society's ills. It is, however, a true statement. Explain the phenomenon from which this phrase is derived.

_____

_____

_____

_____

_____

_____

_____

_____

46. Briefly explain how the hormones auxin and cytokinin interact to regulate root and stem branching.

_____

_____

_____

_____

_____

_____

_____

_____

_____

47. How does a venus fly trap catch a fly? Briefly explain how the leaves close, trapping the insect inside. Also explain how they open again.

_____

_____

_____

_____

_____

_____

Plants have evolved many ways to protect themselves from herbivores. As an evolutionary response, some herbivores have overcome the plant's defense and may use it for their own survival. Use the Case Study and the Web sites for this chapter to answer the following questions.

48. What are primary compounds in a plant? How do these compare to secondary compounds? What function do secondary compounds serve in a plant?

_____

_____

_____

_____

_____

_____

_____

_____

_____

_____

49. Identify some secondary compounds that have been used or exploited by humans. Has your life been affected by any of these?

_____

_____

_____

_____

_____

_____

_____

_____

50. What is an inducible defense? How do these serve to protect a plant?

_____

_____

_____

_____

_____

_____

_____

_____

51. What is Volicitin and how do plants respond to it? Is a plant's response a generic signal or is it "insect specific"?

_____

_____

_____

_____

_____

_____

52. How can Agronomy researchers and, in turn, farmers use Volicitin to control crop pests and reduce the use of pesticides on crops?

_____

_____

_____

_____

ANSWERS TO EXERCISES

1. phototropism
   hormone
   auxin
2. plant hormones
   gravitropism
   plastids
3. gibberellin
   cytokinin
   ethylene
   abscisic acid
4. florigen
   biological clock
5. short-day plants
   long-day plants
   day-neutral plants
   phytochrome
6. senescence
   abscission layer
7. true

8. true
9. true
10. true
11. false, auxin and cytokinin
12. true
13. false, day length
14. false, when mature
15. false, pigment
16. false, far-red
17. false, phytochrome
18. false, red
19. true
20. false, increases
21. true
22. phototropism
23. gravitropism
24. both
25. phototropism

26. gravitropism
27. phototropism
28. gravitropism
29. gravitropism
30. b
31. a
32. d
33. a
34. e
35. c
36. f
37. a
38. d
39. a
40. b
41. e
42. e
43. f

44. Hormones are chemicals that are produced by cells in one area and are transported to another area, where they exert specific effects.

45. "One bad apple spoils the bunch..." is derived from the fact that ethylene gas is given off from ripe fruit. Since ethylene influences fruit ripening, any fruit nearby will be affected by the gas being released. If an over-ripe apple is in a barrel, surrounded by less-ripe apples, the ethylene released from the over-ripe apple will cause the surrounding apples to ripen quickly. These apples will, in turn, release more ethylene, affecting the apples around them. In a relatively short time, the entire barrel will be over-ripe and spoiled.

46. High concentrations of auxin at the apical meristem suppress lateral bud growth at the top of the plant. As the concentration of auxin decreases down the stem, the inhibitory effects are lessened. Low concentrations of auxin stimulate root growth and branching. At the same time, cytokinin produced by roots moves up the stem, stimulating lateral bud growth. The two balance the growth of the plant. Large shoot systems will produce auxin in sufficient amounts to stimulate root-system expansion. If the shoot system gets too large for the root system to support nutritionally, cytokinin levels will decrease, reducing the rate of expansion of the shoot system until the root system can catch up.

47. Simply, the triggering of the sensory hairs on the leaves causes rapid growth of an outer layer of cells. This causes the leaves to shut, much like the growth of outer stem cells stimulated by auxin causes the stem to bend toward light. Over time, the cells of the second layer of the leaf grow, balancing the size of the cells of the two layers and slowly opening the trap.

48. According to S. P. McLaughlin at the Univ. of Arizona, primary compounds are chemicals found in all plant species. Secondary compounds are chemicals that are directly used in photosynthesis, respiration, growth, reproduction, or other basic metabolic or developmental processes. It was thought that these secondary metabolites were "waste products" of metabolism and that they served no real purpose. There is a great deal of evidence, however, that these compounds defend plants against many herbivores. These compounds may function as toxins, as feeding deterrents, or as digestibility reducers. Plant chemicals exist that affect an insect's nervous system, muscles, respiration, hormonal balance, reproduction, and behavior. Secondary metabolites also function to attract pollinators, provide protection from UV radiation, serve as temporary nutrient storage, regulate phytohormones, increase drought resistance, enable nutrient uptake, and mediate microbial interactions.

49. According to S. P. McLaughlin at the Univ. of Arizona, classes of chemical compounds that act as chemical defenses in different plants include alkaloids, coumarins, glucosinolates, terpenoids, flavenoids, and tannins. Alkaloids are nitrogen-containing cyclic compounds that may be highly toxic and include nicotine and caffeine. You may be familiar with pyrethrins. They are found in several species of chrysanthemums and are used in shampoos to remove fleas and lice. Tannins are very complex compounds that accumulate in the foliage and wood of many perennials. Condensed tannins bind to proteins, interfere with digestive enzymes, and render food proteins undigestible. These are the very properties that make them useful in preserving leathers.

50. S. P. McLaughlin at the Univ. of Arizona teaches that inducible defenses are produced in response to tissue damage. Alkaloids and proteinase inhibitors block digestive enzymes. Therefore, they act as anti-feeding compounds and inhibit insect feeding behavior. Proteinase inhibitors are produced within two hours after the initiation of insect feeding; they accumulate in the vacuoles of undamaged cells. Also, many plants produce jasmonic acid in response to herbivore feeding. The jasmonic acid stimulates the production of a volatile ester, methyl jasmonate, which is released and can be sensed by neighboring plants. So, maybe plants do talk to each other...

51. The Agricultural Research Service of the U.S. Department of Agriculture reports that volicitin is an insect chemical that prompts corn seedlings to send out a distress signal. Volicitin is secreted in the saliva of beet armyworm caterpillars and other similar pests that feed on crops. Researchers have found that volicitin caused damaged plants to give off chemicals that acted like distress signals. The plant chemicals attracted a beneficial wasp, *Cotesia marginiventris*, that attacked the beet armyworm caterpillars.

52. According to the Agricultural Research Service of the U.S. Department of Agriculture, the discovery could help plant breeders develop new crop varieties with enhanced chemical defense systems. Plants would be better able to attract beneficial insects that could fend off the attacking pests. This would cut down on pest damage and allow farmers to reduce pesticide applications to control crop pests.

Chapter 26: Homeostasis and the Organization of the Animal Body

OVERVIEW

This chapter presents an overview of the mechanisms that enable animals (including you) to regulate and control bodily functions, even when they live in harsh environments or are exposed to harsh conditions such as the subzero temperatures and days without food and water that Wayne Crill survived even though he was critically injured. These mechanisms include ways to maintain a stable internal environment as well as organs and organ systems that have specialized functions within that internal environment.

1) How Do Animals Maintain Internal Constancy?

Animals maintain internal consistency (**homeostasis**), in part, through negative feedback. **Negative feedback** systems counteract the effects of the external environment so that the body system is returned to its original condition. An example of a negative feedback response is your body regulating its internal temperature. When specific nerve endings sense a change in your body's temperature, the message is sent to the hypothalamus in your brain. The hypothalamus turns on the mechanisms needed to bring your body temperature back to its set point (98.6° F / 37° C). When your body temperature has reached its set point, your hypothalamus turns off the mechanisms used to restore your body temperature to normal.

Homeostasis is also maintained through positive feedback. **Positive feedback** systems reinforce changes, thus the change continues in a controlled but self-limiting chain reaction. The initiation of childbirth is such a mechanism. The initial contractions of the uterus begin a sequence of events that result in stronger and stronger contractions. This is a self-limiting event that ends with the birth of the child and the placenta. Numerous negative and positive feedback systems act together to maintain internal consistency. In order to achieve this, animals have evolved complex mechanisms to coordinate the systems by communicating among them with chemical signals.

2) How Is the Animal Body Organized?

Animal bodies are made up of cells organized into **tissues**, either epithelial, connective, muscle, or nerve. These tissues combine to make up **organs**, each with specific functions. The organs, in turn, combine to form **organ systems**.

Epithelial tissues form the **membranes** that cover your body and line your organs. They provide a barrier that controls what substances cross into or out of your organs. Epithelial tissues withstand a great deal of wear and tear; so they must constantly be replaced with new cells produced by mitosis. A second function of epithelial tissue is to serve as **glands**. **Exocrine glands** are connected to the outer surface of an organ or your body by a duct. **Endocrine glands** are no longer attached to outer surfaces; these glands typically produce and secrete hormones.

Cells forming **connective tissues** tend to be surrounded by an extracellular substance produced by the cells. This substance is often complexed with protein fibers of **collagen**. Immediately beneath the epithelial tissue of your skin is a layer of connective tissue, the **dermis**. **Tendons** function to connect your muscles to bones while **ligaments** attach your bones to other bones. The ends of your bones, the pads between your vertebrae, and the structures of your ears, nose, and respiratory passages are made of **cartilage**. **Bone**, itself, is a connective tissue that contains deposits of calcium phosphate. Your fat cells make up the connective tissue called **adipose tissue** that specializes in high energy storage. **Blood** and **lymph** are also included in connective tissue since they are primarily composed of extracellular fluid.

Muscle tissue cells are specialized to contract and relax. **Skeletal muscle** typically contracts in response to a conscious decision to move a portion of your skeleton. **Cardiac muscle** is found only in your heart and contracts and relaxes spontaneously as your heart beats. **Smooth muscle**, found in your digestive tract, bladder, large blood vessels, and the uterus of women, produces contractions that are slow, rhythmic, and involuntary.

Nerve tissue cells are called **neurons** and are specialized to generate and conduct electrical signals. The **dendrites** of your neurons receive a stimulus from the external environment or from other neurons. The **cell body** controls the functions of the neuron. **Axons** of your neurons conduct the electrical signal to target cells, and **synaptic terminals** transmit the signal to the target cells, either other neurons, muscle cells, or cells of a gland. **Glial cells** allow neurons to function properly by surrounding and protecting them.

When two or more tissue types function together, an organ is formed. By definition, then, the skin is a typical organ. The **epidermis** is the outer layer of your skin. It is covered and protected by a layer of dead epithelial cells containing the protein **keratin**. Beneath the epidermis is the connective tissue layer, the dermis. The dermis has a dense network of arterioles and capillaries, lymph vessels, and nerve endings. The dermis also has **hair follicle**s, the glands that produce hair, sweat glands to cool your skin and remove waste products, and **sebaceous glands** to produce oil to moisturize your epithelium.

KEY TERMS AND CONCEPTS

Fill-In: From the following list of key terms fill in the blanks in the following statements.

adipose tissue	collagen	exocrine glands	lymph	positive feedback
axon	connective tissue	fat	membrane	sebaceous glands
blood	dendrite	glands	negative feedback	skeletal muscle
bones	dermis	glial	nerve tissue	smooth muscle
cardiac muscle	endocrine glands	hair follicles	neurons	synaptic terminals
cartilage	epidermis	keratin	organ system	tendons
cell body	epithelial tissue	ligaments	organs	tissues

1. _____ refers to your body's ability to maintain a relatively constant internal environment. This consistency is maintained, in part, through _____, which responds to counteract change, and _____, which responds to enhance change.

2. The cells of multicellular organisms, including you, are organized into _____ that have a specific function. These organized cells then form _____, such as your lungs, heart, or stomach. Several different organs involved in a similar process form an _____.

3. Your body's cavities are lined with _____, creating a _____, which controls substances moving into and out of an organ.

4. Epithelial tissue also forms _____ that secrete substances. _____ remain connected to epithelial tissue by way of a duct. _____ separate from the epithelium and tend to secrete hormones into surrounding extracellular fluid.

5. The most diverse tissue in structure and function is _____. Most of this tissue has a network of fibrous strands of the protein _____. The connective

tissue located under your skin's epidermis is the _____.

6. The human body shape is maintained, in part, by the internal skeleton. Connective tissue plays a part in that _____ connect your muscles to your bones while _____ connect your bones to other bones. Your _____ themselves are connective tissue stiffened by calcium phosphate deposits.

7. Adipose tissue is composed of _____ cells, which specifically store energy. And the most unusual connective tissues are the fluids _____ and _____.

8. Your muscle tissue is specialized for contraction and exists as three types. _____ contracts when you tell it to by a conscious thought. _____ is found only in your heart and contracts spontaneously. _____ also contracts involuntarily and is found in your gastrointestinal tract.

9. Your brain and spinal cord are composed of _____. Nerve cells, or _____, generate and conduct electrical impulses. The _____ receives signals, the _____ maintains and repairs the neuron, and the _____ conducts the impulse to a target cell.

10. An electrical signal is actually transmitted to other cells at the _____. _____ allow the neuron to function optimally.

11. Your skin is an organ and has an outer layer, the _____, that is protected with dead cells containing the protein _____. The connective tissue of the skin, the dermis, contains glands that produce your hair. These glands are the _____. Also, the _____ secrete oil to keep your skin moist and lubricated.

Key Terms and Definitions

adipose tissue (a´-di-pōs): tissue composed of fat cells.

axon: a long extension of a nerve cell, extending from the cell body to synaptic endings on other nerve cells or on muscles.

blood: a fluid consisting of plasma in which blood cells are suspended; carried within the circulatory system.

bone: a hard, mineralized connective tissue that is a major component of the vertebrate endoskeleton; provides support and sites for muscle attachment.

cardiac muscle (kar´-dē-ak): the specialized muscle of the heart, able to initiate its own contraction, independent of the nervous system.

cartilage (kar´-teh-lij): a form of connective tissue that forms portions of the skeleton; consists of chondrocytes and their extracellular secretion of collagen; resembles flexible bone.

cell body: the part of a nerve cell in which most of the common cellular organelles are located; typically a site of integration of inputs to the nerve cell.

collagen (kol´-uh-jen): a fibrous protein in connective tissue such as bone and cartilage.

connective tissue: a tissue type consisting of diverse tissues, including bone, fat, and blood, that generally contain large amounts of extracellular material.

dendrite (den´-drīt): a branched tendril that extends outward from the cell body of a neuron; specialized to respond to signals from the external environment or from other neurons.

dermis (dur´-mis): the layer of skin beneath the epidermis; composed of connective tissue and containing blood vessels, muscles, nerve endings, and glands.

endocrine gland: a ductless, hormone-producing gland consisting of cells that release their secretions into the extracellular fluid from which the secretions diffuse into nearby capillaries.

epidermis (ep-uh-der´-mis): in animals, specialized epithelial tissue that forms the outer layer of skin.

epithelial tissue (eh-puh-thē´-lē-ul): a tissue type that forms membranes that cover the body surface and line body cavities, and that also gives rise to glands.

exocrine gland: a gland that releases its secretions into ducts that lead to the outside of the body or into the digestive tract.

gland: a cluster of cells that are specialized to secrete substances such as sweat or hormones.

glial cell: a cell of the nervous system that provides support and insulation for neurons.

hair follicle: a gland in the dermis of mammalian skin, formed from epithelial tissue, that produces a hair.

homeostasis (hōm-ē-ō-stā´sis): the maintenance of a relatively constant environment required for the optimal functioning of cells, maintained by the coordinated activity of numerous regulatory mechanisms, including the respiratory, endocrine, circulatory, and excretory systems.

keratin (ker´-uh-tin): a fibrous protein in hair, nails, and the epidermis of skin.

ligament: a tough connective tissue band connecting two bones.

lymph (limf): a pale fluid, within the lymphatic system, that is composed primarily of interstitial fluid and lymphocytes.

membrane: in multicellular organism, a continuous sheet of epithelial cells that covers the body and lines body cavities.

negative feedback: a situation in which a change initiates a series of events that tend to counteract the change and restore the original state. Negative feedback in physiological systems maintains homeostasis.

nerve tissue: the tissue that make up the brain, spinal cord, and nerves; consists of neurons and glial cells.

neuron (noor´-on): a single nerve cell.

organ: a structure (such as the liver, kidney, or skin) composed of two or more distinct tissue types that function together.

organ system: two or more organs that work together to perform a specific function; for example, the digestive system.

positive feedback: a situation in which a change initiates events that tend to amplify the original change.

sebaceous gland (se-bā´-shus): a gland in the dermis of skin, formed from epithelial tissue, that produces the oily substance sebum, which lubricates the epidermis.

skeletal muscle: the type of muscle that is attached to and moves the skeleton and is under the direct, normally voluntary, control of the nervous system; also called *striated muscle*.

smooth muscle: the type of muscle that surrounds hollow organs, such as the digestive tract, bladder, and blood vessels; normally not under voluntary control.

synaptic terminal: a swelling at the branched ending of an axon; where the axon forms a synapse.

tendon: a tough connective tissue band connecting a muscle to a bone.

tissue: a group of (normally similar) cells that together carry out a specific function; for example, muscle; may include extracellular material produced by its cells.

THINKING THROUGH THE CONCEPTS

True or False: Determine if the statement given is true or false. If it is false, change the <u>underlined</u> word(s) so that the statement reads true.

12. _____   Cells are the building blocks of <u>organs</u>.

13. _____   The kidney is an example of an <u>organ system</u>.

14. _____   The outer layer of your skin is composed of <u>epithelial tissue</u>.

15. _____   The epidermis is replaced <u>two times a year</u>.

16. _____   Salivary glands are an example of <u>exocrine glands</u>.

17. _____   <u>Endocrine glands</u> are connected to the epithelium by ducts.

18. _____   Blood is a <u>connective</u> tissue.

19. _____   <u>Cartilage</u> forms the internal structure of the ear.

20. _____   <u>Smooth</u> muscle is found only in the heart.

21. _____   Skeletal muscle contractions are <u>voluntary</u>.

22. _____   Blood vessels contain <u>muscle tissue</u>.

23. _____   <u>Dendrites</u> conduct an electrical signal to a target cell.

24. _____   An animal's bladder contracts using <u>smooth</u> muscle.

25. _____   Hair follicles are glands found in the <u>dermis</u>.

Matching: Tissues and their functions or characteristics

26. _____   support and bind other tissues

27. _____   forms tendons and ligaments

28. _____   tissue can contract and relax

29. _____   lines respiratory passages

30. _____   makes up brain and spinal cord

31. _____   cells may secrete mucus

32. _____   forms lining of stomach

33. _____   transmits electrical signals

34. _____   blood is an example

35. _____   forms glands

36. _____   forms lining of mouth

37. _____   includes glial cells and neurons

38. _____   may be under conscious or unconscious control

Choices:

a. epithelial

b. connective

c. muscle

d. nerve

Identify: Determine whether the following statements refer to **exocrine** glands, **endocrine** glands, or **both**.

39. _____   remain connected to epithelium by a duct

40. _____   salivary glands

41. _____   most produce hormones

42. _____   derived from epithelial tissue

43. _____   secrete product into extracellular fluid

44. _____   sweat glands

Short Answer:

45. Identify the three types of muscle found in the animal body.

   a. _____   b. _____   c. _____

46. Provide an example of an organ system listing the organs within the system.

   _____

   _____

   _____

   _____

   _____

   _____

47. Complete the following table relating the major vertebrate organ systems.

Organ system	Major structures	Physiological role
respiratory system		
	lymph, lymph nodes and vessels, white blood cells	
excretory system		
		supplies body with nutrients for growth and maintenance
	smooth muscle, cardiac muscle, skeletal muscle	
endocrine system		
		provides support, protects organs, muscle attachment sites
reproductive system		
circulatory system		
	brain, spinal cord, peripheral nerves	

APPLYING THE CONCEPTS

These practice questions are intended to sharpen your ability to apply critical thinking and analysis to the biological concepts covered in this chapter.

48. The term "homeostasis" actually means unchanging. Do you think this is an accurate description of the state of the body's internal environment? Why or why not?

_____

_____

_____

_____

_____

_____

49. Explain why your home thermostat represents a type of negative feedback mechanism.

_____

_____

_____

_____

_____

Use the Case Study and Web sites provided for this chapter to answer the following questions.

50. When Wayne Crill fell 1,000 feet, several of his body's feedback mechanisms were triggered as he lay critically injured. What type of feedback mechanisms (negative or positive) did his body use to keep him alive? Several of Wayne's systems used these feedback mechanisms to maintain homeostatic control. Identify a few of the systems you think would have been involved.

_____

_____

_____

_____

_____

_____

51. What is shock? From which of the four types of shock would Wayne Crill most likely have suffered?

_____

_____

_____

_____

_____

52. How should the rescue team have treated Wayne Crill for shock?

_____
_____
_____
_____
_____
_____
_____
_____
_____
_____

53. Would Wayne Crill have suffered from hypothermia? If so, what would have caused it?

_____
_____
_____
_____
_____

54. How should the rescue team have treated Wayne Crill for hypothermia?

_____
_____
_____
_____
_____
_____
_____
_____
_____
_____
_____

55. Wayne Crill's condition and situation were extreme. But, even as you wait for a campus bus or walk to your car on a cold, drizzly day, hypothermia could set in. How can you avoid hypothermia?

_____

_____

_____

_____

_____

_____

ANSWERS TO EXERCISES

1. Homeostasis
   negative feedback
   positive feedback
2. tissues
   organs
   organ system
3. epithelial tissue
   membrane
4. glands
   Exocrine glands
   Endocrine glands
5. connective tissue
   collagen
   dermis
6. tendons
   ligaments
   bones
7. fat
   blood
   lymph

8. Skeletal muscle
   Cardiac muscle
   Smooth muscle
9. nervous tissue
   neurons
   dendrite
   cell body
   axon
10. synaptic terminals
    Glial cells
11. epidermis
    keratin
    hair follicles
    sebaceous glands
12. false, tissues
13. false, organ

14. true
15. false, two times
    per month
16. true
17. false, Exocrine
18. true
19. true
20. false, Cardiac
21. true
22. true
23. false, axon
24. true
25. true
26. b
27. b
28. c

29. a
30. d
31. a
32. a
33. d
34. b
35. a
36. a
37. d
38. c
39. exocrine
40. exocrine
41. endocrine
42. both
43. endocrine
44. exocrine
45. a. skeletal,
    b. smooth,
    c. cardiac

46. Respiratory system: nose, trachea, lungs, or gills; Lymphatic/Immune system: lymph, lymph nodes and vessels, white blood cells; Excretory system: kidneys, ureters, bladder, urethra; Digestive system: mouth, esophagus, stomach, small and large intestines, digestive glands; Muscular system: smooth muscle, cardiac muscle, skeletal muscle; Endocrine system: hormone secreting glands: hypothalamus, pituitary, thyroid, pancreas, adrenals; Skeletal system: bones, cartilage, tendons, ligaments; Reproductive system: testes, seminal vesicles, penis, ovaries, oviducts, uterus, vagina, mammary glands; Circulatory system: heart, vessels, blood; Nervous system: brain, spinal cord, peripheral nerves.

47.

Organ system	Major structures	Physiological role
respiratory system	nose, trachea, lungs, or gills	gas exchange between blood and environment
lymphatic/immune system	lymph, lymph nodes and vessels, white blood cells	carries fat, excess fluids to blood, destroys invading microbes
excretory system	kidneys, ureters, bladder, urethra	filters cellular waste, toxins, and excess water and nutrients; stabilizes bloodstream
digestive system	mouth, esophagus, stomach, small & large intestines, digestive glands	supplies body with nutrients for growth and maintenance
muscular system	smooth muscle, cardiac muscle, skeletal muscle	movement through digestive tract, large blood vessels; begins and carries out heart contractions; moves skeleton
endocrine system	hormone secreting glands: hypothalamus, pituitary, thyroid, pancreas, adrenals	regulates physiological processes along with nervous system
skeletal system	bones, cartilage, tendons, ligaments	provides support, protects organs, muscle attachment sites
reproductive system	testes, seminal vesicles, penis, ovaries, oviducts, uterus, vagina, mammary glands	produces sperm, inseminates female, produces egg cells, nurtures developing offspring
circulatory system	heart, vessels, blood	transports nutrients, gases, hormones, wastes; temperature control
nervous system	brain, spinal cord, peripheral nerves	senses environment, directs behavior, controls physiological processes along with endocrine system

48. The term "homeostasis" is used to describe the body's ability to maintain an internal constancy even though the animal's body is exposed to an external environment that is continually changing. However, the term suggests a static (unchanging) state. In fact, this is not an accurate description. The body's internal environment is dynamic (always changing). Chemical and physical changes are constantly occurring. At the same time, the changes occur within certain parameters, enabling cells to function. Thus, dynamic equilibrium is a more appropriate term.

49. Negative feedback mechanisms respond to counteract a change. If the temperature in your home cools beyond a certain set point, the sensor on the thermometer will detect the change and signal the heating device to turn on. Once the set-point temperature has been reached, the thermometer again detects the change and signals the heating device to turn off. This system turns on to counteract a drop in temperature. When the temperature has been restored, it turns off.

50. Wayne Crill's body would have primarily used negative feedback mechanisms to keep him alive. Such mechanisms would have included, among others, internal body temperature since he was exposed to subzero temperatures, blood oxygen content since he was in a low oxygen atmosphere, water balance and blood glucose levels since he would have had nothing to eat or drink for the 63+ hours before he was rescued.

51. According to the Virtual Naval Hospital, shock is the failure of the circulatory system to maintain enough oxygen-rich blood flowing to the vital organs of the body. Wayne Crill most likely would have suffered from hypovolemic shock. This type of shock is caused by a decreased amount of blood or fluids in the body. This decrease results from injuries that produce internal and external bleeding, fluid loss due to burns, and dehydration due to severe vomiting and diarrhea.

52. According to the Virtual Naval Hospital, the rescue team would have taken the following steps to treat Wayne Crill for shock. (1) They would have maintained an open airway. (2) They would have controlled his bleeding. (3) They would have positioned him on his back, with his legs elevated 6 to 12 inches. If he was vomiting or bleeding around the mouth, they would have placed him on his side or back with his head turned to the side. If, however, the rescue team had suspected head or neck injuries, they would have kept him lying flat. (4) They would have placed splints on any suspected broken or dislocated bones in the position in which they were found. They would not have attempted to straighten any broken or dislocated bones, because of the high risk of causing further injury. (5) They would have done their best to keep him comfortable and warm enough to maintain normal body temperature. (6) They would have kept him as calm as possible without excessive handling, and they would have given nothing by mouth. If he had complained of thirst, they would have wet his lips with a wet towel.

53. Wayne Crill most likely would have suffered from hypothermia. According to mercyhealthpartners, hypothermia occurs when body temperature falls below 96° F (35.5° C). Normal body temperature is 98.6° F (37° C). Hypothermia is caused by environmental factors such as low temperatures, high and/or cold winds, dampness, and cold water. Major personal risk factors for hypothermia include prolonged exposure to cold, inadequate clothing/cold-weather gear, inadequate nutrition, dehydration, immobilization, and male gender, all of which Wayne Crill would have experienced (and is a male).

54. According to Outdoorplaces.com, if Wayne was suffering from moderate hypothermia, rescuers would get him bundled up and out of the cold. They would cover his neck and head since almost 50 percent of heat loss happens there. Sudden movement and physical activity would be avoided. Rescuers would apply warm bottles of water, or warm rocks to his armpits and groin area (comfortably warm when touched by a hand flat on the stone and held in place). If he was fully conscious, he could sip luke-warm sweetened, non-alcoholic fluids. But, Wayne Crill would probably have been suffering from severe hypothermia. Therefore, the focus of the rescuers would have changed to maintaining his body temperature. Improper warming would create a condition called metabolic acidosis that could cause shock and heart failure. Warming would only have been performed in this case by a medical facility. (In severe cases surgery may be performed to bypass the extremities and warm the core first through bypass.) Rescuers would have been extremely gentle with him. Sudden or rough movements could have pulled very cold blood from the extremities into the warmer core causing further shock. They would not have rubbed his skin or moved his joints; this would cause more harm than good. In severe hypothermia it is best for three people to get under a pile of blankets or in a sleeping bag. Skin on skin contact of the torso works best with a person on each side of the victim. If rescuers would have done this, they would have ignored any pleas to leave Wayne alone or to be allowed to go to sleep. They would not have administered fluids. Maintaining his temperature and stemming further loss was the most important thing.

55. To help prevent hypothermia you should follow some basic rules. According to Outdoorplaces.com, even on short outings, you should wear a hat (almost a full 50 percent of heat loss is through the head) and keep your extremities warm. You should wear proper layers for the conditions. If there is a chance that you may be outside for a prolonged period you shouldn't drink alcoholic beverages, this will only increase heat loss. You should eat carbohydrate-loaded energy foods and drink plenty of warm fluids in cold weather to keep the internal fires going. You should avoid heavy meals that cause blood to pool in the digestive system. You should urinate frequently because you don't want to waste body heat maintaining the temperature of a full bladder. Most of all, you should use common sense.

Chapter 27: Circulation

OVERVIEW

This chapter will take you through the circulatory system, its components and their functions. The focus of the chapter is on the human system; however, parallels to other vertebrate systems can be made. Since the lymphatic system functions so closely with the circulatory system, it is included here.

1) What Are the Major Features and Functions of Circulatory Systems?

Circulatory systems include **blood**, a fluid that transports nutrients and oxygen to your cells and wastes away from your cells; **blood vessels**, which carry the fluid to and away from your cells; and a **heart**, which serves as the pump to circulate the fluid. A circulatory system may be an **open circulatory system**, such as that found in many invertebrates. In open systems, the organs and tissues are bathed in blood in a **hemocoel**. A circulatory system may be a **closed circulatory system**, such as that found in humans and most other vertebrates. In a closed circulatory system, the blood is maintained within vessels and the heart; this is a more efficient way of transporting nutrients and wastes.

Your circulatory system functions to carry oxygen from your respiratory system to your cells and then carry carbon dioxide away from your cells. Your circulatory system carries nutrients to your cells and takes wastes from them to your liver or kidneys and it distributes hormones throughout your body. Your circulatory system also helps to regulate your body temperature and prevents blood loss. Finally, it enhances your immune system by circulating antibodies and white blood cells throughout your body.

2) How Does the Vertebrate Heart Work?

Your heart is the pump behind the circulation of blood through your body. Within your heart, blood collects in **atria** (or one **atrium**). The muscular atria contract, pushing your blood into **ventricles**. When the ventricles contract, blood is pumped through your body. Oxygen and carbon dioxide are exchanged to and from cells through capillaries, thin-walled vessels.

In the four-chambered vertebrate heart in you and other mammals (as well as in birds), deoxygenated blood flows from your body through a **vein** into the right atrium. The right atrium pumps your blood to the right ventricle. When the right ventricle contracts, it pumps your blood through the pulmonary **arteries** to your lungs. The deoxygenated blood picks up oxygen in your lungs and flows back to your heart through pulmonary veins into the left atrium. Then the left atrium pumps the oxygen-rich blood into the left ventricle. The left ventricle contracts and pumps the oxygen-rich blood out through the aorta and to the rest of your body. The alternating contractions of your atria and ventricles generate the **cardiac cycle**. When you have your blood pressure checked, the systolic pressure measures the strength of your ventricular contractions; the diastolic pressure is measured between contractions.

Once your blood is pumped from an atrium to a ventricle, it must be kept from flowing back into the atrium. The **atrioventricular valves** open as blood is pumped through to your ventricles and then are kept shut by pressure of the blood in the ventricle. In turn, back flow of blood leaving your heart is controlled by **semilunar valves** between the right atrium and the pulmonary artery and the left atrium and the aorta.

Contractions of cardiac muscle fibers must be synchronized so that they do not contract independently. Gap junctions between adjacent fibers allow them to communicate contraction signals rapidly and efficiently. The cells are signaled by a special cluster of cardiac muscle cells that serve as a **pacemaker**. The primary pacemaker is found in the wall of the right atrium and is called the **sinoatrial (SA) node**. The atria must contract before the ventricles. In order to control this, the electrical impulse is delayed at a group of cells called the **atrioventricular (AV) node**. If your pacemaker cannot regulate contractions, **fibrillation**, or irregular contractions, occurs. However, if your heart is healthy, your nervous system and hormones are also involved in regulating your heart rate.

3) What Is Blood?

Your blood is composed of **plasma** and specialized cells within the fluid plasma. Dissolved within the plasma are proteins such as albumins, globulins, and fibrinogen, nutrients, hormones, gases, wastes, and salts. The specialized cells include **erythrocytes**, the red blood cells. The red color of your blood is due to an iron-containing pigment, **hemoglobin**. Hemoglobin functions to carry oxygen to cells and carbon dioxide away from cells. Red blood cells are manufactured in bone marrow and have a life span of only 120 days. The hormone **erythropoietin**, made by your kidneys, is responsible for maintaining red blood cell production through a negative feedback mechanism.

Your blood type is determined by proteins (A or B) present or absent on the plasma membrane of your erythrocytes. Another protein, called the **Rh factor,** may also be present on your red blood cell membranes. Determining the presence or absence of these proteins is extremely important in case you need a blood transfusion or an incompatible pregnancy occurs. An Rh-positive child carried by an Rh-negative mother might cause the mother's body to produce antibodies against the Rh factor. This usually does not harm the child. However, any subsequent pregnancies involving an Rh-positive child could result in the child being affected by **erythroblastosis fetalis**. Erythroblastosis fetalis occurs when the mother's antibodies attack the red blood cells of the fetus, causing severe anemia in the fetus.

Additional specialized cells within blood are the **leukocytes**, your white blood cells. Leukocytes are involved in your body's immune system. Some white blood cells travel to injured sites where bacteria have entered your body. Some will differentiate into foreign-cell-"eating" cells that will feed on the bacteria. Once the cells are "full" they will die, creating white pus. **Lymphocytes** are specialized white blood cells that produce antibodies in response to disease and help you become immune to diseases.

Platelets are membrane-enclosed fragments of **megakaryocytes** that reside in the bone marrow. Platelets are involved in **blood clotting**. Clotting begins when platelets encounter an irregular surface, such as a wounded blood vessel. The platelets stick to the irregular surface. In the meantime, production of the enzyme **thrombin** is induced. Thrombin causes the plasma protein fibrinogen to change into **fibrin**. Fibrin fibers form a mesh that traps red blood cells and platelets. Eventually, a dense scab forms over the wound.

4) What Are the Types and Functions of Blood Vessels?

Blood leaves your heart through arteries. The muscular elastic walls of arteries help to maintain a steady flow of blood to smaller vessels. As the diameter of an artery decreases, **arterioles** are formed. Arterioles are capable of responding to electrical and hormonal signals when changes in tissue needs are detected. When the diameter of the arterioles decreases, **capillaries** are formed. Blood flow into your capillaries is controlled by rings of muscle, **precapillary sphincters**, where arterioles become capillaries.

Capillaries function in gas exchange, as well as nutrient and waste exchange, with your body's cells. Pressure within capillaries causes fluid to continuously leak from them into spaces around the capillaries and tissues. This fluid, called **interstitial fluid**, allows the gas, nutrient, and waste exchange to occur. As blood returns to your heart, the capillaries increase in diameter, becoming **venules** and finally veins. Blood always returns to your heart through veins. The walls of your veins are thin and expandable. There is little resistance to blood flow through your veins. Contractions of your skeletal muscles, squeezing your veins, moves blood through them. One-way valves prevent backward flow of blood during the muscle contractions.

5) How Does the Lymphatic System Work with the Circulatory System?

Within the **lymphatic system** are lymph capillaries and vessels, lymph nodes, the thymus, and the spleen. The lymphatic system functions to remove excess interstitial fluid and molecules that have leaked from capillaries. **Lymph** fluid is interstitial fluid collected by lymph capillaries and carried back to the circulatory system in lymph vessels. The lymphatic system also functions to carry fat globules from the small intestine to the bloodstream. Additionally, the lymphatic system functions to carry white blood cells throughout the body to fight off bacteria and viruses. Clusters of connective tissue containing large numbers of lymphocytes are found throughout the body. The largest of these are the **tonsils**. Lymphocytes are produced in **lymph nodes**, in the **thymus**, and in the **spleen**. The spleen also acts as a filter of the blood, as blood flows through the spleen, **macrophages** and lymphocytes remove and destroy foreign material and aged red blood cells.

KEY TERMS AND CONCEPTS

Fill-In: From the following list of key terms, fill in the blanks in the following statements.

arteries	fibrillation	plasma
arterioles	fibrin	platelets
atria	heart	precapillary sphincters
atrioventricular (AV) node	hemocoel	Rh factor
atrioventricular valves	hemoglobin	semilunar valves
blood	interstitial fluid	sinoatrial (SA) node
blood clotting	leukocytes	spleen
blood vessels	lymph nodes	thrombin
capillaries	lymph	tonsils
cardiac cycle	lymphatic system	veins
closed circulatory system	lymphocytes	ventricles
erythroblastosis fetalis	megakaryocytes	thymus
erythrocytes	open circulatory systems	venules
erythropoietin	pacemaker	

1. All circulatory systems consist of _____ to transport, _____ to conduct fluid, and a _____ to pump the fluid throughout the body. A circulatory system may contain a large open space called the _____, in which organs and tissues are bathed in circulating blood. This exists in _____, found in most invertebrates. The _____, found in humans and other vertebrates, maintains the blood within vessels.

2. The _____ of the heart collect blood from the body or from the lungs. When these chambers pump, the blood goes into the _____.

3. Blood flowing toward the heart is always carried in _____, while blood flowing away from the heart is always carried in _____.

4. The alternating pumping of the atria and the ventricles is called the _____.

5. One-way blood flow within and from the heart is controlled by valves. Specifically, one-way blood flow from the atria into the ventricles is controlled by the _____. Blood flowing to the pulmonary artery or the aorta must pass through the _____ from the respective ventricle.

6. The rhythm of the beating heart is coordinated by a natural _____. The specialized cluster of cells in the wall of the right atrium that is primarily responsible for the coordination is the _____. In order for the ventricles to pump a fraction of a second after the atria, the electrical signal is delayed at the _____. When the pumping cycle fails to be coordinated, irregular contractions, or _____, occur.

7. The fluid component of blood is called _____. Within this fluid are specialized cells and _____, the membrane-enclosed fragments of larger cells called _____. The red blood cells of blood are called _____ and are red due to the pigment _____ within them. The number of red blood cells within the body is monitored by a negative feedback system involving the hormone _____ produced in the kidneys.

8. Proteins embedded in the plasma membrane of red blood cells are used to determine individual blood type as A, B, AB, or O. Another protein, the _____ is a concern when a pregnancy occurs involving an Rh-negative woman carrying an Rh-positive fetus. If the mother's body produces antibodies against the blood of the fetus, _____ could affect the newborn child.

9. White blood cells within blood are called _____, which function in the body's immune response. Platelets are intimately involved with the process of _____, which reduces the chances of bleeding to death. In response to a wound, platelets adhere to the injured surface of a blood vessel, in part, triggering production of the enzyme _____. This enzyme catalyzes the conversion of fibrinogen to _____.

10. Arteries branch into smaller vessels called _____, which help regulate the distribution of blood throughout the body. The tiniest of all vessels are the _____ that are responsible for gas and nutrient exchange with the body's cells. There is continuous leakage of fluid from these tiny vessels, forming the _____, which bathes nearly all the body's cells. Blood flow from arterioles into capillaries is regulated by tiny rings of smooth muscle called the _____. Unoxygenated blood in capillaries drains into larger vessels called _____ which empty into large veins.

11. Closely associated with the circulatory system is a second system of vessels, the _____. This system of vessels receives interstitial fluid leaked from capillaries. The fluid is now referred to as _____ and is carried back to the circulatory system. The function of the system is to help the body defend against foreign invaders. Patches of connective tissue with large numbers of _____ are found in linings of the respiratory, digestive, and urinary tracts. The largest patch of this tissue makes up the _____ behind the mouth. Lymph fluid is passed through kidney-bean-shaped structures called _____. Organs that function as part of the lymphatic system include the _____, which produces lymphocytes, and the _____, which filters blood.

KEY TERMS AND DEFINITIONS

angina (an-jī´-nuh): chest pain associated with reduced blood flow to the heart muscle, caused by the obstruction of coronary arteries.

arteriole (ar-tēr´-ē-ōl): a small artery that empties into capillaries. Contraction of the arteriole regulates blood flow to various parts of the body.

artery (ar´-tuh-rē): a vessel with muscular, elastic walls that conducts blood away from the heart.

atherosclerosis (ath´-er-ō-skler-ō´-sis): a disease characterized by the obstruction of arteries by cholesterol deposits and thickening of the arterial walls.

atrioventricular (AV) node (ā´-trē-ō-ventrik´-ū-lar nōd): a specialized mass of muscle at the base of the right atrium through which the electrical activity initiated in the sinoatrial node is transmitted to the ventricles.

atrioventricular valve: a heart valve that separates each atrium from each ventricle, preventing the backflow of blood into the atria during ventricular contraction.

atrium (ā´-trē-um): a chamber of the heart that receives venous blood and passes it to a ventricle.

blood: a fluid consisting of plasma in which blood cells are suspended; carried within the circulatory system.

blood clotting: a complex process by which platelets, the protein fibrin, and red blood cells block an irregular surface in or on the body, such as a damaged blood vessel, sealing the wound.

blood vessel: a channel that conducts blood throughout the body.

capillary: the smallest type of blood vessel, connecting arterioles with venules. Capillary walls, through which the exchange of nutrients and wastes occurs, are only one cell thick.

cardiac cycle (kar´-dē-ak): the alternation of contraction and relaxation of the heart chambers.

closed circulatory system: the type of circulatory system, found in certain worms and vertebrates, in which the blood is always confined within the heart and vessels.

erythroblastosis fetalis (eh-rith´-rō-blas-tō´-sis fē-tal´-is): a condition in which the red blood cells of a newborn Rh-positive baby are attacked by antibodies produced by its Rh-negative mother, causing jaundice and anemia. Retardation and death are possible consequences if treatment is inadequate.

erythrocyte (eh-rith´-rō-sīt): a red blood cell, active in oxygen transport, that contains the red pigment hemoglobin.

erythropoietin (eh-rith´-rō-pō-ē´-tin): a hormone produced by the kidneys in response to oxygen deficiency that stimulates the production of red blood cells by the bone marrow.

fibrillation: rapid, uncoordinated, and ineffective contractions of heart muscle cells.

fibrin (fī´-brin): a clotting protein formed in the blood in response to a wound; binds with other fibrin molecules and provides a matrix around which a blood clot forms.

heart: a muscular organ responsible for pumping blood within the circulatory system throughout the body.

heart attack: a severe reduction or blockage of blood flow through a coronary artery, depriving some of the heart muscle of its blood supply.

hemocoel (hē´-mō-sēl): a blood cavity within the bodies of certain invertebrates in which blood bathes tissues directly; part of an open circulatory system.

hemoglobin (hē´mō-glō-bin): the iron containing protein that gives red blood cells their color; binds to oxygen in the lungs and releases it to the tissues.

hypertension: arterial blood pressure that is chronically elevated above the normal level.

interstitial fluid (in-ter-sti´-shul): fluid, similar in composition to plasma (except lacking large proteins), that leaks from capillaries and acts as a medium of exchange between the body cells and the capillaries.

leukocyte (loo´-kō-sīt): any of the white blood cells circulating in the blood.

lymph (limf): a pale fluid, within the lymphatic system, that is composed primarily of interstitial fluid and lymphocytes.

lymphatic system: a system consisting of lymph vessels, lymph capillaries, lymph nodes, and the thymus and spleen; helps protect the body against infection, absorbs fats, and returns excess fluid and small proteins to the blood circulatory system.

lymph node: a small structure that filters lymph; contains lymphocytes and macrophages, which inactivate foreign particles such as bacteria.

lymphocyte (lim´-fō-sīt): a type of white blood cell important in the immune response.

macrophage (mak´-rō-fāj): a type of white blood cell that engulfs microbes and destroys them by phagocytosis; also presents microbial antigens to T cells, helping stimulate the immune response.

megakaryocyte (meg-a-kar´-ē-ō-sīt): a large cell type that remains in the bone marrow, pinching off pieces of itself that then enter the circulation as platelets.

open circulatory system: a type of circulatory system found in some invertebrates, such as arthropods and mollusks, that includes an open space (the hemocoel) in which blood directly bathes body tissues.

pacemaker: a cluster of specialized muscle cells in the upper right atrium of the heart that produce spontaneous electrical signals at a regular rate; the sinoatrial node.

plaque (plak): a deposit of cholesterol and other fatty substances within the wall of an artery.

plasma: the fluid, noncellular portion of the blood.

platelet (plāt´-let): a cell fragment that is formed from megakaryocytes in bone marrow and lacks a nucleus; circulates in the blood and plays a role in blood clotting.

precapillary sphincter (sfink´-ter): a ring of smooth muscle between an arteriole and a capillary that regulates the flow of blood into the capillary bed.

Rh factor: a protein on the red blood cells of some people (Rh-positive) but not others (Rh-negative); the exposure of Rh-negative individuals to Rh-positive blood triggers the production of antibodies to Rh-positive blood cells.

semilunar valve: a paired valve between the ventricles of the heart and the pulmonary artery and aorta; prevents the backflow of blood into the ventricles when they relax.

sinoatrial (SA) node (sī´-nō-āt´-rē-ul): a small mass of specialized muscle in the wall of the right atrium; generates electrical signals rhythmically and spontaneously and serves as the heart's pacemaker.

spleen: an organ of the lymphatic system in which lymphocytes are produced and blood is filtered past lymphocytes and macrophages, which remove foreign particles and aged red blood cells.

stroke: an interruption of blood flow to part of the brain caused by the rupture of an artery or the blocking of an artery by a blood clot. Loss of blood supply leads to rapid death of the area of the brain affected.

thrombin: an enzyme produced in the blood as a result of injury to a blood vessel; catalyzes the production of fibrin, a protein that assists in blood clot formation.

thymus (thī´-mus): an organ of the lymphatic system that is located in the upper chest in front of the heart and that secretes thymosin, which stimulates lymphocyte maturation; begins to degenerate at puberty and has little function in the adult.

tonsil: a patch of lymphatic tissue consisting of connective tissue that contains many lymphocytes; located in the pharynx and throat.

vein: in vertebrates, a large-diameter, thin-walled vessel that carries blood from venules back to the heart.

ventricle (ven´-tre-kul): the lower muscular chamber on each side of the heart, which pumps blood out through the arteries. The right ventricle sends blood to the lungs; the left ventricle pumps blood to the rest of the body.

venule (ven´-ūl): a narrow vessel with thin walls that carries blood from capillaries to veins.

THINKING THROUGH THE CONCEPTS

True or False: Determine if the statement given is true or false. If it is false, change the underlined word(s) so that the statement reads true.

12. _____ An open circulatory system is an efficient means of transporting nutrients and wastes.

13. _____ A function of the circulatory system is to distribute hormones from their point of production to their target tissue(s).

14. _____ When atria contract, blood is sent throughout the body.

15. _____ Blood flows away from the heart in veins.

16. _____ At a resting heart rate, the cardiac cycle is complete in under one second.

17. _____ The atria are separated from the ventricles by the semilunar valves.

18. _____ The delay between the pumping of the atria and the ventricles is maintained by the atrioventricular node.

19. _____ The atrioventricular node serves as the "pacemaker" of the heart.

20. _____ Fibrillation of the ventricles can be fatal.

21. _____ Heart rate is influenced by the hormone epinephrine.

22. _____ The average human has five to six liters of blood.

23. _____ The protein fibrinogen in erythrocytes carries oxygen to the body's cells.

24. _____ Red blood cells are capable of mitotic division after living 120 days.

25. _____ Both red and white blood cells are produced by cells from bone marrow.

26. _____ White blood cells are part of the circulatory system and the lymphatic system.

27. _____ Platelets are cells derived from megakaryocytes.

28. _____ Interstitial fluid contains water and dissolved nutrients, gases, wastes, and blood proteins.

29. _____ Red blood cells pass through capillaries in pairs.

30. _____ Veins contain one-way valves to prevent backflow of blood.

Identify: After reading "Health Watch: Matters of the Heart," determine whether the following statements refer to **hypertension**, **angina**, **stroke**, **atherosclerosis**, or **heart attack**.

31. _____ high blood pressure

32. _____ blood vessels in brain rupture

33. _____ arteries thicken, lose elasticity

34. _____ insufficient oxygen causes chest pain

35. _____ strain may increase heart size

36. _____ cholesterol, calcium, and fatty plaques form

37. _____ coronary artery is blocked

38. _____ caused by arterioles constricting

39. _____ major cause of death from atherosclerosis

40. _____ artery in the brain is blocked

41. _____ clots may form

42. _____ may be treated using laser revascularization

Label the heart diagram below.

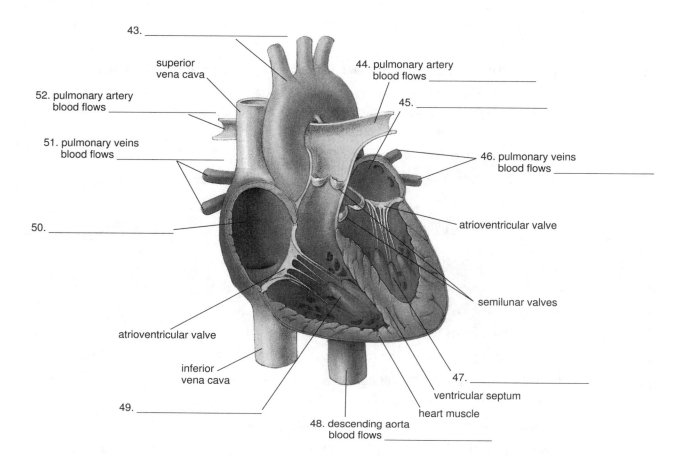

43. _____

superior
vena cava

44. pulmonary artery
blood flows _____

52. pulmonary artery
blood flows _____

45. _____

51. pulmonary veins
blood flows _____

46. pulmonary veins
blood flows _____

atrioventricular valve

50. _____

semilunar valves

atrioventricular valve

inferior
vena cava

47. _____

ventricular septum

49. _____

heart muscle

48. descending aorta
blood flows _____

APPLYING THE CONCEPTS

These practice questions are intended to sharpen your ability to apply critical thinking and analysis to biological concepts covered in this chapter.

53. Identify the three major plasma proteins and their functions.

Plasma Protein	Function

54. Discuss the differences between blood flow in a three-chambered heart and a four-chambered heart. Trace the flow of blood through a four-chambered heart, including its path to and from the lungs.

_____

_____

_____

_____

_____

_____

_____

_____

_____

_____

55. Explain how arterioles function to help regulate your body temperature on extremely cold or extremely hot days.

_____

_____

_____

_____

_____

56. Explain how negative feedback regulates red blood cell production.

_____

_____

_____

_____

57. When you are running, jogging, swimming, or doing other activities, precapillary sphincters to your abdominal organs tighten, reducing blood flow to these organs and supplying your arms and legs with ample blood supply. While you are eating, or after a meal, these sphincters are open to provide adequate blood supply for proper digestion. With this information, why should you rest after lunch on the beach instead of going for a swim?

_____

_____

_____

_____

58. Identify differences between capillaries of your circulatory system and those of your lymphatic system.

_____

_____

_____

_____

_____

_____

Using information from the Case Study and Web sites for this chapter, answer the following questions.

59. What is xenotransplantation? Why are medical researchers exploring its use?

_____

_____

_____

60. The case study presents the use of xenotransplantation to treat severe heart conditions. What xeno-tissues are currently being investigated to treat other life-threatening conditions or diseases?

_____

_____

_____

61. The Case Study discusses the use of pig tissues (hearts and valves) for xenotransplants. Pigs are even being cloned with specific characteristics to reduce rejection by the human body. Why are pigs the preferred animal for use in xenotransplantations?

_____

_____

_____

_____

_____

62. Serious concerns have been raised that the use of xenotransplants poses health risks not only to the transplant recipient, but also to the general public. In fact, health risks to the general public are of such serious concern that some researchers are calling for xenotransplantation research to stop. What are the health risks to the general public?

_____

_____

_____

_____

_____

ANSWERS TO EXERCISES

1. blood
   blood vessels
   heart
   hemocoel
   open circulatory systems
   closed circulatory system
2. atria
   ventricles
3. veins
   arteries
4. cardiac cycle
5. atrioventricular valves
   semilunar valves
6. pacemaker
   sinoatrial (SA) node
   atrioventricular (AV) node
   fibrillation
7. plasma
   platelets
   megakaryocytes
   erythrocytes
   hemoglobin
   erythropoietin
8. Rh factor
   erythroblastosis fetalis
9. leukocytes
   blood clotting
   thrombin
   fibrin

10. arterioles
    capillaries
    interstitial fluid
    precapillary sphincters
    venules
    lymphatic system
11. lymph
    lymphocytes
    tonsils
    lymph nodes
    thymus
    spleen
12. false, closed
13. true
14. false, ventricles
15. false, arteries
16. true
17. false, atrioventricular valves
18. true
19. false, sinoatrial (SA) node
20. true
21. true
22. true
23. false, hemoglobin
24. false,
    are not capable of division
25. true
26. true

27. false, membrane enclosed
    cell fragments
28. true
29. false, single file
30. true
31. hypertension
32. stroke
33. atherosclerosis
34. angina
35. hypertension
36. atherosclerosis
37. heart attack
38. hypertension
39. heart attack
40. stroke
41. atherosclerosis
42. atherosclerosis
43. aorta
44. to left lung
45. left atrium
46. from left lung
47. left ventricle
48. to lower body
49. right ventricle
50. right atrium
51. from right lung
52. to right lung

53.

Plasma Protein	Function
albumins	help maintain osmotic pressure of blood which controls water flow across plasma membranes
globulins	transport nutrients and play a role in the immune system
fibrinogen	important in blood clotting; is converted to fibrin by the enzyme thrombin in response to an injury

54. Three-chambered hearts have two atria and a single ventricle. Deoxygenated blood returns from the body to the right atrium while oxygenated blood returns from the lungs into the left atrium. Both atria empty into the one ventricle. There is some mixing of oxygenated and unoxygenated blood within the ventricle, but for the most part, the two types of blood remain on their respective sides of the ventricle. Mostly oxygenated blood gets pumped throughout the body and unoxygenated blood gets pumped to the lungs. This type of heart works fine for cold-blooded animals; however, warm-blooded birds and mammals require a more efficient method of oxygen delivery to their tissues. A four-chambered heart provides this. In a four-chambered heart, blood returns to the heart to the right atrium through the atrioventricular valve and into the right ventricle. Blood leaving the right ventricle passes through the semilunar valve as it enters the pulmonary artery into the lungs. After being reoxygenated, the blood

leaves the lungs through the pulmonary vein and enters the left atrium. Blood is pumped from the left atrium through the atrioventricular valve and into the left ventricle. Oxygenated blood is pumped through the semilunar valve into the aorta for distribution to the body.

55. Muscles in your arteriole walls are directly controlled by nerves, hormones, and chemicals produced by nearby tissues. Therefore, as the needs of nearby tissues change, arterioles contract and relax in response. Your circulatory system helps to regulate your body's temperature. By relaxing and expanding, the arterioles bring blood flow closer to the surface of your skin; thus, the heat dissipates, cooling your body. During cold weather, arteriole walls constrict, pulling the blood away from your skin's surface to maintain body warmth. Under extremely cold conditions, blood is shunted to your body's internal organs, specifically, to your heart and brain.

56. When low levels of oxygen are detected in the blood stream, your kidneys produce the hormone erythropoietin. Erythropoietin stimulates your bone marrow to rapidly produce new red blood cells. With more red blood cells circulating, the level of oxygen rises, signaling the kidneys to reduce production of erythropoietin. Thus, red blood cell production is controlled by a negative feedback mechanism.

57. After eating, abdominal precapillary sphincters are open to supply the tissues involved in digestion with an adequate supply of blood. Swimming (or other vigorous activities) requires the blood supply to be directed to your arms and legs. This need signals the abdominal precapillary sphincters to close, reducing blood supply to the tissues busy digesting your lunch. The result may be severe abdominal cramps that would inhibit your ability to swim, possibly resulting in drowning.

58. The capillaries of your circulatory system are composed of cells that have plasma membranes with pores that allow only water, dissolved nutrients, hormones, wastes, and small blood proteins to leak out. White blood cells ooze through openings between capillary cells. The capillaries of your circulatory system form a continuous network between the arterial system and the venous system. The capillaries of your lymphatic system contain cells with one-way openings between them. Thus, larger molecules can be carried into them. These capillaries dead end in tissues, serving to collect extra interstitial fluid and its contents.

59. According to the U.S. Food and Drug Administration, xenotransplantation involves transplanting cells, tissues, or organs from one species into another, unrelated species. Medical researchers are investigating this option because there is a severe shortage of human organs available for the large number of patients needing transplants.

60. According to G. Samiotis at Health Canada, xenotransplantation of islet cells from pig pancreas into diabetic patients is being researched. Additionally, such diverse conditions as liver failure and Parkinson's disease may be treated using cells or tissues from other species. In fact, in the United States, controlled clinical trials have already taken place.

61. According to G. Samiotis at Health Canada and F. Hoke for The Scientist, pigs are preferred for several reasons. Pig organs are of a size suitable for adult patients, the animals are available in large numbers, and, as mentioned, pigs can be genetically manipulated to reduce the possibility or severity of transplant rejection. The use of pigs raises fewer ethical concerns since they are already a major food source. The use of baboons is a possibility, especially for liver transplants needed as a result of hepatitis. (Baboon livers are resistant to hepatitis.) Whyfiles.com interviewed N. Fost, medical ethicist at University of Wisconsin-Madison. He states that because of baboon psycho-social interactions, the ethical problems associated with their use is greater.

62. According to G. Samiotis at Health Canada, the primary concerns pertain to xenoses (the transfer of infection, usually viral, from the transplanted organ to the recipient). Of particular concern are Porcine Endogenous Retroviruses, PERV. Pigs carry PERV but are not harmed by them. It is not known if these viruses can be transferred to human tissues from the transplanted organ and then passed from human to human. It is also not known how PERV would affect humans. Researchers are trying to find ways to detect PERV in pig tissues and may be able to produce transgenic pigs without PERV in their tissues.

Chapter 28: Respiration

OVERVIEW

This chapter covers the structures of the respiratory system and their specialized functions. Not all organisms use lungs for gas exchange; these adaptations are reviewed. The focus of the chapter, however, is on vertebrate respiratory systems, specifically of humans.

1) Why Exchange Gases?

Your cells require that oxygen be available for cellular respiration. Remember from Chapter 8, that cellular respiration uses O_2 and releases CO_2 as it produces energy (ATP) for your cells. Organisms use gas exchange to supply their cells with O_2 and to remove the CO_2 that has built up as a waste product.

2) What Are Some Evolutionary Adaptations for Gas Exchange?

All respiratory systems use diffusion for gas exchange. The surface across which diffusion will occur must be moist and must have a large surface area that is in contact with the environment. Some aquatic animals do not have specialized respiratory structures. Oxygen and carbon dioxide simply diffuse across their bodies. This works well for extremely small animals, thin flat animals, or for those that do not move quickly. Some aquatic animals circulate water from their environment through their bodies. Other animals use a large surface area of moist skin and their circulatory system to distribute gases. Most animals, however, have specialized respiratory structures that work closely with the circulatory system to exchange gases between cells and the environment. Gas exchange occurs by diffusion and **bulk flow**. Bulk flow occurs when fluids or gases move across relatively large spaces from an area of high pressure to an area of low pressure.

Many larger aquatic animals utilize **gills** to exchange O_2 and CO_2 with the environment. Gills have a dense network of capillaries just beneath their outer membrane, allowing gas exchange to occur. Fish have an **operculum**, or protective flap, covering their gills. The operculum serves to streamline the body and deter predators from eating the gills. Terrestrial animals evolved respiratory structures that are protected from drying out and have internal support. The respiratory structures of terrestrial animals are internal to reduce the amount of water lost in keeping the respiratory surface moist. The complex system of internal tubes called **tracheae** are the respiratory structures found in insects. (The tubes also support the exoskeleton.) Air, which may be actively pumped, enters the tracheae at valved openings along the abdomen called **spiracles**. Most terrestrial vertebrates primarily respire using saclike **lungs**. Amphibians, however, may also use their skin for supplemental gas exchange, as long as it remains moist.

3) How Does the Human Respiratory System Work?

Your respiratory system consists of a **conducting portion**, a series of passageways that carries air, and a **gas-exchange portion**, where O_2 and CO_2 are exchanged with the blood in sacs within the lung. Air flowing through the conducting portion passageways first enters your nose or mouth and passes through your **pharynx** to your **larynx**. Your **vocal cords** are located in your larynx. When your vocal cords partially block the larynx, exhaled air causes them to vibrate, producing sounds. From your larynx, air passes through your **trachea**. In your chest, the trachea divides into branches, forming two **bronchi**, one going into each lung. Within your lungs each **bronchus** branches many times forming smaller tubes called **bronchioles**. The bronchioles terminate in microscopic **alveoli**, where gas exchange will occur. The alveoli function to increase the gas-exchange surface area of lung tissue, and are infused with capillaries. Gases dissolve in the water covering each **alveolus** and diffuse through alveolar and capillary membranes. Oxygen in the inhaled air diffuses into the oxygen-poor blood returning from your body. In turn, CO_2 in your blood diffuses into the alveoli. Oxygen-rich blood is then transported to your cells and tissues.

Inside your blood cells, the iron-containing protein **hemoglobin** binds loosely with up to four oxygen molecules. This process maintains a low oxygen concentration in your blood to facilitate diffusion from inhaled air. Oxygenated blood appears bright red due to the shape of hemoglobin when it is bound to oxygen. Roughly 20% of CO_2 binds to hemoglobin, while 70% reacts with water to form bicarbonate ions that diffuse in your plasma along with the remaining CO_2 molecules. This process provides the mechanism needed to maintain a low CO_2 gradient in your blood to facilitate CO_2 diffusion into your bloodstream from cells.

Air enters your respiratory system actively through **inhalation**. During inhalation, your chest cavity is enlarged by the contraction of **diaphragm** muscles drawing your diaphragm downward. Muscles in your rib cage also contract, moving your ribs up and out. Since your lungs are held in a vacuum, the expansion of your chest causes your lungs to expand as well, drawing in air. **Exhalation** of air occurs passively as your muscles relax and your chest cavity decreases in size again. Your lungs are protected within your chest cavity by the rib cage, your diaphragm, and your neck muscles and connective tissues. An airtight seal between your chest wall and your lungs is enhanced by **pleural membranes**. The inhalation-exhalation process occurs without conscious thought through muscle contractions stimulated by the **respiratory center** of your brainstem. Your breathing rate is determined by a combination of sensors detecting high CO_2 levels, low O_2 levels, and an increase in activity level.

KEY TERMS AND CONCEPTS

Fill-In: From the following list of key terms, fill in the blanks in the following statements.

alveoli	exhalation	larynx	respiratory center
bronchi	gas-exchange portion	lungs	spiracles
bronchiole	gills	operculum	trachea
bulk flow	hemoglobin	pharynx	tracheae
conducting portion	inhalation	pleural membranes	vocal cords
diaphragm			

1. Fluids or gases moving through large spaces move by _____ from areas of high pressure to areas of low pressure. Diffusion, in contrast, is the movement of individual molecules.

2. The respiratory structures of aquatic animals are the _____. In fish, these structures are covered by a flap called the _____. The respiratory structures of terrestrial animals are internal. Insects respire using _____. Air enters through openings called _____ along the sides of the abdomen of the insect. Other terrestrial animals use _____, which are usually moist sacs.

3. The respiratory system is divided into two parts. The _____ carries air to the _____ , where gases are exchanged.

4. After air enters the nose or mouth, it passes through the _____ behind the mouth. The air then passes through the _____ which is protected by the epiglottis. The _____ are found here, allowing sounds to be made during exhalation. Semicircular bands of cartilage form the _____ . Within the chest, this tube splits into two tubes called _____. Inside the lungs, each tube repeatedly splits into smaller _____. Microscopic chambers within the lung, called _____, increase the surface area of lung tissue immensely.

5. The iron-containing protein _____ binds oxygen within red blood cells.

6. The process of breathing occurs with the help of the muscular _____. As these muscles contract and the rib cage expands, air is drawn into the lungs. This is called _____. Air is passively pushed out during _____ when the muscles relax and the chest decreases in size again. The rhythm of breathing is maintained in an area of the brain stem called the _____.

7. A double layer of membranes, the _____, help to provide an airtight seal between the lungs and the chest wall.

KEY TERMS AND DEFINITIONS

alveolus (al-vē´-ō-lus; pl., alveoli): a tiny air sac within the lungs, surrounded by capillaries, where gas exchange with the blood occurs.

bronchiole (bron´-kē-ōl): a narrow tube, formed by repeated branching of the bronchi, that conducts air into the alveoli.

bronchus (bron´-kus): a tube that conducts air from the trachea to each lung.

bulk flow: the movement of many molecules of a gas or fluid in unison from an area of higher pressure to an area of lower pressure.

chronic bronchitis: a persistent lung infection characterized by coughing, swelling of the lining of the respiratory tract, an increase in mucus production, and a decrease in the number and activity of cilia.

conducting portion: the portion of the respiratory system in lung-breathing vertebrates that carries air to the lungs.

diaphragm (dī´-uh-fram): in the respiratory system, a dome-shaped muscle forming the floor of the chest cavity that, when it contracts, pulls itself downward, enlarging the chest cavity and causing air to be drawn into the lungs.

emphysema (em-fuh-sē´-muh): a condition in which the alveoli of the lungs become brittle and rupture, causing decreased area for gas exchange.

exhalation: the act of releasing air from the lungs, which results from a relaxation of the respiratory muscles.

gas-exchange portion: the portion of the respiratory system in lung-breathing vertebrates where gas is exchanged in the alveoli of the lungs.

gill: in aquatic animals, a branched tissue richly supplied with capillaries around which water is circulated for gas exchange.

hemoglobin (hē´mō-glō-bin): the iron containing protein that gives red blood cells their color; binds to oxygen in the lungs and releases it to the tissues.

inhalation: the act of drawing air into the lungs by enlarging the chest cavity.

larynx (lar´-inks): that portion of the air passage between the pharynx and the trachea; contains the vocal cords.

lung: a paired respiratory organ consisting of inflatable chambers within the chest cavity in which gas exchange occurs.

operculum: an external flap, supported by bone, that covers and protects the gills of most fish.

pharynx (far´-inks): in vertebrates, a chamber that is located at the back of the mouth and is shared by the digestive and respiratory systems; in some invertebrates, the portion of the digestive tube just posterior to the mouth.

pleural membrane: a membrane that lines the chest cavity and surrounds the lungs.

respiratory center: a cluster of neurons, located in the medulla of the brain, that sends rhythmic bursts of nerve impulses to the respiratory muscles, resulting in breathing.

spiracle (spi´-ruh-kul): an opening in the abdominal segment of insects through which air enters the tracheae.

trachea (trā´-kē-uh): in birds and mammals, a rigid but flexible tube, supported by rings of cartilage, that conducts air between the larynx and the bronchi; in insects, an elaborately branching tube that carries air from openings called *spiracles* near each body cell.

vocal cord: one of a pair of bands of elastic tissue that extend across the opening of the larynx and produce sound when air is forced between them. Muscles alter the tension on the vocal cords and control the size and shape of the opening, which in turn determines whether sound is produced and what its pitch will be.

THINKING THROUGH THE CONCEPTS

True or False: Determine if the statement given is true or false. If it is false, change the <u>underlined</u> word(s) so that the statement reads true.

8. _____ Very small animals may exchange O_2 and CO_2 through their <u>skin</u>.

9. _____ Gases must be <u>dissolved in water</u> when they diffuse in or out of cells.

10. _____ <u>Diffusion</u> is the movement of gases through relatively large spaces from an area of high pressure to an area of low pressure.

11. _____ Gills are extensively branched or folded to <u>increase</u> surface area.

12. _____ The first lungs may have arisen as extensions of the <u>digestive tract</u> in freshwater fish.

13. _____ As some animals develop, <u>both gills and lungs</u> are produced as they go through their life stages.

14. _____ Birds acquire <u>CO_2</u> as they inhale and exhale.

15. _____ The <u>tongue</u> functions to guard the opening of the larynx.

16. _____ The respiratory tract is lined with <u>mucus</u>, which serves to trap bacteria and debris from inhaled air.

17. _____ The internal lung surface area equals about 75 m^2 due to the structure of <u>alveoli</u>.

18. _____ Each hemoglobin molecule can bind with <u>eight oxygen molecules</u>.

19. _____ Because of the function of hemoglobin, blood carries <u>less</u> oxygen compared with oxygen simply dissolved in plasma.

20. _____ Hemoglobin binds to CO <u>more strongly</u> than it binds to O_2.

21. _____ The enzyme carbonic anhydrase converts <u>CO_2 and water</u> to the bicarbonate ion in plasma.

22. _____ Inhalation is a <u>passive</u> process.

23. _____ Breathing is controlled by the respiratory center in the <u>cerebellum</u>.

Identify: Do these statements about the results of smoking describe **atherosclerosis**, **emphysema**, or **chronic bronchitis**?

24. _____ causes heart attacks

25. _____ mucus is produced in large quantities

26. _____ breathing is labored

27. _____ action of cilia decreases

28. _____ air flow to alveoli decreases

29. _____ arterial walls are thick with fatty deposits

30. _____ alveoli are destroyed

31. _____ lungs resemble blackened Swiss cheese

32. _____ number of cilia decreases

33. _____ lung tissue becomes brittle

34. _____ respiratory tract lining is swollen

APPLYING THE CONCEPTS:

These practice questions are intended to sharpen your ability to apply critical thinking and analysis to biological concepts covered in this chapter.

35. Outline the general path of gas exchange through the body. Identify whether movement occurs through bulk flow or diffusion in each step.

_____

_____

_____

_____

_____

_____

36. Explain how the respiratory center regulates breathing rate.

_____

_____

_____

37. Why is the respiratory center less sensitive to O_2 concentrations than CO_2 concentrations in the blood?

_____

_____

_____

_____

Use the Case Study and the Web sites for this chapter the answer the following questions.

38. How many teenagers in the U.S. begin smoking each day? Many say they will stop "when the time comes." What are the chances that this will happen?

_____

_____

_____

_____

39. What is ETS? Why is it classified as a group A carcinogen?

_____

_____

_____

_____

_____

40. How does ETS affect adults? Children?

_____

_____

_____

_____

_____

_____

_____

41. What percentage of a cigarette's output is visible? How many different chemical compounds are included in cigarette smoke? Where do these chemicals come from?

_____

_____

_____

_____

_____

_____

_____

42. Some cigarettes advertise that they are "low tar." What is tar? Would any level be healthy?

_____

_____

_____

43. How can lead in cigarette smoke affect children exposed to ETS?

_____

_____

_____

44. What is N-N-K? How does N-N-K affect non-smokers?

_____

_____

45. What are TRAP values? How does exposure to ETS affect your TRAP values?

_____

_____

_____

_____

ANSWERS TO EXERCISES

1. bulk flow
2. gills
   operculum
   tracheae
   spiracles
   lungs
3. conducting portion
   gas-exchange portion
4. pharynx
   larynx
   vocal cords
   trachea
   bronchi
   bronchioles
   alveoli
5. hemoglobin
6. diaphragm

inhalation
exhalation
respiratory center
7. pleural membranes
8. true
9. true
10. false, bulk flow
11. true
12. true
13. true
14. false, O_2
15. false, epiglottis
16. true
17. true
18. false, 4 oxygen molecules
    or 8 oxygen atoms

19. false, 70 times more
20. true
21. true
22. false, active
23. false, brain stem
24. atherosclerosis
25. chronic bronchitis
26. emphysema
27. chronic bronchitis
28. chronic bronchitis
29. atherosclerosis
30. emphysema
31. emphysema
32. chronic bronchitis
33. emphysema
34. chronic bronchitis

35. Air or water containing O_2 is moved past a respiratory surface by <u>bulk flow</u>. The flow may be facilitated by muscular breathing movements. O_2 and CO_2 are exchanged through the respiratory surface by <u>diffusion</u>; O_2 <u>diffuses</u> into the capillaries of the circulatory system and CO_2 <u>diffuses</u> out. The gases are transported in the circulatory system from the respiratory system to the tissues by <u>bulk flow</u> of the blood as it is pumped throughout the body by the heart. The gases are exchanged between the tissues and the circulatory system by <u>diffusion</u>; O_2 <u>diffuses</u> out of the capillaries and CO_2 <u>diffuses</u> in. The gases are transported back to the respiratory system by <u>bulk flow</u> of the blood. At the respiratory surface, O_2 <u>diffuses</u> into the capillaries of the circulatory system and CO_2 <u>diffuses</u> out again.

36. The respiratory center, located in the brainstem, receives signals from CO_2 receptor neurons. The CO_2 receptors are sensitive to CO_2 levels in the bloodstream. If CO_2 levels rise only 0.3% above the body's acceptable level, the receptors signal the center to increase breathing rate and the depth of breaths.

37. The respiratory center is less sensitive to the O_2 concentration in the bloodstream because under normal conditions, there is an overabundance of O_2. A relatively small drop in O_2 levels will not present a problem to tissues in need of O_2. If, however, a sudden drop in O_2 levels occurs, receptors in the aorta and carotid arteries will stimulate the respiratory center.

38. Each day in the U.S., 3,000 teenagers will begin smoking and about 1,000 of them will eventually die from a smoking-related illness. Really, the only time to quit is before you start, since nicotine is a powerfully addictive drug. In fact, the percentage of people who begin using it and become addicted is about the same as for cocaine and heroin use. Only 20% of regular smokers successfully quit.

39. According to the U.S. Environmental Protection Agency, ETS is Environmental Tobacco Smoke, commonly known as secondhand smoke. Non-smokers are exposed to the smoke exhaled by smokers and smoke given off by the burning end of cigarettes ("passive smoking"). ETS is classified as a Group A carcinogen under EPA's carcinogen assessment guidelines. This classification is reserved for those compounds or mixtures which have been shown to cause cancer in humans.

40. According to the U.S. Environmental Protection Agency, in the U.S., ETS is responsible for approximately 3,000 lung cancer deaths annually in nonsmoking adults. In children, ETS exposure increases the risk of lower respiratory tract infections such as bronchitis and pneumonia. ETS exposure increases chronic middle ear disease. ETS exposure in children irritates the upper respiratory tract and causes some loss of lung function. ETS exposure increases the frequency of episodes and severity of symptoms in asthmatic children. ETS exposure increases the risk for new cases of asthma in children.

41. Health Canada reports that a burning cigarette emits solid particles, gases and liquids. Only the solid particles, about 5–8 percent of the cigarette's output are visible. The list of ingredients in tobacco smoke includes more than 4,000 different chemical compounds, including toxic heavy metals and pesticides. About half the compounds are found naturally in the green tobacco leaf and half are created by chemical reactions when tobacco is burned. Some chemicals are introduced by the manufacturer during the curing process, others are added by manufacturer to give a distinctive flavor or quality to the cigarette.

42. According to Health Canada, tar is a sticky, black residue containing hundreds of chemicals, some of which are classified as hazardous waste.

43. According to Health Canada, lead is a heavy metal. Severe lead poisoning can cause birth defects and learning disabilities in children. Studies have shown that children who live with a smoking parent have more lead in their blood than children living near a lead smelter.

44. R. Atkins of the American Chemical Society reports that N-N-K is a carcinogen. People exposed to passive smoke are more prone to the particular lung cancer caused by N-N-K.

45. According to T. Kuusi and M. Valkonen in the American Heart Association Journal, after spending only 30 minutes in a smoke-filled room, blood levels of antioxidants of study participants was reduced. TRAP (total peroxyl radical trapping potential of serum) measures the ability of all blood antioxidants to rid the body of free radicals. Passive smoking caused a 31% drop in TRAP values. This was enough to change cholesterol metabolism, increasing atherosclerosis development.

Chapter 29: Nutrition and Digestion

OVERVIEW

This chapter provides an overview of the nutrients animals, including you, need and how they get those nutrients from their food. **Digestion** is the physical grinding and chemical breakdown of food. Since the energy an animal needs for body maintenance, growth, and reproduction amounts to the energy that remains after eating and digestion is complete, animals that have evolved the most efficient means of digesting their food will survive. Many different mechanisms accomplishing this have arisen; however, this chapter will focus on the relatively unspecialized, but well-orchestrated, **digestive system** of humans that allows us to ingest and digest a wide variety of food items.

1) What Nutrients Do Animals Need?

The **nutrients** animals (including humans) need include lipids, carbohydrates, and proteins to provide energy for cellular metabolism and to provide the chemical building blocks for making complex molecules. Vitamins and minerals are needed as well, for specific metabolic reactions. How you get these nutrients into your body and process them into a usable form is called **nutrition**.

Fats, carbohydrates, and proteins are used for energy. The energy available in food is measured in **calories** (1000 calories = 1 **Calorie**). In order for you to synthesize the fats your body requires, you must get **essential fatty acids** from your food. Carbohydrates are synthesized from fats, amino acids, or other carbohydrates in your diet and are stored as **glycogen** in your liver and muscles. Your body also uses a small amount of protein for energy. When this protein is broken down, the waste product **urea** is produced. The protein is replaced by protein eaten in your diet. Of the 20 amino acids that your body needs to make proteins, your liver can synthesize 11. The other 9 amino acids, called **essential amino acids**, must be acquired through your diet. **Minerals**, elements, and small inorganic molecules, must be obtained from your food or dissolved in drinking water. **Vitamins** are essential organic compounds required in small amounts for the proper functioning of the human body. Vitamins are either fat-soluble or water-soluble.

The availability and abundance of food in countries like the United States makes it seem unlikely that individuals would be malnourished. However, many people make poor nutritional choices. Therefore, the U.S. government has set up recommendations and guidelines, including the Food Guide Pyramid. Making poor nutritional choices also sets the stage for many individuals in the U.S. to be overweight. To determine if your weight is appropriate for your height, you can calculate your **body mass index** (BMI). A BMI between 20 and 24 is considered healthy.

2) How Is Digestion Accomplished?

Before any animal can begin digestion, food items must be ingested; food is put into its digestive tract, usually through an opening called the **mouth**. The food is physically broken down into smaller pieces and exposed to chemical breakdown by digestive enzymes. The resulting small molecules are absorbed by cells. What could not be used or broken down is eliminated from the body. The way in which ingestion, digestion, absorption, and elimination occurs differs, depending on the animal.

Digestion may occur, as in sponges, within cells. **Intracellular digestion** occurs after microscopic food particles have been engulfed by cells. The food particles are enclosed in a **food vacuole**. The vacuole fuses with **lysosomes** containing digestive enzymes, and the food is broken down and transported to the cell cytoplasm. **Extracellular digestion** occurs in larger animals that have evolved specialized digestive chambers, such as a **gastrovascular cavity**. Food is both ingested and eliminated through a single opening to the cavity; thus, one meal must be eaten, completely digested, and eliminated before another can begin.

Evolution of a digestive tube allows animals to eat more frequently. In the specialized digestive tube of earthworms, a *pharynx* connects the mouth with the *esophagus*. Beyond the esophagus is the *crop*, which stores food particles. Slowly, the food is passed on to the muscular *gizzard*, where the food is ground using sand particles. The food is passed on to the intestines for chemical digestion and is absorbed by the cells of the intestinal lining. Organisms that ingest plant material as their primary energy source have evolved digestive systems specialized to digest cellulose. Within the first of several digestive chambers, the rumen, microorganisms produce **cellulase** to break cellulose into its sugar components. After exposure to one round of cellulase, the *cud*, as the plant material is now called, is regurgitated and reswallowed to the rumen for further exposure to cellulase. Eventually, the cud is passed on to the three other chambers of the system before entering the intestines.

3) How Do Humans and Other Mammals Digest Food?

Humans are omnivores; we eat and digest all types of food items. This contrasts with animals that only eat plants, the herbivores, and the animals that only eat other animals, the carnivores.

Within your mouth, food is ground by your teeth and mixed with saliva by your tongue. This is the beginning of the mechanical and chemical breakdown of food. Saliva contains the enzyme **amylase**, which acts to break down starch to sugar. As you swallow, your tongue pushes your food into your **pharynx**. Your **epiglottis** covers the opening of your trachea as food is swallowed into your esophagus. The muscles of the **esophagus** contract, pushing the food toward your **stomach**, in a process called **peristalsis**. Once in your stomach, food is mixed and churned by peristalsis and other muscular contractions before it is slowly released into your **small intestine** through the **pyloric sphincter**. While in your stomach, however, the food is mixed with chemicals secreted by cells of your stomach. These chemicals include **gastrin**, a hormone that stimulates the production and secretion of hydrochloric acid by stomach cells, and *pepsin*, a **protease** enzyme that digests proteins into shorter peptides. The churning and mixing that occurs in your stomach converts the food into **chyme**. Chyme is slowly released into your small intestine, where it will be digested into small molecules to be absorbed into your bloodstream.

Digestion in your small intestine is accomplished by secretions from cells of the small intestine, your liver, and your pancreas. Your **liver** functions in digestion by producing **bile**, which is stored in your **gallbladder**. Bile is a mixture of **bile salts**, water, cholesterol, and other salts. Bile salts help digest lipids by acting as detergents, breaking fats into microscopic particles. The microscopic lipid particles are then digested by **lipase** enzymes. Your **pancreas** secretes **pancreatic juice** into your small intestine, which neutralizes the acid in the chyme and uses amylase to digest carbohydrates, lipases to digest lipids, and the proteases *trypsin*, *chymotrypsin*, and *carboxypeptidase* to digest proteins. Your small intestine is the primary site of chemical digestion and the primary site of nutrient **absorption**. As nutrients are being absorbed, additional enzymes such as sucrase, maltase, and lactase break disaccharides into monosaccharides and lipases and proteases complete the breakdown of lipids and proteins. To increase the absorptive surface area of your small intestine, it is extensively folded and covered internally by **villi**. Each **villus** is covered by **microvilli**. Each villus contains numerous blood capillaries and one **lacteal**, a lymph capillary. **Segmentation movements** move the chyme around in your small intestine, so that nutrients contact the absorptive surface of the villi. Peristalsis moves what is left after absorption into your **large intestine**, where water is absorbed to form **feces**. The large intestine consists of your **colon** and your **rectum**.

The digestion process is regulated by both your nervous system and hormones. Your nervous system is stimulated by the sight, smell, taste, or even the thought of food, as well as by the process of chewing it. Your nervous system readies your mouth and stomach for the arrival of food. Your stomach begins to secrete acid, stimulating its cells to produce the hormone *gastrin*. Gastrin stimulates the production of more acid. As chyme enters the small intestine, its acidity stimulates the production of **secretin**. Secretin is released into your bloodstream to stimulate your pancreas to dump sodium bicarbonate into your small intestine, neutralizing the acidity. Chyme in your small intestine also causes the hormone **cholestokinin** to be produced. This stimulates your pancreas to release additional digestive enzymes into your small intestine and stimulates your gall bladder to contract and dump bile into your small intestine. When sugars and fatty acids are detected in chyme, your small intestine secretes **gastric inhibitory peptide** to reduce the rate at which your stomach releases chyme. This increases the length of time chyme spends in your small intestine for nutrient absorption.

KEY TERMS AND CONCEPTS

Fill-In: From the following list of key terms, fill in the blanks of the following statements.

absorption	esophagus	lacteal	pharynx
amylase	essential amino acids	large intestine	protease
bile salts	essential fatty acids	lipase	pyloric sphincter
calorie	extracellular digestion	liver	rectum
Calorie	food vacuole	lysosome	segmentation movement
cellulase	gallbladder	microvilli	small intestine
chyme	gastrin	minerals	stomach
colon	gastrovascular cavity	mouth	urea
crop	gizzard	pancreas	villi
digestion	glycogen	pancreatic juice	vitamins
epiglottis	intracellular digestion	peristalsis	

1. _____ is the physical grinding and chemical breakdown of food.

2. A _____ is the amount of energy needed to raise the temperature of one gram of water by one degree Celsius. A _____ is 1,000 of these units and is the term most commonly used to express the energy content in foods. Animals store energy as the carbohydrate _____ in the liver and muscles. A small amount of protein is metabolized for energy, creating the waste product _____.

3. When digestion occurs within a cell, it is termed _____. This type of digestion encloses food items in a membrane bound _____. The food item is digested when enzymes from a _____ are released into the vacuole.

4. When digestion occurs by enzymes outside the cell, it is termed _____, which usually occurs in a digestive sac with a single opening. This sac is called the _____.

5. An earthworm's tubular digestive system includes a thin-walled storage organ called a _____ and a muscular _____ where food is ground.

6. The human diet requires _____ to build lipids, _____ to build proteins, carbohydrates, _____ and _____.

7. The enzyme _____ breaks starch into simple sugars, while the enzymes _____ breaks fats into glycerol and fatty acid molecules and _____ breaks proteins into peptides.

8. Food enters the digestive tract through an opening called the _____. After the food is chewed, it passes through the _____ on its way to the esophagus. The _____ prevents the food from entering the trachea while it is being swallowed. As food goes down the _____, smooth muscles contract rhythmically in a process called _____.

9. Food moving from the _____ to the small intestine is called _____ and its entrance into the small intestine is controlled by a ring of muscle, the _____.

10. _____ and the microscopic _____ greatly increase the absorptive surface area of the _____.

11. Each villus contains a lymph capillary called a _____. Once absorption is complete, the contents remaining in the small intestine move into the _____, which is divided into two portions; the first part is the _____, and the last 15 cm is called the _____.

12. Secretion of acid by the stomach is controlled by the hormone _____. Many of the digestive enzymes that are secreted into the small intestine are produced in the _____ and make up the substance called _____.

13. The _____ produces bile, which is stored in the _____. Bile is a complex mixture of _____, water, cholesterol, and other salts.

14. The primary functions of the small intestine are the chemical breakdown of food and _____ of small molecules into the bloodstream. _____ increase the nutrients and other molecules absorbed from chyme while it is in the small intestine.

KEY TERMS AND DEFINITIONS

absorption: the process by which nutrients are taken into cells.

amylase (am´-i-lās): an enzyme, found in saliva and pancreatic secretions, that catalyzes the breakdown of starch.

bile (bīl): a liquid secretion, produced by the liver, that is stored in the gallbladder and released into the small intestine during digestion; a complex mixture of bile salts, water, other salts, and cholesterol.

bile salt: a substance that is synthesized in the liver from cholesterol and amino acids and that assists in the breakdown of lipids by dispersing them into small particles on which enzymes can act.

body mass index (BMI): a number derived from an individual's weight and height used to estimate body fat. The formula is: weight (in kg)/height^2 (in meters^2).

calorie (kal´-ō-rē): the amount of energy required to raise the temperature of 1 gram of water by 1 degree Celsius.

Calorie: a unit of energy, in which the energy content of foods is measured; the amount of energy required to raise the temperature of 1 liter of water 1 degree Celsius; also called a *kilocalorie*, equal to 1000 calories.

cellulase: an enzyme that catalyzes the breakdown of the carbohydrate cellulose into its component glucose molecules; almost entirely restricted to microorganisms.

cholecystokinin (kō´-lē-sis-tō-ki´-nin): a digestive hormone, produced by the small intestine, that stimulates the release of pancreatic enzymes.

chyme (kīm): an acidic, souplike mixture of partially digested food, water, and digestive secretions that is released from the stomach into the small intestine.

colon: the longest part of the large intestine, exclusive of the rectum.

digestion: the process by which food is physically and chemically broken down into molecules that can be absorbed by cells.

digestive system: a group of organs responsible for ingesting and then digesting food substances into simple molecules that can be absorbed and then expelling undigested wastes from the body.

epiglottis (ep-eh-glah´-tis): a flap of cartilage in the lower pharynx that covers the opening to the larynx during swallowing; directs food down the esophagus.

essential amino acid: an amino acid that is a required nutrient; the body is unable to manufacture essential amino acids, so they must be supplied in the diet.

essential fatty acid: a fatty acid that is a required nutrient; the body is unable to manufacture essential fatty acids, so they must be supplied in the diet.

extracellular digestion: the physical and chemical breakdown of food that occurs outside a cell, normally in a digestive cavity.

feces: semisolid waste material that remains in the intestine after absorption is complete and is voided through the anus. Feces consist of indigestible wastes and the dead bodies of bacteria.

food vacuole: a membranous sac, within a single cell, in which food is enclosed. Digestive enzymes are released into the vacuole, where intracellular digestion occurs.

gallbladder: a small sac, next to the liver, in which the bile secreted by the liver is stored and concentrated. Bile is released from the gallbladder to the small intestine through the bile duct.

gastric inhibitory peptide: a hormone, produced by the small intestine, that inhibits the activity of the stomach.

gastrin: a hormone, produced by the stomach, that stimulates acid secretion in response to the presence of food.

gastrovascular cavity: a saclike chamber with digestive functions, found in simple invertebrates; a single opening serves as both mouth and anus, and the chamber provides direct access of nutrients to the cells.

glycogen (glī´-kō-jen): a long, branched polymer of glucose that is stored by animals in the muscles and liver and metabolized as a source of energy.

intracellular digestion: the chemical breakdown of food within single cells.

lacteal (lak-tēl´): a single lymph capillary that penetrates each villus of the small intestine.

large intestine: the final section of the digestive tract; consists of the colon and the rectum, where feces are formed and stored.

lipase (lī´-pās): an enzyme that catalyzes the breakdown of lipids such as fats.

liver: an organ with varied functions, including bile production, glycogen storage, and the detoxification of poisons.

lysosome (lī´-sō-sōm): a membrane-bound organelle containing intracellular digestive enzymes.

microvillus (mī-krō-vi´-lus; pl., microvilli): a microscopic projection of the plasma membrane of each villus; increases the surface area of the villus.

mineral: an inorganic substance, especially one in rocks or soil.

mouth: the opening of a tubular digestive system into which food is first introduced.

nutrient: a substance acquired from the environment and needed for the survival, growth, and development of an organism.

nutrition: the process of acquiring nutrients from the environment and, if necessary, processing them into a form that can be used by the body.

pancreas (pān´-krē-us): a combined exocrine and endocrine gland located in the abdominal cavity next to the stomach. The endocrine portion secretes the hormones insulin and glucagon, which regulate glucose concentrations in the blood. The exocrine portion secretes enzymes for fat, carbohydrate, and protein digestion into the small intestine and neutralizes the acidic chyme.

pancreatic juice: a mixture of water, sodium bicarbonate, and enzymes released by the pancreas into the small intestine.

peristalsis: rhythmic coordinated contractions of the smooth muscles of the digestive tract that move substances through the digestive tract.

pharynx (far´-inks): in vertebrates, a chamber that is located at the back of the mouth and is shared by the digestive and respiratory systems; in some invertebrates, the portion of the digestive tube just posterior to the mouth.

protease (prō´-tē-ās): an enzyme that digests proteins.

pyloric sphincter (pī-lor´-ik sfink´-ter): a circular muscle, located at the base of the stomach, that regulates the passage of chyme into the small intestine.

rectum: the terminal portion of the vertebrate digestive tube, where feces are stored until they can be eliminated.

secretin: a hormone, produced by the small intestine, that stimulates the production and release of digestive secretions by the pancreas and liver.

segmentation movement: a contraction of the small intestine that results in the mixing of partially digested food and digestive enzymes. Segmentation movements also bring nutrients into contact with the absorptive intestinal wall.

small intestine: the portion of the digestive tract, located between the stomach and large intestine, in which most digestion and absorption of nutrients occur.

stomach: the muscular sac between the esophagus and small intestine where food is stored and mechanically broken down and in which protein digestion begins.

urea (ū-rē´-uh): a water-soluble, nitrogen-containing waste product of amino acid breakdown; one of the principal components of mammalian urine.

villus (vi´-lus; pl., villi): a fingerlike projection of the wall of the small intestine that increases the absorptive surface area.

vitamin: one of a group of diverse chemicals that must be present in trace amounts in the diet to maintain health; used by the body in conjunction with enzymes in a variety of metabolic reactions.

THINKING THROUGH THE CONCEPTS

True or False: Determine if the statement given is true or false. If it is false, change the underlined word(s) so that the statement reads true.

15. _____ An advantage of the evolution of a digestive tract is that it enables an organism to eat frequently.

16. _____ The nematode has a complex digestion tube that is unspecialized along its length.

17. _____ Animals with tubular digestive systems use intracellular digestion to break down their food.

18. _____ The primary function of the stomach is water reabsorption.

19. _____ Bacteria in the large intestine serve no purpose to the human body.

20. _____ The secretion of saliva, stimulated by the sight or thought of food, or the presence of food in the mouth, is controlled by the nervous system.

21. _____ Mucus produced by the stomach serves to neutralize the acid conditions.

22. _____ Within the esophagus, the enzyme pepsin breaks proteins into shorter polypeptides.

23. _____ As food moves into the small intestine, the pancreas is stimulated to produce sodium hydroxide to neutralize the acidic chyme.

24. _____   Bile salts <u>emulsify</u> fats.

25. _____   Absorption of nutrients and small food molecules occurs in the <u>large intestine</u>.

26. _____   The purpose of villi and microvilli is to increase the surface area of the <u>stomach</u>.

27. _____   The <u>vitamins of the B-complex and vitamin C</u> are used to aid enzymes in chemical reactions.

28 _____   Fat soluble vitamins may be <u>toxic</u> if taken in high doses.

Matching: Match the following enzymes with their substrate.

29. _____   cellulase   Choices:

30. _____   pepsin   a. proteins

31. _____   lipase   b. disaccharides

32. _____   lactase   c. individual amino acids

33. _____   chymotrypsin   d. peptides

34. _____   amylase   e. lipids

35. _____   peptidases   f. starches

36. _____   sucrase   g. cellulose

37. _____   trypsin

38. _____   maltase

39. Complete the table below by identifying the functions of the structures listed.

Area of the Digestive Tract	Function
mouth	
esophagus	
stomach	
small intestine	
liver	
pancreas	
large intestine	

Identify: Determine whether the following vitamins are **fat** soluble or **water** soluble.

40. _____ vitamin A   43. _____ vitamin D

41. _____ vitamin B-complex 44. _____ vitamin E

42. _____ vitamin C   45. _____ vitamin K

Matching: To which hormone are the following statements related?

46. _____ stimulates hydrochloric acid   Choices:
                    production by stomach cells
                                                             a. gastrin
47. _____ stimulates release of pancreatic
                    digestive enzymes   b. secretin

48. _____ stimulates bicarbonate release   c. cholecystokinin
                    into small intestine
                                                             d. gastric inhibitory peptide
49. _____ inhibits movement of chyme into
                    the small intestine

50. _____ stimulates contraction of the gall bladder

51. _____ is regulated by acid production

52. _____ release is stimulated by the presence of fatty
                    acids in chyme

53. _____ release is stimulated by acidic chyme in the
                    small intestine

Short Answer:

54. List the five tasks a digestive system must accomplish, regardless of its level of complexity.

    _____

    _____

    _____

    _____

    _____

55. What are the two primary functions of a digestive tract?

    _____

    _____

56. Where are trypsin, chymotrypsin, and carboxypeptidase found? When are they active? What is their
    function?

    _____

    _____

    _____

APPLYING THE CONCEPTS

These practice questions are intended to sharpen your ability to apply critical thinking and analysis to biological concepts covered in this chapter.

57. Consider the chemicals secreted by the stomach: hydrochloric acid and protein digesting pepsin. Why does the stomach not digest itself?

_____

_____

_____

_____

_____

_____

58. Some people might react strongly if they knew that their digestive system supported a healthy population of bacteria. How would you convince them that the bacteria play a very important role in their health?

_____

_____

_____

_____

_____

_____

_____

_____

_____

59. If you are being treated for a bacterial infection with an antibiotic, a side effect you may experience is diarrhea. Using the information you now have concerning the large intestine, what would you determine to be the reason? How might eating yogurt with live bacterial cultures counteract this side effect?

_____

_____

_____

_____

_____

_____

_____

_____

Use the Case Study and the Web sites for this chapter to answer the following questions.

60. What is recommended for individuals whose BMI indicates they are overweight or obese and have two or more risk factors?

_____

_____

_____

_____

_____

_____

61. In genetic studies involving adoptees and/or twins, is obesity genetic, or is it environmentally (upbringing) caused?

_____

_____

_____

_____

_____

_____

_____

_____

_____

_____

62. What is Leptin? Which body cells produce Leptin?

_____

_____

_____

_____

63. Based on studies involving mice, what would mice that do not produce Leptin look like? Why? What if they were given injections of Leptin?

_____

_____

_____

_____

ANSWERS TO EXERCISES

1. Digestion
2. calorie
   Calorie
   glycogen
   urea
3. intracellular digestion
   food vacuole
   lysosome
4. extracellular digestion
   gastrovascular cavity
5. crop
   gizzard
6. essential fatty acids
   essential amino acids
   vitamins
   minerals
7. amylase
   lipase
   protease
8. mouth
   pharynx
   epiglottis
   esophagus

   peristalsis
9. stomach
   chyme
   pyloric sphincter
10. Villi
    microvilli
    small intestine
11. lacteal
    large intestine
    colon
    rectum
12. gastrin
    pancreas
    pancreatic juice
13. liver
    gallbladder
    bile salts
14. absorption
    Segmentation movements
15. true
16. false, simple
17. false, extracellular digestion

18. false, store food
19. false, synthesize vitamins
    B_{12}, thiamin, riboflavin, K
20. true
21. false, protects stomach cells
22. false, stomach
23. false, sodium bicarbonate
24. true
25. false, small intestine
26. false, small intestine
27. true
28. true
29. g
30. a
31. e
32. b
33. a
34. f
35. d
36. b
37. a
38. b

39.

Area of the Digestive Tract	Function
mouth	mechanical (chewing) and chemical (amylase) breakdown of food
esophagus	transfers food from mouth to stomach
stomach	stores food, mechanical breakdown of food, secretes enzymes
small intestine	digests food into small molecules, absorbs food molecules
liver	produces bile to emulsify fats in small intestine
pancreas	produces pancreatic juice to digest carbohydrates, fats, and protein, and to neutralize chyme
large intestine	absorbs water and salts from chyme, absorbs vitamins made by bacteria

40. fat	42. water	44. fat	46. a	48. b	50. c	52. d
41. water	43. fat	45. fat	47. c	49. d	51. a	53. b

54. Ingestion of food item; mechanical breakdown (food is physically broken into smaller pieces); chemical breakdown (food is exposed to digestive enzymes); absorption of small molecules; elimination of wastes.

55. Digestion of food; absorption of nutrients.

56. All three enzymes are in pancreatic juice secreted by the pancreas into the small intestine. All three are proteolytic enzymes that digest proteins and large peptides into small peptide chains.

57. Thick mucus secreted by cells of the stomach lining protects the stomach cells from the acidic environment. The enzyme pepsin is produced in an inactive form, pepsinogen, that is converted to pepsin in the acid environment of the stomach. Again, the mucus lining the stomach would protect the cells from being digested by pepsin. However, cells lining the stomach are damaged or partially digested and must be replaced every few days. In extreme cases, an ulcer forms where the lining and deeper tissues of the stomach have been digested.

58. The bacteria living in the large intestine synthesize the vitamins B_{12}, thiamin, riboflavin, and K, which are absorbed by the large intestine. The normal diet does not provide these vitamins in sufficient amounts. You could include in your argument that without the production of these vitamins, a person could be affected by the following vitamin deficiency symptoms:
B_{12} — pernicious anemia, neurological disorders
Thiamin — muscle weakness, peripheral nerve changes, edema, heart failure (Beriberi)
Riboflavin — red, cracked lips, eye lesions
K — bleeding, internal hemorrhages.

59. An antibiotic that you would be taking for a general bacterial infection would most likely be a broad-spectrum antibiotic. That is, it acts against a wide range of bacteria. This would include the bacteria active in your large intestine. The antibiotic does not distinguish between "good" and "bad" bacteria. With the environment of your large intestine disrupted, diarrhea may result. Eating yogurt containing "active cultures" (live *Acidophilus* bacteria) while taking the antibiotic regimen may help restore the bacterial population in your large intestine, reducing the side effects of the antibiotic.

60. According to the National Institutes of Health, for people who are considered obese (BMI greater than or equal to 30) or those who are overweight (BMI of 25 to 29.9) and have two or more risk factors, the guidelines recommend weight loss. Even a small weight loss (just 10 percent of your current weight) will help reduce your risk of developing diseases associated with obesity.

61. According to the Canadian Medical Association's Task Force on the Periodic Health Examination, the most convincing evidence in the cause of obesity has come from recent studies involving adoptees and twins. A Danish study involving 540 adoptees showed a strong statistical relationship between the BMIs of the adoptees and their biologic parents but no statistical relationship with the BMIs of their adoptive parents. The researchers came to the conclusion that family environment alone has no apparent effect on obesity. The evidence from studies involving twins is even more convincing. BMI or skinfold thickness values between sets of identical and fraternal twins that were reared either together or apart were evaluated. These comparisons allowed investigators to determine the effects of genetic factors, shared environmental factors, and unshared individual environmental factors. The results indicated that obesity is genetic.

62. According to L. Austgen and R.A. Bowen, Colorado State Univ. and C. A. Muller-Rideau, Univ. of Tennessee-Knoxville, leptin is a protein hormone with important effects in regulating body weight, metabolism and reproductive function. This protein hormone is controlled by the obese (*ob*) gene. Leptin is produced predominantly by adipocytes (fat cells).

63. According to L. Austgen and R.A. Bowen, Colorado State Univ., mice with the inability to produce Leptin would be obese since leptin plays an important role in the long term regulation of body weight. Mice that were unable to make their own leptin were given daily injections of leptin. The leptin injections caused the mice to dramatically reduce the amount of food they ate. It also resulted in roughly a 50% loss of body weight within a month.

Chapter 30: The Urinary System

OVERVIEW

This chapter presents the role of the urinary system in simple animals, then focuses on the form and function of the urinary system in vertebrates; humans in particular. The urinary system of animals plays a very important role in the maintenance of homeostasis. It removes excess water, salts, nutrients, and minerals in a precise and regulated manner. From simple animals to the more complex vertebrates, the structures of the urinary system function in rather similar ways.

1) How Does Excretion Occur in Simpler Invertebrate Animals?

Excretory systems of all animals play a very important role in the maintenance of a stable internal environment, or maintaining **homeostasis**. They must collect and filter body fluids, retain nutrients, and remove wastes. The first and most simple structures specialized for excretion are called **protonephridia** and are found in flatworms. This system uses **flame cells**, single-celled bulbs that filter water and dissolved wastes. More advanced excretory structures are found in other invertebrates. These kidney-like **nephridia** consist of a funnel-shaped **nephrostome** with cilia that conduct body fluids along a long, twisted tube. As the fluid passes through the tube, nutrients and salts are reabsorbed and water and wastes form urine. The urine is released from the body through the **excretory pore**.

2) What Are the Functions of Vertebrate Urinary Systems?

In vertebrates, the excretory system is more often called the urinary system. And, in vertebrates, the **kidneys** collect the fluid plasma from blood, and reabsorb nutrients and water while removing toxins, waste products, and excess water, salts, nutrients, and minerals. What is removed is channeled and stored until it leaves the body during the process of **excretion**. **Urea** is produced when the body digests protein and removes the amino functional group ($-NH_2$) from an amino acid; the amino group is released as **ammonia** (NH_3). Ammonia is highly toxic to mammals, so it is converted to water-soluble urea in the liver. Urea leaves the body as **urine**. However, birds and reptiles must conserve water, therefore, they convert urea to *water-insoluble* **uric acid**.

3) How Does the Human Urinary System Function?

Blood to be filtered by your urinary system enters both of your kidneys through the **renal artery**. Once filtration is complete, your blood leaves your kidneys through the **renal vein**. The urine that is produced is transported from each kidney to your **bladder** through a tube called the **ureter**. The ureter is muscular and uses peristaltic contractions to move the urine to your bladder. Urine leaves your body through the **urethra**.

Your kidneys are complex. A hollow inner chamber, called the *renal pelvis*, funnels urine to the ureter. Outside the renal pelvis is the *renal medulla* covered by the *renal cortex*. **Nephrons** fill the renal cortex and may extend into the medulla. Each nephron has three parts. A dense collection of capillaries, called a **glomerulus,** that filters the blood is surrounded by a cup-shaped **Bowman's capsule**. A long, twisted **tubule** carries what was removed away from the glomerulus. The tubule, in turn, is also divided into three parts; the **proximal tubule**, the **loop of Henle**, and the **distal tubule**. The distal tubule leads to the **collecting duct**.

Blood flowing through the glomerulus undergoes **filtration**. The fluid **filtrate**, containing water and dissolved substances, is collected in the Bowman's capsule and transported through the tubule. The filtered blood is now very thick and concentrated; it moves through very porous capillaries that wind around the tubule. When the capillaries come in contact with the tubule, water and any nutrients in the filtrate are reabsorbed by the capillaries. This is called **tubular reabsorption** and occurs in the proximal tubule using active transport, osmosis, and diffusion. Any wastes still in the capillary blood move into the filtrate in the distal tubule by **tubular secretion**. These wastes are removed from your body by excretion.

The loop of Henle enables the filtrate (now called urine) to become concentrated. This can occur because the interstitial fluid surrounding the loop has a greater osmotic concentration than the fluid in the loop. Therefore, water flows out of the loop, wastes stay in the loop. Then, as the collecting duct passes through the concentrated interstitial fluid, additional water moves out of the duct by osmosis. The amount of water that is reabsorbed into your blood is controlled by the level of **antidiuretic hormone** (**ADH**) in your blood. ADH makes the cells of the distal tubule more permeable to water and is regulated according to the volume of blood in your body.

The kidneys are involved in regulating your blood pressure. If your blood pressure is too low, your kidneys will secrete **renin** into your bloodstream. Renin acts like an enzyme and catalyzes the production of **angiotensin**. Angiotensin causes your arterioles to constrict, elevating your blood pressure again. The rate of filtration by your kidneys is decreased because your arterioles are constricted. This reduces the amount of water removed from your blood, increasing your blood volume and your blood pressure.

The kidneys are also involved in regulating the oxygen content in your blood. If your blood is low in oxygen, your kidneys release the hormone **erythropoietin**. Erythropoietin stimulates blood cell production by your bone marrow.

KEY TERMS AND CONCEPTS

Fill-In: From the following list of key terms, fill in the blanks in the following statements.

ammonia	filtration	renal arteries
angiotensin	flame cell	renal cortex
antidiuretic hormone (ADH)	glomerulus	renal medulla
bladder	hemodialysis	renal pelvis
Bowman's capsule	homeostasis	renal veins
collecting duct	kidney	renin
dialysis	loop of Henle	tubular reabsorption
distal tubule	nephridia	tubular secretion
erythropoietin	nephrons	urea
excretion	nephrostome	ureter
excretory pore	protonephridia	urethra
filtrate	proximal tubule	urine

1. _____ is defined as the maintenance of a stable environment. The _____ play a role in maintaining this stability. The wastes that the urinary system filters from the blood are removed from the body through a process called _____.

2. During amino acid digestion, the amine group may be removed forming the highly toxic chemical _____. This toxic chemical is taken to the liver where it is converted to _____. Eventually, the waste is excreted from the body as _____.

3. The simplest structures specialized for excretion are the _____ found in flatworms.

This structure uses single celled bulbs with cilia called _____. Some invertebrates use more advanced, kidney-like structures called_____. Coelomic fluid is conducted into the funnel-shaped _____. After collecting in a bladderlike sac, the urine leaves the body through the _____.

4. In the human, a pair of kidneys receives blood to be filtered through the _____. After it has been filtered, the blood leaves the kidneys through the _____. Urine leaves each kidney through a long tube called a _____ and is stored in the _____. Urine leaves the body by way of the _____.

5. Within the human kidney is a hollow collecting chamber called the _____. Outside this chamber is the fan-shaped _____ covered by the _____. Microscopic structures, the _____ act as filters.

6. Each nephron contains a mass of capillaries, the _____ which is surrounded by a cup-like structure, the _____. A long twisted tubule is subdivided into the _____, the _____, and the _____, which leads to the _____.

7. Within the glomerulus, water and dissolved substances are removed from the blood in a process called _____. The fluid that is removed is called the _____.

8. Reabsorption of H_2O and nutrients from the filtrate by the proximal tubule is called _____. When the distal tubule removes wastes not filtered out initially, _____ is occurring. The amount of water reabsorbed by the proximal tubule is regulated by the hormone _____.

9. The kidneys help regulate blood pressure by releasing the hormone _____ into the bloodstream if blood pressure falls. This stimulates the release of an arteriole constricting hormone _____. The kidneys also aid in increasing blood oxygen levels by releasing the hormone _____ when low oxygen levels are detected.

10. Individuals with kidney damage are treated with an artificial kidney. The process, called _____ uses an artificial semipermeable membrane to passively filter substances from the blood. Since it is blood that is being filtered, the process is specifically called _____.

KEY TERMS AND DEFINITIONS

ammonia: NH_3; a highly toxic nitrogen-containing waste product of amino acid breakdown. In the mammalian liver, it is converted to urea.

angiotensin (an-jē-ō-ten´-sun): a hormone that functions in water regulation in mammals by stimulating physiological changes that increase blood volume and blood pressure.

antidiuretic hormone (an-tē-dī-ū r-et´-ik; ADH): a hormone produced by the hypothalamus and released into the bloodstream by the posterior pituitary when blood volume is low; increases the permeability of the distal tubule and the collecting duct to water, allowing more water to be reabsorbed into the bloodstream.

bladder: a hollow muscular storage organ for storing urine.

Bowman's capsule: the cup-shaped portion of the nephron in which blood filtrate is collected from the glomerulus.

collecting duct: a conducting tube, within the kidney, that collects urine from many nephrons and conducts it through the renal medulla into the renal pelvis. Urine may become concentrated in the collecting ducts if ADH is present.

distal tubule: in the nephrons of the mammalian kidney, the last segment of the renal tubule through which the filtrate passes just before it empties into the collecting duct; a site of selective secretion and reabsorption as water and ions pass between the blood and the filtrate across the tubule membrane.

erythropoietin (eh-rith´-rō-pō-ē´-tin): a hormone produced by the kidneys in response to oxygen deficiency that stimulates the production of red blood cells by the bone marrow.

excretion: the elimination of waste substances from the body; can occur from the digestive system, skin glands, urinary system, or lungs.

excretory pore: an opening in the body wall of certain invertebrates, such as the earthworm, through which urine is excreted.

filtrate: the fluid produced by filtration; in the kidneys, the fluid produced by the filtration of blood through the glomerular capillaries.

filtration: within Bowman's capsule in each nephron of a kidney, the process by which blood is pumped under pressure through permeable capillaries of the glomerulus, forcing out water, dissolved wastes, and nutrients.

flame cell: in flatworms, a specialized cell, containing beating cilia, that conducts water and wastes through the branching tubes that serve as an excretory system.

glomerulus (glō-mer´-ū-lus): a dense network of thin-walled capillaries, located within the Bowman's capsule of each nephron of the kidney, where blood pressure forces water and dissolved nutrients through capillary walls for filtration by the nephron.

hemodialysis (hē-mō-dī-al´-luh-sis): a procedure that simulates kidney function in individuals with damaged or ineffective kidneys; blood is diverted from the body, artificially filtered, and returned to the body.

homeostasis (hōm-ē-ō-stā´sis): the maintenance of a relatively constant environment required for the optimal functioning of cells, maintained by the coordinated activity of numerous regulatory mechanisms, including the respiratory, endocrine, circulatory, and excretory systems.

kidney: one of a pair of organs of the excretory system that is located on either side of the spinal column and filters blood, removing wastes and regulating the composition and water content of the blood.

loop of Henle (hen´-lē): a specialized portion of the tubule of the nephron in birds and mammals that creates an osmotic concentration gradient in the fluid immediately surrounding it. This gradient in turn makes possible the production of urine more osmotically concentrated than blood plasma.

nephridium (nef-rid´-ē-um): an excretory organ found in earthworms, mollusks, and certain other invertebrates; somewhat resembles a single vertebrate nephron.

nephron (nef´-ron): the functional unit of the kidney; where blood is filtered and urine formed.

nephrostome (nef´-rō-stōm): the funnel-shaped opening of the nephridium of some invertebrates such as earthworms; coelomic fluid is drawn into the nephrostome for filtration.

protonephridium (prō-tō-nef-rid´-ē-um; pl., protonephridia): an excretory system consisting of tubules that have external opening but lack internal openings; for example, the flame-cell system of flatworms.

proximal tubule: in nephrons of the mammalian kidney, the portion of the renal tubule just after the Bowman's capsule; receives filtrate from the capsule and is the site where selective secretion and reabsorption between the filtrate and the blood begins.

renal artery: the artery carrying blood to each kidney.

renal vein: the vein carrying cleansed blood away from each kidney.

renin: an enzyme that is released (in mammals) when blood pressure and/or sodium concentration in the blood drops below a set point; initiates a cascade of events that restores blood pressure and sodium concentration.

tubular reabsorption: the process by which cells of the tubule of the nephron remove water and nutrients from the filtrate within the tubule and return those substances to the blood.

tubular secretion: the process by which cells of the tubule of the nephron remove additional wastes from the blood, actively secreting those wastes into the tubule.

tubule (toob´-ū l): the tubular portion of the nephron; includes a proximal portion, the loop of Henle, and a distal portion. Urine is formed from the blood filtrate as it passes through the tubule.

urea (ū-rē´-uh): a water-soluble, nitrogen-containing waste product of amino acid breakdown; one of the principal components of mammalian urine.

ureter (ū´-re-ter): a tube that conducts urine from each kidney to the bladder.

urethra (ū-rē´-thruh): the tube leading from the urinary bladder to the outside of the body; in males, the urethra also receives sperm from the vas deferens and conducts both sperm and urine (at different times) to the tip of the penis.

uric acid (ū r´-ik): a nitrogen-containing waste product of amino acid breakdown; a relatively insoluble white crystal excreted by birds, reptiles, and insects.

urine: the fluid produced and excreted by the urinary system of vertebrates; contains water and dissolved wastes, such as urea.

THINKING THROUGH THE CONCEPTS

True or False: Determine if the statement given is true or false. If it is false, change the underlined word(s) so that the statement reads true.

11. _____   Urea is formed when underlined carbohydrates are digested.

12. _____   Flame cells function to heat up body fluids.

13. _____   In the earthworm body, each segment basically has its own pair of nephridia.

14. _____   Urine is carried to the bladder in the ureter.

15. _____   Urination is under voluntary control of the external sphincter.

16. _____   An adult bladder can hold 0.5 L of urine.

17. _____   Within the kidney, the loop of Henle extends into the renal medulla.

18. _____   Filtration in the glomerulus occurs because of a difference in diameter between the arterioles bringing blood in and the arterioles taking blood out.

19. _____   Tubular reabsorption occurs in the <u>distal tubule</u>.

20. _____   During tubular secretion, wastes that are actively secreted into the tubule may include <u>drugs</u>.

21. _____   Each drop of blood passes through a kidney <u>10 times</u> a day.

22. _____   ADH is released when receptors in the <u>hypothalamus</u> detect an inappropriate osmotic level in the blood.

23. _____   A long loop of Henle is found in the kidneys of <u>aquatic</u> animals.

Identify: Determine whether the following refer to tubular **reabsorption** or tubular **secretion**.

24. _____   involves the proximal tubule

25. _____   involves the distal tubule

26. _____   salts, amino acids, and glucose are actively transported into the blood

27. _____   water leaves the tubule by osmosis

28. _____   wastes are secreted into the tubule

29. _____   urea remains in the tubule

30. _____   wastes become more concentrated

Using the diagram of the nephron, answer the questions below.

31. _____   identify structure (a)

32. _____   what is the function of (a)?

33. _____   identify structure (b)

34. _____   what is the function of (b)?

35. _____   identify structure (c)

36. _____   if (c) was extremely long, in what sort of habitat would this animal live?

37. _____   if (c) was rather short, in what sort of habitat would this animal live?

38. _____   identify structure (d)

39. _____   what is the function of (d)?

40. _____   identify structure (e)

41. _____   what is the function of (e)?

42. Trace the path of urine from its production in the kidneys to its point of excretion from the body.

Kidneys → _____ → _____ → _____ → excretion

APPLYING THE CONCEPTS

These practice questions are intended to sharpen your ability to apply critical thinking and analysis to the biological concepts covered in this chapter.

43. Identify the six ways the mammalian urinary system helps maintain homeostasis.

_____

_____

_____

_____

_____

_____

44. How do the loop of Henle and the collecting duct concentrate urine?

_____

_____

_____

_____

45. How does negative feedback regulate the water content of your blood?

_____

_____

_____

_____

_____

_____

_____

_____

_____

46. Why does beer (or any alcohol) in your system make you urinate more frequently?

_____

_____

_____

Using the Case Study and Web sites provided for this chapter, answer the following questions.

47. What is ESRD? What are the warning signs of ESRD?

_____

_____

_____

_____

48. Glomerulonephritis is a painless inflammation of the glomerulus. This results in high blood pressure which may lead to kidney failure. Why would glomerulonephritis lead to high blood pressure?

_____

_____

_____

_____

49. What special dietary changes must a person undergoing hemodialysis make? How do these changes compare to diets of persons undergoing peritoneal dialysis?

_____

_____

_____

_____

_____

_____

_____

_____

_____

_____

_____

_____

50. Is dialysis a cure for ESRD? Why or why not?

_____

_____

_____

_____

_____

_____

_____

_____

ANSWERS TO EXERCISES

1. Homeostasis
   kidneys
   excretion
2. ammonia
   urea
   urine
3. protonephridia
   flame cells
   nephridia
   nephrostome
   excretory pore
4. renal arteries
   renal veins
   ureter
   bladder
   urethra
5. renal pelvis
   renal medulla
   renal cortex
   nephrons
6. glomerulus
   Bowman's capsule
   proximal tubule
   loop of Henle
   distal tubule
   collecting duct

7. filtration
   filtrate
8. tubular reabsorption
   tubular secretion
   ADH, antidiuretic hormone
9. renin
   angiotensin
   erythropoietin
10. dialysis
    hemodialysis
11. false, protein
12. false, filter
13. true
14. true
15. true
16. true
17. true
18. true
19. false, proximal tubule
20. true
21. false, about 350 times
22. true
23. false, desert

24. reabsorption
25. secretion
26. reabsorption
27. reabsorption
28. secretion
29. reabsorption
30. reabsorption
31. glomerulus
32. filtration
33. proximal tubule
34. reabsorption
35. loop of Henle
36. desert
37. aquatic
38. distal tubule
39. tubular secretion
40. collecting duct
41. concentrate urine

42. Kidneys → ureter → bladder → urethra → excretion.

43. Regulation of blood ion levels. Regulation of blood water content. Maintenance of blood pH. Retention of important nutrients. Secretion of hormones. Elimination of cellular waste products.

44. The loop of Henle concentrates urine because the interstitial fluid surrounding the loop has a greater osmotic concentration gradient than the fluid in the loop. Therefore, water flows out of the loop, wastes stay in the loop. Then as the collecting duct passes through the concentrated interstitial fluid, additional water moves out of the duct by osmosis.

45. If receptors in the hypothalamus of your brain detect a decrease in the osmotic content of your blood (not as much "stuff" is dissolved in your blood — or there is too much water), it signals your anterior pituitary to stop releasing ADH. The increase of ADH in your blood causes your distal tubules and collecting ducts to be less permeable to water (water will not flow out of them). This increases the osmotic content of your blood (there's more "stuff" dissolved in it — less water). Receptors in your hypothalamus now sense an increase in the osmotic content. Your hypothalamus tells your anterior pituitary to release ADH again. The increased ADH in your blood increases the permeability of your distal tubules and collecting ducts, so more water flows out of them, increasing the water content of your blood. This cycle continues as your body continuously regulates the osmotic content of your blood.

46. Consumption of alcohol causes a decrease in the osmotic content of your blood. This results in a decrease of ADH production and your distal tubules and collecting ducts become less permeable to water (water will not flow out of them). More dilute urine flows into the collecting ducts, into your ureters, and fills your bladder.

47. According to the American Kidney Fund, ESRD is End Stage Renal Disease. According to the American Kidney Foundation the warning signs include: swelling of body parts, especially eyes, ankles, or wrists; burning or abnormal discharge during urination; changes in the frequency of urination — especially at night; lower back pain; bloody, foamy, or coffee-colored urine; or high blood pressure.

48. According to the Kidney Patient Guide, glomerulonephritis is an inflammation of the glomerulus. Therefore, blood would not be filtered properly and would keep a high osmotic content. This would result in water being kept in the bloodstream increasing blood pressure. The increased blood pressure may result in damage to the capillaries of the kidney.

49. The National Kidney and Urologic Diseases Information Clearinghouse suggests that a person under-going hemodialysis should eat balanced amounts of foods high in protein such as meat and chicken (animal protein is better used by the body than protein found in plant foods). Their potassium intake should be limited. Too much or too little potassium can be harmful to the heart. Fluid intake should be limited. Fluids build up quickly when your kidneys are not working. Too much fluid makes tissues swell and can cause high blood pressure and heart trouble. Salt should be avoided. Salt increases thirst and causes water retention. Milk, cheese, nuts, dried beans, and soft drinks should be limited because they contain the mineral phosphorus. Too much phosphorus in the blood causes calcium to be pulled from your bones. Calcium supplements may be needed.

   The National Kidney and Urologic Diseases Information Clearinghouse also states that a diet for peritoneal dialysis is slightly different than a diet for hemodialysis. These patients may be able to have more salt and fluids and may eat more protein. The potassium restrictions may be different and they may need to cut back on the number of calories eaten because the sugar in the dialysate may cause weight gain.

50. The National Kidney and Urologic Diseases Information Clearinghouse warns that dialysis is not a cure for ESRD. Hemodialysis and peritoneal dialysis are treatments that try to replace failed kidneys. These treatments help increase quality and length of life, but they are not cures. While patients with ESRD are living longer than ever, ESRD can cause problems over the years including bone disease, high blood pressure, nerve damage, and anemia (having too few red blood cells).

Chapter 31: Defenses Against Disease: The Immune Response

OVERVIEW

This chapter will present the ways in which your body defends itself against bacteria, fungi, protists, and environmental toxins. Your body is designed so that it is difficult for foreign substances to enter it. But, when they do, your body has a well-developed system to detect and destroy invaders. It is able to determine foreign cells from body cells and it is able to remember invaders that have been present before. Occasionally, the system fails, however. Examples of what occurs when an immune system breaks down are discussed.

1) How Does the Body Defend Against Invasion?

The human body is designed to keep invaders from entering. Your **skin** provides a barrier to **microbes**. The outer layer of skin is dry and has no nutrients, and the dead cells slough off. Your sweat and sebaceous glands produce natural antibiotics. Microbes have a difficult time actually getting into your body through your skin. However, you have natural openings into your body. **Mucous membranes** lining the digestive and respiratory tracts trap microbes; and they contain antibiotic substances, as well. Cilia of cells lining these tracts move the mucous and trapped particles up and out of their respective areas. The mucous is then either coughed or sneezed out, or swallowed.

When microbes do gain entry, your body has a line of defense to remove them. **Phagocytic cells** such as the white blood cells called **macrophages**, ingest microbes by phagocytosis. **Natural killer cells** destroy your body's cells that have been attacked by viruses. When viruses enter a cell they usually leave viral proteins on the outside of the cell. Natural killer cells know that those proteins do not belong to you and respond by killing the infected cell. Natural killer cells also destroy cancerous cells.

When your skin is cut, an **inflammatory response** is elicited. The damaged cells release **histamine**, relaxing smooth muscle of arterioles and making capillaries leak. This increases blood flow to the area, causing it to become red, swollen, and warm. Phagocytes brought to your wound by the increased blood flow ingest microbes infecting the injury. If the phagocytes cannot remove all of the microbes, the microbes may infect larger areas of the body, initiating a **fever**. A fever is part of the body's natural defense against infection. It increases phagocyte activity while reducing microbial reproduction. Fever also induces the production of virus-fighting **interferon**. Macrophages, responding to an infection, release **endogenous pyrogens** that stimulate the hypothalamus to reset the body's thermostat and to set in motion the responses that lead to an increase in body temperature.

You have just read about the *non-specific defenses* your body uses to fight off *any* foreign substance. These defenses usually work well, but when they fail, your body uses a highly *specific* set of defenses; your **immune system**.

2) What Are the Key Characteristics of the Immune Response?

Your **immune response** is specifically directed toward a particular invading organism. *Lymphocytes* called **B cells** and **T cells** recognize foreign organisms, initiate an attack against them, and remember the organisms in order to combat future attacks.

B cells produce proteins called **antibodies**. Antibodies may remain attached to the surface of the B cell or they may be secreted into the bloodstream. T cells produce proteins called **T-cell receptors** that remain attached to the T cells.

Antibodies are made of two heavy peptide chains and two light peptide chains. Both the heavy and the light chains have a **constant region** and a **variable region**. The variable regions are very specific and will only react to specific **antigens**. So, the antibodies on the membrane of a B cell act as receptors, projecting outward, ready to bind to the right *anti*body *gen*erating molecules (antigens). Molecules that your body responds to as antigens include proteins, polysaccharides, and glycoproteins.

When an antigen binds to an antibody receptor on a B cell, a response is stimulated in the B cell. The antibodies that circulate in the bloodstream neutralize or destroy antigens in the blood. There are five classes of antibodies. The constant regions of each class determine how that antibody will act against invaders. It may determine that the antibody will serve as a receptor on the surface of a cell, or it may bind to **complement** proteins in the blood promoting microbe destruction.

T-cell receptors function only as receptors on the surface of T cells. When they bind to a specific antigen, they cause a response in the T cell.

The plasma membranes of your body's cells contain proteins and polysaccharides that indicate to your antibodies that these cells belong to your body. These proteins form the **major histocompatibility complex** (**MHC**) and are unique to you (and perhaps to your identical twin). During embryological development, immune cells that react to "self" cells are destroyed so that they do not destroy cells with your body's own antigens.

When your body is threatened by an infection, two types of immunity are triggered. **Humoral immunity** involves the B cells and the antibodies secreted into the bloodstream. When antibodies on B cells bind to antigens from an invading organism, the B cell is induced to divide rapidly, producing genetically identical cells. This is referred to as **clonal selection**. The daughter clones differentiate into either **plasma cells** or **memory cells**. Plasma cells produce large quantities of their specific antibodies and release them into the bloodstream. Memory cells are involved in future immunity. They will be activated if you are reinfected with the same organism in the future.

Antibodies carry out their effect using four mechanisms. (1) Antibodies may neutralize a toxic antigen (bind to the molecule) preventing damage to the body. (2) They may coat the surface of a microorganism promoting phagocytosis by white blood cells. (3) An antibody may bind to two microbes at a time. Because additional antibodies also bind with more than one organism, clumping occurs, enhancing phagocytosis. (4) Antibodies circulating in the blood may attach to microbe antigens with their variable regions while their stems attach to proteins of the **complement system**, enhancing phagocytosis.

Cell-mediated immunity involves four types of T cells that attack organisms inside your body's cells. When receptors on **cytotoxic T cells** bind to antigens on an infected cell, the cytotoxic T cells produce and release proteins that disrupt the plasma membrane of the infected cell. When antigens bind to receptors on the surface of **helper T cells**, the helper T cells produce and then release chemicals that act as hormones, stimulating both cytotoxic T cells and B cells to divide and differentiate. Once an infection has been successfully fought, **suppressor T cells** shut off the immune response of both cytotoxic T cells and B cells. However, some cytotoxic T cells and helper T cells stay in your system and act as **memory T cells**, ready for the next time you encounter the antigen.

When your body is infected with an organism that has antigens your immune system has responded to before, memory cells specific to that antigen will recognize the organism. The memory cells multiply rapidly and stimulate the production of huge populations of plasma cells and cytotoxic T cells. This second round of defense occurs much more quickly than the first and it is why you can become "immune" to certain diseases.

3) How Does Medical Care Augment the Immune Response?

The use of *antibiotics* serves to compromise the ability of microbes to grow and reproduce. With their population growth in check, the immune system can fight the infection and stands a better chance of winning. Antibiotics are effective against bacteria, fungi, and protists; but not viruses. When you overuse or improperly use antibiotics, you help the already strong qualities of natural selection present in bacteria. This has given rise to many antibiotic resistant strains of microbes. New anti-viral drugs, called *neuraminidase inhibitors*, stop a virus from entering your cells. These new drugs may prove to be very helpful to your immune system fighting off viral infections.

By injecting dead or weakened disease agents into a healthy individual, the individual does not acquire the disease. However, the immune system responds to the presence of the foreign antigens. The response includes the production of a large number of memory cells. Thus, the practice of giving **vaccinations** gives you immunity against potentially deadly diseases.

4) What Happens When the Immune System Malfunctions?

Yes, unfortunately, your immune system can malfunction. **Allergies** are a common example of such a malfunction. When pollen or another foreign substance, such as poison ivy toxin, enters your bloodstream and is identified by a B cell as an antigen, the B cell multiplies. The proliferation of B cells produces plasma cells that produce antibodies against the pollen antigen. When the pollen antigen binds to antibodies on **mast cells** in the respiratory or digestive tract, the mast cells are stimulated to produce and release histamine. The release of histamine induces the typical "hay fever" symptoms. **Autoimmune diseases** occur when your immune system does not recognize your body's own cells. Thus, a portion of the body is attacked. This is the cause of some forms of anemia, juvenile-onset diabetes, multiple sclerosis, and rheumatoid arthritis.

Children born with **severe combined immune deficiency (SCID)** are unable to produce sufficient numbers of immune cells, if any. For the first few months of life, the children are protected from disease by immunity conferred by their mother during pregnancy. Bone marrow transplants, to supply the child with healthy white blood cell-producing marrow, is one form of therapy offered to these children.

Other individuals lose the effectiveness of their immune system. Individuals with **acquired immune deficiency syndrome (AIDS)** are infected with the **human immunodeficiency virus** either type 1 or 2 (HIV-1 or HIV-2). These viruses attack helper T cells. Thus, both the humoral and cell-mediated responses are severely hindered, and these individuals are susceptible to a number of otherwise easily combated diseases. Both HIV-1 and HIV-2 viruses are **retroviruses**. That is, they are RNA viruses that transcribe their RNA into DNA using their own enzyme called **reverse transcriptase**. The viral DNA is then inserted into the DNA of the helper T cells. When the gene is activated in the helper T cell, HIV particles are manufactured and released into the bloodstream, killing the helper T cell in the process.

When the immune system is overwhelmed, its effectiveness is severely reduced. This is, in part, what happens when the body develops cancer. A **tumor** occurs when a population of cells grows at an abnormal rate. If the tumor is benign, the growth remains local. If the tumor is malignant, the growth is uncontrolled and increasingly uses the body's nutrient and energy supplies. **Cancer** is the disease resulting from uncontrolled growth of a malignant tumor. When cancerous cells are formed, changes often occur in their plasma membrane proteins. This usually enables natural killer cells and cytotoxic T cells to destroy them before they can multiply. However, the cancerous cells may multiply faster than the immune system can destroy them, they may suppress the immune system, or they may be resistant to immune attack. Any or all result in a cancerous growth.

KEY TERMS AND CONCEPTS

Fill-In: From the following list of key terms fill in the blanks in the following statements:

acquired immune deficiency syndrome (AIDS)
allergy
antibodies
antigens
autoimmune disease
B cells
cancer
cell-mediated immunity
clonal selection
complement
complement system
constant region
cytotoxic T cells
endogenous pyrogens
fever
helper T cells
histamine
human immunodeficiency virus (HIV)
humoral immunity
immune response

inflammatory response
interferon
macrophages
major histocompatibility complex (MHC)
mast cells
memory cells
mucous membranes
natural killer
phagocytic
plasma cells
retrovirus
reverse transcriptase
severe combined immune deficiency (SCID)
skin
suppressor T cells
T cells
T-cell receptors
tumor
vaccination
variable region

1. The _____ and _____ provide your body barriers to invasion by foreign substances.

2. Your body's nonspecific defenses use _____ cells, specifically _____ to destroy microbes by phagocytosis, and _____ cells to destroy virus infected cells.

3. An injury provokes the _____. Injured cells release _____ making capillaries leaky and smooth muscles relax. If a major infection becomes established, the body produces a _____ to slow microbial growth and reproduction and to help fight viral infections by increasing the production of _____. Macrophages, producing the hormones _____ signal the hypothalamus to "turn up" the body's thermostat.

4. When nonspecific defenses fail, the body initiates the highly specific _____. This response involves two specific lymphocytes. The _____ produce protein _____ on their plasma membranes or release them into the bloodstream. The _____ produce _____ on their cell surfaces.

5. Antibodies consist of a _____ that is similar within an antibody class, and a _____ that differs even within an antibody class. Antibodies may be attached to the plasma membrane of a B cell or they may circulate in the bloodstream. In both roles, the variable regions respond and bind to _____ produced by a foreign substance. The constant region may bind to the B cell plasma membrane or it may bind to _____ proteins in your blood.

6. Your immune system is able to differentiate "self" cells from "non-self" cells because of plasma membrane proteins making up the _____.

7. In order to attack microbes before they enter cells, the body uses B cells and antibodies to provide _____. When antibodies bind to antigens, B cells are stimulated to divide rapidly. Since the new cells are identical, this is called _____. The daughter cells will differentiate into _____ and _____.

8. Antibodies in the blood may defend the body against foreign substances by neutralizing a toxic antigen, by identifying a microbe to be engulfed, thereby promoting phagocytosis, by causing microbes to clump together, or by triggering blood proteins of the _____ which causes a series of reactions.

9. T cells are used in _____ to attack cells infected by an invader. This type of response involves three cell types: _____ that disrupt the plasma membrane of infected cells, _____ that assist other immune cells, and _____ that shut off the immune system when an infection has been fought.

10. A _____ is an injection of weakened or killed microbes causing an immune response. This decreases the likelihood of an individual contracting a severe case of the disease.

11. When the immune system recognizes pollen or mold spores as antigens, an _____ has developed to those substances. When the antigen binds to antibodies on _____ in the respiratory tract, histamine is released.

12. When the immune system does not recognize cells as belonging to "self," and "anti-self" antibodies are produced, the result is an _____. If a child is born with little or no ability to produce immune cells, that child has _____. However, due to the _____, many individuals in the world are now combating the effects of _____. This virus is a _____, which uses its RNA and the enzyme _____ to make DNA and inserts it into the genome of helper T cells.

13. A population of cells that are no longer regulated and are growing at an abnormal rate is called a _____. This growth is benign if it is localized or it is malignant if it grows uncontrollably. _____ is the disease caused by a malignant growth.

KEY TERMS AND DEFINITIONS

acquired immune deficiency syndrome (AIDS): an infectious disease caused by the human immunodeficiency virus (HIV); attacks and destroys T cells, thus weakening the immune system.

allergy: an inflammatory response produced by the body in response to invasion by foreign materials, such as pollen, that are themselves harmless.

antibody: a protein, produced by cells of the immune system, that combines with a specific antigen and normally facilitates the destruction of the antigen.

antigen: a complex molecule, normally a protein or polysaccharide, that stimulates the production of a specific antibody.

autoimmune disease: a disorder in which the immune system produces antibodies against the body's own cells.

B cell: a type of lymphocyte that participates in humoral immunity; gives rise to plasma cells, which secrete antibodies into the circulatory system, and to memory cells.

cancer: a disease in which some of the body's cells escape from normal regulatory processes and divide without control.

cell-mediated immunity: an immune response in which foreign cells or substances are destroyed by contact with T cells.

clonal selection: the mechanism by which the immune response gains specificity; an invading antigen elicits a response from only a few lymphocytes, which proliferate to form a clone of cells that attack only the specific antigen that stimulated their production.

complement: a group of blood-borne proteins that participate in the destruction of foreign cells to which antibodies have bound.

complement system: a series of reactions in which complement proteins bind to antibody stems, attracting to the site phagocytic white blood cells that destroy the invading cell that triggers the reactions.

constant region: the part of an antibody molecule that is similar in all antibodies.

cytotoxic T cell: a type of T cell that, upon contacting foreign cells, directly destroys them.

endogenous pyrogen: chemical, produced by the body, that stimulates the production of fever.

fever: an elevation in body temperature caused by chemicals (pyrogens) that are released by white blood cells in response to infection.

helper T cell: type of T cell that helps other immune cells recognize and act against antigens.

histamine: a substance released by certain cells in response to tissue damage and invasion of the body by foreign substances; promotes the dilation of arterioles and the leakiness of capillaries and triggers some of the events of the inflammatory response.

human immunodeficiency virus (HIV): a pathogenic retrovirus that causes acquired immune deficiency syndrome (AIDS) by attacking and destroying the immune system's T cells.

humoral immunity: an immune response in which foreign substances are inactivated or destroyed by antibodies that circulate in the blood.

hybridoma: a cell produced by fusing an antibody-producing cell with a myeloma cell; used to produce monoclonal antibodies.

immune response: a specific response by the immune system to the invasion of the body by a particular foreign substance or microorganism, characterized by the recognition of the foreign substance by immune cells and its subsequent destruction by antibodies or by cellular attack.

immune system: cells such as macrophages, B cells, and T cells and molecules such as antibodies that work together to combat microbial invasion of the body.

inflammatory response: a nonspecific, local response to injury to the body, characterized by the phagocytosis of foreign substances and tissue debris by white blood cells and by the walling off of the injury site by the clotting of fluids that escape from nearby blood vessels.

interferon: a protein released by certain virus-infected cells that increases the resistance of other, uninfected, cells to viral attack.

macrophage (mak´-rō-fāj): a type of white blood cell that engulfs microbes and destroys them by phagocytosis; also presents microbial antigens to T cells, helping stimulate the immune response.

major histocompatibility complex (MHC): proteins, normally located on the surfaces of body cells, that identify the cell as "self"; also important in stimulating and regulating the immune response.

mast cell: a cell of the immune system that synthesizes histamine and other molecules used in the body's response to trauma and that are a factor in allergic reactions.

mechanoreceptor: a receptor that responds to mechanical deformation, such as that caused by pressure, touch, or vibration.

memory B cell: a type of white blood cell that is produced as a result of the binding of an antibody on a B cell to an antigen on an invading microorganism. Memory B cells persist in the bloodstream and provide future immunity to invaders bearing that antigen.

memory T cell: a type of white blood cell that is produced as a result of the binding of a receptor on a T cell to an antigen on an invading microorganism. Memory T cells persist in the bloodstream and provide future immunity to invaders bearing that antigen.

microbe: a microorganism.

monoclonal antibody: an antibody produced in the laboratory by the cloning of hybridoma cells; each clone of cells produces a single antibody.

mucous membrane: the lining of the inside of the respiratory and digestive tracts.

natural killer cell: a type of white blood cell that destroys some virus-infected cells and cancerous cells on contact; part of the immune system's nonspecific internal defense against disease.

phagocytic cell (fa-gō-sit´-ik): a type of immune system cell that destroys invading microbes by using phagocytosis to engulf and digest the microbes.

photoreceptor: a receptor cell that responds to light; in vertebrates, rods and cones.

plasma cell: an antibody-secreting descendant of a B cell.

retrovirus: a virus that uses RNA as its genetic material. When it invades a eukaryotic cell, a retrovirus "reverse transcribes" its RNA into DNA, which then directs the synthesis of more viruses, using the transcription and translation machinery of the cell.

reverse transcriptase: an enzyme found in retroviruses that catalyzes the synthesis of DNA from an RNA template.

severe combined immune deficiency (SCID): a disorder in which no immune cells, or very few, are formed; the immune system is incapable of responding properly to invading disease organisms, and the individual is very vulnerable to common infections.

skin: the tissue that makes up the outer surface of an animal body.

suppressor T cell: a type of T cell that depresses the response of other immune cells to foreign antigens.

T cell: a type of lymphocyte that recognizes and destroys specific foreign cells or substances or that regulates other cells of the immune system.

T-cell receptor: a protein receptor, located on the surface of a T cell, that binds a specific antigen and triggers the immune response of the T cell.

tumor: a mass that forms in otherwise normal tissue; caused by the uncontrolled growth of cells.

vaccination: an injection into the body that contains antigens characteristic of a particular disease organism and that stimulates an immune response.

variable region: the part of an antibody molecule that differs among antibodies; the ends of the variable regions of the light and heavy chains form the specific binding site for antigens.

THINKING THROUGH THE CONCEPTS

True or False: Determine if the statement given is true or false. If it is false, change the <u>underlined</u> word(s) so that the statement reads true.

14. _____ The skin provides an <u>excellent</u> breeding ground for microbes.

15. _____ Mucous membranes <u>secrete enzymes</u> to destroy microbes.

16. _____ A <u>fever</u> is actually a defense mechanism against disease.

17. _____ <u>Natural killer cells</u> recognize and destroy cancerous cells.

18. _____ Chemicals produced by wounded cells serve to <u>attract</u> phagocytic cells to an injury.

19. _____ Interferon <u>decreases</u> a cell's resistance to viral attack.

20. _____ <u>Antibodies</u> are formed from a wide array of genes, some of which mutate easily.

21. _____ Antibodies <u>provide no protection against</u> poisons such as snake venom.

22. _____ Antibodies <u>are custom made to fit a specific antigen</u>.

23. _____ B cells, differentiated into <u>plasma cells</u>, release antibodies into the bloodstream.

24. _____ <u>Helper T cells</u> turn the immune system off after an infection has been eradicated from the body.

25. _____ Humans continue to suffer from the flu because <u>our immune system is unable to produce memory cells for the flu virus</u>.

26. _____ Antibiotics help fight diseases caused by <u>viruses</u>.

27. _____ An anthrax vaccine has been produced using <u>synthetic antigens</u>.

28. _____ <u>Mast cells</u> produce histamine in response to an allergen antigen.

29. _____ Autoimmune diseases can be <u>cured</u> with today's medical technology.

Identify: Determine whether the following statements refer to **B cells**, **T cells**, or **both**.

30. _____ precursor cells produced in bone marrow

31. _____ differentiate in bone marrow

32. _____ differentiate in thymus

33. _____ long lived, can provide future immunity

34. _____ secrete antibodies into the bloodstream

35. _____ destroy cells infected with viruses

36. _____ shut down the immune response when appropriate

37. _____ activate both B and T cells

38. _____ provide humoral immunity

39. _____ provide cell-mediated immunity

Label the antibody diagram below.

40. _____

41. _____

42. _____

43. _____

44. _____

45. _____

Short Answer:

46. Identify the body's nonspecific defenses against microbial invasion.

_____

_____

47. Explain the four ways antibodies destroy extracellular microbes and molecules.

_____

_____

_____

_____

Complete the following questions specifically about AIDS.

48. How do HIV-1 and HIV-2 affect the immune system? Once they are in a cell, how are more viruses made?

_____

_____

_____

49. If AIDS does not directly kill its victims, what does cause them to die?

_____

_____

_____

_____

50. How do drugs like AZT and protease inhibitors function against HIV? Are they a cure?

_____

_____

_____

_____

_____

_____

51. Why is an effective vaccine against HIV so difficult to produce?

_____

_____

_____

_____

_____

_____

_____

52. Where are the most AIDS cases found in the world and who is most at risk of being infected with HIV?

_____

_____

_____

_____

_____

APPLYING THE CONCEPTS

These practice questions are intended to sharpen your ability to apply critical thinking and analysis to biological concepts covered in this chapter.

53. Antihistamines are often taken by individuals suffering from allergies. Using what you now know about the role of histamine in the immune response, how would antihistamines decrease your allergy symptoms? Would taking antihistamines lessen your symptoms from a cold virus?

_____

_____

_____

_____

_____

54. Why will your body reject an organ received during a transplant?

_____

_____

_____

_____

_____

_____

55. How does the immune system combat the cells that do become cancerous?

_____

_____

_____

56. Why is it common for cancer patients to experience hair loss, nausea, and severe dry mouth during chemotherapy treatments?

_____

_____

_____

_____

ANSWERS TO EXERCISES

1. skin
   mucous membranes
2. phagocytic
   macrophages
   natural killer
3. inflammatory response
   histamine
   fever
   interferon
   endogenous pyrogens
4. immune response
   B cells
   antibodies
   T cells
   T-cell receptors
5. constant region
   variable region
   antigens
   complement
6. major histocompatibility complex (MHC)

7. humoral immunity
   clonal selection
   plasma cells
   memory cells
8. complement system
9. cell-mediated immunity
   cytotoxic T cells
   helper T cells
   suppressor T cells
10. vaccination
11. allergy
    mast cells
12. autoimmune disease
    severe combined immune deficiency (SCID)
    human immunodeficiency virus (HIV)
    acquired immune deficiency syndrome (AIDS)
    retrovirus
    reverse transcriptase
13. tumor
    Cancer

14. false, poor
15. true
16. true
17. true
18. true
19. false, increases
20. true
21. false, neutralize
22. false, fit reasonably well to specific antigens
23. true
24. false, suppressor T cells
25. false, the flu virus mutates frequently
26. false, bacteria, fungi, and protists
27. true
28. true
29. false, suppressed

30. both
31. B cells
32. T cells
33. both
34. B cells
35. T cells
36. T cells
37. T cells
38. B cells
39. T cells
40. light chain
41. heavy chain
42. constant regions
43. variable regions
44. antigen
45. antigen binding site

46. phagocytic cells and natural killer cells, the inflammatory response, and fever

47. Antibodies destroy extracellular microbes and molecules by binding to and neutralizing a toxic antigen; coating a microbe, thus promoting phagocytosis; clumping of microbes, and by inducing complement reactions with the proteins of the complement system, promoting phagocytosis.

48. HIV-1 and HIV-2 infect and destroy helper T cells. Once in a helper T cell, the RNA uses its reverse transcriptase enzyme to create a DNA fragment molecule. The DNA fragment is inserted into the helper T cell's genome, where it will eventually be used to make more HIV viruses.

49. Because the helper T cell population declines in number as the cells are destroyed, there are fewer and fewer cells to assist other immune cells in their defenses. The hormone-like chemicals that helper T cells release to trigger immune cell division and differentiation, antibody production, and cytotoxic cell development are not circulated. Thus, the person with an active HIV infection becomes increasingly susceptible to other diseases. It is one of these "opportunistic" diseases that will cause the patient's death.

50. AZT inhibits reverse transcriptase slowing the production of DNA from the viral RNA. The hope is that DNA will never be properly formed from the RNA, thus new virus will not be made. Unfortunately, AZT does not successfully block the reverse transcription. Protease inhibitors inactivate an enzyme responsible for assembling new viral particles. Protease inhibitors are most effective when used in combination with reverse transcriptase inhibitors. Thus the term "cocktail" of drugs has come into common use. No, these drugs are not a cure. That search continues.

51. An effective vaccine against HIV has been difficult to develop for several reasons. One reason is that antibodies naturally produced against HIV seem to offer little protection against infection. This indicates that a vaccine would need to function in a way that produced more effective immune responses than normal HIV does. Another reason is that HIV mutates at a phenomenal rate. Therefore, a vaccine developed for one strain may have absolutely no effect on the person vaccinated. To further complicate the situation, HIV can exist in different forms within the same person.

52. Most (95%) individuals infected with HIV live in developing countries. Sub-Saharan Africa has the highest rate of infection with approximately 10% of the adult population infected. Worldwide, transmission through heterosexual sex is the most common avenue of infection. Individuals most at risk of being infected are those individuals not practicing "safe sex," regardless of the gender of the individuals involved.

53. Histamines are produced when an allergen antigen binds to antibodies on mast cells. The result of histamine circulating in the bloodstream is increased mucus production and the inflammatory response increasing discomfort. Antihistamines would inhibit the production of histamine or block its effect, thus reducing the production of mucus and inflammation. Since the cold virus does not typically stimulate mast cells to produce histamine, congestion symptoms due to a cold will not be alleviated by taking antihistamines.

54. The cell plasma membranes contain large proteins and polysaccharides that indicate to the immune system that they belong to "self." These proteins make up the major histocompatibility complex. They are unique to each individual (and an identical twin). If you receive an organ transplant, a "match" will be made in an attempt to reduce the number of different proteins. Since no match will be perfect (except from an identical twin), the body will naturally mount an immunological attack against the foreign cells, rejecting the transplanted organ.

55. Natural killer cells and cytotoxic cells detect the protein changes that occur on the plasma membranes of cancerous cells. They destroy nearly all cancerous cells that occur in our bodies before they have a chance to proliferate

56. Chemotherapy agents target cells that are rapidly growing and frequently dividing. Therefore, any cell in the body that tends to be replaced frequently would be affected by the chemotherapy. Hair follicles and cells lining the stomach and the mouth are body cells that divide and grow rapidly. Therefore, they tend to be damaged along with any cancerous cells.

Chapter 32: Chemical Control of the Animal Body: The Endocrine System

OVERVIEW

This chapter covers how your body's internal chemistry is controlled by hormones. The authors discuss the four classes of animal hormones and the endocrine glands that produce them. Hormone regulation by negative feedback is outlined, and the major mammalian endocrine glands are presented in detail.

1) What Are the Characteristics of Animal Hormones?

A **hormone** is a chemical secreted by cells in one part of your body, is transported in your bloodstream to other parts of your body, where it affects particular **target cells**. Hormones are released by the cells of endocrine glands and endocrine organs located throughout your body. Together, these glands, organs, and target cells make up your **endocrine system**. Hormones can be grouped into four general classes. **Peptide hormones** are made of chains of amino acids. *Amino acid based hormones* are formed from single amino acids (for example, adrenaline is made from tyrosine). **Steroid hormones** are derived from cholesterol. **Prostaglandins** are made from fatty acids.

Hormones function by binding to specific receptors on target cells that respond to particular hormone molecules. Receptors for hormones are found either on the plasma membrane or inside the cell, usually within the nucleus. Most peptide and amino acid based hormones cannot cross the plasma membrane. Therefore, these hormones react with protein receptors on the outer surface of the plasma membrane. Hormones bound to a receptor on the outer cell surface usually trigger the release of a chemical inside the cell. This chemical is called a **second messenger** and it starts a series of biochemical reactions. **Cyclic AMP** (cAMP) is a common second messenger.

Steroid hormones are lipid-soluble and can cross the plasma membrane. Inside the cell, they bind to protein receptors. These intracellular receptors are usually in the nucleus. The receptor-hormone complex binds to DNA and stimulates particular genes to become active.

Hormones are regulated by feedback mechanisms. In animals, the secretion of a hormone usually stimulates a response in target cells that will inhibit further secretion of the hormone. This mechanism is called negative feedback. For example, loss of water through perspiration triggers the pituitary gland to produce *antidiuretic hormone* (ADH). ADH causes your kidneys to reabsorb water and to produce very concentrated urine. Drinking water replaces what has been lost and triggers your pituitary gland to turn off ADH secretion when water content in your blood returns to normal. In a few cases, positive feedback controls hormone release. For example, contractions of the uterus early in childbirth cause the release of the hormone *oxytocin* which stimulates stronger contractions of the uterus.

2) What Are the Structures and Functions of the Mammalian Endocrine System?

Mammals have both **exocrine glands** and **endocrine glands**. Exocrine glands produce secretions that are released outside your body or into your digestive tract through tubes or openings called **ducts**. Exocrine glands include your sweat and sebaceous (oil) glands in your skin, lacrimal (tear) glands of your eyes, mammary glands, and glands that produce digestive secretions. Endocrine glands are ductless glands that secrete hormones into capillaries of your bloodstream.

The **hypothalamus** is a part of your brain that contains **neurosecretory cells**. These cells make and store peptide hormones and release them when stimulated. Your hypothalamus controls the secretions of your **pituitary gland**. Your pituitary gland hangs below your hypothalamus and has two lobes: the **anterior pituitary** and the **posterior pituitary**. Specifically, your hypothalamus produces **releasing hormones** and **inhibiting hormones**. These peptide hormones regulate your anterior pituitary.

Your anterior pituitary produces and releases a variety of peptide hormones, four of which regulate hormone release in other endocrine glands. **Follicle-stimulating hormone (FSH)** and **luteinizing hormone (LH)** stimulate production of sperm and testosterone in males, and eggs, estrogen, and progesterone in females. **Thyroid-stimulating hormone (TSH)** stimulates your thyroid gland to release its hormones, and **adrenocorticotropic hormone (ACTH)** stimulates release of hormones from the adrenal cortex of your adrenal glands. The other four hormones of the anterior pituitary do not act on other endocrine glands. **Prolactin** stimulates development of mammary glands during pregnancy. **Endorphins** inhibit the perception of pain by binding to receptors in your brain. **Melanocyte-stimulating hormone (MSH)** stimulates production of the skin pigment melanin. **Growth hormone (GH)** regulates your body's growth. If you produce too little GH, the result may be pituitary *dwarfism*. If you produce too much GH, the result may be *gigantism*. Growth hormone can now be made by genetic engineering.

Your posterior pituitary releases two peptide hormones that are actually produced by cells in your hypothalamus. **Antidiuretic hormone (ADH)** helps prevent dehydration. It increases the permeability of the collecting ducts of your kidney nephrons. This causes water to be reabsorbed from your urine and kept in your body. Alcohol inhibits the release of ADH. **Oxytocin** causes uterine contractions during childbirth and triggers the "milk letdown reflex" in nursing mothers by causing breast muscle tissue to contract during breast-feeding. Oxytocin also aids in male ejaculation in some animals.

Your **thyroid gland** produces two major hormones. **Thyroxine**, an iodine-containing modified amino acid, raises the metabolic rate of most of your cells by stimulating the synthesis of enzymes that break down glucose and provide energy. Thyroxine helps regulate your body temperature and helps you respond to stress. In juvenile animals, thyroxine helps regulate growth by stimulating both metabolic rate and development of the nervous system. Too little thyroxine early in life causes *cretinism*. A diet deficient in iodine causes thyroid enlargement. The enlarged thyroid bulges from the neck producing a condition called **goiter**. Using iodized salt prevents this condition. **Calcitonin**, a peptide hormone, helps control the concentration of calcium in your blood by regulating calcium release from your bones. If calcium levels in your blood are too high, calcitonin is released to inhibit the release of calcium from your bones.

Behind your thyroid are four small **parathyroid glands**. Your parathyroid glands secrete the hormone **parathormone** in response to low levels of calcium in your blood. When blood calcium levels are low, parathormone triggers your bone cells to release calcium. Between the functions of calcitonin and parathormone, the calcium content of your blood is regulated.

Your **pancreas** functions as an exocrine gland by releasing digestive secretions from the pancreatic duct into your small intestine and it functions as an endocrine gland when clusters of **islet cells** produce the peptide hormones **insulin** and **glucagon**. Insulin and glucagon regulate carbohydrate and fat metabolism. Insulin *reduces* blood glucose levels by causing cells to take in glucose for metabolism or storage as glycogen or fat. Glucagon *increases* blood glucose levels by activating a liver enzyme to break down glycogen. Glucagon also promotes lipid breakdown, releasing fatty acids for metabolism. A deficiency in insulin production, release, or entrance into target cells results in **diabetes mellitus**. Diabetes is a condition in which blood glucose levels are high and fluctuate wildly with sugar intake. Human insulin for injection by diabetic individuals can now be made by genetic engineering.

Your **testes** or **ovaries** are endocrine organs that produce steroid hormones. In males, the testes make a group of hormones called **androgens**, which includes **testosterone**, and in females, the ovaries make **estrogen** and **progesterone**. Sex hormones influence development, pregnancy, menstrual cycles, and the physiological changes of puberty that result in reproductive capacity and secondary sex characteristics.

Your **adrenal glands** sit on top of your kidneys and have two parts that secrete different hormones. The interior **adrenal medulla** contains secretory cells that originated from and are controlled by your nervous system. This area produces two amino acid based hormones: **epinephrine** (*adrenaline*) and

norepinephrine (*noradrenaline*). These two hormones prepare your body for emergency action; they increase heart and respiratory rates, increase blood glucose levels, and direct blood flow away from digestion and toward muscles and your brain. The outer layer of each adrenal gland forms the **adrenal cortex**, which secretes three types of steroid hormones called **glucocorticoids**. These hormones help control glucose metabolism. Stressful stimuli (such as trauma, infection, or hot or cold temperatures) stimulate your hypothalamus to secrete releasing hormones, which cause the anterior pituitary to release ACTH. ACTH, in turn, stimulates the adrenal cortex to release glucocorticoid hormones. The adrenal cortex also secretes **aldosterone**, which regulates the sodium content of your blood. If your blood sodium levels fall, the release of aldosterone causes your kidneys and sweat glands to retain sodium. When blood sodium levels rise to normal, aldosterone secretion is stopped. The adrenal cortex also secretes small amounts of testosterone in males and females. Adrenal medulla tumors in females can lead to excessive testosterone release, causing masculinization of women.

Prostaglandins are unique hormones. Every cell in your body may produce them. Some prostaglandins cause muscle contractions enhancing uterine contractions (and menstrual cramps) and constricting umbilical arteries, shutting off maternal blood flow to a newborn. Among many other functions, some prostaglandins cause inflammation and stimulate pain receptors.

Your **pineal gland** lies between the two hemispheres of your brain and produces the amino acid based hormone **melatonin**. Melatonin is secreted in a daily rhythm regulated in mammals by the eyes. Melatonin appears to regulate the seasonal reproductive cycles of many mammals. It may influence sleep-wake cycles in humans.

Your **thymus** is located in your chest cavity and produces white blood cells and the hormone **thymosin**. Thymosin stimulates the production of T cells, which play a vital role in your immune system.

Your kidneys produce the hormones **erythropoietin** to increase red blood cell production and **renin** in response to low blood pressure. Renin functions as an enzyme catalyzing the formation of **angiotensin** from blood proteins. Angiotensin constricts arterioles and signals your adrenal cortex to produce aldosterone to retain sodium in the blood, regulating your blood pressure Your heart produces a peptide hormone called **atrial natriuretic peptide (ANP)**. ANP causes a reduction in blood pressure by decreasing the release of ADH and aldosterone. Your stomach and small intestine also produce peptide hormones. **Gastrin, secretin,** and **cholecystokinin** help regulate digestion.

A newly discovered hormone, **leptin**, is released by fat cells. Researchers hypothesize that leptin signals the body that it has fat reserves. This should help regulate how much you eat. Leptin may also stimulate capillary growth and the immune system. Leptin production is required for puberty to begin.

KEY TERMS AND CONCEPTS

Fill-In: From the following list of terms, fill in the blanks below.

adrenocorticotropic	gigantism	melatonin	prostaglandins
amino acid based	glucagon	oxytocin	second messenger
antidiuretic	growth	pancreas	steroid
cyclic AMP	hormone	peptide	target
endocrine	insulin	pineal	thymosin
endorphins	luteinizing	pituitary dwarfism	thymus
exocrine	melanocyte-stimulating	prolactin	thyroid-stimulating
follicle-stimulating			

1. A _____ is a chemical secreted by cells in one part of your body that is transported in your bloodstream to other parts of your body, where it affects particular _____ cells.

2. Hormones can be grouped into four general classes. _____ hormones are made of chains of amino acids. _____ are formed from single amino acids. Most _____ hormones are derived from cholesterol. _____ are made from fatty acids and may be produced by every cell in your body.

3. Hormones binding to a receptor on the outer cell surface trigger the release, inside the cell, of a chemical called the _____ which initiates a cascade of biochemical reactions. _____ is a common second messenger.

4. Mammals have both _____ glands, which produce secretions released outside the body or into the digestive tract through tubes or openings called ducts, and _____ glands, which are ductless glands that secrete hormones into capillaries of the bloodstream.

5. The anterior pituitary produces and releases several peptide hormones, four of which regulate hormone release in other endocrine glands. _____ hormone and _____ hormone stimulate production of sperm and testosterone in males and eggs, estrogen, and progesterone in females. _____ hormone stimulates release of thyroid gland hormones. _____ hormone stimulates release of hormones from the adrenal cortex.

6. Four hormones of the anterior pituitary do not act on other endocrine glands. _____ helps stimulate development of the mammary glands during pregnancy, and _____ inhibit the perception of pain by binding to brain receptors. _____ hormone stimulates production of the skin pigment melanin, and _____ hormone regulates body growth: too little causes _____ and too much can cause _____.

7. The posterior pituitary releases two types of peptide hormones produced by cells in the hypothalamus: _____ hormone, which helps prevent dehydration, and _____, which causes uterine contractions and triggers the "milk letdown reflex" in the breasts of nursing mothers.

8. In the _____, clusters of islet cells produce the peptide hormones _____, which reduces blood glucose levels, and _____, which increases blood glucose levels by activating a liver enzyme that breaks down glycogen.

9. The _____ gland lies between the two hemispheres of the brain and produces the amino acid derived hormone _____, which is secreted in a daily rhythm.

10. The _____ is in the chest cavity and produces white blood cells and the hormone _____, which stimulates production of T cells which play a role in the immune system.

Key Terms and Definitions

adrenal cortex: the outer part of the adrenal gland, which secretes steroid hormones that regulate metabolism and salt balance.

adrenal gland: a mammalian endocrine gland, adjacent to the kidney; secretes hormones that function in water regulation and in the stress response.

adrenal medulla: the inner part of the adrenal gland, which secretes epinephrine (adrenaline) and norepinephrine (noradrenaline).

adrenocorticotropic hormone (a-drēn-ō-kor-tik-ō-trō´-pik; ACTH): a hormone, secreted by the anterior pituitary, that stimulates the release of hormones by the adrenal glands, especially in response to stress.

aldosterone: a hormone, secreted by the adrenal cortex, that helps regulate ion concentration in the blood by stimulating the reabsorption of sodium by the kidneys and sweat glands.

androgen: a male sex hormone.

angiotensin (an-jē-ō-ten´-sun): a hormone that functions in water regulation in mammals by stimulating physiological changes that increase blood volume and blood pressure.

anterior pituitary: a lobe of the pituitary gland that produces prolactin and growth hormone as well as hormones that regulate hormone production in other glands.

antidiuretic hormone (an-tē-dī-ū r-et´-ik; ADH): a hormone produced by the hypothalamus and released into the bloodstream by the posterior pituitary when blood volume is low; increases the permeability of the distal tubule and the collecting duct to water, allowing more water to be reabsorbed into the bloodstream.

atrial natriuretic peptide (ā´-trē-ul nā-trē-ū-ret´-ik; ANP): a hormone, secreted by cells in the mammalian heart, that reduces blood volume by inhibiting the release of ADH and aldosterone.

calcitonin (kal-si-tōn´-in): a hormone, secreted by the thyroid gland, that inhibits the release of calcium from bone.

cholecystokinin (kō´-lē-sis-tō-ki´-nin): a digestive hormone, produced by the small intestine, that stimulates the release of pancreatic enzymes.

cyclic AMP: a cyclic nucleotide, formed within many target cells as a result of the reception of amino acid derivatives or peptide hormones, that causes metabolic changes in the cell; often called a second messenger.

diabetes mellitus (dī-uh-bē´-tēs mel-ī´-tus): a disease characterized by defects in the production, release, or reception of insulin; characterized by high blood glucose levels that fluctuate with sugar intake.

duct: a tube or opening through which exocrine secretions are released.

endocrine disruptors: environmental pollutants that interfere with endocrine function, often by disrupting the action of sex hormones.

endocrine gland: a ductless, hormone-producing gland consisting of cells that release their secretions into the extracellular fluid from which the secretions diffuse into nearby capillaries.

endocrine system: an animal's organ system for cell-to-cell communication, composed of hormones and the cells that secrete them and receive them.

endorphin (en-dor´-fin): one of a group of peptide neuromodulators in the vertebrate brain that, by reducing the sensation of pain, mimics some of the actions of opiates.

environmental estrogens: chemicals in the environment that mimic some of the effects of estrogen in animals.

epinephrine (ep-i-nef´-rin): a hormone, secreted by the adrenal medulla, that is released in response to stress and that stimulates a variety of responses, including the release of glucose from skeletal muscle and an increase in heart rate.

erythropoietin (eh-rith´-rō-pō-ē´-tin): a hormone produced by the kidneys in response to oxygen deficiency that stimulates the production of red blood cells by the bone marrow.

estrogen: in vertebrates, a female sex hormone, produced by follicle cells of the ovary, that stimulates follicle development, oogenesis, the development of secondary sex characteristics, and growth of the uterine lining.

exocrine gland: a gland that releases its secretions into ducts that lead to the outside of the body or into the digestive tract.

follicle-stimulating hormone (FSH): a hormone, produced by the anterior pituitary, that stimulates spermatogenesis in males and the development of the follicle in females.

gastrin: a hormone, produced by the stomach, that stimulates acid secretion in response to the presence of food.

glucagon (gloo´-ka-gon): a hormone, secreted by the pancreas, that increases blood sugar by stimulating the breakdown of glycogen (to glucose) in the liver.

glucocorticoid (gloo-kō-kor´-tik-oid): a class of hormones, released by the adrenal cortex in response to the presence of ACTH, that make additional energy available to the body by stimulating the synthesis of glucose.

goiter: a swelling of the neck caused by iodine deficiency, which affects the functioning of the thyroid gland and its hormones.

growth hormone: a hormone, released by the anterior pituitary, that stimulates growth, especially of the skeleton.

hormone: a chemical that is synthesized by one group of cells, secreted, and then carried in the bloodstream to other cells, whose activity is influenced by reception of the hormone.

hypothalamus (hī-pō-thal´-a-mus): a region of the brain that controls the secretory activity of the pituitary gland; synthesizes, stores, and releases certain peptide hormones; directs autonomic nervous system responses.

inhibiting hormone: a hormone, secreted by the neurosecretory cells of the hypothalamus, that inhibits the release of specific hormones from the anterior pituitary.

insulin: a hormone, secreted by the pancreas, that lowers blood sugar by stimulating the conversion of glucose to glycogen in the liver.

islet cell: a cluster of cells in the endocrine portion of the pancreas that produce insulin and glucagon.

leptin: a peptide hormone. One of the functions of leptin, which is released by fat cells, is to help the body monitor its fat stores and regulate weight.

luteinizing hormone (LH): a hormone, produced by the anterior pituitary, that stimulates testosterone production in males and the development of the follicle, ovulation, and the production of the corpus luteum in females.

melanocyte-stimulating hormone (me-lan´-ō-sīt): a hormone, released by the anterior pituitary, that regulates the activity of skin pigments in some vertebrates.

melatonin (mel-uh-tōn´-in): a hormone, secreted by the pineal gland, that is involved in the regulation of circadian cycles.

neurosecretory cell: a specialized nerve cell that synthesizes and releases hormones.

norepinephrine (nor-ep-i-nef-rin´): a neurotransmitter, released by neurons of the parasympathetic nervous system, that prepares the body to respond to stressful situations; also called *noradrenaline.*

ovary: in animals, the gonad of females; in flowering plants, a structure at the base of the carpel that contains one or more ovules and develops into the fruit.

oxytocin (oks-ē-tō´-sin): a hormone, released by the posterior pituitary, that stimulates the contraction of uterine and mammary gland muscles.

pancreas (pǎn´-krē-us): a combined exocrine and endocrine gland located in the abdominal cavity next to the stomach. The endocrine portion secretes the hormones insulin and glucagon, which regulate glucose concentrations in the blood. The exocrine portion secretes enzymes for fat, carbohydrate, and protein digestion into the small intestine and neutralizes the acidic chyme.

parathormone: a hormone, secreted by the parathyroid gland, that stimulates the release of calcium from bones.

parathyroid gland: one of a set of four small endocrine glands, embedded in the surface of the thyroid gland, that produces parathormone, which (with calcitonin from the thyroid gland) regulates calcium ion concentration in the blood.

peptide hormone: a hormone consisting of a chain of amino acids; includes small proteins that function as hormones.

pineal gland (pī-nē´-al): a small gland within the brain that secretes melatonin; controls the seasonal reproductive cycles of some mammals.

pituitary gland: an endocrine gland, located at the base of the brain, that produces several hormones, many of which influence the activity of other glands.

posterior pituitary: a lobe of the pituitary gland that is an outgrowth of the hypothalamus and that releases antidiuretic hormone and oxytocin.

progesterone (prō-ge´-ster-ōn): a hormone, produced by the corpus luteum, that promotes the development of the uterine lining in females.

prolactin: a hormone, released by the anterior pituitary, that stimulates milk production in human females.

prostaglandin (pro-stuh-glan´-din): a family of modified fatty acid hormones manufactured by many cells of the body.

releasing hormone: a hormone, secreted by the hypothalamus, that causes the release of specific hormones by the anterior pituitary.

renin: an enzyme that is released (in mammals) when blood pressure and/or sodium concentration in the blood drops below a set point; initiates a cascade of events that restores blood pressure and sodium concentration.

second messenger: an intracellular chemical, such as cyclic AMP, that is synthesized or released within a cell in response to the binding of a hormone or neurotransmitter (the first messenger) to receptors on the cell surface; brings about specific changes in the metabolism of the cell.

secretin: a hormone, produced by the small intestine, that stimulates the production and release of digestive secretions by the pancreas and liver.

steroid hormone: a class of hormone whose chemical structure (four fused carbon rings with various functional groups) resembles cholesterol; steroids, which are lipids, are secreted by the ovaries and placenta, the testes, and the adrenal cortex.

target cell: a cell on which a particular hormone exerts its effect.

testis (pl., testes): the gonad of male mammals.

testosterone: in vertebrates, a hormone produced by the interstitial cells of the testis; stimulates spermatogenesis and the development of male secondary sex characteristics.

thymosin: a hormone, secreted by the thymus, that stimulates the maturation of cells of the immune system.

thymus (thī´-mus): an organ of the lymphatic system that is located in the upper chest in front of the heart and that secretes thymosin, which stimulates lymphocyte maturation; begins to degenerate at puberty and has little function in the adult.

thyroid gland: an endocrine gland, located in front of the larynx in the neck, that secretes the hormones thyroxine (affecting metabolic rate) and calcitonin (regulating calcium ion concentration in the blood).

thyroid-stimulating hormone (TSH): a hormone, released by the anterior pituitary, that stimulates the thyroid gland to release hormones.

thyroxine (thī-rox´-in): a hormone, secreted by the thyroid gland, that stimulates and regulates metabolism.

THINKING THROUGH THE CONCEPTS

True or False: Determine if the statement given is true or false. If it is false, change the <u>underlined</u> word(s) so that the statement reads true.

11. _____   Endocrine glands usually <u>have</u> ducts.

12. _____   The action of hormones is dependent on <u>gland</u> cells.

13. _____   <u>Endocrine</u> glands produce chemicals that exert their effects outside the body of the animal producing them.

14. _____   <u>Peptide</u> hormones are able to enter cells easily.

15. _____   Animals regulate most hormone release through <u>positive</u> feedback.

16. _____   An increase in antidiuretic hormone will <u>decrease</u> blood pressure.

17. _____ The <u>anterior</u> pituitary is more like a part of the brain than an endocrine gland.

18. _____ The <u>posterior</u> pituitary produces most of the body's hormones.

19. _____ The breakdown of glycogen into glucose is favored by the presence of <u>insulin</u>.

20. _____ Insulin <u>increases</u> blood glucose levels.

21. _____ Epinephrine and norepinephrine prepare the body for <u>emergency action</u>.

22. _____ Adrenaline causes blood to flow <u>toward</u> the stomach.

23. _____ The adrenal medulla secretes a <u>female</u> hormone.

Identification: Determine whether the following statements refer to **endocrine** glands or **exocrine** glands.

24. _____ clusters of hormone-producing cells embedded in capillaries

25. _____ secrete substances into ducts leading to the outside of the body

26. _____ sweat glands

27. _____ secrete substances into extracellular spaces surrounding capillaries

28. _____ mammary glands

29. _____ salivary glands

30. _____ lacrimal (tear) glands

31. _____ ductless glands

Matching: Pituitary hormones. Some questions may have more than one correct answer.

32. _____ affects gamete release in both sexes

33. _____ produced by the posterior pituitary

34. _____ regulates growth of bones

35. _____ reduces dehydration

36. _____ stimulates development of the mammary glands during pregnancy

37. _____ causes uterine contractions during childbirth

38. _____ stimulates release of hormone by the thyroid gland

39. _____ allows milk to flow from the breasts during lactation

40. _____ causes release of hormones from the cortex of the adrenal glands

41. _____ allows ejaculation to occur in males of some animals

Choices:

a. ACTH

b. ADH

c. FSH

d. LH

e. growth hormone

f. oxytocin

g. prolactin

h. TSH

Matching: Adrenal gland hormones. Some questions may have more than one correct answer.

42. _____   produced by the adrenal medulla

43. _____   release is stimulated by ACTH

44. _____   help control glucose metabolism

45. _____   male sex hormone

46. _____   acts like glucagon

47. _____   overproduction, caused by certain tumors,
can result in "bearded ladies"

Choices:

a. epinephrine

b. norepinephrine

c. glucocorticoids

d. testosterone

48. List the major glands and/or organs of the endocrine system.

_____

_____

_____

_____

_____

_____

APPLYING THE CONCEPTS

These practice questions are intended to sharpen your ability to apply critical thinking and analysis to biological concepts covered in this chapter.

49. Antidiuretic hormone (ADH) acts on very specific target cells in the kidney, while the hormone insulin affects every cell in the body. Why are some hormones very specific as to which cells they affect and other hormones have a very general effect on many target organs?

_____

_____

_____

_____

_____

50. Explain why complications from diabetes mellitus include high blood pressure, heart disease, blindness, and kidney failure.

_____

_____

_____

_____

_____

51. Explain how calcitonin and parathormone work "together" to regulate your blood calcium levels.

_____

_____

_____

_____

_____

52. How is the pancreas both an exocrine gland and an endocrine gland?

_____

_____

_____

Use the Case Study and the Web sites for this chapter to answer the following questions.

53. What are "anabolic-androgenic" steroids? What are their primary medical uses?

_____

_____

_____

_____

_____

_____

54. What are the side effects related to anabolic steroid use? What are the side effects specific to men? What are the side effects specific to women?

_____

_____

_____

_____

_____

_____

_____

_____

55. When did American athletes begin using anabolic steroids? Who was Dr. John Ziegler?

_____

_____

_____

_____

_____

56. Instead of anabolic steroid use, some athletes might choose to use human Growth Hormone. What might the athlete gain by using GH? What are the side effects of using GH?

_____

_____

_____

_____

57. Do over the counter (OTC) drugs have actual performance enhancing effects? Why could they really decrease an athlete's performance? What are these drugs used to treat? What are their side effects?

_____

_____

_____

_____

_____

_____

_____

ANSWERS TO EXERCISES

1. hormone
   target
2. Peptide
   Amino acid based
   steroid
   Prostaglandins
3. second messenger
   cyclic AMP
4. exocrine
   endocrine
5. Follicle-stimulating
   luteinizing
   Thyroid-stimulating
   Adrenocorticotropic
6. Prolactin
   endorphins
   Melanocyte-stimulating
   growth
   pituitary dwarfism
   gigantism
7. antidiuretic
   oxytocin

8. pancreas
   insulin
   glucagon
9. pineal
   melatonin
10. thymus
    thymosin
11. false, do not have
12. false, target
13. false, exocrine
14. false, steroid
15. false, negative
16. false, increase
17. false, posterior
18. false, anterior
19. false, glucagon
20. false, decreases
21. true
22. false, away from
23. false, male
24. endocrine
25. exocrine

26. exocrine
27. endocrine
28. exocrine
29. exocrine
30. exocrine
31. endocrine
32. c, d
33. b, f
34. e
35. b
36. g
37. f
38. h
39. f
40. a
41. f
42. a, b
43. c
44. a, b, c
45. d
46. c
47. d

48. hypothalamus, pituitary, thyroid, parathyroids, pancreas, sex organs, adrenals, (pineal, thymus, kidneys, heart, stomach, small intestine, fat cells)

49. An organ can respond to the presence of a hormone only if it has the proper receptor molecule in its cell membrane or internally. Only kidney cells have the specific receptor for ADH, while most cells have receptors for insulin.

50. Diabetes mellitus causes the body to rely on fat metabolism resulting in high levels of lipids in the blood. In severe cases, fat deposits in the blood vessels. The presence of the fat deposits results in high blood pressure and heart disease. Deposits in vessels on the eye may damage the retina, causing blindness. Deposits located in the small vessels of the kidney may result in kidney failure.

51. Calcitonin helps control calcium levels in your blood by regulating calcium release from your bones. If calcium levels in your blood are too high, calcitonin is released to inhibit release of calcium from your bones. The parathyroid glands secrete parathormone in response to low levels of calcium in your blood. When blood calcium levels are low, parathormone triggers your bone cells to release calcium. Between calcitonin and parathormone, the calcium content of your blood is regulated.

52. Your pancreas functions as an exocrine gland by releasing digestive secretions from the pancreatic duct into your small intestine and it functions as an endocrine gland when clusters of islet cells produce the peptide hormones insulin and glucagon.

53. According to the National Institute on Drug Abuse, "Anabolic-androgenic" steroids are commonly called "anabolic steroids," synthetic substances related to male sex hormones. They increase growth of skeletal muscle (anabolic effects) and development of male sexual characteristics (androgenic effects). Anabolic steroids were developed in the late 1930s to treat hypogonadism (the testes do not produce sufficient testosterone for normal growth, development, and sexual functioning). The primary medical uses of these compounds are to treat delayed puberty, some types of impotence, and wasting of the body caused by HIV infection or other diseases.

54. According to C. Freudenrich of HowStuffWorks, the side effects from anabolic steroids include jaundice and liver damage (these substances are broken down in the liver), mood swings, and depression and aggression because they act on various centers of the brain. In males, excessive concentrations interfere with normal sexual function and cause baldness, infertility, and breast development. In females, excessive concentrations cause male characteristics to develop and interfere with normal female functions. They stimulate hair growth on the face and body, interfere with the menstrual cycle, possibly leading to infertility, thicken the vocal cords (the voice deepens) possibly permanently, and if pregnant, interfere with the developing fetus.

55. The American Chemical Society reports that the use of anabolic steroids by American athletes can be traced to the 1950s. Dr. John Ziegler, physician, scientist, and leading U.S. weight lifter, first introduced anabolic steroids to American athletes.

56. According to C. Freudenrich of HowStuffWorks, excessive GH levels increase muscle mass by stimulating protein synthesis, strengthen bones by stimulating bone growth and reduce body fat by stimulating the breakdown of fat cells. Use of GH has become increasingly popular because it is difficult to detect. Side effects include overgrowth of hand, feet, and face (acromegaly) because of the increased muscle and bone development in these parts, enlarged internal organs, especially heart, kidneys, tongue and liver, and heart problems.

57. According to "the physician and sportsmedicine," some over the counter drugs have actual performance-enhancing side effects. However, many actually decrease performance, primarily because of adverse cardiovascular effects and impaired judgment. Some of these drugs are used for weight loss (phenylpropanolamine hydrochloride) and to treat asthma and upper respiratory infections (ephedrine, phenylephrine, and epinephrine). Their primary effect is autonomic cardiac stimulation as well as an amphetamine-like central nervous system (CNS) stimulation. Possible side effects at commonly used dosages include tachycardia (rapid heart beat), headache, dizziness, hypertension, anorexia (hence their use by athletes in appearance sports), irritability, anxiety, mania, and, at high doses, psychosis.

Chapter 33: The Nervous System and the Senses

OVERVIEW

This chapter covers the structure and function of nerve cells, the nature of resting potentials in nerves, and how action potentials (nerve signals) are generated and conducted. The authors discuss the human nervous system, including the central nervous system (brain and spinal cord), the peripheral nervous system, and neurotransmitters. The ways in which animals perceive and respond to nervous stimulation are discussed, and the structures and functions of the major sense organs are covered. Learning, memory, and retrieval are briefly discussed.

1) What Are the Structures and Functions of Neurons?

An individual **neuron** (nerve cell) has four functions besides normal metabolism. A neuron receives information, it integrates information and produces an appropriate output signal, it conducts the signal to its terminal endings, and it transmits the signal to other nerve cells, glands, or muscles. A typical vertebrate neuron has four structural regions. **Dendrites** receive signals from other neurons or from the environment. Dendrites of brain and spinal cord neurons respond to specific chemicals released by other neurons. Dendrites of *sensory neurons* have specialized membranes to respond to specific stimuli such as pressure, odor, light, or heat. Dendrites receive electrical signals that travel to the neuron's **cell body**. The cell body performs typical metabolic activities. The cell body also integrates all the signals received from dendrites. A long, thin fiber called an **axon** extends from the cell body. An axon may be up to 1 m in length (~3 ft). Axons are usually bundled together into **nerves**. Some axons are wrapped with insulation, called **myelin**, which allows very rapid conduction of the electrical signal.

If the signals received from dendrites are sufficiently positive, the cell body will produce an **action potential**, an electrical signal. Axons carry action potentials to **synaptic terminals** located at the far ends of each axon. Signals are conducted to other cells from the synaptic terminals. Most synaptic terminals contain a **neurotransmitter**. A neurotransmitter is a chemical that is released in response to an action potential reaching the synaptic terminal. The site where synaptic terminals communicate with other cells is called the **synapse**.

2) How Is Neural Activity Produced and Transmitted?

Researchers have measured a difference in electrical charges between the outside and the inside of an unstimulated, inactive neuron. In other words, neurons have an electrical difference, or electrical *potential*, across their plasma membranes. This is called the **resting potential** and is always negative inside the cell. The resting potential ranges between -40 and -90 mV (millivolts). If a neuron is stimulated, the inner negative potential can be altered. If the inside of the neuron becomes sufficiently *less* negative, it reaches a **threshold**. The threshold is usually 15 millivolts less negative than the resting potential. Once threshold is reached, an *action potential* is triggered. This causes the neuron's potential inside to *rise* rapidly to about +50 mV. (This action potential will last a few milliseconds, and then the cell's normal resting potential will be restored.) The positive charge of the action potential flows rapidly down the axon to synaptic terminals where the signal is communicated to another cell at a synapse.

The signals transmitted at synapses are called **postsynaptic potentials (PSPs)**. The neuron that releases neurotransmitter into the synapse is the **presynaptic neuron**. If the cell on the other side of the synapse is another neuron, it is called the **postsynaptic neuron**. A tiny gap separates the synaptic terminal of the presynaptic neuron from the postsynaptic neuron. When an action potential reaches a synaptic terminal, the inside of the terminal becomes positively charged. This causes storage vesicles in the synaptic terminal to release a neurotransmitter into the gap. The neurotransmitter molecules rapidly diffuse across the gap and bind briefly to membrane receptors of the postsynaptic neuron.

The binding of the neurotransmitter molecules causes ion-specific channels in the postsynaptic membrane to open. With the channels open, ions flow across the plasma membrane along their concentration gradients. A small, brief change in electrical charge occurs; this is the postsynaptic potential (PSP). PSPs can be *excitatory* (EPSPs) or *inhibitory* (IPSPs), depending on which channels are opened and which ions flow. EPSPs make the neuron less negative inside so the neuron is more likely to produce an action potential. IPSPs make the neuron more negative and it is less likely to produce an action potential. The PSPs travel to the cell body. The cell body determines whether an action potential will be produced. Keep in mind here, that the dendrites and cell body of a neuron may receive *both* EPSPs and IPSPs from the synaptic terminals of thousands of presynaptic neurons. The cell body must "add up," or integrate, all the PSPs. An action potential will be produced *only* if the EPSPs and IPSPs collectively raise the electrical potential inside the neuron above the threshold.

The nervous system uses over 50 different neurotransmitters. Included in the ever-growing list are acetylcholine, dopamine, and serotonin.

3) What are Some General Features of Nervous Systems?

Information processing requires a nervous system to perform four basic operations.

(1) A nervous system must determine the type of stimulus. The type of stimulus is distinguished by wiring patterns in the brain. Since all action potentials are the same, the nervous system monitors which neurons are firing. For instance, the brain interprets optic nerve action potentials as light, and olfactory nerve action potentials as odors, etc.

(2) A nervous system must signal the intensity of a stimulus. All action potentials are of the same magnitude and duration; therefore the **intensity**, or strength, of a stimulus is determined by the number of neurons firing or the frequency of action potentials generated. In response to an intense stimulus, each neuron will produce many action potentials quickly and a large number of neurons will respond. Thus, a loud noise stimulates a larger number of auditory nerves to fire more rapidly than does less intense sound.

(3) A nervous system must integrate information from many sources. The brain must filter all the stimuli it is constantly receiving. Information from many sources is processed through **convergence**. During convergence, many neurons funnel their signals to fewer decision-making neurons in the brain.

(4) A nervous system must initiate and direct the response. Complex responses occur through the **divergence** of signals. During divergence, electrical signals from a relatively small number of decision-making cells in the brain flow to many different neurons that control muscle or glandular activity.

Neuron-to-muscle pathways direct most behaviors and are composed of four elements. **Sensory neurons** respond to a stimulus. **Association neurons** receive signals from many sources and activate motor neurons. **Motor neurons** receive instructions from association neurons and activate muscles or glands. **Effectors**, usually muscles or glands, perform the response directed by the nervous system.

In animals, the simplest behavior is the **reflex**. A reflex is produced by neurons in the spinal cord. It is an involuntary response that does not require interaction with the brain. The knee-jerk and pain-withdrawal reflexes are good examples. However, more complex behaviors (interpreting the pain) rely on neural pathways that are interconnected, integrating the stimuli.

There are only two designs for nervous systems within the animal kingdom. In radially symmetrical animals, a network of neurons called a **nerve net** is integrated with the animal's tissues. Clusters of neurons, called **ganglia**, exist in the net but do not resemble a brain. Over time, evolution of bilaterally symmetrical animals led to a concentration of more complex ganglia at one end of the animal, the head. The evolutionary trend toward development of a brain is called **cephalization**.

4) How Is the Human Nervous System Organized?

The human nervous system has two parts. The **central nervous system (CNS)** includes your **brain** and your **spinal cord**. The **peripheral nervous system (PNS)** includes **peripheral nerves** that connect the CNS to the rest of your body. Peripheral nerves contain the axons of the sensory neurons that bring sensory information *to* your CNS from all parts of your body. Peripheral nerves also contain axons of motor neurons that carry signals *from* your CNS to your organs and muscles.

The motor portion of your peripheral nervous system has two parts. The **somatic nervous system** controls your skeletal muscles and voluntary muscle movement. Its motor neurons are located in the gray matter of your spinal cord; their axons go directly to the muscles they control. The **autonomic nervous system** controls your involuntary responses. Its motor neurons form synapses with your heart, smooth muscle, and glands. It is controlled by both the *medulla* and the *hypothalamus* of your brain.

Your autonomic nervous system has two divisions. The **sympathetic division** prepares your body for a stressful or high energy activity. The responses involved are referred to as "fight or flight" responses and include increased heart rate, decreased digestive processes, dilated pupils, and increased lung capacity. The **parasympathetic division**, in contrast, allows your body to "rest and ruminate." It allows your digestive system to become active and your heart rate to slow; it is involved in leisurely activities.

Your central nervous system receives and processes information, generates thoughts, and directs responses. The CNS contains up to 100 billion association neurons, all of which are protected in three ways. Your brain is encased in your *skull* and your spinal cord is surrounded by your *vertebral column*. Beneath these bones, a triple layer of connective tissue (the **meninges**) covers your CNS and **cerebro-spinal fluid** flows between the layers of the meninges providing an additional cushion of protection. Finally, the capillaries providing oxygen and nutrients to your brain are much less permeable than capillaries in other parts of your body. They form the **blood-brain barrier**, helping to keep dangerous chemicals and microbes out of your brain tissues.

Your spinal cord is a cable of axons protected by your vertebral column. Nerves, carrying axons of sensory neurons and motor neurons, come out from between vertebrae and merge to form the peripheral nerves of your spinal cord (part of your PNS). In the center of your spinal cord is **gray matter**. Gray matter is made up of the cell bodies of neurons that control voluntary muscles and the autonomic nervous system, and neurons that communicate with the brain and other parts of the spinal cord. The gray matter is surrounded by **white matter**. This area is white because of the myelin-coated axons of neurons that extend up or down your spinal cord. These axons carry sensory information from your organs, muscles, and skin *to* your brain. They also carry motor signals *from* your brain to your PNS.

The spinal cord also contains the neural pathways for simple behaviors such as reflexes. A reflex response may involve the following scenario: If you touch a hot burner on the stove, the signal travels along the dendrite of a sensory neuron to its cell body in a **dorsal root ganglion** on spinal nerves just outside your spinal cord. An axon from the sensory neuron would synapse with an association neuron whose cell body is located in the gray matter of your spinal cord. The association neuron would send the signal to a motor neuron, resulting in the withdrawal of your hand from the burner. At the same time, the signal would travel along the axon of the association neuron to your brain, alerting it to the pain.

Your brain consists of many parts specialized for specific functions. Embryologically, the vertebrate brain begins as a simple tube that develops into three parts. Simplistically, the **hindbrain** controls automatic behaviors such as breathing and heart rate, the **midbrain** controls vision, and the **forebrain** controls smell. However, the mammalian brain is much more complex. The human *hindbrain* includes the **medulla**, which controls automatic functions such as your breathing, heart rate, blood pressure, and swallowing; the **pons**, which influences transitions between sleep and wakefulness and between stages of sleep, as well as the rate and pattern of your breathing, and the **cerebellum**, which coordinates movements of the body and is involved in learning and memory storage for behaviors.

The *midbrain* in humans is small. It contains an auditory relay center and a center that controls reflex movements of your eyes. It also contains another relay center, the **reticular formation**. The reticular formation receives information from almost every sense, every part of your body, and from many areas of your brain. It filters sensory inputs before they reach the conscious regions of your brain. It also plays a role in sleep and arousal, emotion, muscle tone, and certain movements and reflexes.

The large human *forebrain* is called the **cerebrum**. It includes the *thalamus*, the *limbic system*, and the *cerebral cortex*. The **thalamus** carries sensory information to and from other forebrain regions. The **limbic system** produces our most basic and primitive emotions, drives, and behaviors, such as fear, thirst, pleasure, sexual response, and memory formation. The limbic system includes three areas of your brain. The **hypothalamus** contains neurons and neurosecretory cells that release hormones into your blood, control the release of hormones from your pituitary gland, and direct the activities of your autonomic nervous system. The **amygdala** contains neurons that produce the sensations of pleasure, sexual arousal, or punishment when stimulated. The **hippocampus** plays a role in emotions and in the formation of the long-term memory needed for learning.

The **cerebral cortex** forms the outer layer of your forebrain. It is the largest part of your brain. It is divided into the **cerebral hemispheres**, which communicate through axons making up the **corpus callosum**. To accommodate some 100 billion neurons, the cortex forms folds called **convolutions** to increase its area. Cortex neurons receive and process sensory information, store some of it as memory for future use, direct voluntary movements, and are responsible for thinking. The cerebral cortex is divided anatomically into four regions: the *frontal*, *parietal*, *occipital*, and *temporal lobes*. Damage to the cortex from trauma, stroke, or a tumor results in specific deficits depending on the area(s) damaged.

5) How Does the Brain Produce the Mind?

The "mind-brain" problem has occupied philosophers and neurobiologists for generations. Studies of accident, stroke, or surgery patients have revealed that the "left brain" and the "right brain" are specialized for different functions. These studies indicate that if your are right-handed, your left hemisphere functions in speech, reading, writing, language comprehension, mathematical ability, and logical problem solving, and your right hemisphere functions in musical skills, artistic ability, facial recognition, spatial visualization, and the ability to recognize and express emotions. However, this left-right dichotomy is not always so cleancut.

While the mechanics of learning and memory are poorly understood, researchers do know that memory may be brief or long lasting and that learning occurs in two phases. An initial **working memory** is a temporary electrical or biochemical activity in the brain. This allows you to remember a phone number or room number long enough to make the call or find the room. Working memory may be converted to **long-term memory**, but that involves actual changes to brain structure, such as the formation of new, long-lasting neural connections or the strengthening of existing ones. So, if you want to *learn* the information you are studying, you can't just read about it, you have to work at it to make the necessary structural changes in your brain. You also have to be able to retrieve the information you've learned. Learning and memory storage, and memory retrieval may be controlled by separate regions of your brain. Learning and memory may be controlled by your hippocampus, while retrieval of those memories seems to be controlled by your temporal lobes. If you tap or rub the side of your head when you are trying to remember something, you are tapping or rubbing near your temporal lobe.

6) How Do Sensory Receptors Work?

Generally, a **receptor** is a structure that changes when acted on by a stimulus. All receptors convert signals from one form to another. A **sensory receptor** is usually a neuron specialized to produce an electrical signal in response to specific stimuli. Stimulation of a sensory receptor causes a **receptor potential**, an electrical signal whose size is proportional to the strength of the stimulus. However, intensity is still conveyed to the nervous system by the frequency, not the size, of the action potentials.

7) How Is Sound Sensed?

Sound is produced by any vibrating object. The structure of your ear allows it to capture, transmit, and convert sound into electrical signals. Your **outer ear** consists of the **external ear** and the **auditory canal**. The auditory canal conducts sound waves to your **middle ear**. Your middle ear consists of the **tympanic membrane** (eardrum), three tiny bones (the hammer, the anvil, and the stirrup), and the **Eustachian tube**. The eustachian tube connects to your pharynx and equalizes air pressure between your middle ear and the atmosphere. This is why your ears "pop" when you go up or down in elevation. In your middle ear, sound vibrates the tympanic membrane, which in turn vibrates the hammer, anvil, and stirrup bones. These bones transmit vibrations to your **inner ear**. The fluid-filled, hollow bones of your inner ear form a spiral-shaped **cochlea** and other structures that detect movement of your head and the pull of gravity. Sounds enter your inner ear when the stirrup vibrates a membrane that covers a hole in the cochlea.

The central canal of the cochlea contains the **basilar membrane**. On top of the basilar membrane are the sound receptors, the *hair cells*. Protruding into the central canal is the gelatinous **tectorial membrane** in which the hairs are embedded. You perceive sound when the oval window passes vibrations to the fluid in the cochlea. This vibrates the basilar membrane, causing it to move up and down. The moving of the basilar membrane bends the hairs embedded in the tectorial membrane, producing receptor potentials in the hair cells. The hair cells release neurotransmitters onto neurons of the auditory nerve, whose action potentials travel to the brain. The cochlea allows us to perceive *loudness*. The magnitude of sound vibrations bends the hair cell hairs in proportion to the loudness. The cochlea also allows us to perceive *pitch*. The frequency of sound vibrations is determined by the area of the basilar membrane that vibrates. High pitched sounds vibrate the membrane closest to the oval window, low pitched sounds vibrate the membrane further in the cochlea.

8) How Is Light Sensed?

All forms of vision use *photoreceptors* containing colored receptor molecules called **photopigments**. Photopigments absorb light and are chemically changed in the process. The chemical change alters ion channels in the plasma membrane, producing a receptor potential.

Even though all light receptors function in the same way, not all eyes evolved in the same way. Arthropods have **compound eyes** made up of many light sensing **ommatidia**. Images from each **ommatidium** come together to give the insect a grainy, but rather real, image of their surroundings.

The mammalian eye collects, focuses, and converts light waves. Incoming light first encounters the transparent **cornea** covering your eyeball. Behind your cornea is a chamber filled with a watery nourishing fluid called **aqueous humor**. Your **iris** controls how much light enters your eye because it is pigmented muscular tissue with a circular opening, the **pupil**. Light strikes a flattened sphere of transparent protein fibers, the **lens**, which is suspended behind the pupil by ligaments and muscles that can change its shape. Behind the lens is a chamber filled with clear jellylike **vitreous humor**. After passing through the vitreous humor, light reaches your **retina**, a multilayered sheet of photoreceptors and neurons. Your retina converts light energy into electrical nerve impulses that are transmitted to your brain. Behind your retina is the **choroid**, a darkly pigmented tissue that absorbs stray light, providing you with clear vision. Surrounding the outer portion of your eyeball is the **sclera**, tough connective tissue forming the white of your eye.

You are able to focus on both distant and nearby objects because your lens is adjustable. Visual images are focused by the lens onto a small area of the retina called the **fovea**. If your eyeball is too long, images are focused in front of the fovea and you are **nearsighted**, unable to focus on distant objects. If your eyeball is too short, images are focused beyond the fovea and you are **farsighted**. External lenses change how light reaching your eye is bent, allowing your lens to focus the light onto your fovea.

Light reaching your retina is captured by photoreceptors called **rods** and **cones** present in the retina. The signal is then processed by several layers of neurons. The retinal cells closest to the vitreous humor are **ganglion cells**, whose axons make up your **optic nerve**. Ganglion cell axons must pass through the retina to reach the brain. The spot at which the axons pass through the retina is called the **optic disc** or your **blind spot**.

Cones and rods differ in distribution and light sensitivity. Although cones are located throughout your retina, they are concentrated in the fovea. Human eyes have three types of cones. Each cone has a slightly different photopigment that is most strongly stimulated by a particular wavelength of light (red, green, or blue). Your brain distinguishes color according to the relative intensity of stimulation of different cones. Rods dominate the peripheral portions of your retina. Rods are more sensitive to light than cones but cannot distinguish color. They are largely responsible for our vision in dim light.

Where an animal's eyes are placed in their head evolved as their lifestyle evolved. Predators and omnivores (such as humans) have both eyes facing forward. We have slightly different, but extensively overlapping, visual fields (**binocular vision**) that allows for depth perception. Most herbivores, however, have one eye on each side of their head. Herbivores have no depth perception but they have a nearly 360-degree field of view, an advantage when watching for predators.

9) How Are Chemicals Sensed?

The ability of animals to sense chemicals enables them to find food, avoid poisons, locate homes, and find mates. Terrestrial vertebrates have two chemical senses: **olfaction** (smell) for airborne molecules and **taste** for molecules dissolved in water or saliva. Olfactory receptors are nerve cells with hairlike dendrites. They are found in epithelial tissue of the upper portion of each of your nasal cavities. Odor molecules dissolve in the mucus covering the receptors, allowing you to smell. Taste receptors are located in clusters called **taste buds** on your tongue. Each taste bud has up to 80 taste receptor cells surrounded by supporting cells in a small pit. The four (maybe five) major types of taste receptors are sweet, sour, salty, and bitter. A fifth, *umami*, responds to glutamate used in MSG to enhance flavors of foods. However, what we call taste is mostly olfactory receptors responding to odor molecules from our food.

Pain is actually a specialized chemical sense. Most pain is caused by tissue damage. When cells are damaged, their contents flow into the extracellular fluid. Potassium ions from the cell contents stimulate **pain receptors**. Damaged cells also release enzymes that produce *bradykinin*, another stimulus that activates pain receptors. Your pain receptors signal specific brain cells. This enables you to determine what hurts.

Other animals have specialized senses that have evolved in response to more diverse habitats. Bats navigate through darkness and hunt their prey using **echolocation**. Bats (and porpoises) send out pulses of ultrasonic noise frequencies that bounce off nearby objects. The pattern of the returning frequencies convey a great deal of accurate information to the bat (or the porpoise). Some fish use **electrolocation** to navigate. They send out high-frequency electric signals from their tails, creating an electric field. Electroreceptors located on the sides of their body detect distortions in the field. The impulses are interpreted by their brain.

KEY TERMS AND CONCEPTS

Fill-In: From the following list of terms, fill in the blanks below.

action potential	cornea	neurotransmitters
auditory	dendrites	olfactory
axon	fovea	optic disc
blind spot	inner	oval
bradykinin	iris	peripheral nervous system (PNS)
brain	lens	pons
cell body	limbic system	resting potential
central nervous system (CNS)	medulla	rod
cerebellum	motor	sensory
cerebral cortex	myelin	spinal cord
cerebrum	nerve	taste buds
cone	neuron	thalamus

1. A single nerve cell is called a _____. The part of a nerve cell containing the cellular organelles is the _____. The long process that carries impulses away from the cell body to synaptic endings on other nerve cells or on muscles is the _____. Branched tendrils that respond to signals from the external environment or from other neurons are _____. A bundle of axons bound together in a sheath is a _____.

2. _____ covers many axons and insulates them.

3. Chemicals released into synapses by the synaptic terminals of presynaptic neurons are called _____.

4. The _____ of a neuron is always negative. Once the cell becomes more negative and threshold is reached, an _____ is generated.

5. _____ neurons respond to a stimulus; and _____ neurons activate muscles or glands.

6. The human nervous system has two parts: the _____, consisting of the brain and spinal cord; and the _____ which links the brain and spinal cord to the rest of the body.

7. The _____ is the part of your central nervous system that is enclosed within the skull, and the _____ is protected by your vertebral column.

8. The hindbrain of humans develops into the _____ which controls automatic activities such as breathing, swallowing, heart rate, and blood pressure; the _____ which coordinates movements of the body; and the _____ is just above the medulla and contains neurons that influence sleep and your rate and pattern of breathing.

9. The human forebrain is called the _____ and includes the _____, which carries sensory information; the _____, which is responsible for our primitive emotions and behaviors; and the _____, which is involved in memory and thinking.

10. The _____ canal is within the outer ear and conducts sound from the external ear to the eardrum. The _____ ear is composed of the bony, fluid-filled tubes of the cochlea. The membrane-covered entrance to the cochlea is the _____ window, which vibrates in response to vibrations of the stirrup.

11. The clear outer covering of the eye in front of the pupil and iris is the _____. The _____ is the pigmented muscular tissue of the eye that surrounds and controls the size of the pupil. A flexible or movable _____ is used to focus both close and distant objects.

12. A photoreceptor cell in the retina that is sensitive to dim light but not color is a _____. The region of your retina where images are focused is the _____. The area where ganglion cell axons pass through your retina is the _____, also called the _____.

13. Your sense of smell involves _____ receptors, and your sense of taste involves clusters of receptors called _____.

14. _____ is a chemical made during tissue damage. It binds to receptor molecules on pain nerve endings, causing the sensation of pain.

Key Terms and Definitions

action potential: a rapid change from a negative to a positive electrical potential in a nerve cell. This signal travels along an axon without a change in intensity.

amygdala (**am-ig´-da-la**): part of the forebrain of vertebrates that is involved in the production of appropriate behavioral responses to environmental stimuli.

aqueous humor (**ā´-kwē-us**): the clear, watery fluid between the cornea and lens of the eye.

association neuron: in a neural network, a nerve cell that is postsynaptic to a sensory neuron and presynaptic to a motor neuron. In actual circuits, there may be many association neurons between individual sensory and motor neurons.

auditory canal (**aw´-di-tor-ē**): a canal within the outer ear that conducts sound from the external ear to the tympanic membrane.

autonomic nervous system: the part of the peripheral nervous system of vertebrates that synapses on glands, internal organs, and smooth muscle and produces largely involuntary responses.

axon: a long extension of a nerve cell, extending from the cell body to synaptic endings on other nerve cells or on muscles.

basilar membrane (**bas´-eh-lar**): a membrane in the cochlea that bears hair cells that respond to the vibrations produced by sound.

binocular vision: the ability to see objects simultaneously through both eyes, providing greater depth perception and more accurate judgment of the size and distance of an object from the eyes.

blind spot: the area of the retina at which the axons of the ganglion cell merge to form the optic nerve; the blind spot of the retina.

blood–brain barrier: relatively impermeable capillaries of the brain that protect the cells of the brain from potentially damaging chemicals that reach the bloodstream.

brain: the part of the central nervous system of vertebrates that is enclosed within the skull.

cell body: the part of a nerve cell in which most of the common cellular organelles are located; typically a site of integration of inputs to the nerve cell.

central nervous system: in vertebrates, the brain and spinal cord.

cephalization (**sef-ul-ī-zā´-shun**): the tendency of sensory organs and nervous tissue to become concentrated in the head region over evolutionary time.

cerebellum (**ser-uh-bel´-um**): the part of the hindbrain of vertebrates that is concerned with coordinating movements of the body.

cerebral cortex (**ser-ē´-brul kor´-tex**): a thin layer of neurons on the surface of the vertebrate cerebrum, in which most neural processing and coordination of activity occurs.

cerebral hemisphere: one of two nearly symmetrical halves of the cerebrum, connected by a broad band of axons, the corpus callosum.

cerebrospinal fluid: a clear fluid, produced within the ventricles of the brain, that fills the ventricles and cushions the brain and spinal cord.

cerebrum (**ser-ē´-brum**): the part of the forebrain of vertebrates that is concerned with sensory processing, the direction of motor output, and the coordination of most bodily activities; consists of two nearly symmetrical halves (the hemispheres) connected by a broad band of axons, the corpus callosum.

choroid (**kor´-oid**): a darkly pigmented layer of tissue, behind the retina, that contains blood vessels and pigment that absorbs stray light.

cochlea (kahk´-lē-uh): a coiled, bony, fluid-filled tube found in the mammalian inner ear; contains receptors (hair cells) that respond to the vibration of sound.

compound eye: a type of eye, found in arthropods, that is composed of numerous independent subunits called *ommatidia*. Each ommatidium apparently contributes a piece of a mosaiclike image perceived by the animal.

cone: a cone-shaped photoreceptor cell in the vertebrate retina; not as sensitive to light as are the rods. The three types of cones are most sensitive to different colors of light and provide color vision; see also *rod*.

convergence: a condition in which a large number of nerve cells provide input to a smaller number of cells.

convolution: a folding of the cerebral cortex of the vertebrate brain.

cornea (kor´-nē-uh): the clear outer covering of the eye, in front of the pupil and iris.

corpus callosum (kor´pus kal-ō´-sum): the band of axons that connect the two cerebral hemispheres of vertebrates.

dendrite (den´-drīt): a branched tendril that extends outward from the cell body of a neuron; specialized to respond to signals from the external environment or from other neurons.

divergence: a condition in which a small number of nerve cells provide input to a larger number of cells.

dorsal root ganglion: a ganglion, located on the dorsal (sensory) branch of each spinal nerve, that contains the cell bodies of sensory neurons.

echolocation: the use of ultrasonic sounds, which bounce back from nearby objects, to produce an auditory "image" of nearby surroundings; used by bats and porpoises.

effector (ē-fek´-tor): a part of the body (normally a muscle or gland) that carries out responses as directed by the nervous system.

electrolocation: the production of high-frequency electrical signals from an electric organ in front of the tail of an electric fish; used to detect and locate nearby objects.

Eustachian tube (ū-stā´-shin): a tube connecting the middle ear with the pharynx; allows pressure between the middle ear and the atmosphere to equilibrate.

external ear: the fleshy portion of the ear that extends outside the skull.

farsighted: the inability to focus on nearby objects, caused by the eyeball being slightly too short.

forebrain: during development, the anterior portion of the brain. In mammals, the forebrain differentiates into the thalamus, the limbic system, and the cerebrum. In humans, the cerebrum contains about half of all the neurons in the brain.

fovea (fō´-vē-uh): in the vertebrate retina, the central region on which images are focused; contains closely packed cones.

ganglion (gang´-lē-un): a cluster of neurons.

ganglion cell: a type of cell, of which the innermost layer of the vertebrate retina is composed, whose axons form the optic nerve.

gray matter: the outer portion of the brain and inner region of the spinal cord; composed largely of neuron cell bodies, which give this area a gray color.

hindbrain: the posterior portion of the brain, containing the medulla, pons, and cerebellum.

hippocampus (hip-ō-kam´-pus): the part of the forebrain of vertebrates that is important in emotion and especially learning.

hypothalamus (hī-pō-thal´-a-mus): a region of the brain that controls the secretory activity of the pituitary gland; synthesizes, stores, and releases certain peptide hormones; directs autonomic nervous system responses.

inner ear: the innermost part of the mammalian ear; composed of the bony, fluid-filled tubes of the cochlea and the vestibular apparatus.

intensity: the strength of stimulation or response.

iris: the pigmented muscular tissue of the vertebrate eye that surrounds and controls the size of the pupil, through which light enters.

lens: a clear object that bends light rays; in eyes, a flexible or movable structure used to focus light on a layer of photoreceptor cells.

limbic system: a diverse group of brain structures, mostly in the lower forebrain, that includes the thalamus, hypothalamus, amygdala, hippocampus, and parts of the cerebrum and is involved in basic emotions, drives, behaviors, and learning.

long-term memory: the second phase of learning; a more-or-less permanent memory formed by a structural change in the brain, brought on by repetition.

medulla (med-ū´-luh): the part of the hindbrain of vertebrates that controls automatic activities such as breathing, swallowing, heart rate, and blood pressure.

meninges (men-in´-jēz): three layers of connective tissue that surround the brain and spinal cord.

midbrain: during development, the central portion of the brain; contains an important relay center, the reticular formation.

middle ear: the part of the mammalian ear composed of the tympanic membrane, the Eustachian tube, and three bones (hammer, anvil, and stirrup) that transmit vibrations from the auditory canal to the oval window.

motor neuron: a neuron that receives instructions from the association neurons and activates effector organs, such as muscles or glands.

myelin (mī´-uh-lin): a wrapping of insulating membranes of specialized nonneural cells around the axon of a vertebrate nerve cell; increases the speed of conduction of action potentials.

nearsighted: the inability to focus on distant objects caused by an eyeball that is slightly too long.

nerve: a bundle of axons of nerve cells, bound together in a sheath.

nerve net: a simple form of nervous system, consisting of a network of neurons that extend throughout the tissues of an organism such as a cnidarian.

neuron (noor´-on): a single nerve cell.

neurotransmitter: a chemical that is released by a nerve cell close to a second nerve cell, a muscle, or a gland cell and that influences the activity of the second cell.

olfaction (ōl-fak´-shun): a chemical sense, the sense of smell; in terrestrial vertebrates, the result of the detection of airborne molecules.

ommatidium (ōm-ma-tid´-ē-um): an individual light-sensitive subunit of a compound eye; consists of a lens and several receptor cells.

optic nerve: the nerve leading from the eye to the brain, carrying visual information.

outer ear: the outermost part of the mammalian ear, including the external ear and auditory canal leading to the tympanic membrane.

pain receptor: a receptor that has extensive areas of membranes studded with special receptor proteins that respond to light or to a chemical.

parasympathetic division: the division of the autonomic nervous system that produces largely involuntary responses related to the maintenance of normal body functions, such as digestion.

peripheral nerve: a nerve that links the brain and spinal cord to the rest of the body.

peripheral nervous system: in vertebrates, the part of the nervous system that connects the central nervous system to the rest of the body.

photopigment (fō´-tō-pig-ment): a chemical substance in photoreceptor cells that, when struck by light, changes in molecular conformation.

pons: a portion of the hindbrain, just above the medulla, that contains neurons that influence sleep and the rate and pattern of breathing.

postsynaptic neuron: at a synapse, the nerve cell that changes its electrical potential in response to a chemical (the neurotransmitter) released by another (presynaptic) cell.

postsynaptic potential (PSP): an electrical signal produced in a postsynaptic cell by transmission across the synapse; it may be excitatory (EPSP), making the cell more likely to produce an action potential, or inhibitory (IPSP), tending to inhibit an action potential.

presynaptic neuron: a nerve cell that releases a chemical (the neurotransmitter) at a synapse, causing changes in the electrical activity of another (postsynaptic) cell.

pupil: the adjustable opening in the center of the iris, through which light enters the eye.

receptor: a cell that responds to an environmental stimulus (chemicals, sound, light, pH, and so on) by changing its electrical potential; also, a protein molecule in a plasma membrane that binds to another molecule (hormone, neurotransmitter), triggering metabolic or electrical changes in a cell.

receptor potential: an electrical potential change in a receptor cell, produced in response to the reception of an environmental stimulus (chemicals, sound, light, heat, and so on). The size of the receptor potential is proportional to the intensity of the stimulus.

reflex: a simple, stereotyped movement of part of the body that occurs automatically in response to a stimulus.

resting potential: a negative electrical potential in unstimulated nerve cells.

reticular formation (reh-tik´-ū-lar): a diffuse network of neurons extending from the hindbrain, through the midbrain, and into the lower reaches of the forebrain; involved in filtering sensory input and regulating what information is relayed to conscious brain centers for further attention.

retina (ret´-in-uh): a multilayered sheet of nerve tissue at the rear of camera-type eyes, composed of photoreceptor cells plus associated nerve cells that refine the photoreceptor information and transmit it to the optic nerve.

rod: a rod-shaped photoreceptor cell in the vertebrate retina, sensitive to dim light but not involved in color vision; see also *cone*.

sclera: a tough, white connective tissue layer that covers the outside of the eyeball and forms the white of the eye.

sensory neuron: a nerve cell that responds to a stimulus from the internal or external environment.

sensory receptor: a cell (typically, a neuron) specialized to respond to particular internal or external environmental stimuli by producing an electrical potential.

somatic nervous system: that portion of the peripheral nervous system that controls voluntary movement by activating skeletal muscles.

spinal cord: the part of the central nervous system of vertebrates that extends from the base of the brain to the hips and is protected by the bones of the vertebral column; contains the cell bodies of motor neurons that form synapses with skeletal muscles, the circuitry for some simple reflex behaviors, and axons that communicate with the brain.

sympathetic division: the division of the autonomic nervous system that produces largely involuntary responses that prepare the body for stressful or highly energetic situations.

synapse (sin´-aps): the site of communication between nerve cells. At a synapse, one cell (presynaptic) normally releases a chemical (the neurotransmitter) that changes the electrical potential of the second (postsynaptic) cell.

synaptic terminal: a swelling at the branched ending of an axon; where the axon forms a synapse.

taste: a chemical sense for substances dissolved in water or saliva; in mammals, perceptions of sweet, sour, bitter, or salt produced by the stimulation of receptors on the tongue.

taste bud: a cluster of taste receptor cells and supporting cells that is located in a small pit beneath the surface of the tongue and that communicates with the mouth through a small pore. The human tongue has about 10,000 taste buds.

tectorial membrane (tek-tor´-ē-ul): one of the membranes of the cochlea in which the hairs of the hair cells are embedded. In sound reception, movement of the basilar membrane relative to the tectorial membrane bends the cilia.

thalamus: the part of the forebrain that relays sensory information to many parts of the brain.

threshold: the electrical potential (less negative than the resting potential) at which an action potential is triggered.

tympanic membrane (tim-pan´-ik): the eardrum; a membrane, stretched across the opening of the ear, that transmits vibration of sound waves to bones of the middle ear.

vitreous humor (vit´-rē-us): a clear, jelly-like substance that fills the large chamber of the eye between the lens and the retina.

white matter: the portion of the brain and spinal cord that consists largely of myelin-covered axons and that give these areas a white appearance.

working memory: the first phase of learning; short-term memory that is electrical or biochemical in nature.

THINKING THROUGH THE CONCEPTS

True or False: Determine if the statement given is true or false. If it is false, change the underlined word(s) so that the statement reads true.

15. _____ Dendrites carry an impulse away from the nerve cell body and are the long extensions of a nerve cell.

16. _____ Axons initiate an impulse.

17. _____ Nerves pass on impulses with undiminished intensity.

18. _____ The resting potential inside a nerve cell is always positive within the cell.

19. _____ When the threshold level is reached, an action potential is triggered.

20. _____ The size of the action potential is dependent on the strength of the stimulus.

21. _____ Myelin covers some dendrites.

22. _____ Receptors for neurotransmitters are located on the postsynaptic neuron.

23. _____ The cerebellum controls coordination.

24. _____ The thalamus connects the two sides of the cerebrum.

25. _____ The left side of the brain is associated with creativity.

26. _____   Short-term memory is <u>electrical</u>.

27. _____   Long-term memory is <u>chemical</u>.

28. _____   Structures that change when acted on by stimuli are <u>receptors</u>.

29. _____   High-frequency vibrations of air or water are <u>sounds</u>.

30. _____   The cochlea is part of the human <u>middle</u> ear.

31. _____   The sense of vision involves <u>chemical</u> changes.

32. _____   The fovea contains virtually no <u>cones</u>.

33. _____   The retinas of nocturnal animals are made up of <u>cones</u>.

34. _____   Pain is a special kind of <u>chemical</u> sense.

35. _____   <u>Rods</u> are responsible for color vision.

Identification: Determine whether the following statements refer to the **cell body**, **synaptic terminals**, **dendrites**, or **axons** of neurons.

36. _____   carry action potentials to output terminals

37. _____   the cell's integration center

38. _____   receive information from the environment

39. _____   sites where signals are transmitted to other cells

40. _____   convert environmental information into electrical signals

41. _____   bundled together into nerves

42. _____   initiates action potentials

Using the following terms. label the parts of the diagram below:

axon   dendrites   presynaptic terminals
cell body   myelin   postsynaptic neurons

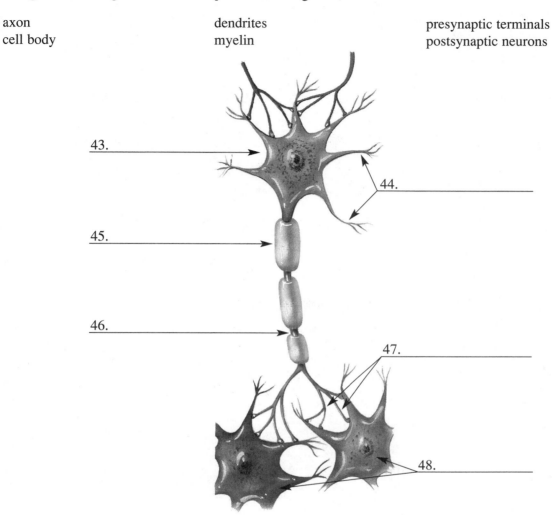

43.

44.

45.

46.

47.

48.

Matching: Nerve cell function.

49. _____ a nerve impulse

50. _____ always negative within a nerve cell

51. _____ a sudden positive charge within a nerve cell

52. _____ occurs when a nerve cell becomes sufficiently less
negative inside

53. _____ ranges between -40 to -90 mV

54. _____ electrical output of a neuron

Choices:

a. threshold

b. action potential

c. resting potential

Matching: The human brain. Some questions may have more than one correct answer.

55. _____ midbrain

56. _____ like an extension of the spinal cord

57. _____ channels sensory information to other forebrain parts

58. _____ hindbrain

59. _____ controls several automatic functions

60. _____ hypothalamus

61. _____ forebrain

62. _____ receives input from all sense organs and "decides" which require attention

63. _____ controls learning, emotions, and the autonomic nervous system

64. _____ largest part of the brain

65. _____ coordinates body movements and body positions

66. _____ controls speech, reading, math ability, and musical skills

Choices:

a. cerebellum

b. cerebrum

c. limbic system

d. medulla

e. reticular formation

f. thalamus

g. pons

Identify: Determine whether the following statements refer to the **CNS** or to the **PNS**.

67. _____ includes the brain

68. _____ includes motor neurons

69. _____ generates thoughts and emotions

70. _____ includes the spinal cord

71. _____ controls reflexes

72. _____ stores memories

73. _____ includes sensory neurons

Multiple Choice: Pick the most correct choice for each question.

74. A rapid change from a negative to a positive electrical potential in a nerve cell is
   a. a reflex
   b. an amygdala
   c. a divergence
   d. an action potential
   e. a synapse

75. The gap between the axon of one neuron and the dendrite of another is called a
   a. synapse
   b. myelin
   c. cell body
   d. convergence
   e. dendritic junction

76. The resting potential of a neuron becomes less negative when
    a. convergence occurs
    b. an IPSP signal is received
    c. it reaches threshold
    d. none of the above, since it cannot become less negative

77. The autonomic nervous system innervates all of the following except
    a. skeletal muscle
    b. heart
    c. stomach
    d. kidney

78. A photoreceptor is a receptor designed for
    a. touch
    b. pleasure
    c. chemicals
    d. pain
    e. light

79. The middle ear contains
    a. the cochlea
    b. three bones (hammer, anvil, and stirrup)
    c. the semicircular canals
    d. receptor cells for hearing

80. The amount of light entering the human eye is regulated by the muscular
    a. cornea
    b. pupil
    c. sclera
    d. iris

81. Completely color-blind animals lack
    a. cones
    b. rods
    c. ommatidia
    d. irises
    e. lenses

APPLYING THE CONCEPTS

These practice questions are intended to sharpen your ability to apply critical thinking and analysis to biological concepts covered in this chapter.

82. The neurotransmitter acetylcholine stimulates postsynaptic neurons to transmit action potentials to skeletal muscles, which then contract. Many chemicals can act on the nervous system and cause problems. Determine the effect each of the following would have on a person. (1) Botulism toxin, from a certain bacterium, inhibits the release of acetylcholine. (2) Curare, used by South American Indians to coat the tip of their arrows, binds to the same receptors that normally bind acetylcholine. (3) Diisopropyl fluorophosphate, a chemical that could be used as a nerve gas during war, blocks the enzyme that breaks down acetylcholine after it has crossed the synapse and becomes bound to postsynaptic neurons (so acetylcholine stays in the synapse).

_____

_____

_____

_____

_____

_____

_____

_____

83. Explain how each of the following medical problems could result in deafness: (1) injury to your auditory nerve; (2) arthritis of your middle ear bones; (3) a punctured eardrum; (4) too much earwax in your auditory canal; and (5) damage to your hair cells from being exposed to loud music.

_____

_____

_____

_____

_____

_____

_____

_____

_____

Use the Case Study and Web sites for this chapter to answer the following questions.

84. Actor Christopher Reeve has been "walking" on a treadmill. Since his accident damaged his spinal cord high in his neck, how is it possible for him to "walk"?

_____

_____

_____

_____

_____

_____

_____

_____

_____

_____

85. Most spinal cord injuries do not result in severing of the spinal cord. How are they damaged instead? How can your body, in responding to the injury, actually cause more harm?

_____

_____

_____

_____

_____

_____

_____

_____

86. What are some complications that can occur as your body continues to heal from a spinal cord injury?

_____

_____

_____

_____

_____

_____

87. In 1990, the FDA approved the use of methylprednisolone in treating spinal cord injuries. How does this drug help in reducing potential damage to the spinal cord?

_____

_____

_____

_____

_____

_____

_____

88. Spinal cord injuries may be complete or incomplete. Explain these terms. The patient with a spinal cord injury may be a quadriplegic or a paraplegic. Explain these terms.

_____

_____

_____

_____

_____

_____

89. What is FES? In what primary areas is it being used?

_____

_____

_____

_____

_____

_____

ANSWERS TO EXERCISES

1. neuron	11. cornea	30. false, inner	56. d
cell body	iris	31. true	57. f
axon	lens	32. false, rods	58. a, d, g
dendrites	12. rod	33. false, rods	59. d
nerve	fovea	34. true	60. c
2. Myelin	optic disc	35. false, cones	61. b, c, f
3. neurotransmitters	blind spot	36. axons	62. e
4. resting potential	13. olfactory	37. cell body	63. c
action potential	taste buds	38. dendrites	64. b
5. Sensory	14. Bradykinin	39. synaptic terminals	65. a
motor	15. false, axons	40. dendrites	66. b
6. central nervous	16. false, The cell body	41. axons	67. CNS
system (CNS)	17. true	42. cell body	68. PNS
peripheral nervous	18. false, negative	43. cell body	69. CNS
system (PNS)	19. true	44. dendrites	70. PNS
7. brain	20. false, independent of	45. myelin	71. PNS
spinal cord	21. false, axons	46. axon	72. CNS
8. medulla	22. true	47. presynaptic terminals	73. PNS
cerebellum	23. true	48. postsynaptic neurons	74. d
pons	24. false,	49. b	75. a
9. cerebrum	corpus callosum	50. c	76. c
thalamus	25. false, right	51. b	77. a
limbic system	26. true	52. a	78. e
cerebral cortex	27. false, structural	53. c	79. b
10. auditory	28. true	54. b	80. d
inner	29. true	55. e	81. a
oval			

82. Acetylcholine stimulates postsynaptic neurons to transmit action potentials to skeletal muscles, which then contract. (1) When botulism toxin inhibits the release of acetylcholine, transmitting cells will not be able to signal receiving cells, and nerve signals will be stopped at the synapse, resulting in paralysis of skeletal muscles. (2) When curare blocks the acetylcholine receptors, skeletal muscles cannot be stimulated to contract, resulting in paralysis. (3) Diisopropyl fluorophosphate prevents acetylcholine from being removed from the receptors where it is bound, so the receiving cells will continue to transmit signals to skeletal muscles, resulting in prolonged contraction of the muscles.

83. (1) If the auditory nerve is injured, the brain will not be able to receive signals that sound is occurring. (2) If the middle ear bones are arthritic, they will be unable to move in response to sound vibrations, preventing them from transmitting vibrations to the cochlea. (3) A punctured eardrum will be unable to pick up vibrations from the air and transmit them to the ear bones. (4) Too much earwax will block sound waves from striking the eardrum and setting it in motion. (5) Damaged hair cells will not produce receptor potentials and thus will not release transmitter onto neurons of the auditory nerve.

84. In addition to simple reflexes, the entire "software" for operating some fairly complex activities exists in the spinal cord. All the neurons and interconnections needed for the basic movements of walking (and running) are contained in the spinal cord, allowing Christopher Reeve to "walk" on a treadmill using a harness. This semi-independent arrangement between brain and spinal cord probably increases speed and coordination, because messages do not have to travel all the way up to the brain and back

down again just to swing one leg forward (in the case of walking). The brain's role in these "semi-automatic" behaviors is to initiate, guide, and modify the activity of spinal motor neurons, based on conscious decisions. This Christopher Reeve will not be able to do.

85. According to D. Barnes of the National Institute of Health's "Research In the News," spinal cord injuries usually crush the axons of nerve cells that are surrounded by the vertebrae. The axons of nerve cells with similar functions run in groups or pathways. Some carry sensory information upward to the brain; others run downward from the brain to control the body's movements. A few or many of these pathways may be damaged. Nevertheless, a person can often recover some functions that were lost because of the initial injury.

D. Barnes also explains that damage that occurs to spinal cord axons within the first few hours after injury is complex and it occurs in stages. Normal blood flow is disrupted, causing oxygen deprivation to some of the tissues of the spinal cord. Bleeding into the injured area leads to swelling, which can further compress and damage spinal cord axons. Free radicals are released into the damaged area. These negatively charged ions can damage cell membranes, killing cells that were not injured initially. Macrophages invading the site to clean up debris may also damage uninjured tissue. Non-neuronal cells may divide too often, forming a scar that slows the regrowth of injured nerve cell axons.

86. According to D. Barnes of the National Institute of Health's "Research In the News," early events after a spinal cord injury can lead to other kinds of damage later on. Within weeks or months, cysts may form at the injury site and fill with cerebrospinal fluid. Typically, scar tissue develops around the cysts, creating permanent cavities that can elongate and cause further damage to nerve cells. Also, axons that were not damaged initially often lose their myelin. Over time, these and other events can contribute to more tissue degeneration and a greater loss of function.

87. D. Barnes of the National Institute of Health's "Research In the News" interviewed W. Young from the New York Medical Center Neurosurgery Research Laboratory. He stated that researchers do not know exactly how methylprednisolone, a synthetic steroid, helps injured spinal cord tissue to recover, but they think the drug has at least two effects. One is that methylprednisolone suppresses immune responses throughout the body. This is beneficial because inflammatory responses at the injury site may cause more damage. Second, methylprednisolone may work to block the formation of free radicals. These charged, highly energetic ions can damage membranes of cells that were not initially injured. The overall effect of methylprednisolone for people with spinal cord injuries seems to be protective. It may be best to give methylprednisolone within three hours after a spinal cord injury occurs and patients may benefit from treatment up to eight hours after the injury.

88. According to I. Carrano of the Helen Hayes Hospital, in a complete injury, a patient has no movement or sensation below the level of the injury. With an incomplete injury there may be some sensation and/or movement spared below the level of injury. Depending on the injury location, a patient may be a quadriplegic (if the injury is in the cervical spine) or a paraplegic (if the injury is in the remaining regions). A quadriplegic is affected in all four limbs. A paraplegic is only affected in the legs and trunk.

89. According to the Spinal Cord Injury Association, functional electrical stimulation (FES) artificially induces neural activity in paralyzed, incontinent, or sensory impaired persons. A cochlear implant for the ear has been developed that allows people with sensorineural deafness to hear certain kinds of sounds, and thus interpret speech. Phrenic nerve stimulator implants help people breathe without a respirator. Spinal cord injury males with sexual dysfunction can induce ejaculation using FES. Electrical stimulation can assist paralyzed individuals regain control of lost bladder function or may even restore some function to paralyzed arms and legs.

Chapter 34: Action and Support: The Muscles and Skeleton

OVERVIEW

This chapter discusses the structures and tissues that allow your body to move, have form, and support itself. Muscle tissue is specialized to contract and relax. Contracting and relaxing muscles move your body, move substances through your organs, and allow your heart to beat. The human body gets its support from a bony skeleton. The muscles attached to your skeleton by connective tissue allow it to move. However, movement would not be possible if the skeleton were not articulated, or jointed. This chapter discusses how these structures function together.

1) How Do Muscles Work?

Muscle tissue is specialized according to its function. **Skeletal muscle**, or **striated muscle**, functions to move your skeleton and is usually under voluntary control. **Cardiac muscle** is found only in your heart. Cardiac muscle is stimulated by nerves and hormones, but it initiates its own contractions. **Smooth muscle** is found lining the walls of organs such as the uterus, your stomach, intestines, and esophagus, as well as the walls of your blood vessels. Smooth muscle contractions are involuntary.

Interactions between microfilaments of the proteins **actin** and **myosin** allow cells to move. These interactions are the ancestors of today's muscle movement.

Muscle cells are referred to as **muscle fibers**. Extending into each muscle fiber are indentations of plasma membrane called **T tubules**. Muscle fibers contain many individual contractile subunits called **myofibrils**. Each myofibril is enclosed by a membrane, the **sarcoplasmic reticulum**, which stores high concentrations of calcium ions. The myofibrils are arranged into subunits called **sarcomeres**. Within each sarcomere, **thick filaments** of myosin lie between **thin filaments** of actin. The thin and thick filaments connect by small branches of myosin called **cross-bridges**. Each sarcomere is attached to fibrous protein bands called **Z lines**. When muscles contract, the thin filaments of actin are pulled past the thick filaments of myosin. This is called the *sliding-filament mechanism.*

Muscle contraction is controlled by the nervous system. The point of attachment between a motor neuron and a muscle cell is called a **neuromuscular junction**. A neuromuscular junction is always excitatory and every action potential from a motor neuron causes an action potential in a muscle fiber. Therefore, all the sarcomeres in the muscle fiber will contract. The strength and degree of contraction is regulated by the number of muscle fibers stimulated and the frequency at which it occurs. Motor neurons form synapses with a group of muscle fibers called a **motor unit**.

When an action potential reaches a muscle fiber, it flows down the T-tubules to the sarcoplasmic reticulum. The sarcoplasmic reticulum releases Ca^{++} into the cytoplasm. Calcium binds to an accessory protein (*troponin*) on the thin filament. With Ca^{++} bound to troponin, the binding site for the myosin cross-bridge is open. Myosin binds to actin and, using energy from ATP, the cross-bridges repeatedly bend, release, and reattach farther along. The thin filaments are pulled past the thick filaments, shortening the sarcomere and contracting the muscle.

The heart is composed of cardiac muscle. It, too, contains sarcomeres of thin and thick filaments. Cardiac muscle fibers initiate their own contractions in the pacemaker, the SA node. Gap junctions allow the action potential to spread from one cardiac muscle cell to another, synchronizing the contraction.

Smooth muscle lines the walls of organs such as the uterus, your stomach, intestines, and esophagus, as well as the walls of your blood vessels. Smooth muscle does not have the regular arrangement of sarcomeres that skeletal and cardiac muscles exhibit, they also do not have a sarcoplasmic reticulum. As in cardiac muscle, gap junctions allow the action potential to spread along the muscle, synchronizing the contraction in slow, sustained, or slow, wavelike movements. Contraction of smooth muscle is involuntary and may be stimulated by stretching, hormonal signals, nervous signals, or a combination of signals.

2) What Does the Skeleton Do?

The **skeleton** is a supporting framework for the body. A **hydrostatic skeleton** consists of a fluid-filled sac. This type of skeleton is found in worms, mollusks, and cnidarians. **Exoskeletons** support the body from the outside. Both thin, flexible and thick, rigid exoskeletons exist to support the bodies of arthropods. **Endoskeletons** support the body from the inside and are found in echinoderms and chordates.

The vertebrate skeleton supports your body and protects your internal organs. It provides a framework for muscle attachment that allows you to move. It produces red blood cells, white blood cells, and platelets. It is your body's storage area for calcium and phosphorous. And, since hearing relies on three tiny bones, your skeleton even allows you to hear. Your skeleton is divided into the **axial skeleton** (your body's axis) and the **appendicular skeleton** (your appendages).

3) Which Tissues Compose the Vertebrate Skeleton?

Cartilage and **bone** are the tissues that make up your skeleton. The cells of both tissues are embedded in a protein matrix called **collagen**. Cartilage is found at the ends of bones at your joints. It forms your ears and nose, the structures of your respiratory system, and it forms "shock absorbing" pads in your knees and the **intervertebral discs** between your vertebrae. The cells of cartilage are called **chondrocytes**. Since no blood vessels are found in cartilage, the chondrocytes metabolize very slowly and rely on diffusion to receive nutrients and remove wastes.

The collagen in bone tissue is hardened by calcium phosphate, forming a dense support structure. Bones have a hard outer covering of **compact bone** and a lightweight, porous interior of **spongy bone**. Bone marrow is found in the cavities of spongy bone. Bone tissue has a healthy network of blood vessels. The cells of bone include **osteoblasts**, **osteocytes**, and **osteoclasts**. Osteoblasts build new bone tissue, trapping themselves inside the hardened matrix they secrete. Osteoblasts mature into osteocytes. Osteoclasts dissolve bone. During the process of forming bone tissue (bone remodeling), **osteons** are created. Osteons are concentric layers of bone embedded with osteocytes. In the center of an osteon, a capillary passes, supplying the cells with nutrients. Bone remodeling slows as bone ages, resulting in more fragile bones.

4) How Does the Body Move?

The body moves through the use of **antagonistic muscles**. When one muscle is actively contracted, another is passively extended. Muscles in vertebrates move the bones of the skeleton at points where two bones meet, the **joints**. Muscles attach to bones by connective tissue forming **tendons**. Bones attach to other bones by connective tissue forming **ligaments**. In **hinge joints**, a bone on one side of the joint moves while the other bone maintains a fixed position. The antagonistic muscle pair involved in moving a hinged joint includes a **flexor** and an **extensor**. Each muscle is attached to an immovable bone. This attachment point is called the **origin**. The end attached to the moveable bone is the **insertion**. **Ball-and-socket joints** consist of a rounded end of one bone that fits into a hollow depression of another. This type of joint allows movement in many directions and involves two pairs of muscles oriented at right angles to each other.

KEY TERMS AND CONCEPTS

Fill-In: From the following list of key terms, fill in the blanks in the following statements.

actin
antagonistic muscles
appendicular skeleton
axial skeleton
ball-and-socket joints
cardiac muscle
cartilage
chondrocytes
collagen
compact bone
cross-bridges

endoskeleton
exoskeleton
extensor
flexor
hinge joints
hydrostatic skeleton
insertion
intervertebral discs
joint
ligaments
motor unit

muscle fibers
myofibrils
myosin
neuromuscular junction
origin
osteoblasts
osteoclasts
osteocytes
osteons
osteoporosis
sarcomeres

sarcoplasmic reticulum
skeletal muscle
skeleton
smooth muscle
spongy bone
striated muscle
tendons
thick filaments
thin filaments
T tubules
Z lines

1. _____ is used to move the skeleton. It is also called _____ because of its appearance under the microscope. _____ is found only in the heart, and _____ lines the walls of hollow organs and blood vessels.

2. Movement within cells occurs because of protein microfilaments made of _____. These microfilaments interact with another protein, _____, to change the shape of the cell.

3. Individual muscle cells are called _____, which contain individual contractile subunits, the _____. Within muscle cells, high concentrations of Ca^{++} are stored in fluid of the _____. Deep indentations of muscle cell plasma membrane called _____ extend into the muscle fiber.

4. In myofibrils, precise arrangement of actin and myosin filaments make up _____. Actin molecules are called _____, and myosin molecules are called _____. The actin and myosin filaments interact using connections called _____. Between adjacent sarcomeres, the thin filaments are attached to _____.

5. Motor neurons form synapses with muscle cells at _____, which are always excitatory. However, most motor neurons synapse with many muscle fibers. This group of fibers is a _____.

6. The supporting structure for the body is the _____. A _____ is a fluid-filled sac that is highly flexible. Arthropods, such as insects and crustaceans, are encased in an _____, while humans and other chordates have an _____.

7. The vertebrate skeleton is subdivided into the _____, containing the skull, vertebral column, and rib cage, and the _____, forming the appendages.

8. The cells making up the skeleton are imbedded in a protein matrix of _____. When the skeleton is formed during embryological development, it is composed of _____. The

cells of this connective tissue are called _____. Adult skeletons still rely on this connective tissue to form the larynx, trachea, and bronchi. It also protects vertebrae by serving as shock-absorbing pads, the _____.

9. Bones have a hard outer covering called _____, with lightweight, porous _____ on the inside.

10. The cells of bone include _____, the bone forming cells, _____, mature bone cells, and _____, the bone dissolving cells. Hard bone is composed of tightly packed units, the _____, which are concentric layers of bone embedded with osteocytes.

11. The skeleton moves because of attached muscle pairs. These are _____ in that when one contracts, the other is extended. Muscles are attached to bones by _____. Bones, however, are attached to other bones by _____. When bones move at areas where two bones attach, a _____ is formed.

12. _____, such as the knee and elbow, are movable in only two dimensions. The pair of muscles controlling this type of joint includes a _____ and a _____. The area where muscle attaches to an immovable bone is called the _____. The area where muscle attaches to a movable bone is called the _____. The hip and shoulder form _____ that allow movement in several directions.

KEY TERMS AND DEFINITIONS

actin (ak´-tin): a major muscle protein whose interactions with myosin produce contraction; found in the thin filaments of the muscle fiber; see also *myosin*.

antagonistic muscles: a pair of muscles, one of which contracts and in so doing extends the other; an arrangement that makes possible movement of the skeleton at joints.

appendicular skeleton (ap-pen-dik´-ū-lur): the portion of the skeleton consisting of the bones of the extremities and their attachments to the axial skeleton; the pectoral and pelvic girdles, the arms, legs, hands, and feet.

axial skeleton: the skeleton forming the body axis, including the skull, vertebral column, and rib cage.

ball-and-socket joint: a joint in which the rounded end of one bone fits into a hollow depression in another, as in the hip; allows movement in several directions.

cardiac muscle (kar´-dē-ak): the specialized muscle of the heart, able to initiate its own contraction, independent of the nervous system.

cartilage (kar´-teh-lij): a form of connective tissue that forms portions of the skeleton; consists of chondrocytes and their extracellular secretion of collagen; resembles flexible bone.

chondrocyte (kon´-drō-sīt): a living cell of cartilage. With their extracellular secretions of collagen, chondrocytes form cartilage.

collagen (kol´-uh-jen): a fibrous protein in connective tissue such as bone and cartilage.

compact bone: the hard and strong outer bone; composed of osteons.

cross-bridge: in muscles, an extension of myosin that binds to and pulls on actin to produce muscle contraction.

endoskeleton (en´-dō-skel´-uh-tun): a rigid internal skeleton with flexible joints to allow for movement.

exoskeleton (ex´-ō-skel´-uh-tun): a rigid external skeleton that supports the body, protects the internal organs, and has flexible joints that allow for movement.

extensor: a muscle that straightens a joint.

flexor: a muscle that flexes (decreases the angle of) a joint.

Haversian system (ha-ver´-sē-un): see *osteon*.

hinge joint: a joint at which one bone is moved by muscle and the other bone remains fixed, such as in the knee, elbow, or fingers; allows movement in only two dimensions.

hydrostatic skeleton (hī-drō-stat´-ik): a body type that uses fluid contained in body compartments to provide support and mass against which muscles can contract.

insertion: the site of attachment of a muscle to the relatively movable bone on one side of a joint.

intervertebral disc (in-ter-ver-tē´-brul): a pad of cartilage between two vertebrae that acts as a shock absorber.

joint: a flexible region between two rigid units of an exoskeleton or endoskeleton, allowing for movement between the units.

ligament: a tough connective tissue band connecting two bones.

motor unit: a single motor neuron and all the muscle fibers on which it forms synapses.

muscle fiber: an individual muscle cell.

myofibril (mī-ō-fī´-bril): a cylindrical sub-unit of a muscle cell, consisting of a series of sarcomeres; surrounded by sarcoplasmic reticulum.

myosin (mī´-ō-sin): one of the major proteins of muscle, the interaction of which with the protein actin produces muscle contraction; found in the thick filaments of the muscle fiber; see also *actin*.

neuromuscular junction: the synapse formed between a motor neuron and a muscle fiber.

origin: the site of attachment of a muscle to the relatively stationary bone on one side of a joint.

osteoblast (os´-tē-ō-blast): a cell type that produces bone.

osteoclast (os´-tē-ō-klast): a cell type that dissolves bone.

osteocyte (os´-tē-ō-sīt): a mature bone cell.

osteon: a unit of hard bone consisting of concentric layers of bone matrix, with embedded osteocytes, surrounding a small central canal that contains a capillary.

osteoporosis (os´-tē-ō-por-ō´-sis): a condition in which bones become porous, weak, and easily fractured; most common in elderly women.

sarcomere (sark´-ō-mēr): the unit of contraction of a muscle fiber; a subunit of the myofibril, consisting of actin and myosin filaments and bounded by Z lines.

sarcoplasmic reticulum (sark´-ō-plas´-mik re-tik´-ū-lum): the specialized endoplasmic reticulum in muscle cells; forms interconnected hollow tubes. The sarcoplasmic reticulum stores calcium ions and releases them into the interior of the muscle cell, initiating contraction.

skeletal muscle: the type of muscle that is attached to and moves the skeleton and is under the direct, normally voluntary, control of the nervous system; also called *striated muscle*.

skeleton: a supporting structure for the body, on which muscles act to change the body configuration; may be external or internal.

smooth muscle: the type of muscle that surrounds hollow organs, such as the digestive tract, bladder, and blood vessels; normally not under voluntary control.

spongy bone: porous, lightweight bone tissue in the interior of bones; the location of bone marrow.

striated muscle: see *skeletal muscle*.

tendon: a tough connective tissue band connecting a muscle to a bone.

thick filament: in the sarcomere, a bundle of myosin that interacts with thin filaments, producing muscle contraction.

thin filament: in the sarcomere, a protein strand that interacts with thick filaments, producing muscle contraction; composed primarily of actin, with accessory proteins.

T tubule: a deep infolding of the muscle plasma membrane; conducts the action potential inside a cell.

Z line: a fibrous protein structure to which the thin filaments of skeletal muscle are attached; forms the boundary of a sarcomere.

THINKING THROUGH THE CONCEPTS

True or False: Determine if the statement given is true or false. If it is false, change the underlined word(s) so that the statement reads true.

13. _____   Moving <u>food through the digestive system</u> involves muscle action.

14. _____   Muscles are active only during the <u>extension</u> phase.

15. _____   Striated muscle contractions are under <u>conscious control</u>.

16. _____   Cardiac muscle contractions are under <u>conscious control</u>.

17. _____   Smooth muscle contractions are under <u>conscious control</u>.

18. _____   Actin and myosin work <u>against one another</u> to contract a muscle.

19. _____   <u>One</u> muscle cell may be 35 centimeters long.

20. _____   Actin and myosin filaments are arranged in subunits called <u>motor units</u>.

21. _____   An action potential causes the <u>sarcoplasmic reticulum</u> to release Ca^{++}, allowing muscle fibers to contract.

22. _____   Skeletal muscle is called striated muscle because of the appearance of the <u>thick and thin filaments in sarcomeres</u>.

23. _____   <u>ATP</u> provides the energy for myosin cross bridges to move along the thin filaments, contracting the muscle.

24. _____   An action potential causes the contraction of <u>only one muscle cell</u>.

25. _____   <u>Neuromuscular junctions</u> allow action potentials to travel from one cardiac muscle cell to another, synchronizing their contractions.

26. _____   Bones immobilized in a cast <u>remain strong and high in calcium</u>.

27. _____   Bones produce <u>red blood cells</u>.

28. _____   Cartilage is <u>living tissue</u>.

29. _____ Spongy bone provides sites for muscle attachment.

30. _____ Calcium levels in blood need to remain constant. If the blood level drops, calcium is taken from bone and retained in the blood.

31. _____ Ligaments connect muscle to bone.

32. _____ The insertion point of a muscle attaches to the mobile bone on the far side of a joint.

Matching: Correctly match the following characteristics with the appropriate cell type.

33. _____ bone dissolving cells

34. _____ cartilage cell

35. _____ mature bone cell

36. _____ bone forming cell

37. _____ found in osteons

38. _____ relies on diffusion for nutrients

39. _____ capillaries bring nutrients

40. _____ secrete flexible, elastic matrix

41. _____ dissolve cartilage

42. _____ replace cartilage with bone

43. _____ create channels during bone remodeling

44. _____ fill channels during bone remodeling

45. _____ become osteocytes

Choices:

a. chondrocyte

b. osteoblast

c. osteocyte

d. osteoclast

Identify: Determine whether the following bones are part of the **axial** skeleton or the **appendicular** skeleton.

46. _____ skull

47. _____ humerus

48. _____ rib

49. _____ vertebra

50. _____ patella

51. _____ tarsals

52. _____ mandible

53. _____ sternum

54. _____ femur

55. _____ tibia

56. _____ coccyx

57. _____ ulna

Short answer:

58. List the five functions of the vertebrate skeleton.

_____

_____

_____

_____

_____

59. Use the following terms to explain how muscle contraction occurs. Include the differences between skeletal, cardiac, and smooth muscle in how contractions are induced.

accessory proteins	cardiac muscle	sarcoplasmic reticulum	T tubules
action potential	motor neuron	skeletal muscle	thin filament
calcium ions	myosin	smooth muscle	

_____

_____

_____

_____

_____

_____

_____

_____

_____

60. Which bones are involved in transmitting sound?

_____

APPLYING THE CONCEPTS

These practice questions are intended to sharpen your ability to apply critical thinking and analysis to biological concepts covered in this chapter.

61. Damage to a knee often involves loss of cartilage. A new procedure using chondrocyte transplants is being used to repair joints damaged in accidents (but not due to arthritis). This procedure uses cultured chondrocytes from the injured person and places them at the injury site. Why is such a procedure necessary? Why is it unlikely that damaged cartilage will repair itself, while bone readily heals?

_____

_____

_____

62. Using what you have read about bone formation and bone remodeling, what would you predict to be the effect weightlessness has on astronauts in space with respect to their bone strength?

_____

_____

_____

_____

Use the Case Study and Web sites for this chapter to answer the following questions.

63. What are your long bones? Identify the parts of their "anatomy."

_____

_____

_____

_____

_____

_____

_____

_____

_____

_____

_____

_____

64. What is involved in bone remodeling? How often is a healthy thigh bone (femur) closest to the hip replaced?

_____

_____

_____

_____

_____

65. What is needed in a healthy diet to ensure healthy bones or that damaged bones heal properly?

_____

_____

_____

_____

66. How are osteoclasts involved in getting calcium and phosphorous into the bloodstream?

_____

_____

_____

_____

_____

67. What is involved in a bone graft?

_____

_____

_____

_____

_____

68. A new polymer was created in August of 2000. This polymer may replace the use of metal plates and pins for setting severe fractures. What is so unique about this polymer and how will doctors "set" it?

_____

_____

_____

_____

_____

_____

_____

69. News reports about Electric and Magnetic Fields (EMFs) have focused on emanations from power lines, building wiring, and appliances. They have chronicled the continuing controversy over whether these fields have unhealthy effects, such as perturbed sleep patterns, altered heart rhythms, and cancer. While these risks have grabbed headlines, researchers have begun to use EMFs in medicine. How?

_____

_____

_____

_____

_____

_____

ANSWERS TO EXERCISES

1. Skeletal muscle
   striated muscle
   Cardiac muscle
   smooth muscle
2. actin
   myosin
3. muscle fibers
   myofibrils
   sarcoplasmic reticulum
   T tubules
4. sarcomeres
   thin filaments
   thick filaments
   cross bridges
   Z lines
5. neuromuscular junctions
   motor unit
6. skeleton
   hydrostatic skeleton
   exoskeleton
   endoskeleton
7. axial skeleton
   appendicular skeleton
8. collagen
   cartilage
   chondrocytes
   intervertebral discs
9. compact bone
   spongy bone
10. osteoblasts

osteocytes
osteoclasts
osteons
11. antagonistic muscle
    tendons
    ligaments
    joint
12. Hinge joints
    flexor
    extensor
    origin
    insertion
    ball-and-socket joints
13. true
14. false, contraction
15. true
16. false, nerve and hormone control
17. false, unconscious control
18. false, together
19. true
20. false, sarcomeres
21. true
22. true
23. true
24. false, a motor unit
25. false, gap junctions
26. false, weaken and lose calcium
27. true

28. true
29. false, compact bone
30. true
31. false, tendons
32. true
33. d
34. a
35. c
36. b
37. c
38. a
39. c
40. a
41. d
42. b
43. d
44. b
45. b
46. axial
47. appendicular
48. axial
49. axial
50. appendicular
51. appendicular
52. axial
53. axial
54. appendicular
55. appendicular
56. axial
57. appendicular

58. The skeleton provides a rigid framework that supports the body and protects the organs; allows animals to move; produces red blood cells, white blood cells. and platelets; stores calcium and phosphorous, and allows sensory transduction (transmits sound vibrations)

59. A *skeletal* muscle cell is stimulated by a motor neuron, creating an action potential. The action potential travels to the muscle cell's interior by passing down T tubules. When the action potential reaches the sarcoplasmic reticulum, Ca^{++} are released into the cytoplasm of the muscle cell. The Ca^{++} bind to the accessory proteins of thin filaments. This changes the shape of the accessory proteins so that the myosin binding sites are exposed, contracting sarcomeres. Contraction of *cardiac* muscle occurs when an action potential travels through T tubules, releasing Ca^{++} from the sarcoplasmic reticulum. Ca^{++} enter the cell from the extracellular fluid and from the sarcoplasmic reticulum. Gap junctions connect cardiac muscle cells, enabling contractions to be synchronized. *Smooth* muscle cells do not have sarcoplasmic reticula. Ca^{++}, initiating contraction of smooth muscle cells, flow into the cell from the extracellular fluid. Gap junctions between muscle cells synchronize contractions.

60. The bones of the middle ear, the hammer, anvil, and stirrup, are involved in conducting sounds.

61. Damaged cartilage is unlikely to repair itself because it lacks capillaries. It relies on diffusion to provide the necessary nutrients and to remove wastes. This process is very slow and the metabolism of cartilage

is, therefore, very slow also. Bone, on the other hand, is supplied with a network of capillaries and active cells to heal damaged tissue.

62. Normal stress on bone increases its thickness and its strength. Simply immobilizing a bone in a cast decreases its strength. NASA is studying the effects of weightlessness on astronaut's bone strength. Astronauts experiencing extended periods without gravity putting stress on their bones have experienced bone loss.

63. According to Starnet.esc20.net, the long bones include your arms, legs, feet, and hands. The parts of long bones include: the *diaphysis*, the shaft of the long bone; the *medullary cavity*, the cavity in the diaphysis of long bones, filled with yellow marrow (fat); the *epiphysis*, an end of the long bone, filled with spongy bone and red bone marrow; the *epiphyseal plate*, the area of longitudinal bone growth, composed of cartilage which becomes ossified as growth occurs; the *epiphyseal line*, the remnant of the epiphyseal plate in mature bones; the *periosteum*, the connective tissue membrane which surrounds the diaphysis; the *endosteum*, a connective tissue membrane which lines the inner bone surfaces; and *articular cartilage*, the ends of bones are covered with cartilage, which helps protect them and allows for smoother joint movement.

64. I. N. Gurov at the New York State Univ. at Buffalo explains that bone remodeling involves removing and depositing bone inside and outside the bone structure. This process is performed by osteoblasts and osteoclasts. In healthy young adults, the rates of bone deposits and removal is in balance, but the bone remodeling process is not uniform. For example, the femur near the thigh, is replaced every 5–6 months; its shaft is altered much more slowly.

65. According to I. N. Gurov at the New York State Univ. at Buffalo, for optimal bone deposit, a diet rich in proteins, vitamins C (for collagen synthesis), A, and B_{12}, and several minerals (calcium, phosphorus, magnesium, manganese) is essential.

66. According to I. N. Gurov at the New York State Univ. at Buffalo, osteoclasts are specialized bone cells that use enzymes to digest the matrix of bone. They take calcium from your bone by changing calcium salts into a solution that contains phosphate ions. These ions eventually go into your blood to be used by your body. The control of bone remodeling involves hormones (parathormone) produced by your parathyroid glands.

67. According to the National Library of Medicine's Medline Plus, a bone graft involves surgery to place new bone into spaces between or around broken bone or holes in bone. New bone to be grafted can be taken from your own healthy bone (hip bones or ribs) or from frozen, donated bone. An incision is made over the bone defect and the bone graft is shaped and inserted into and around the defect. The graft is held in place with pins, plates, or screws. Bone grafts are used for fractures with bone loss, repair of bone that has not healed properly, and fusion of joints to prevent movement.

68. The American Chemical Society interviewed A. Burkoth of the Univ. of Colorado at Boulder. She stated that the new polymer is unique because it dissolves from the surface inward, so it retains its strength. It can be designed to dissolve over a period of several days to more than a year, depending on the injury. The degradation rate can be timed so it matches the bone's healing rate. This allows for a gradual transfer of load from the degrading polymer to the healing bone. Another advantage is that it can be molded right on the bone. This makes it easier for surgeons to use. To harden it in place, an intense light is applied, causing cross-links between individual molecules to form and stiffen the polymer.

69. According to J. Raloff of Science News, during the past 20 years, the FDA has approved EMF generators for two medical uses. The devices are frequently used to treat bone fractures that have stopped healing. EMF treatment is used to fuse spinal vertebrae in people with intractable back pain. EMF therapy also helps people with established joint disease, such as advanced osteoarthritis in their knees. Two studies have found that EMFs reduce pain and swelling possibly "by changing the chemistry of the joint." Other studies indicate that EMFs can increase a joint's production of natural anti-inflammatory agents.

Chapter 35: Animal Reproduction

OVERVIEW

This chapter presents the mechanisms by which animals reproduce. These mechanisms are the result of the effects of natural selection on generations of organisms. The chapter, however, focuses on human reproduction. Included is a summary of the reproductive organs and the hormones regulating sexual development and functioning. Using the information learned about the human reproductive system, methods to limit or induce fertility are explored.

1) How Do Animals Reproduce?

When reproduction involves gametes produced through meiosis, it is **sexual reproduction**. The offspring produced will have a genome that combines those of the parents. However, some animals produce offspring by mitotic divisions in an area of their body; this is **asexual reproduction**, and the offspring are genetically identical to the parent. Asexual reproduction may involve **budding**. These animals, such as hydra and sponges, produce a **bud** that is a miniature version of the "parent." The bud will eventually break off and continue independently. Reproduction of new individuals by **regeneration** is rare, but does occur and is actually called **fission**. Annelid and flatworm species may spontaneously break into two or more parts. Each part then regrows the missing body parts (regeneration). Some corals divide in half, lengthwise, producing two smaller individuals. Some species can produce eggs that will develop into adults without being fertilized. These adults are haploid and are produced by **parthenogenesis**. **Fertilization** is the union of egg and sperm, forming a diploid zygote.

Species with both male and female individuals are **dioecious**. Females produce **eggs** and males produce **sperm**. In **monoecious** species, an individual produces both egg and sperm. These individuals are **hermaphrodites**.

If the union of egg and sperm occurs outside the body of the parents, **external fertilization** has occurred. This is known as **spawning**. Animals that spawn must synchronize their release of eggs and sperm. Female mussels and sea stars achieve this by releasing **pheromones** into the water when they release their eggs. The males detect this chemical and release sperm into the area. Other species rely on specialized mating rituals to bring males and females together. Frogs assume a specialized mating posture called **amplexus**. During amplexus, the male mounts the female and prods her side, causing her to release eggs. He then fertilizes the eggs by releasing sperm into the water.

If the union of the egg and sperm occur inside the female's body, **internal fertilization** has occurred. Internal fertilization uses **copulation** behavior and increases the chances that the sperm will reach the egg. Some species deposit a packet of sperm, a **spermatophore**, which the female picks up and inserts into her reproductive cavity. Regardless of the method of getting sperm into the female body, fertilization will only occur if the female has released a mature egg. Mature eggs are released during **ovulation**.

2) How Does the Human Reproductive System Work?

The organs that produce your sex cells are called **gonads**. The gonads in a male are the **testes** which are contained outside his body within a sac called the **scrotum**. Within each testis, sperm are produced by the **seminiferous tubules**. **Interstitial cells** between the tubules produce the hormone *testosterone*. Inside the wall of the seminiferous tubules are the **spermatogonia**, the diploid cells from which sperm arise. Also within the wall of the tubules are **Sertoli cells** which regulate **spermatogenesis**, the development of sperm.

During spermatogenesis, spermatogonia grow and differentiate into **primary spermatocytes**. The primary spermatocytes divide by *meiosis I*, producing two **secondary spermatocytes**. Secondary spermatocytes divide by *meiosis II*, each producing two **spermatids**. The spermatids differentiate into sperm nourished by the sertoli cells. (***Four sperm are produced for every spermatogonium undergoing meiosis.)

The organization of sperm cells is unlike that of any other cell. In the head of the sperm is the DNA and an **acrosome**. The acrosome is a specialized lysosome that contains enzymes to digest the protective layers around the egg. Behind the head of the sperm is the midpiece containing many mitochondria to provide energy for the whiplike movement of the sperm's flagellum (tail). Males do not begin producing sperm until the hypothalamus releases **gonadotropin-releasing hormone (GnRH)** at puberty. GnRH stimulates the anterior pituitary to produce **luteinizing hormone (LH)** and **follicle-stimulating hormone (FSH)**. LH stimulates the interstitial cells to produce **testosterone**. Testosterone and FSH stimulate Sertoli cells and spermatogonia to begin spermatogenesis. Testosterone is also needed for the development of secondary sexual characteristics and to maintain an erection of the **penis** for successful intercourse.

Sperm leave the male body through accessory structures of their reproductive system. The seminiferous tubules merge to form a continuous tube, the **epididymis**, which leads to the **vas deferens**. The vas deferens merges with the **urethra** forming a shared path with the urinary system to the tip of the penis. Sperm leaving the body (ejaculation) are mixed in a fluid called **semen**. Semen is formed from three glands, the **seminal vesicles**, the **prostate gland**, and the **bulbourethral glands**.

The gonads in a female are the **ovaries**. Within the ovaries, eggs are formed through **oogenesis**. Oogenesis begins in the ovaries of a developing **fetus** when the **oogonia** are formed. By the end of the third month of fetal development, all of the oogonia have matured into **primary oocytes**. The primary oocytes begin *meiosis I*, but the process is halted at prophase I. It is not until after puberty that development will continue, and then only a few primary oocytes each month will continue oogenesis.

Surrounding each primary oocyte are accessory cells; together, they form a **follicle**. After puberty, pituitary hormones stimulate a few follicles to continue development. Within each follicle, the first meiotic division is completed, forming one **secondary oocyte** and one **polar body**. The polar body contains chromosomes to be discarded. The accessory cells of the follicle secrete **estrogen**. A mature follicle containing a secondary oocyte and the polar body is released from the ovary into the **uterine tube** during *ovulation*. If fertilization occurs, the secondary oocyte will undergo *meiosis II*. (***One egg is produced for every oogonium undergoing meiosis.) Meanwhile, accessory cells left behind in the ovary enlarge, forming the **corpus luteum**. The corpus luteum secretes both estrogen and the hormone **progesterone**. The secondary oocyte (now commonly referred to as the egg) is conducted into a uterine tube by a current created by ciliated **fimbriae**. If fertilization occurs, it usually takes place in the uterine tube, forming a **zygote**. The zygote is carried through the uterine tube by ciliated cells into the uterus. The inner wall of the uterus is well supplied with blood vessels and forms the **endometrium**, the mother's portion of the **placenta**. The outer wall of the uterus is the muscular **myometrium**, which will contract during delivery. The development of the endometrium is stimulated by estrogen and progesterone secreted by the corpus luteum. If the egg is not fertilized, the corpus luteum will disintegrate, causing estrogen and progesterone levels to decrease. When the hormone levels drop, the endometrium disintegrates and is expelled from the uterus during **menstruation** as part of the **menstrual cycle**. At the outer end of the uterus is a ring of connective tissue forming the **cervix**. On the other side of the cervix is the birth canal, or **vagina**.

(See "A Closer Look: Hormonal Control of the Menstrual Cycle." The menstrual cycle is regulated by interactions of hormones from the hypothalamus, pituitary gland, and the ovary. Cells in the hypothalamus spontaneously release GnRH all the time. GnRH stimulates the secretion of FSH and LH by the anterior pituitary. FSH and LH circulating in the blood stream stimulate the development of several follicles in the ovaries as well as estrogen production by the follicles. One or two follicles complete development over a two-week period. As estrogen production increases, follicle development continues, endometrium growth in the uterus is stimulated, and FSH and LH are released in a surge by the hypothalamus and pituitary gland. The sudden increase in LH stimulates the primary oocyte to resume meiosis I, forming the secondary oocyte and first polar body; it initiates ovulation; and it transforms the follicle cells remaining in the ovary into the corpus luteum. The corpus luteum secretes estrogen and progesterone that turns off the release of FSH and LH, continues endometrial growth in the uterus, and turns off GnRH release. If fertilization occurs, the embryo secretes the hormone **chorionic gonadotropin (CG)**. CG prevents the breakdown of the corpus luteum. Therefore, estrogen and progesterone secretion continues, maintaining the endometrial lining that nourishes the developing embryo. If fertilization does not occur, the corpus luteum breaks down, the amount of estrogen and progesterone in the blood drops, the endometrial lining breaks down and is shed, and the production of GnRH resumes.)

Traditionally, in order for a sperm to fertilize an egg, copulation (intercourse, in humans) must occur. With the influences of psychological and physical stimulation, blood flow into tissue spaces of the penis increases. When the tissues swell, the vessels draining blood from the penis become blocked. As pressure increases, the penis becomes erect, so that it can be inserted into the vagina. Further stimulation causes the muscles surrounding the epididymis, vas deferens, and the urethra to contract, resulting in ejaculation. During an ejaculation, 300–400 million sperm are released. Similarly, psychological and physical stimulation increases blood flow into the vagina and external reproductive tissues. The external reproductive tissues include folds of skin called the **labia** and a rounded projection, the **clitoris**. Since the clitoris develops from the same embryological tissue as the tip of the penis, it becomes erect when blood flow into the area increases. Sperm, released into the vagina following an ejaculation, swim up the female reproductive tract from the vagina, through the cervix, into the uterus, and into the uterine tubes. If an egg has been released in the past 24 hours or so, one sperm may succeed in fertilizing it.

The follicle cells that remain around the egg form the **corona radiata**, a barrier between the egg and sperm. Between the corona radiata and the egg is a second barrier, a jellylike **zona pellucida**. The hundreds of sperm actually reaching the egg release enzymes from their acrosomes, weakening the corona radiata and the zona pellucida, allowing one sperm to wiggle into the egg. The plasma membranes of the sperm and egg fuse and the head of the sperm is drawn into the egg's cytoplasm. As this occurs, the egg releases chemicals into the zona pellucida that reinforce it so that no other sperm can enter. ***The egg now begins meiosis II, finally producing a haploid gamete. The two haploid nuclei fuse forming a diploid cell, the zygote. Couples unable to successfully fertilize an egg are increasingly relying on artificial insemination or in vitro fertilization to help them out.

3) How Can People Limit Fertility?

Fertility can be limited by various methods of **contraception**. The most reliable method, of course, is abstinence. **Sterilization** provides a rather permanent means of contraception. Sterilization in men is achieved through a **vasectomy**. This rather minor operation, performed under local anesthetic, cuts the vas deferens. Sperm are still produced but are not released during ejaculation. A vasectomy has no known physical side effects. Sterilization in women is achieved through a **tubal ligation**. This operation, performed under general anesthetic, cuts the uterine tubes. Ovulation still occurs but sperm do not reach the eggs, nor can eggs reach the uterus. More temporary means of contraception prevent ovulation, prevent the sperm from reaching the egg, or prevent a fertilized egg from implanting in the uterus. **Birth control pills** prevent ovulation because the continuous estrogen and prostaglandin that the pill provides suppresses LH release. Barrier methods prevent sperm from reaching the egg. The **diaphragm** and the **cervical cap** fit securely over the cervix, preventing sperm entry. When used in conjunction with **spermicides**, these devices are very effective with no known side effects. As an alternative barrier method, a male can wear a **condom** over his penis. Female condoms are also now available. These cover the inside of the vagina. Less effective methods of contraception include **withdrawal** of the penis from the vagina before ejaculation and **douching** in an attempt to wash sperm from the vagina before they reach the uterus. Both of these methods are extremely unreliable. Another method that has a high failure rate is the **rhythm method**. This method involves abstinence from intercourse just before, during, and after ovulation. Users of this method often have difficulty determining when ovulation occurs each month or are undisciplined in their habits. An alternative method involves using devices, such as the **intrauterine device (IUD)**, that prevent a fertilized egg from implanting in the uterus. The "morning after pill," by providing a large dose of estrogen, essentially has the same effect. When contraception fails, and an unplanned and unwanted pregnancy results, the pregnancy can be terminated by **abortion**. Abortion procedures typically involve dilation of the cervix followed by suction to remove the fetus and placenta. An alternative to the surgical procedure can be used within the first month of pregnancy. The RU-486 drug (mifepristone), in pill form, was approved by the FDA in 2000 for use in the U.S. RU-486 blocks the action of progesterone, ending the pregnancy.

KEY TERMS AND CONCEPTS

Fill-In: From the following list of key terms fill in the blanks in the following statements.

abortion	estrogen	ovaries	seminiferous tubules
acrosome	external fertilization	oviduct	Sertoli cells
amplexus	fimbriae	ovulation	sexual reproduction
asexual reproduction	fission	parthenogenesis	spawning
budding	follicle	penis	spermatogenesis
bulbourethral gland	gonads	pheromone	spermatogonia
cervix	hermaphrodites	placenta	spermatophore
clitoris	internal fertilization	polar body	testes
contraception	interstitial cells	primary oocytes	urethra
copulation	labia	prostate gland	uterine tube
corona radiata	menstrual cycle	regeneration	uterus
corpus luteum	menstruation	scrotum	vagina
dioecious	monoecious	secondary oocyte	vas deferens
eggs	myometrium	semen	zona pellucida
endometrium	oogenesis	seminal vesicles	zygote
epididymis	oogonia		

1. During _____ haploid gametes are produced by meiosis. During _____, repeated mitosis of cells of some part of the body produces an exact copy of the parent.

2. The process of _____ produces a bud off of the body of the adult that breaks off as a new individual. The ability to regrow lost body parts is called _____. Some species are capable of reproducing by this method. Corals divide lengthwise, producing two smaller copies of the parent coral. This type of asexual reproduction is _____. When haploid egg cells develop into adults, as with the development of male honey bees, _____ has occurred.

3. When species have both male and female individuals, the species is _____. Females produce _____ and males produce _____. When individuals of a species produce both egg and sperm, the species is _____ and the individuals are _____.

4. If the union of the egg and sperm occurs outside the body, _____ has taken place. _____ involves egg and sperm being released into water. The sperm then swim to the eggs. The female may communicate to males that eggs are being released by releasing a _____ into the water. Courtship rituals help ensure that both a male and a female are in the area at the right time.

Frogs include a mating posture called _____ in their mating behavior.

5. If sperm are taken into the female's body, _____ occurs. Typically, this involves _____, in which the male directly deposits the sperm into the female's body. Males of some species may package their sperm as a _____, which the female picks up and places in her reproductive cavity. However sperm are to reach the egg, fertilization can only occur if a mature egg has been produced and _____ has occurred.

6. The paired organs that produce sex cells in mammals are called _____. The male structures that produce sperm are the _____ which are enclosed in the _____ outside the body. Coiled within the testes are the _____.

7. The male hormone testosterone is produced by _____. The diploid cells that give rise to sperm are the _____, while the cells that will nourish the sperm are the _____. The process by which sperm are formed is called _____.

8. The human sperm contains enzymes in the specialized lysosome, the _____ and DNA in the head region. Sperm are released from the male body through the _____ during ejaculation.

9. The seminiferous tubules merge to form the _____, which leads to the _____. This tube finally merges with the _____ of the urinary system.

10. The fluid ejaculated from the penis is called _____ and contains sperm mixed with secretions from the _____, the _____, and the _____.

11. The female structures that produce eggs are the _____. The process by which eggs are formed is called _____. The diploid cells that give rise to the eggs are the _____. At three months of development, a female fetus has none of these precursor cells left. They have all matured into _____, each of which will begin meiosis I.

12. Accessory cells surrounding each primary oocyte make up the _____. After puberty, a few primary oocytes will initiate the completion of meiosis I. One oocyte will complete meiosis I producing one _____ and one _____.

13. The accessory cells of the follicle secrete the hormone _____. After being released from an ovary, a secondary oocyte may complete meiosis II in the _____. Accessory cells left behind in the ovary form the _____, which produces both estrogen and progesterone.

14. Each oviduct, also known as the _____, has an open end, almost surrounding the ovary, with ciliated "fingers" called _____.

15. After the egg is fertilized, the _____ is swept along the oviduct to the _____. This organ has a two-layered wall. The inner layer is dense with blood vessels and

forms the _____, the mother's contribution to the _____. If fertilization of the egg does not occur, this layer is shed during _____, as part of the _____.

16. The outer layer of the uterus is muscular. This _____ contracts strongly during birth. The outer end of the uterus contains a ring of connective tissue, the _____. On the other side of the ring of connective tissue is the _____, which serves as the birth canal.

17. Sexual excitement in a female increases blood flow to external reproductive tissues. These include external folds of skin, the _____, and a rounded projection, the _____, which becomes erect.

18. Follicle cells surrounding the ovulated "egg" develop into the _____. Between this layer and the egg is a clear jellylike layer called the _____. Both layers protect the egg.

19. The prevention of pregnancy involves various methods of _____. When these methods fail, a pregnancy may be terminated by _____.

KEY TERMS AND DEFINITIONS

abortion: the procedure for terminating pregnancy; the cervix is dilated, and the embryo and placenta are removed.

acquired immune deficiency syndrome (AIDS): an infectious disease caused by the human immunodeficiency virus (HIV); attacks and destroys T cells, thus weakening the immune system.

acrosome (ak´-rō-sōm): a vesicle, located at the tip of an animal sperm, that contains enzymes needed to dissolve protective layers around the egg.

amplexus (am-plek´-sus): in amphibians, a form of external fertilization in which the male holds the female during spawning and releases his sperm directly onto her eggs.

asexual reproduction: reproduction that does not involve the fusion of haploid sex cells. The parent body may divide and new parts regenerate, or a new, smaller individual may form as an attachment to the parent, to drop off when complete.

birth control pill: a temporary contraceptive method that prevents ovulation by providing a continuing supply of estrogen and progesterone, which in turn suppresses LH release; must be taken daily, normally for 21 days of each menstrual cycle.

bud: in animals, a small copy of an adult that develops on the body of the parent and eventually breaks off and becomes independent; in plants, an embryonic shoot, normally very short and consisting of an apical meristem with several leaf primordia.

budding: asexual reproduction by the growth of a miniature copy, or bud, of the adult animal on the body of the parent. The bud breaks off to begin independent existence.

bulbourethral gland (bul-bō-ū-rē´-thrul): in male mammals, a gland that secretes a basic, mucus-containing fluid that forms part of the semen.

cervical cap: a birth control device consisting of a rubber cap that fits over the cervix, preventing sperm form entering the uterus.

cervix (ser´-viks): a ring of connective tissue at the outer end of the uterus, leading into the vagina.

chlamydia (kla-mid´-ē-uh): a sexually transmitted disease, caused by a bacterium, that causes inflammation of the urethra in males and of the urethra and cervix in females.

chorionic gonadotropin (CG): a hormone, secreted by the chorion (one of the fetal membranes), that maintains the integrity of the corpus luteum during early pregnancy.

clitoris: an external structure of the female reproductive system; composed of erectile tissue; a sensitive point of stimulation during sexual response.

condom: a contraceptive sheath worn over the penis during intercourse to prevent sperm from being deposited in the vagina.

contraception: the prevention of pregnancy.

copulation: reproductive behavior in which the penis of the male is inserted into the body of the female, where it releases sperm.

corona radiata (kuh-rō´-nuh rā-dē-a´-tuh): the layer of cells surrounding an egg after ovulation.

corpus luteum (kor´-pus loo´-tē-um): in the mammalian ovary, a structure that is derived from the follicle after ovulation and that secretes the hormones estrogen and progesterone.

crab lice: an arthropod parasite that can infest humans; can be transmitted by sexual contact.

diaphragm (dī´-uh-fram): in a reproductive sense, a contraceptive rubber cap that fits snugly over the cervix, preventing the sperm from entering the uterus and thereby preventing pregnancy.

dioecious (dī-ē´-shus): pertaining to organisms in which male and female gametes are produced by separate individuals rather than in the same individual.

douching: washing the vagina; after intercourse, an attempt to wash sperm out of the vagina before they enter the uterus; an ineffective contraceptive method.

egg: the haploid female gamete, normally large and nonmotile, containing food reserves for the developing embryo.

embryo: in animals, the stages of development that begin with the fertilization of the egg cell and end with hatching or birth; in mammals in particular, the early stages in which the developing animal does not yet resemble the adult of the species.

endometrium (en-dō-mē´-trē-um): the nutritive inner lining of the uterus.

epididymis (e-pi-di´-di-mus): a series of tubes that connect with and receive sperm from the seminiferous tubules of the testis.

estrogen: in vertebrates, a female sex hormone, produced by follicle cells of the ovary, that stimulates follicle development, oogenesis, the development of secondary sex characteristics, and growth of the uterine lining.

external fertilization: the union of sperm and egg outside the body of either parent.

fertilization: the fusion of male and female haploid gametes, forming a zygote.

fetus: the later stages of mammalian embryonic development (after the second month for humans), when the developing animal has come to resemble the adult of the species.

fimbria (**fim´-brē-uh**; pl., **fimbriae**): in female mammals, the ciliated, fingerlike projections of the oviduct that sweep the ovulated egg from the ovary into the oviduct.

fission: asexual reproduction by dividing the body into two smaller, complete organisms.

follicle: in the ovary of female mammals, the oocyte and its surrounding accessory cells.

follicle-stimulating hormone (**FSH**): a hormone, produced by the anterior pituitary, that stimulates spermatogenesis in males and the development of the follicle in females.

genital herpes: a sexually transmitted disease, caused by a virus, that can cause painful blisters on the genitals and surrounding skin.

genital warts: a sexually transmitted disease, caused by a virus, that forms growths or bumps on the external genitalia, in or around the vagina or anus, or on the cervix in females or penis, scrotum, groin, or thigh in males.

gonad: an organ where reproductive cells are formed; in males, the testes, and in females, the ovaries.

gonadotropin-releasing hormone (**GnRH**): a hormone produced by the neurosecretory cells of the hypothalamus, which stimulates cells in the anterior pituitary to release FSH and LH. GnRH is involved in the menstrual cycle and in spermatogenesis.

gonorrhea (**gon-uh-rē´-uh**): a sexually transmitted bacterial infection of the reproductive organs; if untreated, can result in sterility.

hermaphrodite (**her-maf´-ruh-dīt´**): an organism that possesses both male and female sexual organs.

internal fertilization: the union of sperm and egg inside the body of the female.

interstitial cell (**in-ter-sti´-shul**): in the vertebrate testis, a testosterone-producing cell located between the seminiferous tubules.

intrauterine device (**IUD**): a small copper or plastic loop, squiggle, or shield that is inserted in the uterus; a contraceptive method that works by irritating the uterine lining so that it cannot receive the embryo.

labium (pl., **labia**): one of a pair of folds of skin of the external structures of the mammalian female reproductive system.

luteinizing hormone (**LH**): a hormone, produced by the anterior pituitary, that stimulates testosterone production in males and the development of the follicle, ovulation, and the production of the corpus luteum in females.

menstrual cycle: in human females, a complex 28-day cycle during which hormonal interactions among the hypothalamus, pituitary gland, and ovary coordinate ovulation and the preparation of the uterus to receive and nourish the fertilized egg. If pregnancy does not occur, the uterine lining is shed during menstruation.

menstruation: in human females, the monthly discharge of uterine tissue and blood from the uterus.

monoecious (**mon-ē´-shus**): pertaining to organisms in which male and female gametes are produced in the same individual.

myometrium (**mī-ō-mē´-trē-um**): the muscular outer layer of the uterus.

oogenesis: the process by which egg cells are formed.

oogonium (**ō-ō-gō´-nē-um**; pl., **oogonia**): in female animals, a diploid cell that gives rise to a primary oocyte.

ovary: in animals, the gonad of females; in flowering plants, a structure at the base of the carpel that contains one or more ovules and develops into the fruit.

oviduct: see *uterine tube.*

ovulation: the release of a secondary oocyte, ready to be fertilized, from the ovary.

parthenogenesis (**par-the-nō-jen´uh-sis**): a specialization of sexual reproduction, in which a haploid egg undergoes development without fertilization.

penis: an external structure of the male reproductive and urinary systems; serves to deposit sperm into the female reproductive system and delivers urine to the exterior.

pheromone (**fer´-uh-mōn**): a chemical produced by an organism that alters the behavior or physiological state of another member of the same species.

placenta (**pluh-sen´-tuh**): in mammals, a structure formed by a complex interweaving of the uterine lining and the embryonic membranes, especially the chorion; functions in gas, nutrient, and waste exchange between embryonic and maternal circulatory systems and secretes hormones.

polar body: in oogenesis, a small cell, containing a nucleus but virtually no cytoplasm, produced by the first meiotic division of the primary oocyte.

primary oocyte (**ō´-ō-sīt**): a diploid cell, derived from the oogonium by growth and differentiation that undergoes meiosis, producing the egg.

primary spermatocyte (**sper-ma´-tō-sīt**): a diploid cell, derived from the spermatogonium by growth and differentiation, that undergoes meiosis, producing four sperm.

progesterone (**prō-ge´-ster-ōn**): a hormone, produced by the corpus luteum, that promotes the development of the uterine lining in females.

prostate gland (**pros´-tāt**): a gland that produces part of the fluid component of semen; prostatic fluid is basic and contains a chemical that activates sperm movement.

regeneration: the regrowth of a body part after loss or damage; also, asexual reproduction by means of the regrowth of an entire body from a fragment.

rhythm method: a contraceptive method involving abstinence from intercourse during ovulation.

scrotum (**skrō´-tum**): the pouch of skin containing the testes of male mammals.

secondary oocyte (**ō´-ō-sīt**): a large haploid cell derived from the first meiotic division of the diploid primary oocyte.

secondary spermatocyte (**sper-ma´-tō-sīt**): a large haploid cell derived by meiosis I from the diploid primary spermatocyte.

semen: the sperm-containing fluid produced by the male reproductive tract.

seminal vesicle: in male mammals, a gland that produces a basic, fructose-containing fluid that forms part of the semen.

seminiferous tubule (**sem-i-ni´-fer-us**): in the vertebrate testis, a series of tubes in which sperm are produced.

Sertoli cell: in the seminiferous tubule, a large cell that regulates spermatogenesis and nourishes the developing sperm.

sexually transmitted disease (**STD**): a disease that is passed from person to person by sexual contact.

sexual reproduction: a form of reproduction in which genetic material from two parent organisms is combined in the offspring; normally, two haploid gametes fuse to form a diploid zygote.

spawning: a method of external fertilization in which male and female parents shed gametes into the water, and sperm must swim through the water to reach the eggs.

sperm: the haploid male gamete, normally small, motile, and containing little cytoplasm.

spermatid: a haploid cell derived from the secondary spermatocyte by meiosis II; differentiates into the mature sperm.

spermatogenesis: the process by which sperm cells form.

spermatogonium (pl., **spermatogonia**): a diploid cell, lining the walls of the seminiferous tubules, that gives rise to a primary spermatocyte.

spermatophore: in a variation on internal fertilization in some animals, the males package their sperm in a container that can be inserted into the female reproductive tract.

spermicide: a sperm-killing chemical; used for contraceptive purposes.

sterilization: a generally permanent method of contraception in which the pathways through which the sperm (vas deferens) or egg (oviducts) must travel are interrupted; the most common form of contraception.

syphilis (si´-ful-is): a sexually transmitted bacterial infection of the reproductive organs; if untreated, can damage the nervous and circulatory systems.

testis (pl., **testes**): the gonad of male mammals.

testosterone: in vertebrates, a hormone produced by the interstitial cells of the testis; stimulates spermatogenesis and the development of male secondary sex characteristics.

trichomoniasis (trik-ō-mō-nī´-uh-sis): a sexually transmitted disease, caused by the protist *Trichomonas*, that causes inflammation of the mucous membranes that line the urinary tract and genitals.

tubal ligation: a surgical procedure in which a woman's oviducts are cut so that the egg cannot reach the uterus, making her infertile.

urethra (ū-rē´-thruh): the tube leading from the urinary bladder to the outside of the body; in males, the urethra also receives sperm from the vas deferens and conducts both sperm and urine (at different times) to the tip of the penis.

uterine tube: also called the oviduct, the tube leading out of the ovary to the uterus, into which the secondary oocyte (egg cell) is released.

uterus: in female mammals, the part of the reproductive tract that houses the embryo during pregnancy.

vagina: the passageway leading from the outside of a female mammal's body to the cervix of the uterus.

vas deferens (vaz de´-fer-enz): the tube connecting the epididymis of the testis with the urethra.

vasectomy: a surgical procedure in which a man's vas deferens are cut, preventing sperm from reaching the penis during ejaculation, thereby making him infertile.

withdrawal: the removal of the penis from the vagina just before ejaculation in an attempt to avoid pregnancy; an ineffective contraceptive method.

zona pellucida (pel-oo´-si-duh): a clear, noncellular layer between the corona radiata and the egg.

zygote (zī´-gōt): in sexual reproduction, a diploid cell (the fertilized egg) formed by the fusion of two haploid gametes.

THINKING THROUGH THE CONCEPTS

True or False: Determine if the statement given is true or false. If it is false, change the underlined word(s) so that the statement reads true.

20. _____ Meiosis occurs in asexual reproduction.

21. _____ A flatworm reproducing by regeneration would produce identical "offspring."

22. _____ Male honey bees develop from unfertilized eggs.

23. _____ Some species of fish do not produce males; all members are female.

24. _____ Due to the evolution of sexual reproduction, genetic variation occurs, enabling the action of natural selection to take place.

25. _____ Species with individuals of both sexes are monoecious.

26. _____ In order for external fertilization to be successful, egg and sperm must be released at the same time and in the same area.

27. _____ Copulation occurs during external fertilization.

28. _____ The testes are outside the body, keeping them 4° C warmer than the body's core temperature.

29. _____ Primary spermatocytes are diploid.

30. _____ After meiosis I, spermatids are formed.

31. _____ Oogenesis begins at puberty.

32. _____ Follicle-stimulating hormone functions in males with testosterone to initiate spermatogenesis.

33. _____ A method of contraception that could be developed would block FSH production, resulting in an infertile male who would not be impotent.

34. _____ In males, the reproductive tract merges with the urinary tract.

35. _____ The egg is, in actuality, a primary oocyte.

36. _____ Fertilization occurs in the vagina.

37. _____ Every month the uterus prepares for implantation.

38. _____ The presence of a <u>corpus luteum</u> maintains the endometrium.

39. _____ The embryo secretes CG which <u>maintains the pregnancy</u>.

40. _____ <u>External</u> fertilization increases the chances for transmission of disease.

41. _____ Birth control pills prevent s<u>perm from uniting with an egg</u>.

42. _____ Douching after intercourse is an <u>effective</u> method of contraception.

43. _____ Condom use not only prevents pregnancy, but it also <u>reduces</u> disease transmission.

44. _____ Successful use of the <u>diaphragm</u> requires diligent observations of changes in body temperature and mucus production from the cervix.

45. _____ <u>Condoms</u> are available for women.

Identify: Using the terms provided, label the structures in the diagrams below.

acrosome   DNA in head   midpiece/mitochondria   zona pellucida
corona radiata   flagellum (tail)   secondary oocyte

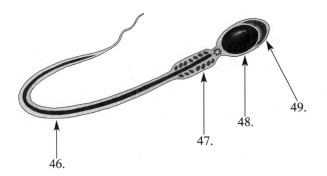
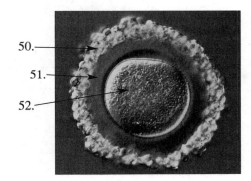

46. _____   50. _____

47. _____   51. _____

48. _____   52. _____

49. _____

Identify: Using the terms provided, label the structures in the diagrams below.

cervix   ovary   scrotum   urethra   vagina
epididymis   penis   seminal vesicle uterine tube   vas deferens
fimbriae   prostate gland testis   uterus

53. _____   60. _____
54. _____   61. _____
55. _____   62. _____
56. _____   63. _____
57. _____   64. _____
58. _____   65. _____
59. _____   66. _____

Matching: Birth control methods.

67. _____ IUD   Choices:

68. _____ condom   a. sterilization

69. _____ cervical cap   b. barrier method

70. _____ birth control pill   c. prevent ovulation

71. _____ tubal ligation   d. prevent implantation

72. _____ spermicides alone   e. ineffective

73. _____ douching

74. _____ vasectomy

75. _____ withdrawal

76. _____ rhythm method

Identify: Determine whether these sexually transmitted diseases are caused by a **bacterium**, **virus**, **protistan**, or an **arthropod**.

77. _____ gonorrhea

78. _____ AIDS

79. _____ trichomoniasis

80. _____ genital herpes

81. _____ syphilis

82. _____ crab lice

83. _____ chlamydia

Short answer:

84. Trace the path of spermatogenesis from spermatogonium to mature sperm. Include where in the pathway meiosis I and meiosis II occur.

Spermatogonium → _____ → _____ → _____ →

_____ → _____ → _____

85. What is the purpose of the polar bodies?

_____

_____

_____

_____

86. Complete the following table by identifying the function of the given hormone and whether or not it is active in **males**, **females**, or **both**.

Hormone	Function	Active in males, females, or both
Gonadotropin -releasing hormone (GnRH)		
Lutenizing hormone (LH)		
Follicle -stimulating hormone (FSH)		
Testosterone		
Estrogen		
Progesterone		
Chorionic gonadotropin (CG)		

APPLYING THE CONCEPTS

These practice questions are intended to sharpen your ability to apply critical thinking and analysis to the biological concepts covered in this chapter.

87. If a couple is unsuccessful in becoming pregnant, a physician may suggest a sperm count be conducted. Why would the number of sperm produced affect the success of attempts to become pregnant?

_____

_____

_____

_____

_____

_____

Use the Case Study and the Web sites for this chapter to answer the following questions.

88. What is a marsupium? Do all marsupials have a marsupium?

_____

_____

_____

_____

_____

89. What is embryonic diapause? What is its possible function?

_____

_____

_____

_____

_____

_____

_____

90. Monotremes have been classified as an order within the class Mammalia. However, they share several
    characteristics with reptiles. What reproductive structure(s) do they share with reptiles? With mammals?

_____

_____

_____

91. Do monotremes lay their eggs in a nest as reptiles do?

_____

_____

_____

_____

92. Reproduction in the nine-banded armadillo is marked by two distinct and apparently unrelated
    phenomena: delayed implantation and the phenomenon of specific polyembryony, which results in the
    normal formation of identical quadruplets. What would be the function of delayed implantation?

_____

_____

_____

_____

93. When do Emperor penguins mate? How do the males help with the incubation of the egg?

_____

_____

_____

_____

ANSWERS TO EXERCISES

1. sexual reproduction
   asexual reproduction
2. budding
   regeneration
   fission
   parthenogenesis
3. dioecious
   eggs
   sperm
   monoecious
   hermaphrodites
4. external fertilization
   Spawning
   pheromone
   amplexus
5. internal fertilization
   copulation
   spermatophore
   ovulation
6. gonads
   testes
   scrotum
   seminiferous tubules
7. interstitial cells
   spermatogonia
   Sertoli cells
   spermatogenesis
8. acrosome
   penis
9. epididymis
   vas deferens
   urethra
10. semen
    seminal vesicles
    prostate gland
    bulbourethral gland
11. ovaries
    oogenesis

    oogonia
    primary oocytes
12. follicle
    secondary oocyte
    polar body
13. estrogen
    oviduct
    corpus luteum
14. uterine tube
    fimbriae
15. zygote
    uterus
    endometrium
    placenta
    menstruation
    menstrual cycle
16. myometrium
    cervix
    vagina
17. labia
    clitoris
18. corona radiata
    zona pellucida
19. contraception
    abortion
20. false, sexual
21. true
22. true
23. true
24. true
25. false, dioecious
26. true
27. false, internal fertilization
28. false, cooler
29. true
30. false, meiosis II
31. false, spermatogenesis
32. true

33. true
34. true
35. false, secondary oocyte
36. false, oviduct (fallopian tube)
37. true
38. true
39. true
40. false, internal fertilization
41. false, ovulation
42. false, very ineffective
43. true
44. false, rhythm method
45. true
46. flagellum (tail)
47. midpiece/mitochondria
48. DNA in head.
49. acrosome
50. corona radiata
51. zona pellucida
52. secondary oocyte
53. testis
54. epididymis
55. vas deferens
56. seminal vesicle
57. prostate gland
58. urethra
59. penis
60. scrotum
61. vagina
62. fimbriae
63. ovary
64. uterine tube
65. uterus
66. cervix
67. d
68. b
69. b
70. c

71. a
72. e
73. e
74. a
75. e

76. e
77. bacterium
78. virus
79. protistan

80. virus
81. bacterium
82. arthropod
83. bacterium

84. Spermatogonium → primary spermatocyte → meiosis I → secondary spermatocyte → meiosis II → spermatid → sperm

85. Polar bodies serve to hold discarded chromosomes. Meiosis produces haploid cells. Since the egg requires as much of the cytoplasm of the dividing cells as possible and "extra" chromosomes must be discarded to produce the haploid state, polar bodies are the result.

86.

Hormone	Function	Active in males, females, or both
Gonadotropin-releasing hormone (GnRH)	stimulates anterior pituitary to produce LH and FSH	both
Lutenizing hormone (LH)	stimulates interstitial cells to produce testosterone; initiates follicle development; surge triggers meiosis I to continue, ovulation, development of corpus luteum	both
Follicle-stimulating hormone (FSH)	stimulates Sertoli cells and spermatagonia, causes spermatogenesis; initiates follicle development	both
Testosterone	stimulates Sertoli cells and spermatagonia, causes spermatogenesis, triggers development of secondary sex characteristics, sexual drive, required for intercourse	males
Estrogen	stimulates endometrium production; decrease triggers menstruation; increase causes increase in LH and FSH	females
Progesterone	maintains endometrium production; decrease triggers menstruation and increases production of GnRH	females
Chorionic gonadotropin (CG)	prevents breakdown of corpus luteum	females

87. The number of sperm produced is important since a large number of sperm are needed to digest the corona radiata and the zona pellucida around the egg. If an insufficient number of sperm is produced to break down the barriers, no sperm can enter the egg to fertilize it.

88. According to the people at Pawprint, most female marsupials have a pouch in which they carry their newborns. Called a marsupium, this pouch is often considered the defining characteristic of a marsupial. But, some marsupials, such as South American opossums, don't have a pouch. Others develop pouches only during the reproductive season. And, in a few cases, the marsupium is simply a protective fold of skin covering the mammae, not a "true" pouch.

89. According to the people at Pawprint, many marsupials, including kangaroos, wallabies and gliders, have a unique ability known as embryonic diapause. Following the birth of one litter into the "pouch," many female marsupials can rebreed almost immediately. When this occurs, the development of the newly fertilized egg stops at approximately 100 cells. The embryo remains in this state of diapause until the first litter is nearing the end of its pouch life, dies, or is abandoned for some reason. Growth of the embryo will then resume.

90. D. Blomstrom of the Geobopological Survey reports that both monotremes and reptiles have a cloaca and lay shelled eggs that are hatched outside the body of the mother. Both monotremes and mammals are covered with hair and nurse their young with milk secreted by specialized glands.

91. According to the Pelican Lagoon Research Centre, on Kangaroo Island, So. Australia, monotremes do not lay their eggs in a nest as reptiles do. Instead, the female lays one single egg directly into a pouch about 24 days after mating. The egg is about the size and texture of a small green grape. The pouch of an echidna is different than a kangaroo pouch. It is formed by pulling two longitudinal muscles on either side of the stomach together. Both males and females can form a pouch. When the mother has a young, the pouch is fleshier because she is lactating.

92. According to the Texas Parks and Wildlife, Nongame and Urban Program, delayed implantation may, in part, account for the successful invasion of the armadillo into temperate regions. Without this characteristic of the reproductive cycle, the young would be born at the beginning of winter, when their chance of survival would be greatly reduced.

93. J. Eisman at the Univ. of Michigan, Museum of Zoology reports that reproduction occurs in the dead of winter, in the middle of the polar night. Emperor pairs gather near a solid iceberg to each lay a single egg. There are no preparations and no nest. To keep the egg off the ice, the male Emperor places it on his feet, holds it between his legs, and protects it with a fold of skin covered with feathers at the base of his abdomen.

Chapter 36: Animal Development

OVERVIEW

This chapter presents the processes that occur after a zygote has been produced. As the development of an embryo progresses, the cells differentiate so that the specialized body functions can take place. This chapter presents information relating to the indirect development of animals that undergo metamorphosis; however, it provides more detail relating to the direct development humans and other mammals follow. Information concerning the changes that occur in the mother's body through pregnancy is also presented.

1) How Do Indirect and Direct Development Differ?

An organism in its early stages of development is an embryo. A developing embryo is nourished by a protein and lipid rich yolk. The amount of yolk present in an egg is determined by the way in which an animal develops. **Indirect development** occurs in animals that hatch from an egg in a form different from that of the adult. The maturation process involves drastic changes of the form of the animal. Amphibians and invertebrates, such as echinoderms and insects, undergo indirect development. The eggs that are laid by these animals are numerous, but they contain only a small amount of **yolk**. Since the amount of yolk is small, the embryo within an egg develops rapidly into a sexually immature **larva**. The larva will eventually undergo **metamorphosis** and become a sexually mature adult.

Direct development occurs in animals that are born resembling miniature adults. The maturation process does not involve drastic changes in body form. These young may be provided large amounts of nourishing yolk within their eggs or they may be nourished within the mother's body before birth. Since both development strategies demand a large input of energy from the mother, a small number of offspring is produced.

The successful transition from aquatic to terrestrial life occurred in animals after the **amniotic egg** evolved in reptiles. Reptiles, birds, and mammals produce amniotic eggs consisting of four **extraembryonic membranes**. The **chorion** is the outermost membrane and conducts gas exchange through the shell. The **amnion** is a membrane that encloses the embryo in a watery environment. The **allantois** surrounds wastes, and the **yolk sac** stores food for the developing embryo. Although the mammalian egg does not produce much yolk, the yolk sac membrane still exists.

2) How Does Animal Development Proceed?

Animal development begins with a series of mitotic divisions called **cleavage**. Although the number of cells increases during cleavage, the size of the overall structure does not increase. As daughter cells are produced, gene-regulating substances are distributed among them. As cleavage continues, a solid ball of cells about the same size as the zygote, called a **morula**, forms. The morula progresses to the development of a hollow ball of cells called a **blastula**. Further divisions of the blastula result in the formation of the **blastopore**. During the formation of the blastopore, cell movement, called **gastrulation**, occurs. Three embryonic tissue layers form during gastrulation. The **endoderm** forms from the cells lining the blastopore. This eventually becomes the digestive tract. The **ectoderm** forms from the cells lining the outside of the gastrula. These cells give rise to the epidermis and the nervous system. Cells that move to the area between the endoderm and ectoderm are called the **mesoderm**. Mesoderm cells develop into muscles, the skeleton, including the **notochord**, and the circulatory system. The resulting embryo is specifically called a **gastrula**.

Through a process called induction, chemicals produced by cells influence the development of other cells. **Organogenesis**, the process of organ development, is controlled by induction. Such signals may include "survival" signals or "death" signals. Some cells will die unless they receive a survival signal from surrounding cells, while other cells will live unless they receive a death signal from surrounding cells. In this way, motor neurons synapse with muscle cells properly and separate fingers and toes form in humans. Development does not stop at birth, however. The development of mature reproductive organs, for example, occurs at a genetically controlled age. The genes regulating this development may be triggered by environmental or social cues.

3) How is Development Controlled?

Development is the process an organism goes through from a fertilized egg to maturity to death. During this process, cells that begin as identical cells from the zygote become specialized in form and function. The specialization that occurs is called **differentiation**. Remember that all cells contain the same genes. But, different cells use different genes. So, differentiation is directed by the genes that are used or activated, transcribed, and translated in a particular cell. The genes that are used are controlled by regulatory molecules. Regulatory molecules bind to a gene, either blocking or promoting transcription. In some animals, the regulatory molecules are concentrated in the cytoplasm of the egg. When the fertilized egg divides, the daughter cells that are produced receive different regulatory molecules. The specialization pathway that each daughter cell will take is determined by which regulatory molecules it received.

In vertebrates, cells differentiate because of chemical signals received from other body cells during gastrulation. The signals may be nutrients, hormones, or neurotransmitters. This process is called **induction**. In an experiment in the 1920's, researchers took cells from the dorsal lip of the blastopore of amphibian embryos. They transplanted the dorsal lip cells to another embryo (a host). The transplanted cells caused (induced) the host to produce a second embryo. This told researchers that the dorsal lip cells control the fate of surrounding cells. In the control experiment, non-dorsal lip cells were transplanted. These transplanted cells differentiated according to the area into which they were transplanted. This gave researchers more information about induction. In fact, research is ongoing in this area, including research about the chemicals that direct cell migration during development and the chemical gradients of *morphogens* that cause body parts to form properly.

4) How Do Humans Develop?

Human development reflects our evolutionary ancestry. The fertilized egg undergoes cleavage as it travels through the uterine tubes to the uterus. It is a **blastocyst**, rather than a blastula, that **implants** into the uterine wall. The blastocyst has a thick **inner cell mass** on one side of its hollow ball structure. The inner cell mass will develop into the embryo and three of the extraembryonic membranes. The thin outer wall becomes the fourth extraembryonic membrane, the embryo's portion of the placenta, called the chorion. The innercell mass grows and separates, forming two fluid-filled sacs. The double cell layer between the sacs is the **embryonic disc**. One of the sacs is enclosed by the amnion and contains the amniotic fluid. The amnion grows around the embryo so that it is maintained within the watery environment all animal embryos need. The second sac is the yolk sac, but it contains no yolk.

As gastrulation begins, the layers of the embryonic disc separate slightly, forming the **primitive streak**. This corresponds to the blastopore in other animals. The mesoderm of the embryo forms when cells migrate through the primitive streak to the interior of the embryo. The cell layer above the primitive streak forms the ectoderm. The cell layer below the primitive streak forms the endoderm. As the embryo grows, the endoderm forms a tube that later becomes the digestive tract. The notochord, formed from mesoderm, causes the ectoderm to form a groove. The groove closes over, creating the **neural tube**, the predecessor to the brain and spinal cord. Continued development generates a beating heart and rapid growth of the brain. By the second month, all of the major organs have developed, including the gonads. Sex hormones are secreted by the testes or ovaries, which influence the development of embryonic organs and certain regions of the brain. The embryo is called a **fetus** at the end of the second month.

The development of the **placenta** begins as the embryo burrows into the endometrium. The outer cells of the embryo, the chorion sends fingerlike projections, called **chorionic villi**, into the endometrium, intricately linking the two. The placenta secretes estrogen to stimulate the growth of the uterus and mammary glands, and progesterone to stimulate the mammary glands and prevent premature uterine contractions. The placenta also regulates the exchange of substances between the mother and the fetus, although their blood never actually mixes. The membranes of the fetal capillaries and the chorionic villi are very selective with respect to the substances that are exchanged; however, some disease-causing organisms and damaging chemicals can pass through.

The next seven months of fetal development involves, for the most part, growth of the already-formed structures. The brain and spinal cord grow and the fetus begins to respond to stimuli. The respiratory, digestive, and urinary tracts enlarge and begin functioning. During the last two months, the fetus positions itself head downward, resting against the cervix in preparation for birth.

Birth of the fetus is initiated by stretching of the uterus and by hormones from both the fetus and the mother causing **labor**, a positive feedback reaction. The contractions of the uterine smooth muscle are triggered by the stretching caused by the growing fetus. Toward the end of development, the fetus produces steroid hormones that increase estrogen and prostaglandin production by the uterus and placenta. The increased hormone production, in combination with the stretching, increases the contractions of the uterus. As the baby's head pushes against the cervix, it dilates. This stretching signals the hypothalamus to release oxytocin. With oxytocin and prostaglandins circulating in the blood, the uterus contracts even more intensely, pushing the baby from the birth canal (the vagina).

At birth, the baby suddenly must breath in oxygen and exhale carbon dioxide, it must attempt to regulate its body temperature, and it must rely on suckling to receive nourishment. Suddenly, it is out of the safe environment of the uterus and out in the world. Soon after the baby's birth, contractions begin again to expel the placenta, the afterbirth. The umbilical cord produces prostaglandins that cause muscles around fetal umbilical vessels to contract, shutting off blood supply between the mother and the baby. The baby's circulatory system takes over.

During the pregnancy, changes in the mother's breasts have occurred to prepare for nursing the baby. Increased amounts of estrogen and progesterone in her blood stimulate her **mammary glands** to grow and develop the capacity to produce milk. **Lactation**, the production of milk, is promoted by the hormone prolactin. The substance first available to a nursing infant is **colostrum**. Colostrum is high in protein and antibodies from the mother. Mature milk, high in fats and lactose and lower in protein, gradually replaces the colostrum.

As cells age they function less efficiently. They may be programmed to divide a certain number of times and then die. Or, a cell's longevity may depend on the length of *telomers* on its chromosomes or its ability to repair damage done to its DNA. Normally, cells die and as cells die, the organism ages. Cancer cells, however, are not normal. They have not only lost the mechanism controlling division, but they also have an indefinite life span.

KEY TERMS AND CONCEPTS

Fill-In: From the following list of key terms, fill in the blanks in the following statements.

allantois	ectoderm	larva
amnion	embryo	mesoderm
amniote egg	embryonic disc	metamorphosis
blastocyst	endoderm	morula
blastopore	extraembryonic membrane	neural tube
blastula	fetal alcohol syndrome (FAS)	notochords
chorion	fetus	organogenesis
chorionic villus	gastrula	placenta
cleavage	gastrulation	primitive streak
development	indirect development	yolk
differentiation	induction	yolk sac
direct development	inner cell mass	

1. _____ is the process of an organism developing from a fertilized egg, through adulthood, to death. Throughout this process cells specialize by _____.

2. Animal development begins with an egg containing _____, rich in protein and lipids needed by the _____.

3. Animals going through _____ experience drastic changes from a sexually immature _____ to a sexually mature adult. A caterpillar will undergo _____ to become a butterfly. Animals progressing through _____ are born as miniature versions of the adult.

4. Evolution of the _____ enabled animals to live away from standing water. This structure contains four _____. The _____ allows gas exchange through the shell. The _____ surrounds the embryo, maintaining it in a watery environment. The _____ isolates wastes from the embryo. The _____ stores food for the developing embryo.

5. A series of mitotic divisions initiated after the zygote forms is called _____. After several division cycles, a solid ball of cells forms, called the _____. With continued divisions, a cavity develops, and this hollow ball of cells is called a _____.

6. An indentation forms in the blastula, called the _____. Three embryonic tissues form as cells migrate through the enlarging indentation. Cells lining the indentation develop the _____, eventually becoming the digestive tract. Cells to the outside form the _____, eventually becoming the epidermis and the nervous system. Cells that migrate between the two layers form the _____, eventually becoming the muscles and skeleton, including the _____.

7. The movement of the cells is called _____ and the structure that is formed is the _____.

8. Chemical messages received from other cells often determine the fate of cell development. This developmental process is _____. The formation of organs from rearrangement of cells is _____.

9. During human development, instead of a blastula forming, a _____ forms. The thick _____ will become the _____, and the thin outer wall will become the chorion.

10. The cells destined to become the embryo grow and split so that two fluid-filled sacs are separated by a double layer of cells, the _____. The double layer of cells splits apart slightly, forming the _____ (blastopore in other animals).

11. During the third week of development, the _____ is generated, the precursor of the brain and spinal cord. At the end of the second month, the embryo is called a _____.

12. The chorion from the embryo penetrates the endometrium of the uterus using _____. The chorion, interacting with the endometrium, generates the _____, which is selective as to which substances pass into the fetus's bloodstream. Not all harmful substances are blocked however. For instance, alcohol readily passes from the mother's bloodstream into that of the fetus, often strongly affecting the fetus. Alcoholic mothers (perhaps even mothers who drink occasionally) give birth to children displaying _____.

KEY TERMS AND DEFINITIONS

allantois (al-an-tō´-is): one of the embryonic membranes of reptiles, birds, and mammals; in reptiles and birds, serves as a waste-storage organ; in mammals, forms most of the umbilical cord.

amnion (am´-nē-on): one of the embryonic membranes of reptiles, birds, and mammals; encloses a fluid-filled cavity that envelops the embryo.

amniote egg (am-nē-ōt´): the egg of reptiles and birds; contains an amnion that encloses the embryo in a watery environment, allowing the egg to be laid on dry land.

blastocyst (blas´-tō-sist): an early stage of human embryonic development, consisting of a hollow ball of cells, enclosing a mass of cells attached to its inner surface, which becomes the embryo.

blastopore: the site at which a blastula indents to form a gastrula.

blastula (blas´-tū-luh): in animals, the embryonic stage attained at the end of cleavage, in which the embryo normally consists of a hollow ball with a wall one or several cell layers thick.

chorion (kor´-ē-on): the outermost embryonic membrane in reptiles, birds, and mammals; in birds and reptiles, functions mostly in gas exchange; in mammals, forms most of the embryonic part of the placenta.

chorionic villus (kor-ē-on-ik; pl., chorionic villi): in mammalian embryos, a fingerlike projection of the chorion that penetrates the uterine lining and forms the embryonic portion of the placenta.

cleavage: the early cell divisions of embryos, in which little or no growth occurs between divisions; reduces the cell size and distributes gene-regulating substances to the newly formed cell.

colostrum (kō-los´-trum): a yellowish fluid, high in protein and containing antibodies, that is produced by the mammary glands before milk secretion begins.

development: the process by which an organism proceeds from fertilized egg through adulthood to eventual death.

differentiation: the process whereby relatively unspecialized cells, especially of embryos, become specialized into particular tissue types.

direct development: a developmental pathway in which the offspring is born as a miniature version of the adult and does not radically change in body form as it grows and matures.

ectoderm (ek´-tō-derm): the outermost embryonic tissue layer, which gives rise to structures such as hair, the epidermis of the skin, and the nervous system.

embryonic disc: in human embryonic development, the flat, two-layered group of cells that separates the amniotic cavity from the yolk sac.

embryonic stem cell: a cell derived from an early embryo that is capable of differentiating into any of the adult cell types.

endoderm (en´-dō-derm): the innermost embryonic tissue layer, which gives rise to structures such as the lining of the digestive and respiratory tracts.

extraembryonic membrane: in the embryonic development of reptiles, birds, and mammals, either the chorion, amnion, allantois, or yolk sac; functions in gas exchange, provision of the watery environment needed for development, waste storage, and storage of the yolk, respectively.

fetal alcohol syndrome (FAS): a cluster of symptoms, including retardation and physical abnormalities, that occur in infants born to mothers who consumed large amounts of alcoholic beverages during pregnancy.

fetus: the later stages of mammalian embryonic development (after the second month for humans), when the developing animal has come to resemble the adult of the species.

gastrula (gas´-troo-luh): in animal development, a three-layered embryo with ectoderm, mesoderm, and endoderm cell layers. The endoderm layer normally encloses the primitive gut.

gastrulation (gas-troo-la´-shun): the process whereby a blastula develops into a gastrula, including the formation of endoderm, ectoderm, and mesoderm.

implantation: the process whereby the early embryo embeds itself within the lining of the uterus.

indirect development: a developmental pathway in which an offspring goes through radical changes in body form as it matures.

induction: the process by which a group of cells causes other cells to differentiate into a specific tissue type.

inner cell mass: in human embryonic development, the cluster of cells, on one side of the blastocyst, that will develop into the embryo.

labor: a series of contractions of the uterus that result in birth.

lactation: the secretion of milk from the mammary glands.

larva (lar´-vuh): an immature form of an organism with indirect development prior to metamorphosis into its adult form; includes the caterpillars of moths and butterflies and the maggots of flies.

mammary gland (mam´-uh-rē): a milk-producing gland used by female mammals to nourish their young.

mesoderm (mēz´-ō-derm): the middle embryonic tissue layer, lying between the endoderm and ectoderm, and normally the last to develop; gives rise to structures such as muscle and skeleton.

metamorphosis (met-a-mor´-fō-sis): in animals with indirect development, a radical change in body form from larva to sexually mature adult, as seen in amphibians (tadpole to frog) and insects (caterpillar to butterfly).

morula (mor´-ū-luh): in animals, an embryonic stage during cleavage, when the embryo consists of a solid ball of cells.

neural tube: a structure, derived from ectoderm during early embryonic development, that later becomes the brain and spinal cord.

notochord (nōt´-ō-kord): a stiff but somewhat flexible, supportive rod found in all members of the phylum Chordata at some stage of development.

organogenesis (or-gan-ō-jen´-uh-sis): the process by which the layers of the gastrula (endoderm, ectoderm, mesoderm) rearrange into organs.

placenta (pluh-sen´-tuh): in mammals, a structure formed by a complex interweaving of the uterine lining and the embryonic membranes, especially the chorion; functions in gas, nutrient, and waste exchange between embryonic and maternal circulatory systems and secretes hormones.

primitive streak: in reptiles, birds, and mammals, the region of the ectoderm of the two-layered embryonic disc through which cells migrate, forming mesoderm.

yolk: protein-rich or lipid-rich substances contained in eggs that provide food for the developing embryo.

yolk sac: one of the embryonic membranes of reptilian, bird, and mammalian embryos; in birds and reptiles, a membrane surrounding the yolk in the egg; in mammals, forms part of the umbilical cord and the digestive tract but is empty.

THINKING THROUGH THE CONCEPTS

True or False: Determine if the statement given is true or false. If it is false, change the underlined word(s) so that the statement reads true.

13. _____   Cells specialize because underlined unnecessary genes are lost.

14. _____   Chemical messages from other cells change how a cell differentiates.

15. _____   Animals whose young will develop indirectly produce eggs with a large amount of yolk.

16. _____   Larvae are sexually mature.

17. _____   Human eggs contain a large amount of yolk.

18. _____   During cleavage, embryonic cells do not grow.

19. _____   The presence of a large yolk may actually hinder cleavage.

20. _____   Gastrulation involves the migration of cells, forming three layers.

21. _____   Cells transplanted to a new region of the embryo differentiate according to the area into which they were transferred.

22. _____   Distinct fingers develop because webbing cells receive a "death signal" from other cells.

23. _____   Cell differentiation and development stops after birth.

24. _____   Cell death is preprogrammed.

25. _____   Long-lived cells can divide damaged DNA.

26. _____ The neural tube is the precursor structure to the <u>digestive tract</u>.

27. _____ The <u>X chromosome</u> determines if the gonads will develop into ovaries or testes.

28. _____ Testosterone and estrogen affect certain areas of the <u>brain</u>.

29. _____ The <u>amnion</u> and the endometrium create the placenta.

30. _____ The kidneys, in a fetus, <u>are functional</u> before birth.

31. _____ Most drugs <u>can pass</u> through the placenta to the fetus.

32. _____ Women who smoke are <u>more likely</u> to have a miscarriage than women who do not smoke.

Matching: Adult tissue types and the embryonic layers from which they developed.

33. _____ epidermis of skin   Choices:

34. _____ muscle   a. endoderm

35. _____ skeleton   b. mesoderm

36. _____ digestive tract   c. ectoderm

37. _____ nervous system

38. _____ circulatory system

Identify: **Can** or **cannot** the following substances cross the placental barrier from the mother's circulatory system to that of the fetus?

39. _____ alcohol   45. _____ syphilis

40. _____ most cells   46. _____ oxygen

41. _____ nicotine   47. _____ large proteins

42. _____ carbon monoxide   48. _____ HIV

43. _____ Thalidomide   49. _____ German measles

44. _____ aspirin   50. _____ nutrients

Short answer:

51. The cells undergoing cleavage do not increase in size. Why not?

_____

_____

_____

_____

52. Place the following terms in the proper developmental order.

blastula   zygote   fetus   embryo   gastrula   morula

_____

53. Complete the following table comparing the extraembryonic membranes produced in reptilian and human eggs.

Extraembryonic membrane	Structure and location in reptilian egg	Function in reptilian egg	Structure and location in human egg	Function in human egg
Chorion				
Amnion				
Allantois				
Yolk sac				

APPLYING THE CONCEPTS

These practice questions are intended to sharpen your ability to apply critical thinking and analysis to the biological concepts covered in this chapter.

54. Identify the two ways cells initiate differentiation.

_____

_____

55. Compare indirect development with direct development. Identify two types of animals that develop indirectly and two that develop directly.

_____

_____

_____

_____

_____

_____

_____

_____

_____

_____

56. Explain how the birth process is controlled by a positive feedback mechanism.

_____

_____

_____

_____

_____

_____

_____

Use the Case Study and the Web sites for this chapter to answer the following questions.

57. Warnings from the Surgeon General about the risks of drinking while pregnant are placed on all bottles of beer, wine, or liquor. A primary risk is that the infant will be born with Fetal Alcohol Syndrome (FAS). What are the symptoms of FAS?

_____

_____

_____

_____

_____

_____

_____

_____

_____

_____

58. Why does alcohol damage a fetus?

_____

_____

_____

_____

59. How much alcohol is safe for a woman to drink while pregnant?

_____

_____

60. What does the term "teratogenic" mean?

_____

61. What are the three ways that drugs can impact a pregnancy?

_____
_____
_____

62. When is a fetus most vulnerable to harm from drugs?

_____
_____

63. What is Accutane? How risky is its use during pregnancy?

_____
_____
_____
_____
_____
_____
_____
_____
_____
_____

64. Alcohol and drugs are not the only substances that threaten the health of a developing embryo/fetus.
    Women with occupational exposure to organic solvents also risk harm to the developing child. What is
    the possible damage to the fetus?

_____
_____
_____
_____

ANSWERS TO EXERCISES

1. Development
   differentiation
2. yolk
   embryo
3. indirect development
   larva
   metamorphosis
   direct development
4. amniote egg
   extraembryonic membranes

   chorion
   amnion
   allantois
   yolk sac
5. cleavage
   morula
   blastula
6. blastopore
   endoderm
   ectoderm

   mesoderm
   notochord
7. gastrulation
   gastrula
8. induction
   organogenesis
9. blastocyst
   inner cell mass
10. embryo
    embryonic disc

primitive streak
11. neural tube
    fetus
12. chorionic villi
    placenta
    Fetal Alcohol
    Syndrome (FAS)
13. false, because of
    gene-regulating
    substances
14. true
15. false, small amount
    of yolk

16. false, sexually
    immature
17. false, no yolk
18. true
19. true
20. true
21. true
22. true
23. false, continues
24. true
25. false, repair
26. false, brain and
    spinal cord

27. false, Y chromosome
28. true
29. false, chorion
30. true
31. true
32. true
33. c
34. b
35. b
36. a
37. c
38. b

39. can
40. cannot
41. can
42. can
43. can
44. can
45. can
46. can
47. cannot
48. can
49 can
50. can

51. The zygote is a relatively large cell, containing a great deal of cytoplasm. In order for body cells to be an appropriate size for efficient functioning, the first divisions during development produce smaller and smaller cells with less cytoplasm.

52. zygote → morula → blastula → gastrula → embryo → fetus

53.

Extraembryonic membrane	Structure and location in reptilian egg	Function in reptilian egg	Structure and location in human egg	Function in human egg
Chorion	lines inside of shell	regulates exchange of O_2, CO_2, H_2O	forms placenta with endometrium	regulates exchange of gases, nutrients, and wastes
Amnion	surrounds embryo	encloses embryo in watery environment	surrounds embryo	encloses embryo in watery environment
Allantois	surrounds wastes, connects to embyro's circulation	stores wastes	provides umbilical cord vessels	carries blood between embryo and placenta
Yolk sac	surrounds yolk	yolk nourishes embryo; becomes digestive tract	empty	forms digestive tract

54. Cells differentiate as directed by genes that are turned on by gene-regulating substances in the cytoplasm or by chemicals released by other cells.

55. Indirect development occurs in animals that hatch from an egg in a form different from that of the adult. The maturation process involves drastic form changes in the animal. Amphibians and invertebrates, such as echinoderms and insects, undergo indirect development. Eggs that are laid by these animals are numerous but contain a small amount of yolk. Since the amount of yolk is small, the embryo within the egg rapidly develops into a sexually immature larva. The larva eventually undergoes metamorphosis and become a sexually mature adult. Direct development occurs in animals that are born resembling miniature adults such as reptiles, birds, and mammals. The maturation process does not involve drastic body form changes. The young may be provided with large amounts of nourishing yolk in their eggs or they may be nourished in the mother's body before birth. Both developmental strategies demand a large input of energy from the mother, so a small number of offspring is produced.

56. A positive feedback reaction is defined as a situation in which a change initiates events that tend to amplify the original change. Birth of the fetus is initiated by stretching of the uterus and by hormones from both the fetus and the mother, causing labor. The contractions of the uterine smooth muscle are triggered by the stretching caused by the growing fetus. Toward the end of development, the fetus

produces steroid hormones that increase estrogen and prostaglandin production by the uterus and placenta. The increased hormone production, in combination with the stretching, increases the contractions of the uterus. As the baby's head pushes against the cervix, it dilates. This stretching signals the hypothalamus to release oxytocin. With oxytocin and prostaglandins circulating in the blood, the uterus contracts even more intensely, pushing the baby from the birth canal (the vagina). The changes that occur in the uterus cause further changes to occur until the process reaches a climax, the birth.

57. According to The Nemours Foundation's Kidshealth.com, FAS is a pattern of physical, developmental, and functional abnormalities in a child resulting from a woman's drinking alcohol during pregnancy. Characteristics of children with FAS include: low birth weight, small head circumference, failure to thrive, developmental delay, organ dysfunction, facial abnormalities, including smaller eye openings, flattened cheekbones, an indistinct or underdeveloped groove between the nose and the upper lip, epilepsy, poor coordination or fine motor skills, poor socialization skills (such as difficulty building and maintaining friendships and relating to groups), lack of imagination or curiosity, learning difficulties (including poor memory, inability to understand concepts such as time and money, poor language comprehension, and poor problem-solving skills), and behavioral problems (including hyperactivity, inability to concentrate, social withdrawal, stubbornness, impulsiveness, and anxiety).

58. According to the March of Dimes, when a pregnant woman drinks, alcohol passes quickly across the placenta to her fetus. The unborn baby's immature body breaks down alcohol much more slowly than an adult's. As a result, the alcohol level of the fetus's blood may be higher and can remain elevated longer than in the mother's blood.

59. According to the March of Dimes, no level of drinking has been proven safe. A woman should stop drinking immediately if she even suspects she could be pregnant and abstain from all alcohol if attempting to become pregnant.

60. According to the Victorian Department of Human Services' Australian Drug Foundation, drugs that can cause birth defects are said to be "teratogenic."

61. According to the Victorian Department of Human Services' Australian Drug Foundation, the three ways that drugs can impact a pregnancy include: damage to the baby by interfering with normal development, damage to the placenta, compromising the baby's life support, and initiation of contractions.

62. According to the Victorian Department of Human Services' Australian Drug Foundation, the greatest danger to the baby is between days 17 and 57, when the major systems of the body are developing, often before a women realizes she is pregnant.

63. According to the March of Dimes, accutane is a prescription medication used to treat severe nodular (cystic) acne that has not responded to other treatments. Accutane is a retinoid, which is related to vitamin A. There is an extremely high risk of fetal malformations if a woman becomes pregnant while taking Accutane, even if she is taking a small amount of the drug for a short period of time. Birth defects associated with Accutane include: hydrocephaly (enlargement of the fluid-filled spaces in the brain), microcephaly (small head), mental retardation, ear and eye abnormalities, cleft lip and palate and other facial abnormalities, and heart defects. There also is a high risk of miscarriage. Accutane can cause birth defects in the early weeks after conception when a woman often does not know she is pregnant. Babies without obvious malformations may have mental retardation or learning disabilities.

64. S. Khattak et. al. investigated the impact of exposure to organic solvents during pregnancy. Occupational exposure to organic solvents during pregnancy is associated with an increased risk of major fetal malformations. This risk appears to be increased among women who report symptoms associated with organic solvent exposure. More of these exposed women had a previous miscarriage while working with organic solvents than did control women.

Chapter 37: Animal Behavior

OVERVIEW

This chapter considers various aspects of behavior. The authors cover instinctive (innate) and learned behaviors, as well as the mechanisms of communication, competitive behaviors within species, cooperative behavior in various animal societies, and the study of human behavior.

1) How Do Innate and Learned Behaviors Differ?

Behavior is any observable activity in a living animal. The study of animal behavior is called **ethology**. **Innate** (often referred to as instinctive) behaviors can be performed in reasonably complete form even the first time an animal encounters a specific stimulus. But, the animal has to be at the right age and motivational state. Scientists can determine innate behaviors by depriving an animal the chance to learn the behavior. For instance, red squirrels raised in a bare cage could not have learned to bury nuts. After eating a few, however, they will carry one "away" and make burying motions around it. Human newborns display innate behavior when they turn their head toward the side their mouth is touched in order to suckle.

The capacity to make changes in behavior based on experience is called **learning**. A common form of learning is **habituation**. Habituation occurs when there is a decline in response to a repeated stimulus. The ability to habituate prevents an animal from wasting energy and attention on unimportant stimuli. Humans habituate to many stimuli such as background noises in the city. Additionally, in a noisy room, you can still have a conversation with a friend. **Trial-and-error** learning is a more complex form of learning. This allows behavior to be modified by experience. A frog captures a bee only once and then it learns that they, and others that look similar, hurt. Researchers experiment with trial-and-error learning by using **operant conditioning**. Operant conditioning involves receiving a reward (or punishment) after doing a certain behavior. If an animal hits a lever and food appears, the animal learns to hit the lever to gain food. On the other hand, some animals can solve a problem without prior experience, this is called **insight learning**. Learning, however, may be governed by innate controls. **Imprinting** is a special form of learning. It occurs when learning is rigidly programmed to occur only at a certain critical period of development, the **sensitive period**. For example, many birds learn to follow the animal or object that they most frequently encounter (usually the mother) during the period from about 13 to 16 hours after hatching.

Traditionally, innate behaviors were seen as rigidly controlled by genetic factors, and learned behaviors were seen as determined exclusively by an animal's environment. However, all behavior arises out of an interaction between genes and the environment.

2) How Do Animals Communicate?

Communication is the production of a signal by one organism that causes another organism to change its behavior in a way beneficial to one or both. Communication may occur by visual displays, by sound, by an emitted chemical, or by touch. Most communication takes place between members of the same species to resolve conflicts over food, space, or mates with minimal damage.

For animals with well-developed eyes, visual communication is most effective over short distances. Visual signals may be *active* or *passive*. When an animal makes a specific movement or posture, it is sending an active visual signal. By contrast, the mere size, shape, or color of an animal serves as a passive visual signal. Active and passive signals can be combined, as is seen in many courtship rituals. Visual signals can be advantageous: They are instantaneous and quiet and therefore unlikely to alert distant predators, and they can be rapidly revised. However, relying on visual signals can also be a disadvantage. Visual signals are ineffective in darkness (unless you are a female firefly) or dense vegetation and they are limited to close distances.

Communication by sound is effective over longer distances. Sound can be transmitted instantaneously through darkness, dense forests, and water. Sound can travel over long distances and can be quickly varied to convey rapidly changing messages. Sound is one of the most important methods of communication.

Chemical messages persist longer but are hard to vary. **Pheromones** are chemical substances produced by an individual that influence the behavior of others of its own species. Chemicals may carry messages over long distances for long periods of time, take very little energy to produce, and may not be detected by other species. Fewer messages are communicated with chemicals than with sight or sound, and pheromone signals lack the diversity and gradation of auditory or visual signals. Pheromones act in two ways. Some pheromones cause an immediate, observable behavior in the animal that detects them; others stimulate a physiological change in the animal that detects them. This change is usually a change in its reproductive state. For example, queen bees produce a primer pheromone called **queen substance**, which is eaten by hivemates and prevents other females from becoming sexually mature. Sex pheromones have been used in pesticides to attract Japanese beetles or Gypsy moths. These pheromones either disrupt their ability to mate or they lure them into traps.

Communication by touch helps establish social bonds among group members. Touch often involves kissing, nuzzling, patting, petting, and grooming. Touch can also influence human well-being.

3) How Do Animals Compete for Resources?

Competition for resources underlies many forms of social interaction. Aggressive behavior helps secure resources such as food, space, or mates. **Aggression** is antagonistic but usually harmless behavior, typically involving symbolic displays or rituals between members of the same species. During aggressive displays, animals may exhibit weapons such as claws or fangs and often behave in ways that make them appear larger; they may stand upright, fluff their feathers or fur, and extend their ears or fins, and they may emit intimidating sounds. Actual fighting tends to be a last resort.

Dominance hierarchies reduce aggressive interactions. In a dominance hierarchy, each animal establishes a rank that determines its access to resources. Although aggressive encounters occur frequently while the hierarchy is being established, disputes are minimized after each animal learns its place; after the "pecking order" has been established.

Animals may defend territories that contain resources. **Territoriality** is the defense of an area where important resources, such as places to mate and raise young, feed, or store food, are located. Territorial behavior is most commonly seen in adult males. Territories are normally defended against members of the same species who compete most directly for the resources being protected. Once a territory is established through aggressive interaction, relative peace prevails as boundaries are recognized and respected ("good fences make good neighbors"). For males of many species, successful territorial defense has a direct impact on reproductive success. Territories are advertised through sight, sound, and smell.

4) How Do Animals Find Mates?

Before mating can occur, animals must identify one another as belonging to the same species, as members of the opposite sex, and as being sexually receptive. This commonly involves social interactions between potential mates called courtship behavior. Chemical signals (pheromones) bring potential mates together then vocal and visual signals are used to encode gender, species, and individual quality. The intertwined functions of sex recognition and species recognition, advertisement of individual quality, and synchronization of reproductive behavior commonly require a complex series of signals, both active and passive, by both sexes.

5) What Kind of Societies Do Animals Form?

Social behavior within animal societies requires cooperative interactions and such group living has both disadvantages and advantages. Some <u>disadvantages</u> are (1) increased competition for limited resources, (2) increased risk of infection, (3) increased risk that offspring will be killed, and (4) increased risk of being detected by predators. The <u>benefits</u> include (1) increased protection from predators; (2) increased hunting efficiency; (3) advantages from division of labor; and (4) an increased likelihood of finding mates. If a species has evolved social behavior, then the "pros" must outweigh the "cons." Some types of animals cooperate on the basis of changing needs; for instance, coyotes are solitary when food is abundant, but they hunt in packs when food is scarce.

Some cooperative societies are based on behavior that seems to sacrifice the individual for the good of the group; for example, worker ants die in defense of their nest. These behaviors characterize altruistic behavior that decreases the reproductive success of one individual to the benefit of another. Although **altruism** seems inconsistent with the concept of "survival of the fittest," the individual is actually promoting the survival of some of its own genes by maximizing the survival of its close relatives. This phenomenon is called **kin selection**. Kin selection is exhibited in the extreme by the evolution of the complex societies of honeybees and naked mole rats.

The most difficult of all animal societies to explain are those of the bees, ants, and termites. In these societies, most individuals never breed but labor slavishly to feed and protect the offspring of a different individual. Honeybees form complex societies, with one reproductive queen, male drones, and sterile female workers. The sterile female workers bring food to other bees, construct and clean the hive, and forage for pollen and nectar. They communicate the location of these resources to other workers by performing a **waggle dance**. Pheromones play a major role in regulating the lives of social insects. (Remember the *queen substance* mentioned earlier.)

With the exception of human society, vertebrate societies are not as complex as insect societies. Bullhead catfish illustrate a simple vertebrate society based almost entirely on pheromones. Naked mole rats form a complex and fascinating mammalian society not unlike that of an ant or termite colony.

Many animals have been observed at play which has fascinated and puzzled many researchers. The elements of play are: (1) actions seem to lack any clear immediate function, (2) actions are abandoned in favor of feeding, courtship, or escaping from danger, (3) actions are seen more frequently in young than in adults, (4) actions typically involve movements borrowed from other behaviors such as stalking, (5) actions use a great deal of energy, and (6) actions are potentially dangerous. Play activities may enable young animals to practice behaviors they will need as adults.

6) Can Biology Explain Human Behavior?

Because humans are animals whose behaviors have an evolutionary history, the techniques of ethology can be used to understand human behavior. But, human ethologists cannot experiment with people as animal ethologists do with animals. Instead, researchers have tried to determine genetic components of human behavior.

The behavior of newborns is thought to have a large innate component. The rhythmic movement of a baby's head searching for its mother's breast is an innate behavior. Sucking, smiling, walking movements when the body is supported, and grasping with the hands and feet are also innate actions.

Researchers have tried to identify innate behaviors by studying diverse human cultures, including those that are more isolated. Simple behaviors, such as the facial expressions for pleasure and rage and the "eye flash" greeting, are universal and therefore may be innate. In addition, people may respond to pheromones. For example, studies indicate that pheromones may synchronize the menstrual cycles among female roommates and close friends.

Genetic components of behavior have been revealed through comparisons of identical and fraternal twins. When twins are reared together, environmental influences on behavior are very similar for each member of the pair, so behavioral differences between twins must have a large genetic component. If a particular behavior is heavily influenced by genetic factors, we would expect to find similar expression of that behavior in identical twins but not in fraternal twins. In fact, in tests that measure many aspects of personality, identical twins are about twice as similar in personality as are fraternal twins. Such studies have shown significant genetic components for traits such as activity level, alcoholism, anxiety, sociability, intelligence, dominance, and even political attitudes. The field of human behavioral genetics is controversial, because it challenges the long-held belief that environment is the most important determinant of human behavior.

KEY TERMS AND CONCEPTS

Fill-In: From the following list of terms, fill in the blanks below.

altruism	dominance hierarchy	imprinting	pheromone
behavior	ethology	innate	sensitive period
communication	habituation	learning	territoriality

1. _____ is any observable activity in a living animal. Specifically, _____ is the study of behavior.

2. _____ behaviors can be performed in reasonably complete form even the first time an animal of the right age and motivational state encounters a particular stimulus. However, the ability to make changes in behavior based on experience is called _____.

3. _____ is a decline in response to a repeated stimulus. And, _____ is a special form of learning that is rigidly programmed to occur only at a certain critical period of development, called the _____.

4. _____ is the production of a sound, movement, or chemical by one organism that causes another organism to change its behavior. Specifically, a _____ is a chemical substance produced by an individual that influences the behavior of others of its species.

5. In a _____, each animal establishes a rank that determines its access to resources while reducing aggressive interactions.

6. _____is the defense of an area where important resources, such as places to mate and raise young, feed, or store food, are located.

7. _____ is behavior that decreases the reproductive success of one individual to the benefit of another.

Key Terms and Definitions

aggression: antagonistic behavior, normally among members of the same species, often resulting from competition for resources.

altruism: a type of behavior that may decrease the reproductive success of the individual performing it but benefits that of other individuals.

behavior: any observable activity of a living animal.

communication: the act of producing a signal that causes another animal, normally of the same species, to change its behavior in a way that is beneficial to one or both participants.

dominance hierarchy: a social arrangement in which a group of animals, usually through aggressive interactions, establishes a rank for some or all of the group members that determines access to resources.

ethology (ē-thol´-ō-jē): the study of animal behavior in natural or near-natural conditions.

habituation (heh-bich-oo-ā´-shun): simple learning characterized by a decline in response to a harmless, repeated stimulus.

imprinting: the process by which an animal forms an association with another animal or object in the environment during a sensitive period of development.

innate (in-āt´): inborn; instinctive; determined by the genetic makeup of the individual.

insight learning: a complex form of learning that requires the manipulation of mental concepts to arrive at adaptive behavior.

kin selection: a type of natural selection that favors a certain allele because it increases the survival or reproductive success of relatives that bear the same allele.

learning: an adaptive change in behavior as a result of experience.

operant conditioning: a laboratory training procedure in which an animal learns to perform a response (such as pressing a lever) through reward or punishment.

pheromone (fer´-uh-mōn): a chemical produced by an organism that alters the behavior or physiological state of another member of the same species.

queen substance: a chemical, produced by a queen bee, that can act as both a primer and a pheromone.

sensitive period: the particular stage in an animal's life during which it imprints.

territoriality: the defense of an area in which important resources are located.

trial-and-error learning: a process by which adaptive responses are learned through rewards or punishments provided by the environment.

waggle dance: a symbolic form of communication used by honeybee foragers to communicate the location of a food source to their hivemates.

THINKING THROUGH THE CONCEPTS

True or False: Determine if the statement given is true or false. If it is false, change the underlined word(s) so that the statement reads true.

8. _____ Behavior has some underlined genetic basis.

9. _____ Trial-and-error behaviors lead to learning.

10. _____ The more complex the animal, the more it relies on instinct.

11. _____ In insect societies, almost all behavior is regulated by pheromones.

12. _____ Habituation is characterized by a crucial time during which it can become part of an animal's behavior.

13. _____ If an animal assumes a particular posture to communicate with another animal, this behavior is considered active.

14. _____ Sound and visual signals can be varied.

15. _____ Aggressive behavior occurs during defense of territories.

16. _____ Most aggressive encounters between members of the same species are real.

17. _____ Territories are commonly laid out by females.

18. _____ Territories are usually defended against invasion by members of the same species.

19. _____ Established territories promote conflict among members of the same species.

20. _____ The majority of encounters between highly social animals are competitive and aggressive.

21. _____ Suckling by babies is an innate behavior.

22. _____ Menstrual cycles of roommates are often unsynchronized.

23. _____ Women exposed to early cycle secretions had shorter menstrual cycles than usual.

Identification: Determine whether the following statements refer to **habituation**, or **imprinting**.

24. _____ decline in response to a harmless, repeated stimulus

25. _____ a strong association learned during a sensitive period in life

26. _____ primarily instinctive

27. _____ humans ignoring night sounds while asleep

28. _____ the "following response" of young birds

29. _____ not easily altered by later experiences

30. _____ young birds ignoring a goose flying overhead

Matching: Modes of communication. Some questions may have more than one correct answer.

31. _____ dog urine

32. _____ easily alerts predators

33. _____ uses pheromones

34. _____ may be active or passive

35. _____ persists after the animal has departed

36. _____ can establish social bonds among group members

37. _____ limited to close-range communications

38. _____ ignored by other species

39. _____ grooming in primates

Choices:

a. visual

b. sound

c. chemical

d. touch

Identify: Determine whether the following statements refer to **aggression**, **dominance hierarchies**, or **territoriality**.

40. _____ behavior that makes the animal appear larger

41. _____ defense of an area where important resources are located

42. _____ alpha members of a wolf pack

43. _____ used to establish rank that determines social status

44. _____ harmless symbolic displays or rituals for resolving conflicts without fighting

45. _____ adult males defend an area against members of the same species

46. _____ scent-marking boundaries with pheromones

47. _____ exhibiting fangs, claws, or teeth

48. _____ pecking orders in chickens

49. _____ "good fences make good neighbors"

50. _____ the sheep with the biggest horns gets most access to necessary resources

Multiple Choice: Pick the most correct choice for each question.

51. Most communication occurs
    a. aggressively
    b. as imprinted behaviors
    c. as altruistic behaviors
    d. between members of the same species
    e. between members of different species

52. Chemicals produced by an individual that influence the behavior of members of the same species are called
    a. hormones
    b. pheromones
    c. stimuli
    d. enzymes

53. The dominance hierarchy within a group of animals functions to
    a. eliminate competition
    b. limit population numbers
    c. minimize aggression
    d. increase competition
    e. ensure reproduction

54. An adaptive change in behavior as a result of experience is called
    a. instinct
    b. learning
    c. an innate behavior
    d. kin selection

Short Answer.

55. List three advantages and three disadvantages of visual communication.

    Advantages:

    _____

    _____

    _____

    Disadvantages:

    _____

    _____

    _____

56. List two advantages of sound communication over visual communication.

    _____

    _____

APPLYING THE CONCEPTS

This practice question is intended to sharpen your ability to apply critical thinking and analysis to biological concepts covered in this chapter.

57. How can studies of twins help human ethologists determine the genetic components of human behavior? Specifically, how would instances of identical twins separated at birth aid in the study?

_____

_____

_____

_____

_____

_____

_____

_____

Use the Case Study and the Web sites for this chapter to answer the following questions.

58. What happens to male threespine sticklebacks (*Gasterosteus aculeatus*) courtship behavior when a predator is threatening? What happens to female mate preference when her first choice is threatened by a predator?

_____

_____

_____

_____

_____

59. To which set of stimuli does a female green swordtail (*Xiphophorus helleri*) respond most obviously: moving food particles, an empty aquarium, or the same male engaging in an active courtship display, performing similar levels of feeding activity, and remaining inactive?

_____

_____

_____

_____

60. Yellow-shouldered widowbirds, *Euplectes macrourus*, are sexually dimorphic (male and female look different), polygynous (mate with more than one female) birds in which males have long tails. Do females prefer males with long tails?

_____

_____

_____

61. In Trinidadian guppies, which gender chooses their mate? What is the basis for their selection? Is this always the case?

_____

_____

_____

_____

62. Why might a more drab male be chosen as a mate?

_____

_____

_____

ANSWERS TO EXERCISES

1. Behavior
   ethology
2. Innate
   learning
3. Habituation
   imprinting
   sensitive period
4. Communication
   pheromone
5. dominance hierarchy
6. Territoriality
7. Altruism
8. true
9. true
10. false, more simple
11. true
12. false, imprinting
13. true
14. true
15. true
16. false, symbolic or harmless
17. false, males
18. true
19. false, reduce
20. false, cooperative
21. true
22. false, synchronized
23. true
24. habituation
25. imprinting
26. imprinting
27. habituation
28. imprinting
29. imprinting
30. habituation
31. c
32. b
33. c
34. a
35. c
36. d
37. a, d
38. c
39. d
40. aggression
41. territoriality
42. dominance hierarchies
43. aggression
44. aggression
45. territoriality
46. territoriality
47. aggression
48. dominance hierarchies
49. territoriality
50. dominance hierarchies
51. d
52. b
53. c
54. b

55. Advantages: instantaneous, rapidly revised, quiet, and unlikely to attract predators. Disadvantages: ineffective in darkness, ineffective in dense vegetation, limited to close-range communication, may not be noticed by distracted animals.

56. Sound communication is effective in darkness and through dense forests, in water, and over long distances.

57. Genetic components of behavior have been revealed through comparisons of identical and fraternal twins. When twins are reared together, environmental influences on behavior are very similar for each member of the pair, so behavioral differences between twins must have a large genetic component. If a particular behavior is controlled by genetic factors, we would expect to find that similar expression of behavior in identical twins but not in fraternal twins. In fact, in tests that measure many aspects of personality, identical twins are about twice as similar in personality as are fraternal twins. Such studies have shown a significant genetic component for traits such as activity level, alcoholism, sociability, intelligence, anxiety, dominance, and even political attitudes. Through anecdotal observations reunited twins separated at birth have been found to be as similar in personality as twins raised together. This finding indicates that environment had little to do with development of their personalities.

58. According to A. MacAulay of Mount Allison Univ., New Brunswick, when male threespine sticklebacks, *Gasterosteus aculeatus*, were "attacked" by a heron model overhead, they reduced the frequency of their courtship displays. When females were shown a pair of same sized nesting males with different amounts of red nuptial coloration, they chose the brightly colored male over the drabber. However, after an "attack" on the initially preferred male only (which decreased its courtship rate) the females switched their initial mating preference to the drabber male.

59. G. G. Rosenthal et. al. report that female responsiveness differed before, during, and after the stimulus for the three stimuli showing a male, but not for the two controls. Analyses of female behavior during the stimuli reveal that they "preferred" sequences of courting males to all other stimuli.

60. U. M. Savalli from the Univ. of Kentucky, reports that tail length did not appear to affect female choice, since there was no statistical correlation between natural tail length and male attractiveness (measured by the number of females nesting in his territory).

61. L. E. Dugatkin and J. G. Godin report in Scientific American that female Trinidadian guppies do the choosing when it comes to selecting a mate. Generally speaking, female guppies choose males that are brighter or more orange in color But even guppies are prone to social pressure. If, for example, an older female appears to choose a drabber male, a young female may go against her instincts and copy the older female.

62. According to L. E. Dugatkin and J. G. Godin in Scientific American, male guppies inspect predators; female guppies inspect the males. When a predator approaches a school of guppies, a pair of males may swim over to inspect the potential threat. Such bold behavior may be attractive to females, since they tend to choose the male that swims closest to the predator for a mate. Although the bravest males are often the most colorful, females will choose a less colorful male if he seems to be more courageous than other males.

Chapter 38:   Population Growth and Regulation

OVERVIEW

This chapter examines the factors that control the size and rate of growth of populations. It covers how the environment plays a role in controlling populations and how individual interactions among members of the same species, as well as among members of different species, influence population size. How a population grows may depend on how its members are distributed within a given area, or it may depend on the number of offspring that survive to reach maturity. These factors are applied to the human population as well.

Ecology involves the study of the relationships between living things (**biotic**) and their nonliving (**abiotic**) environment. Abiotic components include soil, water, and weather. An **ecosystem** includes the abiotic environment and all the organisms within a defined area. Within ecosystems, **communities** are made up of all the interacting populations of organisms.

1) How Do Populations Grow?

A **population** is defined as the members of a species that live in an area that allows interbreeding. The size of a population changes depending on the number of births and deaths, the number leaving (**emigration**), and the number coming in (**immigration**). If life in the ecosystem is ideal, the population will increase according to its **biotic potential**, that is, according to its maximum rate. However, the environment cannot sustain a population at this rate since resources, such as food and space, are limited and organisms interact with one another. Therefore, the population's size is limited according to **environmental resistance**. Environmental resistance may decrease birth rates and increase death rates.

The rate at which a population size changes, its **growth rate** (r), is determined by: $b - d = r$ (number of births - number of deaths = rate of growth). However, if the number of individuals that are new to a population within a certain time period is in question, then the rate of growth (r) is multiplied by the number of members in the population at the beginning of the time period (N): therefore, rN = population growth within a given time period. If a population grows at an ever-accelerating rate, then the population is experiencing **exponential growth**. Exponential growth is graphed as a **J-curve**, because that is what it looks like. Using r, you can determine the time it takes for a population to double in size (its **doubling time**).

The biotic potential of an organism is determined by the age it first reproduces, the frequency at which reproduction occurs, and the average number of offspring produced each time. Biotic potential also depends on the length of the organism's reproductive life span and the death rate of individuals under ideal conditions. The text compares population growth of the bacterium *Staphylococcus* and the golden eagle.

2) How Is Population Growth Regulated?

Populations that grow exponentially may suddenly undergo substantial deaths due to disease, to seasonal changes, or to resource availability. Some populations go through exponential growth and massive die-offs on a cyclical basis; these are referred to as **boom-and-bust cycles**. Other populations that undergo exponential growth may do so because there is no natural predator in the area. This often occurs when **exotic** species have been introduced to an ecosystem. The size of most populations, however, is controlled by the ecosystem. Any given area can support only a certain population size indefinitely. This is the **carrying capacity** of the ecosystem. Population numbers that have reached carrying capacity can be graphed as an **S-curve**. Carrying capacity of an ecosystem is determined by renewable resources such as water and nutrients, and nonrenewable resources such as space.

Population numbers may be affected by the course of nature in ways that may or may not be due to the size of the population. Populations that get too crowded or dense may be adversely affected by **density-dependent** factors such as predation, parasitism, disease, or intense competition. On the other hand, **density independent** factors, such as weather, fire events, or human activities, impact a population regardless of its size.

When an animal kills and eats another organism, predation has occurred, and the animal doing the killing and eating is the **predator**. The organism eaten is the **prey**. Predation is an important mechanism in natural population control. However, predation not only controls the size of the prey populations, but it also serves to control the size of the predator populations. As predators reduce the number of prey available, they are, in effect, reducing their own food resource. This results in a reduction in the predator population. When predator numbers are reduced, the prey population will increase again. Thus, predator populations and prey populations undergo **population cycles**. When an animal feeds on another organism without killing it, the animal is a **parasite**, and the organism on which it is feeding is its **host**.

When population numbers increase, **competition** for the limited resources on which the organisms depend becomes more intense. If the competition occurs among members of different species, **interspecific competition** is occurring. However, if the competition is among members of the same species, the more intense **intraspecific competition** is occurring. In order to survive, organisms have evolved mechanisms to gather the scarce resources. **Scramble competition** results in the survival of the individuals best able to accrue the most resources. **Contest competition** uses social or chemical interactions to gather resources. Territory protection and defense is a type of contest competition.

3) How Are Populations Distributed in Space and Time?

If members of a population cluster together in herds, packs, or flocks, or around an area of a plentiful resource, these organisms are **clumped** in their distribution pattern. Other organisms defend territories that are relatively evenly spaced and thus have **uniform distribution** within an area. Rarely, organisms distribute themselves **randomly** throughout the ecosystem. This may occur when resources are equally available and plentiful throughout the area, perhaps in a rainforest.

Populations can also be distributed with respect to time. Graphing the number of individuals in an age group against their age illustrates a pattern of survivorship, a **survivorship curve**. Different species have different patterns of survivorship. Species that have low offspring mortality and many members who tend to reach maturity before death show a "late loss" curve. Species that have an equal chance of survival as they do death at any age show a "constant loss" when graphed. Species that have high offspring deaths with few members reaching maturity have an "early loss" curve.

4) How Is the Human Population Changing?

The human population grew slowly for over one million years. During that time, fire was discovered, tools and weapons were fashioned, shelters were built, and clothing was made to protect individuals. Each of these "inventions" led to a cultural revolution as the populations adapted to these innovations. The domestication of crops and animals led to an agricultural revolution, providing a more dependable food supply. Once advances in medicine and health care occurred, the human death rate was reduced dramatically. This industrial-medical revolution led to an increase in population. In developed countries, the industrial-medical revolution also led to reduced birth rates, stabilizing their population growth. In developing countries, however, reduced birth rates have not occurred, primarily due to social traditions and a lack of access to education and contraceptives. The differences in population growth between developing and developed countries can be illustrated using **age structure** diagrams. These diagrams graph the distribution of males and females in each age group. If there is a large portion of the population below reproductive years, even if people entering their reproductive years had only the number of children needed to replace themselves [**replacement level fertility (RLF)**], the population would continue to grow due to the number of people entering their reproductive years. The United States is experiencing rapid growth because the "baby boom" generation has reached their reproductive years. However, continued immigration to the United States also contributes to the population growth.

KEY TERMS AND CONCEPTS

Fill-In: From the following list of key terms, fill in the blanks in the following statements.

age structure	ecology	parasite
agricultural revolution	ecosystem	parasitism
biotic potential	emigration	population
boom-and-bust cycle	environmental resistance	predation
carrying capacity	exponential	predator
clumped distribution	host	randomly distributed
community	immigration	replacement-level fertility
competition	industrial-medical revolution	S-curve
contest competition	interspecific competition	scramble competition
cultural revolution	intraspecific competition	survivorship curve
density dependent	J-curve	uniformly distributed
density independent		

1. _____ is the study of the interrelationships among living organisms and their nonliving environment. The term _____, however, is used to describe all the organisms and their non-living environment within a specified area. All the potentially interbreeding members of a species within an ecosystem are referred to as a _____.

2. The size of a population changes as new members are born, move into the area (called _____), leave the area (called _____), or die. The actual rate at which a population changes in size depends on the _____, which is the maximum rate a population could increase with unlimited resources, and the _____, which includes the limited availability of the resources such as food, water, space, etc. However, to measure the change in population size, the number of deaths per person is subtracted from the number of births per person that occur in a given time period. This is the _____ or (r).

3. When a population's size increases at a continuously accelerating rate, the population is experiencing _____ growth; an increasing number of individuals is added to the population with each generation. When the population numbers are graphed for this type of growth, a J-shaped curve or a _____ is the result.

4. When resources for a species are temporarily abundant, its population may grow exponentially until it is limited by an environmental factor (temperature, for example). This pattern of rapid growth followed by a massive die-off is called a _____. Species that have been introduced to an area from a foreign location are termed _____ species. Populations of these introduced species often exhibit exponential growth since their natural predators are missing from their new ecosystem.

5. In populations where birth rates equal death rates, the population stabilizes. These populations exhibit S-shaped curves or _____ when population growth rates are graphed. Populations often stabilize when the population numbers reach the maximum size that the ecosystem can support indefinitely; the _____ of the ecosystem.

6. Human activities and environmental factors such as weather or forest fires may limit population size regardless of its density. These are examples of _____ factors. In contrast, parasites, disease, competition for limited resources, and predation are examples of _____ factors that affect population size more intensely as population density increases.

7. Animals capturing, killing, and eating other organisms is called _____. This act naturally helps control population size. The animal that captures, kills, and eats the organism (the prey) is referred to as the _____. As prey population numbers increase, the predator population numbers will increase. Then, as more predators capture and eat more prey, the prey numbers decline. This will cause the predator population numbers to decrease as well. This effect, the _____, is always out of phase.

8. If an organism feeds within or on a larger organism but does not kill it outright, this is called _____. The organism feeding is the _____, while the organism that is being fed upon is the _____.

9. When resources in an area are limited, the organisms using the resources are in _____ for the resources. If the organisms vying for the resources belong to different species, then _____ is occurring. If the organisms vying for the resources belong to the same species, then _____ is occurring and its effect is more intense. One way that some species have evolved to deal with the intensity of this type of competition involves a "free-for-all," the winner of _____ gets the resource. Other species defend territories when population numbers exceed available resources. These organisms engage in _____ where only the most fit are able to defend their territory.

10. The populations of species in any given area may be distributed in specific patterns. If the members of a population live in groups such as in herds or flocks or along resource lines, the population has a _____. If the organisms live a rather consistent distance from each other, then these organisms are _____. Rarely, if the members of the population do not form social groups or do not use territorial spacing, their dispersal may not have any observable pattern and they are _____.

11. Over time, populations tend to show patterns in the numbers that die or survive. The pattern that any given species exhibits is called its _____.

12. Humans have developed ways to increase the carrying capacity of the ecosystem of which they are a part. By using fire and tools and developing protective shelter and clothing, a _____ occurred, allowing previously uninhabitable areas to be habitable. As farming and animal husbandry evolved, a dependable food supply came about with an _____. The human population grew slowly until advances were made that reduced the number of deaths. This _____ began during the mid-eighteenth century and continues today.

13. A diagram of a population representing the number of individuals in specific age categories is the _____ of the population.

14. When individuals in a population of reproductive age bear only the number of children needed to replace themselves, then _____ is occurring.

Key Terms and Definitions

abiotic (ā-bī-ah´-tik): nonliving; the abiotic portion of an ecosystem includes soil, rock, water, and the atmosphere.

age structure: the distribution of males and females in a population according to age groups.

biotic (bī-ah´-tik): living.

biotic potential: the maximum rate at which a population could increase, assuming ideal conditions that allow a maximum birth rate and minimum death rate.

boom-and-bust cycle: a population cycle characterized by rapid exponential growth followed by a sudden massive die-off, seen in seasonal species and in some populations of small rodents, such as lemmings.

carrying capacity: the maximum population size that an ecosystem can support indefinitely; determined primarily by the availability of space, nutrients, water, and light.

clumped distribution: the distribution characteristic of populations in which individuals are clustered into groups; may be social or based on the need for a localized resource.

community: all the interacting populations within an ecosystem.

competition: interaction among individuals who attempt to utilize a resource (for example, food or space) that is limited relative to the demand for it.

contest competition: a mechanism for resolving intraspecific competition by using social or chemical interactions.

density-dependent: referring to any factor, such as predation, that limits population size more effectively as the population density increases.

density-independent: referring to any factor that limits a population's size and growth regardless of its density.

doubling time: the time it would take a population to double in size at its current growth rate.

ecology (ē-kol´-uh-jē): the study of the inter-relationships of organisms with each other and with their nonliving environment.

ecosystem (ē´kō-sis-tem): all the organisms and their nonliving environment within a defined area.

emigration (em-uh-grā´shun): migration of individuals out of an area.

environmental resistance: any factor that tends to counteract biotic potential, limiting population size.

exotic/exotic species: a foreign species introduced into an ecosystem where it did not evolve; such species may flourish and out-compete native species.

exponential growth: a continuously accelerating increase in population size.

growth rate: a measure of the change in population size per individual per unit of time.

host: the prey organism on or in which a parasite lives; is harmed by the relationship.

immigration (im-uh-grā´-shun): migration of individuals into an area.

interspecific competition: competition among individuals of different species.

intraspecific competition: competition among individuals of the same species.

J-curve: the J-shaped growth curve of an exponentially growing population in which increasing numbers of individuals join the population during each succeeding time period.

parasite (par´-uh-sīt): an organism that lives in or on a larger prey organism, called a *host*, weakening it.

population: all the members of a particular species within an ecosystem, found in the same time and place and actually or potentially interbreeding.

population cycle: out-of-phase cyclical patterns of predator and prey populations.

predator: an organism that kills and eats other organisms.

prey: organisms that are killed and eaten by another organism.

random distribution: distribution characteristic of populations in which the probability of finding an individual is equal in all parts of an area.

replacement-level fertility (RLF): the average birthrate at which a reproducing population exactly replaces itself during its lifetime.

scramble competition: a free-for-all scramble for limited resources among individuals of the same species.

S-curve: the S-shaped growth curve that describes a population of long-lived organisms introduced into a new area; consists of an initial period of exponential growth, followed by decreasing growth rate, and, finally, relative stability around a growth rate of zero.

survivorship curve: a curve resulting when the number of individuals of each age in a population is graphed against their age, usually expressed as a percentage of their maximum life span.

uniform distribution: the distribution characteristic of a population with a relatively regular spacing of individuals, commonly as a result of territorial behavior.

THINKING THROUGH THE CONCEPTS

True or False: Determine if the statement given is true or false. If it is false change the <u>underlined</u> word(s) so that the statement reads true.

15. _____ A <u>population</u> is made up of all the members of a species in a certain area that has the potential to interbreed.

16. _____ The availability of food and space serve to <u>limit</u> the biotic potential of a population.

17. _____ In nature, exponential growth occurs for <u>prolonged</u> periods of time.

18. _____ <u>Carrying capacity</u> of an ecosystem is determined, in part, by renewable resources.

19. _____ The effect of <u>density-dependent</u> factors is unaffected by population size.

20. _____ As a population increases and becomes more dense, the result is <u>less</u> die-off from disease and parasites.

21. _____ Predators exert an influence on the size of their prey populations; <u>prey populations also affect predator population size</u>.

22. _____ Humans are likely to cause extinctions because impacts such as pollution and habitat alteration are <u>not density dependent</u>.

23. _____ Parasites living in their host <u>quickly</u> weaken and kill their host.

24. _____ Intraspecific competition is <u>less intense</u> then interspecific competition.

25. _____ Populations may exhibit a <u>clumping</u> pattern due to localized resources.

26. _____ A <u>species' population</u> will show a characteristic pattern of survivorship; either "early loss," "constant loss," or "late loss."

27. _____ Humans have found ways to overcome <u>environmental resistance</u>.

28. _____ In age structure diagrams, if the number of children (ages 0–14) exceeds the number of reproducing individuals (ages 15–45), the population is <u>decreasing</u>.

29. _____ Delayed childbearing <u>slows</u> population growth.

30. _____ The U.S. population is currently growing <u>exponentially</u>.

Matching:

31. _____ illustrates exponential growth

32. _____ illustrates a stable population

33. _____ $r = b - d$

34. _____ rN

35. _____ illustrates patterns of death

36. _____ illustrates distribution of males and females

Choices:

a. age structure diagram

b. population growth

c. S-curve

d. growth rate

e. survivorship curve

f. J-curve

Short answer:

37. Population change = (_____ - _____) + (_____ - _____)

38. Identify 3 of the 5 factors influencing the biotic potential of a species.

    _____

    _____

    _____

    _____

39. Briefly discuss three factors that can lead to boom-and-bust cycles. Include examples of the types of organisms that are susceptible to these cycles.

    _____

    _____

    _____

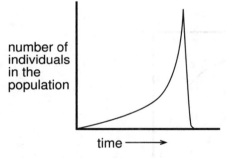

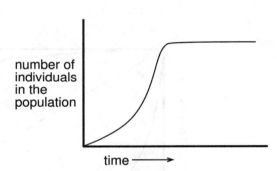

Identify: Use the diagram above to determine whether the characteristics described refer to a **J-shaped** population curve, an **S-shaped** population curve, or **both**.

40. _____   Initial population growth is small.

41. _____   Population growth accelerates with time.

42. _____   Population growth accelerates with time, then levels off.

43. _____   Population grows indefinitely, exceeding carrying capacity.

44. _____   Population is limited at carrying capacity.

45. _____   Population size is probably limited by environmental factors.

46. _____   Density-dependent factors influence population size.

47. _____   Population may suffer a sudden crash.

APPLYING THE CONCEPTS

48. Explain why exotic or introduced species tend to display exponential growth. Discuss the subsequent effects on the ecosystem.

_____

_____

_____

_____

_____

_____

_____

_____

_____

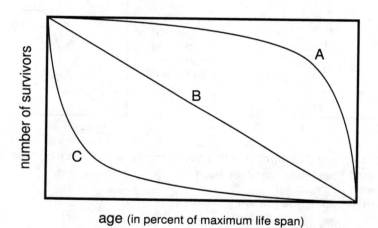

age (in percent of maximum life span)

Identify: Using the survivorship curves **A**, **B**, and **C**, graphed above, identify the following.

49. _____   Which curve shows continued loss?

50. _____   Which curve shows early loss?

51. _____   Which curve shows late loss?

52. _____   Which curve would diagram the survivorship of an invertebrate population?

53. _____   Which curve would diagram the survivorship of a Dall mountain sheep population?

54. _____   Which curve would diagram the survivorship of a population of American Robins?

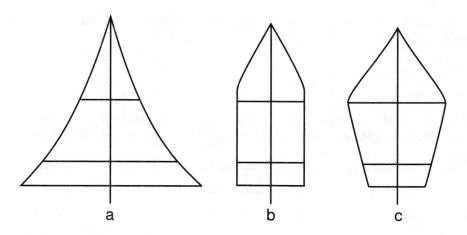

a   b   c

Identify: Using the above age structure diagrams a, b, and c, answer the following.

55. _____   Which diagram shows a growing population?

56. _____   Which diagram shows a shrinking population?

57. _____   Which diagram shows a stable population?

58. _____   Which would diagram the population of a developing country?

59. _____   Which would diagram the population of a country with replacement level fertility?

60. _____   Which would diagram the population of a country with fewer children than reproducing adults?

Use the Case Study and the Web sites for this chapter to answer the following questions.

61. The blacklegged tick *Ixodes scapularis* is a three host tick, i.e. each feeding stage (larva, nymph and adult) requires one vertebrate blood meal for its development. Each stage attaches to a vertebrate host, feeds to repletion (until full), detaches, drops from the host (usually into the leaf litter) and molts to the next stage. How long does it take for *I. scapularis* to complete a full life cycle?

_____

_____

_____

_____

_____

62. What is the typical habitat of the blacklegged tick *Ixodes scapularis*?

_____

_____

_____

_____

_____

_____

63. The white-footed mouse *Peromyscus leucopus* is an important host for larval *I. scapularis*. What is the habitat for the white-footed mouse?

_____

_____

64. The white-tailed deer (*Odocoileus virginianus*) is the mammalian species most commonly infested with the blacklegged tick *I. scapularis*. What is the typical habitat of the white-tailed deer?

_____

_____

_____

_____

_____

65. The gypsy moth, *Lymantria dispar*, is one of North America's most devastating forest pests. What tree species are most vulnerable to the gypsy moth? Where is the gypsy moth most concentrated?

_____

_____

_____

66. How might fungi help control the gypsy moth population?

_____

_____

67. How might nematodes (microscopic roundworms) help control the gypsy moth population?

_____

_____

_____

_____

_____

68. How could researchers place tick collars, similar to those for dogs and cats, on deer? Would this be safe?

_____

_____

_____

_____

_____

ANSWERS TO EXERCISES

1. Ecology
   ecosystem
   population
2. immigration
   emigration
   biotic potential
   environmental resistance
   growth rate
3. exponential
   J-curve
4. boom-and-bust cycle
   exotic
5. S-curve
   carrying capacity
6. density independent
   density dependent
7. predation
   predator
   population cycle
8. parasitism

   parasite
   host
9. competition
   interspecific competition
   intraspecific competition
   scramble competition
   contest competition
10. clumped distribution
    uniformly distributed
    randomly distributed
11. survivorship curve
12. cultural revolution
    agricultural revolution
    industrial-medical revolution
13. age structure
14. replacement-level fertility
15. true
16. true
17. false; short

18. true
19. false; density independent
20. false; more
21. true
22. true
23. false; slowly
24. false; more intense
25. true
26. true
27. true
28. false; increasing
29. true
30. true
31. f
32. c
33. d
34. b
35. e
36. a

37. Population change = (births − deaths) + (immigrants − emigrants).

38. Age at which the organism first reproduces, frequency with which reproduction occurs, average number of offspring produced each time, length of reproductive life span of an organism, death rate of individuals under ideal conditions.

39. Short-lived, rapidly reproducing organisms such as bacteria, algae, and insects have population cycles that are often dependent on amount of rainfall (flood or drought), temperature (intense heat or killing frost), or nutrient availability.

40. b
41. b
42. S

43. J
44. S
45. b

46. S
47. J

48. Introduced foreign (exotic) species often invade new habitats where conditions are favorable, food or nutrients are plentiful, and competition is scarce. Predators, parasites, or disease have little or no effect on the population of the introduced species. This leads to exponential growth of the introduced species which may seriously damage the ecosystem by displacing, out competing, or preying on native species.

49. B
50. C
51. A
52. C

53. A
54. B
55. a
56. c

57. b
58. a
59. b
60. c

61. According to The Tick Research Laboratory of the Department of Fisheries, Animal and Veterinary Science at the Univ. of Rhode Island, the life cycle of *I. scapularis* may range from two to four years and appears to be regulated by host abundance, physiological mechanisms of the tick, and environmental factors. Typically, *I. scapularis* takes about two years to complete one life cycle.

62. According to The Tick Research Laboratory of the Department of Fisheries, Animal and Veterinary Science at the Univ. of Rhode Island, habitats containing dense brush and vegetation such as bayberry, poison ivy, green-briar, and scrub oak appear to be ideal for populations of *I. scapularis*. Feeding nymphs and larvae are abundant in wooded areas, particularly in the leaf litter, rather than in open grasslands. Adult ticks are more common on high shrubs (>1 meter) and in wooded areas. *I. scapularis* may be extremely abundant in heavily forested areas that have a high canopy. They are not usually found in open or grassy areas.

63. According to researchers at the Univ. of Kansas, the white-footed mouse inhabits woodlands, preferably with large climax trees, and dense brushy areas, rarely beyond the forest edge.

64. According to the U.S. Forest Service, white-tailed deer are most frequently found near stream bottoms, swamps, and other riparian areas. They are also found in mixed deciduous and coniferous forests at low to mid elevations with gentle slopes. They prefer dense, coniferous stands near riparian areas with southern exposure. At northern latitudes white-tailed deer need stands of mixed conifer and deciduous trees with partial openings that provide food and protection from winter weather.

65. According to S. Liebhold of the U.S. Forest Service Northeastern Research Station, the gypsy moth feeds on the foliage of hundreds of species of plants in North America but its most common hosts are oaks and aspen. Gypsy moth hosts are located throughout the U.S. but the highest concentrations are in the southern Appalachian Mountains, the Ozark Mountains, and in the northern Lake States.

66. J. Suszkiw reported on P. Allen's work involving fungi infecting black-legged ticks. By exposing black-legged deer ticks to spores of naturally occurring fungi, researchers are hoping the fungus will kill ticks, especially tick eggs and emerging larvae. The fungus *Metarhizium anisopliae* proved the most lethal to the ticks, especially against juvenile forms.

67. D. E. Hill et. al. were interviewed for the U.S. Department of Agriculture's *Agricultural Research* magazine. Researchers hope to biologically control black-legged deer ticks with nematodes. Adult females that have recently fed are most vulnerable to nematode attack because feeding expands their natural body openings. The nematodes may enter the tick's breathing holes (spiracles), mouth, anus, or genital pore. Unfed ticks bat away the nematodes that are trying to get into them. Once infected, death comes quickly. Less than 50 nematodes will cause death within 24 hours. However, it is not actually the worms that kill the ticks. They are killed by a species of symbiotic bacteria that the nematodes carry in their gut. The worms release the bacteria, which liquefy the tick's fat bodies and other tissues.

68. According to L. McGraw at the U.S. Department of Agriculture, pesticide collars are commonly used for controlling ticks and other parasites on domestic animals. Putting collars on deer usually requires trapping or tranquilizing the deer. A new collaring unit, patented by Agricultural Research Service scientists, draws deer to a specially designed feeder filled with corn. To eat, the animal must place its neck near the collaring mechanism, which releases a self-adjusting, flexible collar. ARS researchers have used the collars on captive deer. They have not seen any ticks attached and successfully feeding on the neck and head of collared deer. Without collars, deer typically have hundreds to thousands of ticks feeding on them. The collars contain the pesticide amitraz, approved for livestock but also kills ticks on deer. However, the pesticide is not currently approved for use on deer. Once approved, it would be safe to use during hunting season when most adult blacklegged ticks feed on deer.

Chapter 39: Community Interactions

OVERVIEW

This chapter looks at the how organisms interact with members of their own species, as well as with members of different species. These interactions include mechanisms that have evolved to deal with competition for limited resources, predator/prey relationships, and symbiotic relationships. The species that are present in a community depends on the ecosystem that is present. This chapter also introduces you to how an ecosystem changes over time (succession).

1) Why Are Community Interactions Important?

The interactions among populations within a **community** serve to maintain a balance between available resources (i.e., food, water, shelter) and the number of individuals using them. The interactions among the populations limit population size, but they also lead to changes in characteristics and behaviors, increasing the fitness of the total population (evolution). When changes in one species result in adaptive changes in an interacting species, **coevolution** has occurred.

2) What Are the Effects of Competition Among Species?

Competition among species, or **interspecific competition**, has such a strong effect on the species involved that each evolves ways to reduce any overlap in needs. In other words, each species specializes within the community, developing its own well-defined **ecological niche**, thus partitioning (dividing up) the resources. Therefore, no two species will occupy the exact same niche. This concept is referred to as the **competitive exclusion principle**. Gause showed this in his lab experiments with two species of *Paramecium*. MacArthur investigated Gause's findings by studying six Warbler species that appeared to feed and nest in the same area (a species of spruce). MacArthur realized that the birds were actually using different parts of the tree to feed, they hunted in different ways, and they nested at different times. The birds were using **resource partitioning** to reduce the competition among themselves. Ecologist J. Connell studied the effects of removing one of the competing species from the resource area. He used the **intertidal zone** barnacles *Chthamalus* and *Balanus*. When one of the species was removed from the community, the other expanded its niche because the competition pressure had been reduced. This demonstrated that interspecific competition limits the size and distribution of competing species

3) What Are the Results of Interactions Between Predators and Their Prey?

Predation interactions have intense effects on the species involved. Predators have evolved ways to best capture their prey, while the prey have evolved mechanisms to elude their predators. This coevolution has resulted in some very complex physical characteristics and behaviors. Bats and their moth prey have developed complex "cat and mouse" behaviors, while other species **camouflage** themselves to avoid predators or detection by prey. In contrast to camouflaged species, others stand out with bright coloration. These species advertise their presence, safe in the knowledge that they will be left alone. Their bright colors (or **warning coloration**) warn potential predators that they are poisonous or otherwise distasteful and are to be avoided. Species with common characteristics may share warning patterns as well; for example, stinging insects tend to be bright yellow with black stripes, and poisonous frogs from the tropics display very colorful skin pigments. Some harmless species have evolved to **mimic** their poisonous relatives, thus taking advantage of the effect of the warning pattern on potential predators. Some stingless wasps, for example, are bright yellow with black stripes. Devious predators exists as well: **aggressive mimicry** has evolved among species that resemble harmless species, yet are truly waiting to take a bite out of an unsuspecting prey. Predators, however, may be caught off-guard. Some prey make use of color patterns that mimic a larger organism. These species use their **startle coloration** to scare their predator and make a safe getaway. Some prey species have the ultimate defense: "chemical warfare." Coevolution, however, has also led to a few predator species that are not harmed by the chemical produced and may even use it as its own defense mechanism.

4) What Is Symbiosis?

Within a community, interacting with other species is unavoidable; however, some species have such close interactions that they have developed **symbiotic** relationships. When one species of the relationship benefits and the other is unaffected, the relationship is called **commensalism**. The barnacle-whale relationship is commensalistic. The whale is not harmed by the presence of the barnacle, but the barnacle benefits as it is carried through nutrient-rich water as the whale swims. If one species benefits and the other is harmed, the relationship is called **parasitism**. **Parasites** live in or on their *hosts*. They weaken the host without killing it immediately. If both species benefit, the relationship is called **mutualism**. Nitrogen-fixing bacteria living in nodules on the roots of legumes (beans) capture nitrogen <u>for</u> the plant and the bacteria receive carbohydrates <u>from</u> the plant; therefore, both benefit.

5) How Do Keystone Species Influence Community Structure?

The influence of species on community structure is not necessarily equal. When one species has a role that is out of proportion to its population size, that species is a **keystone species** in the community. The seastar *Pisaster*, lobster off the eastern coast of Canada, and the African elephant are examples of keystone species; they help keep population numbers of other species in balance. Often, a keystone species cannot be identified until it has actually been removed from the community. At this point it may be too late to reduce the impact its absence will have on the community.

6) Succession: How Does a Community Change over Time?

The interactions among members of a community lead to structural changes within that community; changes that are identified as stages in **succession** of the community. **Primary succession** begins with **pioneer** species such as lichen and mosses establishing a hold on bare rock. As soil slowly forms, additional species move into the young community in a recognizable pattern. **Secondary succession** occurs after an established community has been disturbed perhaps by fire, wind storm, or farming.

If left undisturbed, succession will continue to a stable endpoint, the **climax**, determined largely by the geography and climate of the area. If a community is regularly or periodically disturbed, it will be maintained at a succession point below the climax, a **subclimax**. Climax communities covering broad geographical regions are **biomes**. Biomes are distinguished by specific climatic conditions and characterized by specific plant communities.

KEY TERMS AND CONCEPTS

Fill-In: From the following list of key terms fill in the blanks in the following statements.

aggressive mimicry
biomes
camouflaged
climax community
coevolution
commensalism
community
competitive exclusion principle
ecological niche

exotic species
interspecific competition
intertidal zone
keystone species
mimicry
mutualism
parasites
pioneers

primary succession
resource partitioning
secondary succession
startle coloration
subclimax
succession
symbiosis
warning coloration

1. All the interacting populations within an ecosystem make up an ecological _____.

2. In order for _____ to occur, two interacting species serve as agents of natural selection on one another over evolutionary time.

3. When two or more species try to use the same limited resource, each species is harmed. This type of species interaction is called _____.

4. In order to reduce competition among species, each species has evolved its own _____; each species has its own physical environmental factors necessary for its survival, as well as a specific "occupation" within its habitat. With this in mind, the _____ states that no two species can inhabit the same ecological niche; eventually one species would eliminate the other through competition. However, examples of species with very similar niche requirements exist. Through evolutionary adaptations, these organisms have reduced the overlap of their niches. This is referred to as _____.

5. The barnacles *Chthamalus* and *Balanus* live along rocky ocean shores and are exposed to flood and drought conditions as the tide comes in and recedes. This area of the coastline is referred to as the _____.

6. Many animals have evolved colors, patterns, or shapes to avoid a predator or to avoid being noticed by their prey. These animals are _____. Other animals display bright _____ announcing that they are poisonous or otherwise nonpalatable. Some harmless species take full advantage of the brightly colored species by resembling them in appearance. This _____ saves the harmless, tasty species from predation.

7. Predators may also deceive. Through _____, a predator resembles a harmless species, allowing its prey to unknowingly come into range for an attack.

8. Certain moths and caterpillars have evolved "eye-spots" and other color patterns resembling eyes of a larger animal. When prey use _____, the predator is frightened, and the prey escapes.

9. _____ is defined as a close interaction between members of different species for an extended period of time. In a relationship such as _____, one species benefits while the other is unaffected. However, _____ live on or in their hosts, usually harming the host in some way but not killing it. When both species in the relationship benefit from the interaction, the relationship is called _____.

10. One species may play a major role in determining a community's structure. If this _____ is removed, the structure of the community would be dramatically altered.

11. If a community changes structurally over time, _____ has occurred. The changes begin with the first species to invade an area; these species are the _____. As time and climate allow, the community will reach a relatively stable endpoint, the _____.

12. When there has been no previous community in existence, the ecosystem begins with the process of _____. If, however, an ecosystem has been disturbed by fire, storm, or farming, a new ecosystem develops through the process of _____.

13. If an ecosystem is periodically or regularly disturbed, the potential climax community may not be reached, instead, a _____ community is maintained. Large geographical regions consisting of climax communities and characterized by specific plant species make up the _____ of the world.

14. The balance of a community may be severely disrupted when an _____ is introduced to the community.

KEY TERMS AND DEFINITIONS

aggressive mimicry (mim´ik-rē): the evolution of a predatory organism to resemble a harmless animal or part of the environment, thus gaining access to prey.

biome (bī´-ōm): a terrestrial ecosystem that occupies an extensive geographical area and is characterized by a specific type of plant community: for example, deserts.

camouflage (cam´-a-flaj): coloration and/or shape that renders an organism inconspicuous in its environment.

climax community: a diverse and relatively stable community that forms the endpoint of succession.

coevolution: the evolution of adaptations in two species due to their extensive interactions with one another, such that each species acts as a major force of natural selection on the other.

commensalism (kum-en´-sal-iz-um): a symbiotic relationship in which one species benefits while another species is neither harmed nor benefited.

community: all the interacting populations within an ecosystem.

competitive exclusion principle: the concept that no two species can simultaneously and continuously occupy the same ecological niche.

ecological niche (nitch): the role of a particular species within an ecosystem, including all aspects of its interaction with the living and nonliving environments.

exotic/exotic species: a foreign species introduced into an ecosystem where it did not evolve; such species may flourish and outcompete native species.

interspecific   competition: competition among individuals of different species.

intertidal zone: an area of the ocean shore that is alternately covered and exposed by the tides.

keystone species: a species whose influence on community structure is greater than its abundance would suggest.

mimicry (mim´-ik-rē): the situation in which a species has evolved to resemble something else—typically another type of organism.

mutualism (mū´-choo-ul-iz-um): a symbiotic relationship in which both participating species benefit.

parasite (par´-uh-sīt): an organism that lives in or on a larger prey organism, called a *host*, weakening it.

parasitism: a symbiotic relationship in which one organism (commonly smaller and more numerous than its host) benefits by feeding on the other, which is normally harmed but not immediately killed.

pioneer: an organism that is among the first to colonize an unoccupied habitat in the first stages of succession.

primary succession: succession that occurs in an environment, such as bare rock, in which no trace of a previous community was present.

resource partitioning: the coexistence of two species with similar requirements, each occupying a smaller niche than either would if it were by itself; a means of minimizing their competitive interactions.

secondary succession: succession that occurs after an existing community is disturbed—for example, after a forest fire; much more rapid than primary succession.

startle coloration: a form of mimicry in which a color pattern (in many cases resembling large eyes) can be displayed suddenly by a prey organism when approached by a predator.

subclimax: a community in which succession is stopped before the climax community is reached and is maintained by regular disturbances—for example, tallgrass prairie maintained by periodic fires.

succession (suk-seh´-shun): a structural change in a community and its nonliving environment over time. Community changes alter the ecosystem in ways that favor competitors, and species replace one another in a somewhat predictable manner until a stable, self-sustaining climax community is reached.

symbiosis (sim´-bī-ō´sis): a close interaction between organisms of different species over an extended period. Either or both species may benefit from the association, or (in the case of parasitism) one of the participants is harmed. Symbiosis includes parasitism, mutualism, and commensalism.

warning coloration: bright coloration that warns predators that the potential prey is distasteful or even poisonous.

THINKING THROUGH THE CONCEPTS

True or False: Determine if the statement given is true or false. If it is false, change the underlined word(s) so that the statement reads true.

15. _____ The interacting populations of a community influence one another's ability to survive and reproduce, leading to a system that results in coevolution.

16. _____ A species' niche in the environment can be described as its occupation or role in the community.

17. _____ Species may reduce competition by partitioning the available resources resulting in each species occupying a larger niche than if there were no competition.

18. _____ Many poisonous or harmful species display bright coloration to stand out to a predator; in this way, the predator is sure to eat the bright organism.

19. _____ Common coloration patterns of equally poisonous or harmful species aid in a predator learning to avoid these prey items.

20. _____ Prey species have evolved predator avoidance mechanisms, while predator species have responded by evolving deceptive mechanisms to catch their prey.

21. _____ Plants are defenseless against species that feed on them.

22. _____ Parasites, generally, do not kill their hosts.

23. _____ If elephants disappear from the African savanna, the grasslands will eventually succeed to forest. This is because the elephant is a prey species.

24. _____ Following a forest fire, the new community that will develop will do so through primary succession.

25. _____ The organisms that invade bare rock to begin a new community are the biomes of the community.

26. _____ If a farmer allows a field to lie fallow, or abandons the field altogether, secondary succession will quickly establish a new community structure.

27. _____ A pond that is left undisturbed will eventually fill in with silt, forming a marsh; a meadow may eventually be formed as the marsh dries.

28. _____ Climax communities undergo constant change, resulting in diverse populations inhabiting numerous ecological niches.

29. _____ Fields maintained for agriculture represent communities held at a specific subclimax by humans.

Matching: Nature's "Chemical warfare."

30. _____ flowering lupines

31. _____ squid, octopus

32. _____ spiders, snakes

33. _____ bombardier beetle

34. _____ milkweed plants

35. _____ grasses

Choices:

a. produce toxic, distasteful chemicals

b. produce silicon

c. produce a boiling hot, toxic spray

d. produce ink clouds

e. produce a paralyzing venom

f. produce alkaloids

Identify: Determine whether these interactions are examples of **competition**, **parasitism**, **commensalism**, or **mutualism**.

36. _____ Lichens are a growth form that occurs when fungi and algae live together. The fungus absorbs nutrients for the alga while the alga photosynthesizes, providing carbohydrates for the fungus.

37. _____ Bromiliads, "air plants," grow in the notches of tropical trees. The trees are not harmed nor do they benefit from the bromiliads; however, the bromiliads absorb water and nutrients collected in rainwater in the notch.

38. _____ Hyenas and vultures both feed from animal remains after lions have finished feeding.

39. _____ A tick that has imbedded into the hide of a deer feeds off the deer, possibly weakening or infecting the deer.

40. _____ Egrets can often be seen following cattle through a field. As the cattle disturb insects in the grass, the egrets eat the insects; the cattle are unaffected.

41. _____ Roundworms are often found in intestines of feral (wild) cats. The roundworm feeds off the nutrients the cat has ingested, leaving the cat malnourished.

42. _____ An oak seedling and a maple seedling are growing in an opening in the forest. They both require sunlight, water, and nutrients from their environment.

43. _____ Flowers are pollinated by insects. The insects, in turn, receive pollen and nectar for nourishment from the flowers.

Identify: Determine whether the following statements are examples of **primary** succession or **secondary** succession.

44. _____ Large areas of Australia were burned during the southern hemisphere's summer season of 1997. However, new groundcover soon germinated in the open areas and rich nutrients of the fire ashes.

45. _____ When Mt. St. Helen's volcano in Washington state, USA, erupted in 1980, the existing communities were destroyed. Within two years, subalpine flowers could be seen blooming and mountain meadows had formed; the communities were reforming.

46. _____ Tornados during the northern hemisphere's spring season of 1985 leveled acres of virgin forests in northwest Pennsylvania commonwealth, USA. This gave researchers first-hand information on regeneration patterns in forests never touched by human development.

47. _____ As global temperatures rise and glaciers recede, the stratum left behind will be inhabited by lichen, followed by mosses. As plant matter decays and weathering continues, soil will slowly develop which will sustain tundra grasses and wildflowers.

Short Answer:

48. A prey species may avoid being eaten by using body parts that have evolved to camouflage it with its environment. Identify three examples of how a prey species may camouflage itself. A predator may also have evolved camouflage patterns, concealing it from its prey. Identify three examples.

Camouflaged prey: Camouflaged predator:

_____   _____

_____   _____

_____   _____

49. Warning colorations are commonly displayed by poisonous organisms and are often mimicked by equally harmful species and by harmless species as well. Identify three toxic organisms that display warning coloration. Match these organisms with their equally distasteful or harmless mimics.

Poisonous organism: Mimic:

_____   _____

_____   _____

_____   _____

50. What is the key difference between primary succession and secondary succession?

_____

APPLYING THE CONCEPTS

These practice questions are intended to sharpen your ability to apply critical thinking and analysis to the biological concepts covered in this chapter.

51. Kangaroo of Australia fill a niche very similar to that of deer in North America. If a deer population from North America were introduced to Australia, the kangaroo and deer would be in direct competition for many resources. Using what you have learned about competition, the competitive exclusion principle, and introduced species, explain what would happen between the two populations.

_____

_____

_____

_____

_____

52. The organisms in a community are tied together either directly or indirectly. This is illustrated best by studying keystone species. Briefly explain how the removal of a keystone species from an ecosystem affects the balance of the communities found there.

_____

_____

_____

_____

53. The snowberry fly has evolved specific behavioral patterns to ward off its predator, a jumping spider. Explain how the fly's behavior is its defense against predation.

_____

_____

_____

Use the Case Study and the Web sites for this chapter to answer the following questions.

54. How does Tamarask out-compete native desert species? How does it restrict the growth of species near it?

_____

_____

_____

_____

_____

55. Fraser fir (*Abies fraseri*) was a keystone species in the spruce-fir forests of the southern United States. How has the wooly adelgid (*Adelges piceae*) caused the removal of Fraser firs from the forests and what has been the result of its absence?

_____

_____

_____

_____

56. The brown tree snake (*Boiga irregularis*) is an exotic species on the Island of Guam. Besides destroying the native bird population and seriously threatening the survival of any introduced bird species, how else has the snake caused problems on the island?

_____

_____

_____

_____

57. What can you do to prevent or reduce habitat invasion by exotic species?

_____

_____

_____

_____

_____

_____

ANSWERS TO EXERCISES

1. community
2. coevolution
3. interspecific competition
4. ecological niche
   competitive exclusion principle
   resource partitioning
5. intertidal zone
6. camouflaged
   warning coloration
   mimicry
7. aggressive mimicry
8. startle coloration
9. Symbiosis
   commensalism
   parasites
   mutualism
10. keystone species
11. succession
    pioneers
    climax community
12. primary succession
    secondary succession
13. subclimax
    biomes
14. exotic
15. true
16. true
17. false, smaller
18. false, avoid

19. true
20. true
21. false, may produce defense chemicals
22. true
23. false, keystone
24. false, secondary
25. false, pioneers
26. true
27. true
28. false, are stable
29. true
30. f
31. d
32. e
33. c
34. a
35. b
36. mutualism
37. commensalism
38. competition
39. parasitism
40. commensalism
41. parasitism
42. competition
43. mutualism
44. secondary
45. secondary
46. secondary
47. primary

48. Camouflaged prey—dappled fawns; grasshoppers; prey resembling leaves, twigs, thorns, bird droppings; plants resembling rocks. Camouflaged predators—spotted cheetah; striped tiger, frogfish resembling algae covered rocks.

49. Yellow jacket, hornets, bees; coral snake, mountain king snake; monarch butterfly, viceroy butterfly.

50. Primary succession—bare rock, secondary succession—community existed previously

51. The competitive exclusion principle states that if two species with the same niche are placed together and forced to compete for limited resources, one will out-compete the other. The deer, as the introduced species, would have few or no natural predators. This would give the deer a competitive edge. The deer would most likely out-compete and eventually replace the kangaroo in Australia.

52. If the keystone species is removed, the species that it preyed upon will increase in number since the keystone species kept the prey population under control. The prey species may out-compete the other species in the community, possibly eliminating them. With the species distribution altered, the balance of the community has changed.

53. When approached by a predatory jumping spider, the snowberry fly mimics the behavior and appearance of a jumping spider protecting its territory. Seeing this specific behavior pattern, the predator retreats, leaving the fly alone.

54. According to the Nature Conservancy, Tamarasks can out-compete native species for a couple of reasons. Individuals are able to produce up to a half million seeds each year. Then, when the seeds sprout, seedlings may grow 10 feet a year and can survive wherever they can reach water. As the individuals grow, their roots can reach water stored deep in the desert, drying up desert springs and changing natural water flow. They have no natural predators or disease organisms in North America and cattle in the area will not graze on its young seedlings. Additionally, Tamarasks secrete salt from glands in their leaves. This makes the soil around them inhospitable to native species.

55. According to the Nature Conservancy, the wooly adelgid (*Adelges piceae*) is a small, aphid-like insect native to Europe. It sucks the sap from Fraser firs, slowly killing them. As a result, the forests have become warmer and drier. Species that are found only in these Appalachian forests and depend on the cool, damp habitat, or that depend specifically on the Fraser firs, may die out. The effects reach to organisms that rely on these species for food. In the end, the entire ecosystem is affected.

56. According to T. Fritts of the U.S. Geological Survey, since there are no natural predators, the brown tree snake (*Boiga irregularis*) has invaded the island since the 1950's. Population numbers are high in both rural and urban areas. In rural areas, snakes invade farms and feed on poultry and their eggs as well as on small mammals. In residential or urban areas, snakes crawl on electrical lines, causing power outages and damaging electrical lines. Damage to electrical equipment creates an economic burden to both civilian and military activities on Guam. The power outages cause numerous problems including food spoilage. The snakes also invade homes and pose a threat to pets and perhaps, small children.

57. According to A. Copping and S. Smith at the Washington Sea Grant Program, you can help in preventing or reducing invasion by exotic species by: avoiding moving plants and animals when you travel or relocate without first checking with local authorities, never releasing pets, plants, or aquarium animals or plants into the wild, using native plants on your property, and being aware of the ways non-native species are introduced to habitats, educating yourself and others about ecosystems and how the organisms living there affect one another. You can also take part in groups that restore habitat, remove, and report sightings of invasive species.

Chapter 40: How Do Ecosystems Work?

OVERVIEW

This chapter traces the pathways of energy and nutrients through the ecosystem. Energy follows a one-way path through the ecosystem; passing from the sun through the organisms in a community, being lost as heat as it is transferred. Nutrients, on the other hand, cycle through the ecosystem using the natural recycling properties of each community. When the cycles are influenced by man, as is currently the case with the carbon cycle, the natural balance is disturbed. The result has led to acid deposition and global warming.

1) What Are the Pathways of Energy and Nutrients?

Energy flows through communities one-way, from the sun to the organisms in the community. As energy flows, some is lost to the environment as heat. Energy supplied to the community must be continuously replenished from the sun. Nutrients, on the other hand, are recycled. Thus, nutrients remain in the ecosystem.

2) How Does Energy Flow Through Communities?

Plants, algae and a few other protists, and cyanobacteria absorb sunlight energy using light-absorbing pigments within them. Using the process of photosynthesis, sunlight energy is converted to chemical energy and is stored as sugar and structures making up the photosynthetic organism. These organisms are **autotrophs**. Since they produce their "food" themselves, they are also called **producers**. Organisms that feed on other organisms are **heterotrophs**. Since these organisms consume other organisms, they are called **consumers**. The stored energy in photosynthetic organisms is available to the other members of a community. The amount of energy that has been stored is the **net primary productivity**. Net primary productivity is often measured as the **biomass** of the producers that is added to the ecosystem.

As energy flows through the community, it passes from one **trophic level** to the next. Trophic levels begin with producers, then progress to consumers. The number of trophic levels in a system depends on the level of consumers involved. Organisms that feed directly on the producers are **herbivores** and are also **primary consumers**; they form the second trophic level. Organisms that feed on the primary consumers are **carnivores** (flesh-eaters). Carnivores are **secondary consumers**; forming the third trophic level. A fourth trophic level is formed when carnivores eat other carnivores; these are **tertiary consumers**. Some organisms eat both herbivores and carnivores. These organisms, such as humans, are **omnivores**.

When trophic levels are traced to diagram relationships between producers, primary consumers, secondary consumers, and tertiary consumers, a **food chain** has been developed. In nature, however, relationships are not so simple. Instead of a food chain, a **food web** illustrates how food chains merge, forming complex relationships. Rounding out the food web is the group of organisms that break down and decompose plant and animal matter (detritis). **Detritus feeders**, including earthworms, centipedes and millipedes, and vultures, feed on dead organic matter such as fallen leaves, cast-off exoskeletons, carcasses, and bodily wastes. **Decomposers**, such as bacteria and fungi, further break down dead organic matter, releasing any remaining nutrients.

As energy flows through the trophic levels, approximately 10% is transferred from one level to the next. Thus, the energy transfer is very inefficient. **Energy pyramids** are used to diagram the transfer of energy.

The *Earth Watch: Food Chains Magnify Toxic Substances* illustrates that energy is not all that is transferred from one organism to the next. If a caterpillar feeds on leaves sprayed with a pesticide, it may not die from ingesting the chemical. Instead, the chemical may be stored in its body. A bird that feeds on caterpillars may eat several caterpillars with the chemical stored in their bodies. This means that the bird is consuming a great deal more of the pesticide. It is also stored in the bird's body. Any organism that feeds on the bird will ingest a concentrated amount of the chemical each time it feeds on such birds. This organism will store even greater amounts of the pesticide in its body. This is known as **biological magnification**. It occurs because the chemicals used are not **biodegradable**; they are not easily broken down by decomposers into harmless substances.

3) How Do Nutrients Move Within and Among Ecosystems?

Nutrients, as defined here, are elements and small molecules that are used to make the chemical building blocks of life. Molecules needed by organisms in large amounts are **macronutrients**; those needed only in very small amounts are **micronutrients**. Nutrients do not flow through communities, but cycle through. These are the **nutrient cycles**, also referred to as the **biogeochemical cycles**. The nutrients tend to be stored in nonliving, or abiotic, **reservoirs** such as CO_2 and nitrogen in the atmosphere and phosphorous in rocks. Other major reservoirs for CO_2 include the oceans (where CO_2 is dissolved) and the **fossil fuels**, which are the remains of ancient plants and animals transformed by heat and pressure.

The atmosphere is composed of 79% nitrogen gas (N_2). Although plants and animals require nitrogen to make protein, some vitamins, and nucleic acids, they can not use nitrogen in its gaseous form. Atmospheric nitrogen is converted to more usable forms by bacteria and cyanobacteria that conduct **nitrogen fixation**. Plants that belong to the **legume** family play a very important role in nitrogen fixation. Bacteria that can fix nitrogen live in the roots of legumes. Thus, plants such as soy beans and clover are often planted to replace nitrogen in nutrient-poor soils. After cycling through plants and animals, nitrogen is returned to the atmosphere by **denitrifying bacteria**.

Phosphorous is required by plants and animals to make ATP, NADP, nucleic acids, and phospholipids for membranes. Phosphorous is found in rock as phosphate. Phosphorous enters living systems by plants and other producers absorbing phosphate dissolved in water. It is put back into soil by decomposers where it may be reabsorbed by autotrophs or it may become bound as soil sediment becomes rock.

Water is also considered to be a nutrient. It cycles in the **hydrologic cycle**. The ocean serves as the major reservoir for water. Water is evaporated from the oceans and returned to land as precipitation.

4) What Is Causing Acid Rain and Global Warming?

Acid rain, or **acid deposition**, occurs when nitrogen oxide and sulfur dioxide combine with water vapor in the atmosphere. Subsequently, nitric acid and sulfuric acid are formed. The acids fall from the atmosphere in the form of acid rain or as dry particles. Both forms negatively affect any structures or organisms they contact.

From the start of the Industrial Revolution, carbon, stored in fossil fuels, has been released into the atmosphere. In 1850, carbon dioxide made up 0.028% of the atmosphere, now it makes up 0.036%. These numbers seem small, but we can look at the numbers differently to see how much of a change it really is. The CO_2 concentration in the atmosphere has increased from 280 ppm, preindustrial revolution, to over 370 ppm, currently. This is an increase of over 30% since 1850. But, why is this a concern? Carbon dioxide is a **greenhouse gas**. In the atmosphere, it traps heat from both the sun and that radiating from Earth. This provides a natural **greenhouse effect**, keeping our atmosphere warm to sustain life. However, as the CO_2 levels increase, it is like putting a thicker blanket over Earth; leading to warmer temperatures globally. **Deforestation** also contributes to **global warming** since CO_2 stored in cut and burned forests is released into the atmosphere. Additionally, the trees are no longer around to absorb CO_2 from the atmosphere during photosynthesis.

KEY TERMS AND CONCEPTS

Fill-In: From the following list of key terms, fill in the blanks in the following statements.

abiotic
acid deposition
autotrophs
biodegradable
biogeochemical cycles
biological magnification
biomass
carnivores
consumer

decomposers
deforestation
denitrifying bacteria
detritis feeders
energy pyramid
food web
food chain
fossil fuels
global warming

greenhouse effect
greenhouse gas
herbivores
heterotrophs
hydrologic cycle
legumes
macronutrients
micronutrients
net primary productivity

nitrogen fixation
nutrient cycles
primary consumers
producers
reservoirs
secondary consumers
tertiary consumers
trophic level

1. _____ are organisms that produce their own organic material using inorganic materials as their energy source, while organisms that use other living organisms as a source of energy are called _____, and are also known as _____.

2. Plants are _____, using sunlight as their source of energy to make organic material, creating the first trophic level. The amount of energy captured by these organisms is measured as _____. If the amount of energy captured is measured in dry weight of the organisms, the _____ has been determined.

3. Plants as producers form the first _____. Organisms that ingest plant material are the _____ and are also called _____. Animals that eat only other animals are called _____. These animals are the _____, making up the third trophic level, and the _____, the fourth trophic level.

4. Animals such as earthworms, millipedes, and termites, which feed on dead organic matter, are referred to as _____. Fungi and bacteria make up the _____, releasing any remaining nutrients.

5. About10% of biomass energy is passed to the next energy level, forming an _____.

6. Some human-made chemicals are not readily broken down in the environment. These chemicals are not _____. Therefore, their concentration may increase in the organisms as you go up the trophic levels in the environment. This increase is called _____.

7. Nutrients that are required by organisms in large amounts are _____, and those needed only in trace amounts are _____. Within the ecosystem, nutrients are recycled in the _____, also called the _____.

8. In each cycle, large storage areas serve as _____. These storage areas are often nonliving or _____. During the carboniferous period, the remains of ancient plants and animals were covered with sediment. Today, these carbon stores are _____, burned for electricity production.

9. Unusable nitrogen gas is converted to usable ammonia by bacteria living in the roots of plants in the
_____ family. This process is called _____. After passing
through organisms in the ecosystem, nitrogen is returned to the atmosphere by _____.

10. Water evaporating from oceans and falling back to land as rain is part of the _____.
When evaporated water in the atmosphere mixes with nitrogen oxide or sulfur dioxide, nitric acid and
sulfuric acid are formed. With their formation, _____ follows, damaging
ecosystems and human-made structures.

11. Atmospheric CO_2 absorbs heat, creating a natural _____ around Earth. As CO_2
levels increase due to deforestation and fossil fuel burning, the average temperature of Earth is increasing
in an event known as _____.

KEY TERMS AND DEFINITIONS

acid deposition: the deposition of nitric or sulfuric acid, either dissolved in rain (acid rain) or in the form of dry particles, as a result of the production of nitrogen oxides or sulfur dioxide through burning, primarily of fossil fuels.

autotroph (aw´-tō-trōf): "self-feeder"; normally, a photosynthetic organism; a producer.

biodegradable: able to be broken down into harmless substances by decomposers.

biogeochemical cycle: also called a *nutrient cycle*, the process by which a specific nutrient in an ecosystem is transferred between living organisms and the nutrient's reservoir in the nonliving environment.

biological magnification: the increasing accumulation of a toxic substance in progressively higher trophic levels.

biomass: the dry weight of organic material in an ecosystem.

carnivore (kar´-neh-vor): literally, "meat eater"; a predatory organism that feeds on herbivores or on other carnivores; a secondary (or higher) consumer.

consumer: an organism that eats other organisms; a heterotroph.

decomposer: an organism, normally a fungus or bacterium, that digests organic material by secreting digestive enzymes into the environment, in the process liberating nutrients into the environment.

deforestation: the excessive cutting of forests, primarily rain forests in the Tropics, to clear space for agriculture.

denitrifying bacterium (dē-nī´-treh-fī-ing): a bacterium that breaks down nitrates, releasing nitrogen gas to the atmosphere.

detritus feeder (de-trī´-tus): one of a diverse group of organisms, ranging from worms to vultures, that live off the wastes and dead remains of other organisms.

energy pyramid: a graphical representation of the energy contained in succeeding trophic levels, with maximum energy at the base (primary producers) and steadily diminishing amounts at higher levels.

food chain: a linear feeding relationship in a community, using a single representative from each of the trophic levels.

food web: a representation of the complex feeding relationships (in terms of interacting food chains) within a community, including many organisms at various trophic levels, with many of the consumers occupying more than one level simultaneously.

fossil fuel: a fuel such as coal, oil, and natural gas, derived from the remains of ancient organisms.

global warming: a gradual rise in global atmospheric temperature as a result of an amplification of the natural greenhouse effect due to human activities.

greenhouse effect: the process in which certain gases such as carbon dioxide and methane trap sunlight energy in a planet's atmosphere as heat; the glass in a greenhouse does the same. The result, global warming, is being enhanced by the production of these gases by humans.

greenhouse gas: a gas, such as carbon dioxide or methane, that traps sunlight energy in a planet's atmosphere as heat; a gas that participates in the greenhouse effect.

herbivore (erb´-i-vor): literally, "plant eater"; an organism that feeds directly and exclusively on producers; a primary consumer.

heterotroph (het´-er-ō-trōf´): literally, "other-feeder"; an organism that eats other organisms; a consumer.

hydrologic cycle: the water cycle, driven by solar energy; a nutrient cycle in which the main reservoir of water is the ocean and most of the water remains in the form of water throughout the cycle (rather than being used in the synthesis of new molecules).

legume (leg´-ūm): a member of a family of plants characterized by root swellings in which nitrogen-fixing bacteria are housed; includes soybeans, lupines, alfalfa, and clover.

macronutrient: a nutrient needed in relatively large quantities (often defined as making up more than 0.1% of an organism's body).

micronutrient: a nutrient needed only in small quantities (often defined as making up less than 0.01% of an organism's body).

net primary productivity: the energy stored in the autotrophs of an ecosystem over a given time period.

nitrogen fixation: the process that combines atmospheric nitrogen with hydrogen to form ammonium (NH_4^+).

nutrient: a substance acquired from the environment and needed for the survival, growth, and development of an organism.

nutrient cycle: a description of the pathways of a specific nutrient (such as carbon, nitrogen, phosphorus, or water) through the living and nonliving portions of an ecosystem.

omnivore: an organism that consumes both plants and other animals.

primary consumer: an organism that feeds on producers; an herbivore.

producer: a photosynthetic organism; an autotroph.

reservoir: the major source and storage site of a nutrient in an ecosystem, normally in the abiotic portion.

secondary consumer: an organism that feeds on primary consumers; a carnivore.

tertiary consumer (ter´-shē-er-ē): a carnivore that feeds on other carnivores (secondary consumers).

trophic level: literally, "feeding level"; the categories of organisms in a community, and the position of an organism in a food chain, defined by the organism's source of energy; includes producers, primary consumers, secondary consumers, and so on.

THINKING THROUGH THE CONCEPTS

True or False: Determine if the statement given is true or false. If it is false, change the underlined word(s) so that the statement reads true.

12. _____   Almost all of the energy produced from the sun reaches Earth.

13. _____   Ecosystems that have a low producer biomass will have low productivity.

14. _____   Herbivores constitute the first trophic level.

15. _____   Omnivores consume both plant and animal material.

16. _____   A food chain can be very complex since it considers all the feeding relationships between organisms.

17. _____   Vultures are detritus feeders.

18. _____   The transfer of energy between trophic levels is extremely efficient.

19. _____   The most infamous chemical related to biomagnification is DDT.

20. _____   Biogeochemical cycles connect the biotic and abiotic portions of the ecosystem.

21. _____   Plants of the rose family are important because they house nitrogen-fixing bacteria in root nodules that can directly absorb and use nitrogen gas from the air, and directly convert atmospheric nitrogen into fertilizer.

22. _____   Human activities are stabilizing the balance of biogeochemical cycles.

23. _____   In the hydrologic cycle, some of the water enters the living community before it evaporates back into the atmosphere.

24. _____   Acid particles can fall from the atmosphere in dry form.

25. _____   Twenty-five percent of the Adirondack lakes are dead.

26. _____   Acid deposition increases lead and other heavy-metal poisoning in animals.

27. _____   Climate prediction for the future is certain in the face of global warming.

Matching: Members of the ecosystem and their trophic level:

28. _____   bracket fungus   34. _____   mushrooms   Choices:

29. _____   squirrel   35. _____   algae   a. producer

30. _____   sheep   36. _____   shark   b. primary consumer

31. _____   dead leaves   37. _____   bacteria   c. detritus feeder

32. _____   oak tree   38. _____   earthworm   d. detritus

33. _____   hawk   39. _____   snake skin   e. decomposer

                                                                 f. secondary consumer

True or False: Determine if the following statements regarding the phosphorus (P) cycle are true or false.

40. _____   The main source of P is from rock.

41. _____   The P cycle is greatly accelerated by transport through the atmosphere.

42. _____   Organic P is transported through the food chain.

43. _____   P is the limiting nutrient in many ecosystems.

44. _____   Humans have affected the P cycle by inadvertent and undesirable fertilization of waterways, accumulation of phosphorus-rich pollutants in the atmosphere, and depletion of soil supplies by over-cutting and erosion.

Short answer:

45. Define the term abiotic and provide two examples of abiotic reservoirs each for the carbon, nitrogen, and phosphorous cycles.

_____

_____

_____

_____

_____

_____

_____

46. Which of the following derive energy for growth directly from light?

   autotrophs   heterotrophs   tertiary consumers

   decomposers   primary consumers   producers

47. Which of the following derive energy for growth indirectly from light?

   autotrophs   heterotrophs   tertiary consumers

   decomposers   primary consumers   producers

48. Identify six trophic levels. Explain how they interact with one another.

_____

_____

_____

_____

_____

_____

_____

_____

APPLYING THE CONCEPTS

These practice questions are intended to sharpen your ability to apply critical thinking and analysis to the biological concepts covered in this chapter.

49. Assuming typical efficiency of energy transfer from one trophic level to the next, describe how 1000 calories in producer biomass might be converted to carnivores. How many calories will a carnivore receive from the original 1000 calories?

_____

_____

_____

_____

_____

_____

50. Why, do you suppose, has there been no mention in this chapter of quaternary or fourth-level consumers? Energetically, is a fifth trophic level possible? Why or why not?

_____

_____

_____

_____

_____

_____

51. Using the following list, carefully trace the path of a carbon atom through the carbon cycle.

atmosphere	detritus	fungus	primary consumer
cellular respiration	detritus feeder	green leafy plant	secondary consumer
decomposer	fecal matter	plant stem	sugar molecule in plant

_____

_____

_____

_____

_____

_____

_____

_____

_____

52. Briefly describe three ways in which humans have affected the nutrient cycles of carbon, nitrogen, and phosphorus. Do not state three examples for each cycle; give examples of how humans have affected the cycling of nutrients as a whole.

_____

_____

_____

_____

_____

_____

_____

_____

_____

_____

_____

Use the Case Study and the Web sites for this chapter to answer the following questions.

53. A severe population decline of Bald Eagles in the United States was blamed on widespread use of the pesticide DDT. What is DDT? Why was (is) it used?

_____

_____

_____

_____

_____

54. Where is DDT still used? Why might DDT still be detected in foods in countries no longer using DDT?

_____

_____

_____

_____

55. In 2000, the United Nations voted against creating a global ban on DDT. Why was this decision applauded by public health care workers around the world?

_____

_____

_____

_____

_____

ANSWERS TO EXERCISES

1. Autotrophs
   heterotrophs
   consumers
2. producers
   net primary productivity
   biomass
3. trophic level
   herbivores
   primary consumers
   carnivores
   secondary consumers
   tertiary consumers
4. detritis feeders
   decomposers
5. energy pyramid
6. biodegradable
   biological magnification
7. macronutrients
   micronutrients
   nutrient cycles
   biogeochemical cycles

8. reservoirs
   abiotic
   fossil fuels
9. legume
   nitrogen fixation
   denitrifying bacteria
10. hydrologic cycle
    acid deposition
11. greenhouse effect
    global warming
12. false, relatively little
13. true
14. false, second
15. true
16. false, food web
17. true
18. false, inefficient
19. true
20. true
21. false, legume
22. false, threatening

23. true
24. true
25. true
26. true
27. false, uncertain
28. e
29. b
30. b
31. d
32. a
33. f
34. e
35. a
36. f
37. e
38. c
39. d
40. true
41. false
42. true
43. true
44. true

45. Abiotic means nonliving. Carbon reservoirs include gas in the atmosphere, dissolved carbon in the oceans, and fossil fuels. Nitrogen reservoirs include gas in the atmosphere, nitrogen-containing molecules, and wetlands. Phosphorous reservoirs include phosphates in rock, bird guano, and animal teeth and skeletons.

46. autotrophs, producers

47. heterotrophs, primary consumers, decomposers, tertiary consumers

48. Producers capture sunlight energy. Primary consumers feed on producers. Secondary consumers feed on primary consumers. Tertiary consumers feed on secondary consumers. Detritis feeders feed on dead producers, and on waste material from or dead primary, secondary, and tertiary consumers. Decomposers externally digest what remains of producers, and of primary, secondary, and tertiary consumers, and release the remaining nutrients to the environment.

49. 1000 calories exist in producer biomass. If 10% of the energy is lost with each trophic level, 100 calories will be converted to herbivore (primary consumer) biomass. As a carnivore (secondary consumer) feeds on the herbivore, it will receive only 10 calories from the original 1000 in the producer biomass.

50. The existence of a quaternary consumer would indeed be very rare. Using exercise 49 and continuing to assume 10% energy is lost with each trophic level, a tertiary consumer would only receive 1 calorie from the original 1000. If a fourth-level consumer existed, it would receive 0.1 calories from the original 1000. In order to maintain energy demands, this organism would need to consume very large quantities of food.

51. Possible scenario: Atmosphere → green leafy plant → sugar molecule in plant → cell respiration → atmosphere → green leafy plant → plant stem → detritus → detritus feeder → decomposer → atmosphere → green leafy plant → primary consumer → secondary consumer → cell respiration → atmosphere → green leafy plant → primary consumer → secondary consumer → fecal matter → decomposer → atmosphere.

Note* the cycle always starts with and comes back to the atmosphere.

52. The carbon cycle has been affected by the burning of fossil fuels. Massive quantities of carbon were stored in plant material during the carboniferous period, a period that lasted approximately 150 million years. In the roughly 150 years since the beginning of the industrial revolution, humans have returned much of this stored carbon to the atmosphere at a rate that is orders of magnitude greater than the rate at which it was stored. In the meantime, deforestation and the burning of the tropical rainforests have added additional carbon to the atmosphere while at the same time removing areas that could absorb CO_2 from the atmosphere.

The nitrogen cycle has been affected by human-made nitrogen fixation: the production and overuse of fertilizer. Fertilizer runoff from farms into streams and rivers has led to massive die-offs of aquatic ecosystem life. Nitrogen oxides put into the atmosphere by the burning of fossil fuels mixes with water vapor forming nitric acid. Nitric acid is, in part, responsible for damage to crops, aquatic systems, and human-made structures.

The phosphate cycle has been affected, again, by the formation and overuse of fertilizer. Phosphorous-rich soil runoff into waterways stimulates the growth of producers such as algae. When the algae die, their decomposition uses oxygen at a high rate, suffocating fish and other aquatic organisms; ultimately disrupting the aquatic balance.

53. According to the Pesticide Action Network, DDT (dichlorodiphenyltrichloroethane) is an organic insecticide. It kills insects it contacts because it is a nerve poison. Originally, DDT was used to to control typhus since it can be spread by the body louse. It was also used as an agricultural insecticide. DDT is used to combat populations of mosquitos since they can be carriers (vectors) of several diseases, including malaria.

54. According to the Pesticide Action Network, DDT is still used in countries affected by Malaria spread by the *Anopheles* mosquito. DDT may be found in foods in countries no longer using DDT because of illegal use or, even more likely, because contaminated foods are imported from regions where DDT is still used. Additionally, DDT has a reported half-life in the environment of 2–15 years in most soils. That means DDT persists in the environment and may cycle between organisms and the soil.

55. R. Morrison from the Competitive Enterprise Institute interviewed R. Bate, chairman of Save Children From Malaria Campaign. He explained that the use of small amounts of DDT will help keep diseases like malaria under control. It will probably save the lives of millions of people since malaria affects about 500 million people and kills up to 2.5 million each year. When compared to alternative methods of control, DDT is far more effective and less expensive.

Chapter 41: Earth's Diverse Ecosystem

OVERVIEW

This chapter looks at how weather patterns and climate determine the way organisms on Earth are distributed. In terrestrial ecosystems, climate determines which plants inhabit an area. The plants present, in turn, determine the animal life existing there. This chapter outlines the major biomes and their characteristic plant communities. The impact humans have on each biome is addressed. Additionally, since the oceans cover 71% of Earth, aquatic life, both freshwater and marine, is discussed. And again, the impact humans have on aquatic ecosystems is considered.

1) What Factors Influence Earth's Climate?

Earth's climate is determined by the patterns of **weather** that exist in a specific region. The **climate** in a region is influenced by latitude, air currents, ocean currents, the presence of continents, and elevation. *Latitude* is measured as the distance north or south of the equator. The sunlight that heats Earth's surface hits Earth at an angle. The degree of the angle determines the overall temperature for a region. Because of Earth's rotation and temperature differences in air masses, air currents are generated. As warm, moist air rises, it is cooled, causing precipitation. At the same time, ocean currents are formed from the Earth's rotation, wind, and the sun's heating of the water. The presence of continents causes the ocean currents to circulate, forming **gyres**. The circular patterns generate moderate weather patterns on the coastal regions of a land mass. The presence of differing elevations within continents also generates climate variations. As moist air approaches a mountain, it rises, cools, and the moisture condenses, causing precipitation. The now dry air passes over the mountain, absorbing moisture as it flows. This causes a phenomenon called a **rain shadow** on the far side of the mountain.

2) What Conditions Does Life Require?

Four basic resources are required for life. (1) Nutrients are needed to build living tissue, (2) energy is needed to produce the tissues, and in order for metabolic reactions to occur, (3) liquid water must be present, and (4) temperature must be appropriate. Within diverse ecosystems, the organisms have adapted the mechanisms necessary to acquire resources and to survive.

3) How Is Life on Land Distributed?

Basically, temperature and availability of water determine the distribution of terrestrial organisms. Both temperature and water are unevenly distributed, and organisms have adapted various mechanisms to tolerate unfavorable temperatures and drought or flood conditions.

Communities on land are defined by the dominant plants present. Large land areas with similar environmental conditions and characteristic plant communities are called **biomes**. In the tropics, the **tropical rain forest** biome is dominated by a diverse population of huge, broadleaf, evergreen trees that grow well along the equator where temperatures are relatively constant and 250 to 400 cm of rain falls. **Biodiversity** refers to the total number of species within an ecosystem and the resulting complexity of interactions among them. In other words, it defines the biological "richness" of an ecosystem. Rain forests have the highest biodiversity of any ecosystem on Earth. However, the soil is infertile because all the nutrients are tied up in the lush vegetation. Still, humans are deforesting the area for agriculture, inefficient in the infertile soil, causing the loss of countless species and unmeasurable potential. For example, in Malaysia, the sap of a rare species of tree yielded a compound called calanolide A, which offers promise as an anti-AIDS drug.

Tropical deciduous forests are located slightly north and south of the equator where rainfall is not as predictable. During the dry season, trees drop their leaves to reduce water loss. Farther away from the equator, the grasses of the **savanna** dominate. Any trees growing here are scrubby with thorns protecting their sparse foliage. The root systems of the grasses can withstand the severe droughts common to the savanna regions. Human settlements have spread into the savanna, threatening the black rhino and African elephant populations through poaching and habitat conversion to cattle pastures.

Where rainfall decreases below 50 cm a year, the savanna gives way to **desert**. Plants that survive in the deserts have extended root systems to quickly absorb any rain. The absorbed water is stored in fleshy stems covered with wax to prevent evaporation. The area of Earth covered by deserts is increasing as human activities cause **desertification** of fragile habitats. Desertification is caused by deforestation, soil destruction, and over use of once-productive land.

Deserts that merge with a coastal region give rise to unique **chaparral** ecosystems. These ecosystems are found along the Mediterranean Sea and along the southern California coast.

In the centers of Eurasia and temperate North America, **prairies**, or **grasslands**, dominate. Limited rainfall and relatively frequent fires restrict the growth of trees to river banks while promoting the growth of the prairie grasses. Humans have used the fertile soil created by the dense grass growth for intense cereal agriculture or pasture land. Overgrazing on western prairies in the United States has led to desertification while agriculture has left only a few small remnants of original prairie habitat.

As the amount of precipitation increases, trees can take root, shading out the grasses. This gives rise to the **temperate deciduous forests**. Water is unobtainable to the trees during winter, so they drop their leaves to reduce moisture loss. Spring brings short-lived wildflowers to bloom before new leaves form on the trees, shading the forest floor. Along the southeastern coast of Australia, the southwestern coast of New Zealand, and the northwestern coast of North America lie very wet ecosystems called **temperate rain forests**. These rare habitats receive upward of 400 cm of rain a year, giving rise to lush plant growth of evergreen conifers, ferns, and mosses.

In the interior of northern North America and northern Eurasia, harsh temperatures and a short growing season produce the **northern coniferous forests**, or the **taiga**. Humans have *clear-cut* vast areas of the taiga forests for timber; however, due to its harsh climate, much has remained undisturbed. Farther north, the climate becomes even more extreme. Winter temperatures of the **tundra** may drop below -55° C. A **permafrost** layer is impervious to water when temperatures do allow for the top meter or so to thaw. Thus a wet marshy expanse is formed. During the brief summer thaw, small flowering plants quickly grow, bloom, and die back until the next thaw. Fortunately, human impact on the fragile tundra ecosystem is limited to oil drilling sites, pipelines, mines, and military bases.

Temperature and rainfall interact to produce the specific terrestrial biomes. For instance, in the United States, the Sonoran Desert in Arizona, the short-grass prairie of Montana, and the taiga of Alaska all receive annual rainfall levels of 28 cm (12 in). But, because of temperature differences, much different environmental conditions exist and, therefore, much different vegetation dominates each biome.

4) How Is Life in Water Distributed?

Water presents unique characteristics that limit where life can or does exist in aquatic habitats. Water is a great insulator, thus it warms and cools slowly. The sunlight that penetrates into water is quickly absorbed so that depths below 200 m will not receive enough light energy to conduct photosynthesis. Any suspended sediment will reduce the depth of light penetration. Available nutrients settle to the bottom of aquatic ecosystems where light often does not reach, limiting life.

Large freshwater lake ecosystems have distinct life zones. The **littoral zone** occurs where water is shallow, light is abundant, and plants can root and acquire nutrients in the bottom sediments. Among the rooted plants, **plankton** drift. **Phytoplankton**, such as algae and photosynthetic bacteria, and **zooplankton**, such as protozoa and minute crustaceans, add to the diversity of this ecosystem. In deeper water, light is limited to the **limnetic zone** where the aquatic food web is supported by abundant phytoplankton. Light does not reach the deeper depths of the **profundal zone** and the detritus feeders and decomposers living on the lake floor receive nutrients from detritis from the limnetic zone. Lakes that are extremely nutrient-poor are referred to as being **oligotrophic**. These lakes are often formed from deep depressions from glaciers and are fed by mountain streams. Oligotrophic lakes are often clear, deep, and oxygen-rich. Lakes that receive high amounts of nutrients and sediments as run off from surrounding areas are **eutrophic**. Eutrophic lakes are clouded with suspended sediment and phytoplankton. Oxygen content in the profundal zone is often limited as decomposers use any available oxygen while feeding on dead phytoplankton. Human activities increase the rate of lake eutrophication as excess nutrients are carried from farms, sewage, and suburban fertilized lawns. Acid rain falling in the Adirondack mountains of New York state has acidified the lakes to the point that they are unable to support life. The lakes are dead.

Marine ecosystems are also divided into life zones based on depth and availability of light energy. The **photic zone** is the upper region where light energy supports photosynthesis. Below the photic zone is the **aphotic zone** where nutrients settle. Nutrients from the aphotic zone rise to the photic zone during **upwelling** events caused by surface winds displacing surface water. Closer to shore, organisms of the **intertidal zone** experience alternating dehydration, as the tides recede, and re-hydration, as the tides rise again. Bays, coastal wetlands and **estuaries** make up the **near-shore zone**, which is constantly submerged. All of these ecosystems support diverse and abundant life. However, human activities have greatly impacted the fragile coastal ecosystems. By filling in or dredging, humans have removed almost all of the world's wetlands. Equally fragile are **coral reef** communities. Corals are minute animals that live symbiotically with algae, building great structures from calcium carbonate secreted from their bodies. These structures serve as habitat for unique and specialized organisms. However, as sediment and silt from nearby land areas cloud the water around a coral reef, the algae are not able to photosynthesize and provide the energy necessary for the organisms to survive. Ultimately, the reef will die due to human activities.

The open ocean supports free swimming or **pelagic** life forms in the upper photic zone. Here, the food web is supported by phytoplankton. The aphotic zone, however, supports unique life forms of its own. Even the open ocean has been subjected to misuse by humans. The harvesting of fish, shell fish, and mollusks is occurring at such a rate that the populations are unable to be sustained. Additionally, techniques used unintentionally destroy habitat and populations of other organisms. "Tuna-safe" fishing may save numbers of dolphins, but sea turtles and sharks are being destroyed in the process. Trash, dumped by ships or blown in from shore, endangers the lives of many animals in the sea as they mistake it for food. Radioactive wastes have also been dumped in the open ocean under the pretense that it could "do no harm out there." However, unique life forms may live in undiscovered areas of the deep ocean. **Hydrothermal vent communities** represent one such example. Deep in the ocean, cracks in Earth's crust heat surrounding water, supporting an ecosystem containing 284 new species. These organisms do not use the sun's energy for life support. Instead, they use hydrogen sulfide from Earth's crust in a process called *chemosynthesis*.

KEY TERMS AND CONCEPTS

Fill-In: From the following list of key terms, fill in the blanks in the following statements.

aphotic	intertidal zone	photic zone	temperate deciduous forest
chaparral	latitude	phytoplankton	temperate rain forest
climate	limnetic zone	plankton	tropical deciduous forests
coral reef	littoral zone	prairie	tropical rain forest
desert	near-shore zone	profundal zone	tundra
desertification	northern coniferous forest	rain shadow	upwelling
eutrophic	oligotrophic	savanna	vent community
grassland	ozone layer	pelagic	weather
gyres	permafrost	taiga	zooplankton

1. Short-term changes in temperature, precipitation, or cloud cover determine the _____ of an area. Temperature or precipitation patterns over the long term determine the _____ of a region. The equator, at zero degrees _____, has consistently warm temperatures. Ocean currents occur in circular patterns called _____ as determined by the Earth's rotation, wind, warming of the water, and continents.

2. Mountains modify precipitation patterns. As air passes over a mountain it releases precipitation. On the far side of the mountain, then, the air is dry, creating a _____.

3. The plant life in a terrestrial biome may be determined by the amount of rainfall and the frequency with which it occurs. In the _____ temperature and rainfall averages are consistent from year to year and the organisms found there are the most diverse. However, if rainfall is not so consistent and there were distinct wet and dry seasons, the plants would drop their leaves during the dry season. This is more characteristic of the _____. Where grasses are dominant and very tolerant of severe dry seasons, the _____ exists. The use of this biome for cattle grazing is threatening its wildlife. When rainfall averages less than 25–50 cm, water-storing cacti live, making up the flora of the _____. Overuse of land compromised by drought, deforestation, and soil destruction has led to irreversible _____ in many regions.

4. Coastal areas on the margins of deserts result in the unique biome of the _____. In temperate regions, as rainfall increases away from the deserts, grasses and wildflowers grow. This is the _____ biome, more commonly referred to as the _____.

5. In areas that can support tree growth, grasses are shaded out and forests dominate. In regions that have cold winters with below-freezing weather, trees drop their leaves to conserve water. This describes the _____. However, if cold temperatures are moderated and heavy rain fall occurs, as along the Olympic peninsula in Washington state, then a _____ exists.

6. Long, hard, cold winters and short springs of Southern Canada result in heavy growth of evergreen conifers in the _____ or _____ biome. As temperature becomes even colder, the fragile _____ exists as the last biome before the polar ice caps.

7. Aquatic ecosystems are also defined by their unique characteristics. Areas of freshwater lakes are divided into three zones. The _____ has shallow water with diverse communities. Drifting among the plants in this zone are microscopic organisms, the _____. If these organisms are photosynthetic bacteria or protists such as algae, they are called _____. If, instead, the organisms are protozoans and minute crustaceans, they are called _____.

8. Deeper open water that does not allow plants to be rooted to the bottom consists of the upper _____ which allows photosynthesis to occur, and the lower _____ where light does not penetrate to allow photosynthesis to occur.

9. Lakes are also classified according to their nutrient content. _____ lakes are very low in nutrients, having been formed by glaciers and fed by mountain streams. On the other hand, _____ lakes have high sediment rates and high deposition rates of organic and inorganic materials from their surroundings.

10. Ocean ecosystems consist of an upper layer that allows light penetration for photosynthesis, the _____, and the deeper, nonphotosynthetic region, _____. Nutrients from the lower oceanic regions may be brought to the surface as _____ occurs.

11. Aquatic areas along the coast may alternately be wet and dry. This occurs at the _____ as the tide rises and falls. Shallow submerged bays and wetlands make up the _____.

12. Off-shore areas with all the right parameters allow corals and algae to build structures of calcium carbonate. The diverse habitats of these _____ are extremely fragile and sensitive.

13. Life in the open ocean primarily exists in the upper photic zone. The life forms here are free swimming or _____. However, new and unusual life forms have been discovered far from the ocean surface. Deep in the ocean, extremely hot water, heated by Earth's core, supports the _____.

KEY TERMS AND DEFINITIONS

aphotic zone: the region of the ocean below 200 m, where sunlight does not penetrate.

biodiversity: the total number of species within an ecosystem and the resulting complexity of interactions among them.

biome (bī´-ōm): a terrestrial ecosystem that occupies an extensive geographical area and is characterized by a specific type of plant community: for example, deserts.

chaparral: a biome that is located in coastal regions but has very low annual rainfall.

climate: patterns of weather that prevail from year to year and even from century to century in a given region.

coral reef: a biome created by animals (reef-building corals) and plants in warm tropical waters.

desert: a biome in which less than 25 to 50 centimeters (10 to 20 inches) of rain falls each year.

desertification: the spread of deserts by human activities.

estuary: a wetland formed where a river meets the ocean; the salinity there is quite variable but lower than in sea water and higher than in fresh water.

eutrophic lake: a lake that receives sufficiently large inputs of sediments, organic material, and inorganic nutrients from its surroundings to support dense communities; murky with poor light penetration.

grassland: a biome, located in the centers of continents, that supports grasses; also called *prairie*.

gyre (jīr): a roughly circular pattern of ocean currents, formed because continents interrupt the currents' flow; rotates clockwise in the Northern Hemisphere and counterclockwise in the Southern Hemisphere.

hydrothermal vent community: a community of unusual organisms, living in the deep ocean near hydrothermal vents, that depends on the chemosynthetic activities of sulfur bacteria.

intertidal zone: an area of the ocean shore that is alternately covered and exposed by the tides.

limnetic zone: a lake zone in which enough light penetrates to support photosynthesis.

littoral zone: a lake zone, near the shore, in which water is shallow and plants find abundant light, anchorage, and adequate nutrients.

near-shore zone: the region of coastal water that is relatively shallow but constantly submerged; includes bays and coastal wetlands and can support large plants and seaweeds.

northern coniferous forest: a biome with long, cold winters and only a few months of warm weather; populated almost entirely by evergreen coniferous trees; also called *taiga*.

oligotrophic lake: a lake that is very low in nutrients and hence clear with extensive light penetration.

ozone layer: the ozone-enriched layer of the upper atmosphere that filters out some of the sun's ultraviolet radiation.

pelagic (puh-la´-jik): free-swimming or floating.

permafrost: a permanently frozen layer of soil in the arctic tundra that cannot support the growth of trees.

photic zone: the region of the ocean where light is strong enough to support photosynthesis.

phytoplankton (fī´-tō-plank-ten): photosynthetic protists that are abundant in marine and freshwater environments.

plankton: microscopic organisms that live in marine or freshwater environments; includes phytoplankton and zooplankton.

prairie: a biome, located in the centers of continents, that supports grasses; also called *grassland*.

profundal zone: a lake zone in which light is insufficient to support photosynthesis.

rain shadow: a local dry area created by the modification of rainfall patterns by a mountain range.

savanna: a biome that is dominated by grasses and supports scattered trees and thorny scrub forests; typically has a rainy season in which all the year's precipitation falls.

taiga (tī´-guh): a biome with long, cold winters and only a few months of warm weather; dominated by evergreen coniferous trees; also called *northern coniferous forest*.

temperate deciduous forest: a biome in which winters are cold and summer rainfall is sufficient to allow enough moisture for trees to grow and shade out grasses.

temperate rain forest: a biome in which there is no shortage of liquid water year-round and that is dominated by conifers.

tropical deciduous forest: a biome with pronounced wet and dry seasons and plants that must shed their leaves during the dry season to minimize water loss.

tropical rain forest: a biome with evenly warm, evenly moist conditions; dominated by broadleaf evergreen trees; the most diverse biome.

tundra: a biome with severe weather conditions (extreme cold and wind and little rainfall) that cannot support trees.

upwelling: an upward flow that brings cold, nutrient-laden water from the ocean depths to the surface; occurs along western coastlines.

weather: short-term fluctuations in temperature, humidity, cloud cover, wind, and precipitation in a region over periods of hours to days.

zooplankton: nonphotosynthetic protists that are abundant in marine and freshwater environments.

THINKING THROUGH THE CONCEPTS

True or False: Determine if the statement given is true or false. If it is false, change the underlined word(s) so that the statement reads true.

14. _____   The tilt of Earth's axis influences air and ocean currents.

15. _____   The presence of continents affects ocean currents, producing gyres.

16. _____   Deserts are often found in the rain shadow of a mountain range.

17. _____   The specific requirements of life for each organism are the same.

18. _____   Since plants are specifically adapted to their environment, terrestrial communities are defined by the plant life found there.

19. _____   It is possible to harvest products from the tropical rain forests without damaging the ecosystem.

20. _____   Grazing cattle on the range land of the savanna has had little impact on the wildlife there.

21. _____   The total area of land occupied by deserts is decreasing due to human activities.

22. _____   The occurrence of fire plays an important role in the maintenance of healthy prairies.

23. _____   In the temperate rain forest, seedlings often take root in a newly fallen tree, which serves as a nurse log, providing nutrients and protection for the developing plant.

24. _____   Clear-cutting of forests in the tundra has destroyed habitat in large regions.

25. _____   The major factors that determine the quantity and type of life in aquatic ecosystems are energy and nutrients.

26. _____   Wastes from sewer treatment facilities and overuse of fertilizers may lead to excessive amounts of nutrients being washed into aquatic ecosystems.

27. _____   Coral reef ecosystems are ultimately resilient and, therefore, are unaffected by human activities.

28. _____   Fish populations in the ocean have declined dramatically due to overfishing practices.

29. Fill in the table below with characteristics of the biomes listed.

Biome	Precipitation range	Temperature range	Typical plants present	Typical animals present
tropical rain forest				
tropical deciduous forest	NA	NA		NA
savanna		NA		
desert				
chaparral		NA		NA
prairie; grassland				
temperate deciduous forest				
temperate rain forest				NA
taiga; northern coniferous forest				
tundra				

Identify: Determine whether the following characteristics refer to **eutrophic** lakes or **oligotrophic** lakes.

30. _____ receive high inorganic material from area runoff

31. _____ contain very little nutrients

32. _____ are often clear

33. _____ profundal zones often low in oxygen

34. _____ tend to be good trout fishing lakes

35. _____ receive high organic material from area runoff

36. _____ are oxygen rich

37. _____ are murky with a dense phytoplankton population

38. _____ experience seasonal algal blooms

39. _____ limnetic zone may extend to the bottom

Identify: Determine whether the following statements represent **human impacted** ecosystems or **undisturbed** ecosystems.

40. _____ Animals in the highest trophic levels are rare.

41. _____ Species diversity is high.

42. _____ Fueled by sunlight.

43. _____ Nutrients are recycled.

44. _____ Fueled by fossil fuels.

45. _____ Fertilizers, pesticides, and topsoil pollute streams and rivers.

46. _____ Water is stored.

47. _____ Water is polluted.

48. _____ Water is filtered and purified.

49. _____ Natural predators control population growth.

50. _____ Crops planted close together encourages pest outbreaks.

51. _____ Population is expanding exponentially.

52. _____ Populations are relatively stable.

53. _____ Ecosystems are simple.

Short answer:

54. How is climate different from weather?

_____

_____

_____

_____

_____

55. Identify the four fundamental resources necessary for life.

_____

_____

_____

_____

_____

APPLYING THE CONCEPTS

These practice questions are intended to sharpen your ability to apply critical thinking and analysis to the biological concepts covered in this chapter.

56. Discuss how the limnetic zone and profundal zone of fresh water lakes relate to the photic zone and aphotic zone of oceans.

_____

_____

_____

57. Discuss how the Ozone Layer is formed in the stratosphere and its function. How is the layer being destroyed and what are the possible results of this destruction?

_____

_____

_____

_____

_____

_____

_____

_____

58. How can humans reverse the destructive trends of our activities?

_____

_____

_____

_____

_____

_____

_____

Use the Case Study and the Web sites for this chapter to answer the following questions.

59. What is environmental anthropology?

_____
_____
_____
_____
_____
_____
_____

60. Why do environmentalists feel that it is important to protect the Arabuko-Sokoke forest?

_____
_____
_____
_____
_____
_____
_____

61. What have environmental anthropologists accomplished in Project Kipepeo (butterfly)?

_____
_____
_____
_____
_____
_____

62. Is the Arabuko-Sokoke forest safe from destruction?

_____
_____
_____
_____
_____
_____
_____

ANSWERS TO EXERCISES

1. weather
   climate
   latitude
   gyres
2. rain shadow
3. tropical rain forest
   tropical deciduous forest
   savanna
   desert
   desertification
4. chaparral
   grassland
   prairie
5. temperate deciduous forest
   temperate rain forest
6. northern coniferous forest
   taiga

tundra
7. Littoral zone
   plankton
   phytoplankton
   zooplankton
8. limnetic zone
   profundal zone
9. Oligotrophic
   eutrophic
10. photic
    aphotic
    upwelling
11. intertidal zone
    near-shore zone
12. coral reefs
13. pelagic
    vent communities

14. false, rotation of Earth
15. true
16. true
17. false, general requirements
18. true
19. true
20. false, life threatening impact
21. false, increasing
22. true
23. true
24. false, taiga
25. true
26. true
27. false, extremely fragile,
    severely affected
28. true

29.

Biome	Precipitation range	Temperature range	Typical plants present	Typical animals present
tropical rain forest	250 – 400 cm (100 – 160 in)	20 – 25° C (77 – 86° F)	broadleaf evergreens	arboreal monkeys, birds, insects
tropical deciduous forest	NA	NA	deciduous trees	NA
savanna	30 cm (12 in)	NA	scrubby trees, grasses	antelope, buffalo, lions, wildebeest, elephants
desert	< 25 cm (< 10 in)	extreme hot & cold	cacti, succulents	lizards, snakes, kangaroo rats
chaparral	< 75 cm (< 30 in)	NA	small evergreen trees, bushes	NA
prairie; grassland	25 – 75 cm (10 – 30 in)	ave 7° C (45° F), hot summers	grasses	bison, antelope
temperate deciduous forest	75 – 150 cm (30 – 60 in)	periods of subfreezing temperatures	deciduous trees, wildflowers	black bear, deer, wolves shrews, raccoons, birds
temperate rain forest	400 cm (160 in)	moderate	evergreen conifers, mosses, ferns	NA
taiga; northern coniferous forest	25 – 30 cm (10 – 12 in)	ave -4° C (25° F)	evergreen conifers	bison, grizzly bear fox, moose, wolves
tundra	< 25 cm (< 10 in)	below -55° C (-45° F)	perennial wildflowers, dwarf willows	caribou, lemmings, birds mosquitos, snowy owls

30. eutrophic
31. oligotrophic
32. oligotrophic
33. eutrophic
34. oligotrophic
35. eutrophic
36. oligotrophic
37. eutrophic
38. eutrophic
39. oligotrophic
40. human
41. undisturbed
42. undisturbed
43. undisturbed
44. human
45. human
46. undisturbed
47. human
48. undisturbed
49. undisturbed
50. human
51. human
52. undisturbed
53. human

54. Weather refers to short-term fluctuations in temperature, humidity, cloud cover, wind, and precipitation in a region over periods of hours or days. Climate refers to patterns of weather that prevail from year to year and even century to century in a particular region.

55. Nutrients to construct living tissue; energy to support the construction of living tissue; liquid water in which metabolic reactions can occur; appropriate temperature such that the metabolic reactions can occur.

56. In both the limnetic zone of a lake and the photic zone of the ocean, enough sunlight energy penetrates to support photosynthesis. Both areas may provide the energy to support the profundal and the aphotic zones respectively. The profundal and the aphotic zones do not receive enough light energy to support photosynthesis and rely on the limnetic zone or the photic zone for energy.

57. The stratospheric ozone layer is formed by ultraviolet light striking oxygen, forming O_3. Ozone functions to filter damaging UV radiation, greatly reducing the amount that reaches life on Earth. This layer of ozone is destroyed when CFC's (chlorofluorocarbons) are degraded by UV radiation, releasing chlorine atoms. Chlorine reacts with the O_3 molecules, breaking it into O_2 molecules. The chlorine stays in the stratosphere, reacting with ozone molecule after ozone molecule. The result of the destruction of the stratospheric ozone layer is increased UV radiation reaching Earth. This will ultimately result in increased health concerns, such as reduced immune function and increased skin cancer in humans. Globally, it will result in reduced net primary productivity and increased damage to DNA molecules in most organisms.

58. Humans can reverse the destructive trends our activities have on the ecosystem through understanding how healthy ecosystems function, educating ourselves and others about the destruction that is occurring and how it can and should be reversed, and by making a commitment, globally, to the reversal of the destruction. Through the appropriate use of technology, water and air do not have to be polluted, soil does not have to be depleted of nutrients so that artificial fertilizers are needed, and we do not have to rely on fossil fuels to provide energy for our needs. Finally, humans need to stabilize population growth to reduce the expansion of human-dominated ecosystems.

59. According to the Society for Applied Anthropology, environmental anthropology assists communities with making and planning policies pertaining to the environment. To do this, environmental anthropologists must understand the local and regional social and cultural dynamics. They use members of the local communities to identify and solve environmental problems. They must recognize the presence of and consider differences in culturally diverse perceptions, language, values, and behaviors related to the environment.

60. The Kenyan Arabuko Sokoke forest is precious. It is the largest remnant of coastal forest in East Africa and final refuge for endangered birds and mammals that have been displaced by the growing human population. In fact, according to I. Gordon and W. Ayiemba of the Kipepeo Project, it is recognized internationally because of its endemic and endangered species of birds (six species) and mammals (three species).

61. According to I. Gordon and W. Ayiemba of the Kipepeo Project, environmental anthropologists have convinced skeptical local farmers to grow butterflies instead of crops. The communities that participate in the project by raising butterfly pupae for export live on the edge of the forest. There are still three areas on the western edge of the forest that need to be brought into the project and the forest will be surrounded by butterfly farmers. At least 550 local butterfly workers now rely on the Arabuko-Sokoke forest for their livelihood; they make a much better living now than before.

62. The Arabuko-Sokoke forest remains under siege from squatters who want to clear land and establish homes within its confines. But where the farmers gather their butterflies, the forest suffers far less poaching, because the farmers are now reporting poachers rather than joining them. The long-term future of the forest depends on the support of the local community, their leaders, and their politicians for its conservation. But, as one butterfly farmer said, "if they cut the forest, things are going to be very difficult."